Geoffrey Grigson, poet and critic, has been in his time literary editor of a London daily, publisher, and B.B.C. producer. Born and brought up in Cornwall, seventh son of a country clergyman, he was one of the first critics to write with appreciation of the paintings of Ben Nicholson, the sculptures of Henry Moore and the poems of W. H. Auden. Besides *Poems and Poets* (1969), *Collected Poems* (1963), *Sad Grave of an Imperial Mongoose* (1973), *The Contrary View* and *Angles and Circles* (1974) he has written books on the romantic painter Samuel Palmer, on the cave art of the Old Stone Age, on travel, and on natural history. His widely praised *Notes from an Odd Country* (1970) celebrates views of life and art through the window of experiences in France. More information about the meaning of plant names is given in his *Dictionary of English Plant Names* (1974)

Th

Geoffrey Grigson

Englishman's Flora

Illustrated with woodcuts from sixteenth-century herbals

Paladin

Granada Publishing Limited
Published in 1975 by Paladin
Frogmore, St Albans, Herts AL2 2NF

First published in Great Britain by Phoenix House Ltd 1958

Made and printed in Great Britain by
Richard Clay (The Chaucer Press) Ltd
Bungay, Suffolk
Set in Monotype Ehrhardt

For C.M.G.

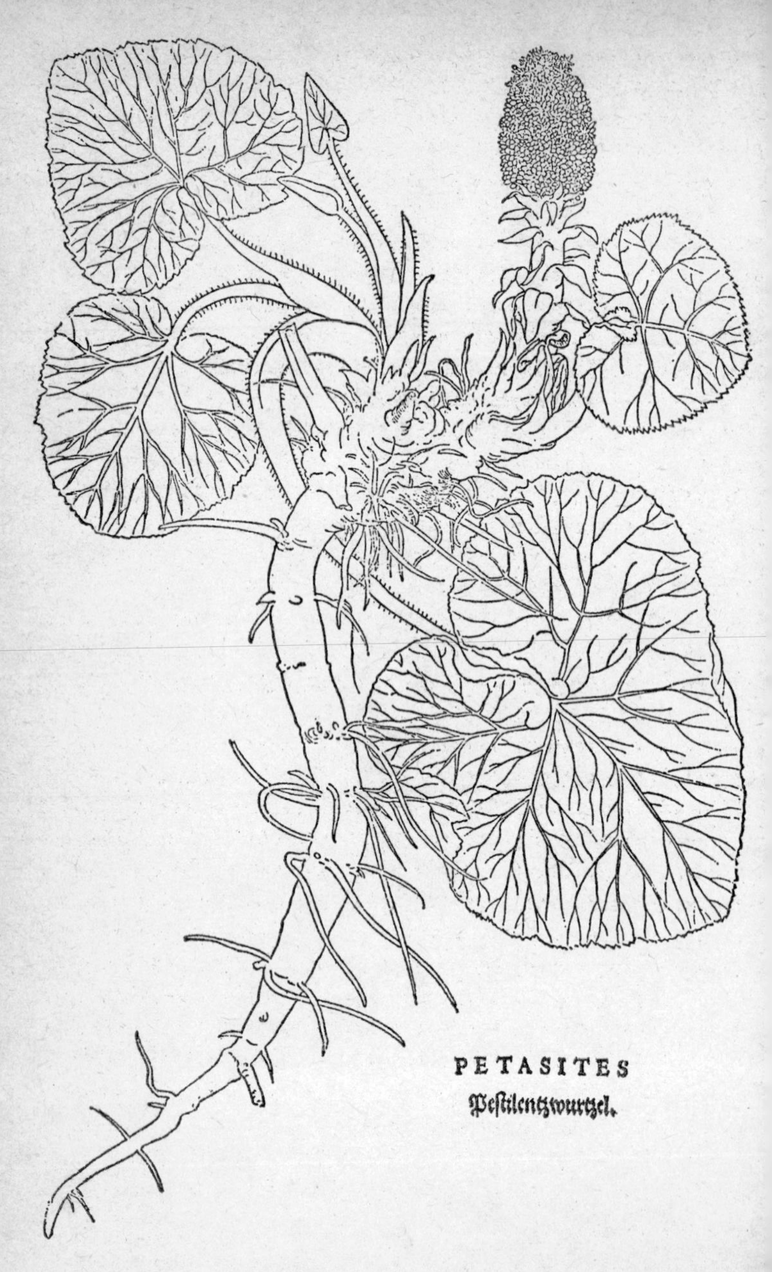

Butterbur (Fuchs, 1542)

. . . Sweetly your time ye spent
Fit, while ye liv'd, for smell or ornament,
And after death for cures.

Contents

Note on the Illustrations

The illustrations are taken from the *Herbarum vivae eicones* of Otto Brunfels (edition of 1532) – Nos. 6, 13, 27, 31, 37, 49, 52, 53, 56, 71, 72, 73, 81; the *De historia stirpium* of Leonhard Fuchs (1542) – Frontispiece, Nos. 4, 5, 7, 9, 14, 15, 23, 25, 26, 38, 39, 41, 42, 43, 57, 60, 61, 64, 67, 69, 74, 76, 77, 78, 79, 83; edition of 1545 – No. 62; Jerome Bock's *Kreuter Buch* (1546 edition) – Nos. 18, 19, 35, 40, 44, 47: Pierandrea Mattioli's *Commentarii 3 in VI libros P. Dioscoridis* (1554 edition) – Nos. 22, 32; Prague edition (1562) – Nos. 28, 29, 30, 55, 65, 70, 80, 82; 1565 edition – Nos. 8, 12, 16, 20, 21, 36, 45, 46, 59; Pierandrea Mattioli's *De plantis epitome utilissima* (1586) – No. 3; Theodor Zwinger's *Theatrum Botanicum* (1744), Nos. 48, 51, 54, 66 (woodcuts) of the sixteenth century); John Gerard's *Herball* (1597) – Nos. 2, 10, 11, 17, 24, 33, 50, 56, 68, 75; Thomas Johnson's edition of Gerard (1633) – Nos. 34, 63.

The woodcuts used by Brunfels were made from drawings by the German master Hans Weiditz (fl. 1520–36), those used by Fuchs from drawings by Albrecht Meyer. Bock's draughtsman was David Kandel (*c.* 1530–*c.* 1587). Drawings by the Italian Giorgio Liberale and the German Wolfgang Meyerpeck were used for the extra woodcuts in the Prague edition of Mattioli.

In these early herbals the woodcuts were meant to be coloured by hand, if bookseller or owner so desired.

Illustrations

Regional Abbreviations

ENGLAND

Beds, Bedfordshire
Berks, Berkshire
Bucks, Buckinghamshire
Cam, Cambridgeshire
Ches, Cheshire
Corn, Cornwall
Cumb, Cumberland
Derb, Derbyshire
Dev, Devon
Dor, Dorset
Dur, Durham
E Ang, East Anglia
E Eng, East of England
Ely, Isle of Ely
Ess, Essex
Glos, Gloucestershire
Hants, Hampshire
Heref, Herefordshire
Herts, Hertfordshire
Hunts, Huntingdonshire
I o W, Isle of Wight
Kent, Kent
Lancs, Lancashire
Leic, Leicestershire
Lincs, Lincolnshire
Mddx, Middlesex
N Eng, North of England
N'hants, Northamptonshire
Norf, Norfolk
Notts, Nottinghamshire
N'thum, Northumberland
Oxf, Oxfordshire
Rut, Rutland
S Eng, South England
Shrop, Shropshire
Som, Somerset
Staffs, Staffordshire
Suff, Suffolk
Surr, Surrey
Suss, Sussex
War, Warwickshire
W Eng, West of England
West, Westmorland
Wilts, Wiltshire
Worc, Worcestershire
Yks, Yorkshire

SCOTLAND

Aber, Aberdeenshire
Angus, Angus
Arg, Argyllshire
Ayr, Ayrshire
Banff, Banffshire
Berw, Berwickshire
Bute, Isle of Bute
Caith, Caithness
Clack, Clackmannanshire
Dumf, Dumfriesshire
Fife, Fifeshire
Heb, Hebrides
Inver, Inverness-shire
I o S, Isle of Skye
Kincard, Kincardineshire
Kinrs, Kinross-shire
Kirk, Kirkcudbrightshire
Lanark, Lanarkshire
Loth, Lothian
Mor, Morayshire
Nairn, Nairnshire
N Scot, North Scotland
Ork, Orkney Islands
Peeb, Peebles
Perth, Perthshire
Renf, Renfrewshire
Ross, Ross and Cromarty
Rox, Roxburghshire
Scot, Scotland
Selk, Selkirkshire
Shet, Shetland Islands
S Scot, South of Scotland
Stir, Stirlingshire
Suth, Sutherland
Wigt, Wigtownshire

WALES

Ang, Anglesey
Breck, Brecknockshire
Caer, Caernarvonshire
Card, Cardiganshire
Carm, Carmarthenshire
Denb, Denbighshire
Flint, Flint
Glam, Glamorgan
Merion, Merionethshire
Mon, Monmouthshire
Mont, Montgomeryshire
Pemb, Pembrokeshire
Rad, Radnor

Ire, Ireland
N Ire, Northern Ireland
I o M, Isle of Man

Foreword

Men and plants are old acquaintances, so kind by kind, family by family, order by order, I have taken those plants, common and less common, and rare, which can be found in the British Isles, and which have their own human dossier (ferns, grasses, sedges, etc., I have had to omit, for reasons of space). What names have these plants been given? Are they native or have they been introduced? If they have been introduced, where did they come from? Or why, or how, were they introduced? If they are natives, when were they first discovered? And what have they meant, generation after generation, to the Englishman in his daily life, in his ceremonies of spring, summer, and winter, his practice of magic and medicine, his feeding of body and of mind?

A delight in plants, above all when their flowers are out (so we talk of 'wild flowers', the part for the whole), seems to us the most natural – and most innocent – and inevitable of emotions. In earlier centuries delight was an ingredient not always present by any means, it was one item in a more practical response. The nature of the response is often revealed in the prosaic names for a beautiful species, or the entire neglect of a beautiful species which did not happen to be useful. Plants (or some plants) were necessary to a degree that we are forgetting. They were not only givers of food, they were nine-tenths of that specialized food we call medicine, they were nine-tenths of all cosmetics and colourings: they possessed, or certain kinds possessed, a powerful *mana*. They could keep malignant influences and the weapons of disease from your body, or they could expel them from your body; they could keep malignant beings from your house, your cow-shed, your milk and butter animals – or your god's temple. It was necessary to have a knowledge of the kinds, of when it was right or wrong to collect them, and of the ways and blends in which they were to be employed. So the physic garden was no fancy, no idle fun, but a need of life. Until the end of the Middle Ages, gardens, first of all, were utilitarian. Delight they gave, but incidentally. And beyond the hedges or the walls of the garden, the countryside, again, was not so enjoyable for itself: it was the world's surface which provided grain and leaf and fruit and timber, and many more plants to be collected for the health and protection of the body and spirit, since the garden could not contain all the plants of virtue. Some were to be found everywhere (the diuretic Dandelion, for example, or the Self-heal for cuts); others, such as the aphrodisiac Orchids, would not endure transplanting.

Then, as we all know, in the sixteenth century botanical studies were intensified. New herbals were compiled, plants were exquisitely and accurately drawn and engraved, critical efforts were made to identify all the plants described by Dioscorides in his *De Materia Medica*. The motives were mixed. Above all, men needed an accurate, reliable report of every plant which could minister to them. Yet the herbals show more than the beginning of a shift of emphasis – a shift away from magic, a shift to the discernment of cause and effect, a shift to the plant considered for itself, either as a part of the knowledge of things or as an object of loveliness. Whether the plant possessed any 'virtues' is no longer quite so important. One herbalist after another included a description and a likeness of the Fritillary (*Fritillaria meleagris*): Gerard sums up their attitude to Fritillaries in his herbal of 1597: 'Of the faculties of these pleasant flowers there is nothing set downe in the ancient or later writers, but [they] are greatly esteemed for the beautifieng of our gardens, and the bosomes of the beautifull.'

Before the printed herbals there was a body of knowledge about plants, mixed with a good deal of surviving magic – with many more of those 'superstitious' practices which were condemned so scornfully and sensibly by popular writers such as Gerard or Parkinson, or even by the astrological yet Protestant Culpepper. I suspect that a fifteenth-century or fourteenth-century housewife knew many more wild and cultivated plants than most country housewives know today. She had to;* and the magic mixed with her botanical knowledge and medicinal knowledge was not surprising. After all, how remarkable that if you eat a plant, it keeps you alive! How remarkable that some plants do the opposite, and do it with dispatch; that others affect the mind narcotically and strangely; that others have local effects on your body inside and out!

Plants which work in special ways must do so by special power; a power one must learn to direct. And where does it come from, this special power, if not from the other beings who live in the world, and may own or live *in*

* So through the eighteenth century and early nineteenth century wild plants must have been far better known in the countryside, since many more kinds were still used, and many kinds were still collected and sold to the apothecaries or in the market.

the plant? Divinities, in classical myth, discover the virtues or reveal them to men. The gods are herbalists, kings delight in medicine. The Welsh have the story of the Doctors of Myddvai, descended from the fairy woman Nelferch – she came out of a lake in the Black Mountains and returned into the lake – who had the fairy woman's power, no doubt, to hurt as well as heal. The first of the Doctors of Myddvai was her son by an earthly father. She instructed him in the virtue of all plants and in the art of healing. The Irish have a tale that the fairies were flung out of Heaven: then God threw plants down after the fairies, instructing them to be good and useful (though matters were confused by the devil, who duplicated many herbs to make them more difficult to recognize). Names of plants and surviving scraps of belief show how many English kinds were associated with elf or fairy, or devil (the Foxglove, for example, or Greater Stitchwort), or with that ultimate being of superstition, the poor witch.

Pre-Christian ceremonies of collecting plants (while they are still under the influence of the moon, still wet with the magic of dew, still untouched by the sun, the collecting of them without iron, naked, on certain nights, at certain times of the year, etc.) were Christianized, up to a point, through the Middle Ages. The plants themselves came under Christian patronage (notice how often plants with names of elf, fairy, devil, witch, snake, etc., also bear names after a saint or the Virgin, or God himself). The necessary incantations and formulae took on a Christian colour. Then comes the drastic change in and after the sixteenth century. Pronouncing an old and more or less Christian formula over plant or patient becomes a fault of superstition, a crime of the craft and subtlety of witches and their master. The power in plants, though not yet subject to exact analysis, must become strictly a portion of the settled and rationally understood beneficence of God, to be employed with the proper soberness. What may seem to us a little strange is the power, nevertheless, which the Doctrine of Signatures wielded over some men in the sixteenth and the 'new' seventeenth century (in which the Royal Society came to birth). It was the old vague notion of sympathetic magic, the yellow plant for the yellow disease and so on, the power of like to affect like; yet in the climate of the time, it was not quite so foolish a notion as we may think. It was now formulated under a precise name, with some show of religious reason or probability, for if a beneficent deity had 'signed' plants with an indication of their faculties, that was merely one of the smaller items of divine kindliness and power, easy to accept when the first chapter of Genesis was still a 'true' picture.

Centuries of belief and practice have left their fossils behind. There are plants we still grow in pleasure gardens from force of habit, because they were grown at one time for physic, or apotropaic power, or some other practical reason. The condition is that the plant, in our eyes, should still be more or less attractive. Hollies and Rowans were apotropaic protectors against evil. We still set them by houses for foliage or berries or both. Monkshood gave a useful supply of poison. We still cultivate it for its pretty flowers. Tansy furnished a supply of bitters, and it is still grown for the scent of the leaves and for the golden buttons. Orpine was useful both in medicine and divination; it survives by its character and colour. Plants introduced from abroad (such as the Greater Celandine and the Evergreen Alkanet and Feverfew, which was once the housewife's aspirin) and then abandoned continue to live in association with men, more or less upon their own terms. Farming and gardening add their plants. Farming profoundly alters a flora by destroying old habitats and supplying new ones, which are 'open', more or less, and are incompletely occupied by crops. Foreign weeds of cultivation come in; or native plants naturally tied to drift margins or to cliffs have a chance to spread and take upon themselves the new nature of weeds. In the age between William Turner and the Chelsea Flower Show, pretty plants from the Mediterranean (Ivy-leaved Toadflax), from the Himalayas (*Impatiens glandulifera*), from China (*Buddleja davidii*), from the wet, foggy Aleutian Islands (*Mimulus guttatus*), have been brought into the gardens and have then broken out. Railways come, with cuttings, embankments, tracks, marshalling yards, and within a hundred years they have helped a plant formerly so scarce as Rosebay Willowherb to become brilliant and sharp throughout the country. Plants, too, have a relation to their discoverers, field botany having its own past, its own adventures, its own heroes, from William Turner and John Ray to a Curtis or a Claridge Druce. Many a wild species can be examined in some rare station, from the clints of Ingleborough to the Bristol Gorge or the black precipices of Snowdon, where it gave pleasure to botanists who have now been dead and gone for centuries.

In all these and other ways a species will have built up its own human dossier, which may stretch back to the remoter antiquity of a neolithic lake dwelling in Switzerland, or to the medical practice of Egypt and Assyria. The dossier, it is true, may be woefully incomplete. Names of one plant and another hint at practices of which there is no full record. Exactly why, for instance, was the picking of some flowers, including Poppies and White Campion and Greater Stitchwort, liable to cause a thunderstorm? There may be no more than clues, but even the clues are fascinating. Introducing a solemn textbook of theoretical botany, two authors have lately remarked of this old *scientia scientiarum* that if, on the one hand, it is a modern experimental discipline, on the other it is a 'storehouse of curious and ancient lore'; or, as a field botanist (Caleb Threlkeld) wrote in 1727, 'Botanick Studies have a native Tendency to the Support, Comfort, and Delight of Mankind'. In the entries which follow, plant by plant, something is given of the curious and ancient lore – though unsystematically, and chiefly from English sources, since the book is about plants in England, or at least in the British Isles (though it is true, of course, that English beliefs are part of a lore about plants spreading right across Europe in space and in time – witness the frequent identity, or parallelism, of plant names). The enjoyment of wild plants, after all – the fullest enjoyment – does involve their history and their associations, which may be added to the pleasures of the hunt, the disciplined identification, the synthesizing, analytic, and aesthetic pleasures of contemplating the plant once found, and once recognized.

* * *

Notes may be useful on a few names which are often repeated, or on books which are often quoted. First of all:

DIOSCORIDES. The *De Materia Medica* of the Greek physician Dioscorides Anazarbeus, which he compiled about A.D. 60. This account of the medicinal plants of Asia Minor was written with an admirable restraint and freedom from wilder fancies. Herbalists of the sixteenth century were immensely stimulated by Dioscorides, partly because of the challenge of identifying the kinds of plants he described so curtly. The pharmacopoeia in the time of Queen Elizabeth was based upon the unquestioned authority of this ancient doctor.

PLINY. The *Natural History*, compiled like a discursive encyclopaedia by Gaius Plinius Secundus, A.D. 23–79, ingenious man of letters and a Roman contemporary of Dioscorides. He was not critical, but he was curious, if credulous. The sixteen books dealing with plants are always entertaining to read, and bring one to the source of many English beliefs. There is a Loeb edition and translation (1951) by Rackham and Jones.

THE PSEUDO-APULEIUS. The *Herbarius* or *Herbarium* of Apuleius Platonicus, another source of English practice and English folklore. This Latin herbal, with a strong magic element, was compiled three or four centuries after the time of Dioscorides and Pliny. The text and translation of an Anglo-Saxon version are given in Cockayne's *Leechdoms, Wortcunning, and Starcraft of Early England*, 1866.

THE *Lacnunga*. An early manuscript of Anglo-Saxon medical and magical practice, partly pre-Christian. It was first edited and translated in Cockayne's great book, which should be read for other medical and herbal manuscripts of the Anglo-Saxon time. Lately it has been re-edited, discussed, and translated by J. H. G. Grattan and Charles Singer, in *Anglo-Saxon Magic and Medicine* (1952). Cockayne is too reckless in identifying plants, Grattan and Singer are too cautious. For Anglo-Saxon knowledge of plants, refer also to *Anglo-Saxon Magic* by G. Storms, 1948 (in which the identifications are sometimes impossible).

Several later manuscripts give an idea of English plant medicines and usages before the Reformation. See, for instance, *Medical Works of the Fourteenth Century*, ed. G. Henslow, 1899; *A Leech Book of the Fifteenth Century*, ed. W. R. Dawson, 1934; the *Liber de Diversis Medicinis*, Early English Text Society, 1938; and Gösta Brodin, *Agnus Castus: A Middle English Herbal*, 1950.

THE PRINTED HERBALS. A modern American edition exists of the *Herball* printed in 1525 by Richard Banckes. Both this and *The Grete Herball* of 1526, which was translated from the French, are brief, readable, and pithy compilations, little touched as yet by the new criticism or scepticism. The first part of Banckes's *Herball* simply puts into print the *Agnus Castus* of the fourteenth century, which exists in many manuscripts and was probably compiled by an Englishman. For an extraordinary change, turn to the works of William Turner (*c.* 1508–79), physician, naturalist, and reformer. There are modern reprints of his first two pamphlets, *Libellus de re herbaria novus*, 1538, and *The names of herbes in Greke, Latin, Englische, Duche and Frenche wyth the commone names that Herbaries and Apotecaris use*, 1548; but not of his *Herball*, published in three parts, in 1551, 1562, and 1568. Nothing about Turner suggests a character either genial or imaginative, everything argues Protestant determination, fierceness, honesty, and force of intellect. Banckes's herbal, the first printed herbal in our language, is mediaeval in tone and substance, Turner exhibits the cool and cautious behaviour of the scientist. Many of the English names we use for plants are Turner's practical, but not always euphonious, or happy, or sensitive, coinage. He is the pioneer and father of English botany. But I do not need to survey the English herbals from Turner to

Gerard and beyond. That has been done well in two books which everyone should read, if they love plants and if the subject entertains them – in *Herbals: Their Origin and Evolution*, 1938, by Agnes Arber, and in C. E. Raven's *English Naturalists from Neckham to Ray* (1947), in which there is the fullest account yet published of the great William Turner.

I do need, though, to say something of the famous herbal, or infamous herbal, as it has now become, of John Gerard, which I have quoted a great deal despite Mrs Arber and Canon Raven. They give Gerard a dressing down, they award him a low mark, he is dismissed (too severely) by Canon Raven as a thief and a rogue. He stole the groundwork of his herbal. Without acknowledgment he used another man's translation of the latest edition of the herbal of Dodoens (of which Henry Lyte, the Somerset squire and amateur, had translated an earlier version in 1578). He made mistakes. He pretended to know more than he did. Of all this I am aware. But I do not apologize for quoting Gerard freely and in preference, though not exclusively so, to others. He had less the temper of a scientist than of an apothecary who delighted in plants. He wrote an awkwardly effective English, which conveyed his delight; and the herbal, for all its faults, for all its origin, contributed immensely and perhaps more than any other volume to the love and appreciation of plants in Great Britain. A great service. If language forgives 'every one by whom it lives', it does so for worse crimes than the one Gerard committed, and he is forgiven assuredly.

Few plant books between Banckes's *Herball* and the time of John Ray fail, indeed, to communicate some delight in the existence of plants. They are compilations (as the latest edition of the *British Pharmacopoeia* must always be a compilation), they are full of repetition and concealed quotation. Subtract the entire contribution out of Dioscorides alone, and many of them would shrink severely. Yet the authors are personal in this or that remark, this or that turn of delightful phrase. Many of the books are scientifically beside the point, and have been overlooked now as a result. Banckes's *Herball* itself, *The Grete Herball*, Lawrence Andrew's translation of Hieronymus Braunschweig (1527), William Bullein's *Booke of Simples* (1562), Thomas Cogan's *Haven of Health* (1584), a handbook for the health of Oxford students – these and others down to the books of William Cole and Robert Turner and Nicholas Culpepper (three who believed in signatures, though they were no fools with the pen), of the tree-loving and salad-loving John Evelyn, and of the sober John Pechey, might all be re-explored from a point of view which is human, and so neither strictly scientific nor herbally whimsical. Pechey's excellent and neglected *Compleat Herbal of Physical Plants*, 1694, marks the end of the period. Here was a cool doctor who knew his profession, knew the old herbals, and had yet taken advantage of the new botanical exploration of a major naturalist, John Ray (1627–1705).

For help on the folklore of plants one or two books I have found most convincing and illuminating: for instance, in the matter of the 'superstitious and vain' ceremonial used in their collection, A. Delatte's *Herbarius: Recherches sur la cérémonial usité chez les anciens pour la cueillette des simples et des plantes magiques* (1938); and for general guidance, Arnold van Gennep's magnificent *Manuel de Folklore Français*, which continually weaves sense out of darkness and always escapes that cross-gartered whimsy which degrades much latter-day English folklore and makes us forget that folklore is properly 'the science of man as a cultural being'. With these two must be mentioned the packed questionnaire and guide book by Seán ó Súilleabháin, *The Handbook of Irish Folklore*, 1942, and the available parts of Heinrich Marzell's *Wörterbuch der Deutschen Pflanzennamen*, which began to appear in 1937, but has unfortunately been left incomplete. Rafael Karsten's *Origins of Religion*, 1935, contains a good chapter on the worship of trees and other plants. But I must not lead readers to expect a formal treatise or a systematic essay, or series of essays, on the folklore of vegetation. I have not the learning for that. I have not altogether practised what I may seem to preach. Like the herbalists themselves, I have taken my plants, as I say, one by one, and I have written my notes about their history (often, as you will detect, they are the plants of dubious standing, dubiously native, or decidedly 'foreign' – Elecampane or Alkanet, Mugwort, or Periwinkle of the magicians). And since, after all, we have an emotion towards the vegetable tuftings and ornaments of the earth, I have allowed myself now and then (though it is unfashionable, through abuse) a raid upon other men's emotion, a reference to a poem or a poet's journal, when the poet has exhibited some attitude of an age, or a sharp and unusual acuity of vision (Gerard Manley Hopkins, for example, on bluebells and daffodils), or an uncalculating love and appreciation such as one finds so refreshingly in older Welsh and Irish poetry or in a few mediaeval English poems or in William Barnes; also some references to painting, mediaeval sculpture, and the visible arts – above all to the Unicorn tapestries, that superlative series aesthetically, botanically, and symbolically, in which the flowers of north-western Europe – Bluebells and Early Purple Orchids and all – have now been transferred, as fresh as when the panels were woven, to the mediaeval treasury of the Cloisters in New York.

This is a book I wanted myself, and a book, it seems, which no one else would write. Perhaps it is a book others have wanted as well, to supplement the increasingly dry or severe paragraphs of the manual of recognition. I hope so. And I hope that follies or failings, misinterpretations or inaccuracies, and certain inconsequent activities of the botanical bower-bird, a certain 'jottedness', a certain delight in glittering objects, or glittering but disparate items of information, will be overlooked, or forgiven, in that spirit.

The plants are arranged in the usual way of a flora, which I think is a gain, since species thus appear in their related groups, Cowslip with Primrose, Deadly Nightshade with Henbane, Marsh Marigold not so far away from the Buttercups, Houseleek near *Sedum acre*, the Bluebell near the Grape Hyacinth, or the Fritillary near Solomon's Seal. The arrangement and scientific names of the plants are taken from Clapham, Tutin, and Warburg's *Flora of the British Isles* (1952). The formula which follows the scientific name (e.g. 112, H 40) may need explanation for those unfamiliar with the geography of English plants, or with Druce's *Comital Flora*, or the new *Flora*, from which, on the whole, the corrected formulae have been taken. England, Scotland, and Wales are divided into 112, and Ireland (Hibernia) into 40 'vice-counties', or botanico-geographical subdivisions. 112, H 40 after a plant implies that it is common and widely distributed, occurring in all the vice-counties, whereas 3, H 2 would indicate a rare plant occurring in only three English and two Irish vice-counties. The formula is a rough and ready index; but observe how the number of vernacular names given to a plant may vary with its distribution and its frequency. Rising above a certain vice-comital figure of distribution, a species may have a great many names, and may often be a general English, Welsh, Scottish, and Irish 'property'; dropping below a certain figure, the plant may be the 'property' only of the field botanist and the inquisitive fellow-traveller; or else it is locally common and remarked upon (like the Fritillary, or the Bath Asparagus, or the Wild Daffodil in Gloucestershire). To the generic and trivial names used in the new standard *Flora*, there is a delightful handbook, A. Gilbert-Carter's *Glossary of the British Flora*, 1950.

The standard English name, the generic and trivial name of each plant, and the formula of distribution, are followed by the English vernacular names, as complete a list as possible from various sources, including county *Floras*, Britten and Holland's pioneer *Dictionary of English Plant-Names*, the *English Dialect Dictionary*, the *Oxford English Dictionary*, etc. If a name is ascribed to a particular county, it does not guarantee that it is still in use in that county, or that it is not spoken in a neighbouring county where no one happens to have recorded it. Sometimes the names are prosaic and disillusioning, often they are widely distributed through Europe or northern Europe, often they show how book learning sinks down into folklore (how much of the folklore of English plants, by the way, comes from literature – and from Greece and Rome). The names are often difficult or impossible to interpret, and I have been chary of the wild guess, following the authority of the *Oxford English Dictionary*, except now and again, when the philologist had not concerned himself to notice botanical improbability, or likelihood, or impossibility. Of course we need an English counterpart to Heinrich Marzell's *Wörterbuch der Deutschen Pflanzennamen*, so far as that remarkable dictionary has been published, or to Eugène Rolland's *Flore populaire* (1896–1914).

Few abbreviations are used except ME and OE for Middle English and Old English, and the county and district abbreviations, the meaning of which is obvious. The English counties are given in consistent order beginning with Cornwall and finishing with Northumberland; after which come the counties of Wales, from Pembrokeshire to Anglesey, followed by Scotland and Scottish counties from the Border to Caithness, followed in turn by Hebrides, Orkney, and Shetland. For Ireland, there are not enough records by county. A few English names taken to the United States have been included. These often show how a plant name recorded at home only from a single county – e.g. Somerset – is now used, or was used until lately, over the vaster areas of the New World into which an Old World species irrupted.

The lists of plant names are fuller than any now in print. No species with any considerable tale of names, which has therefore attracted considerable attention, has been left out.

Numbers in brackets in the text refer to the Bibliography at the end of the book.

The Flora

I. Pinaceae

1. Scotch Fir, Scots Pine. *Pinus sylvestris* ssp. *scotica* (Schott) E. F. Warburg 14

This tall, elegant, splendid tree is the native pine of the Highland forests, or what is left of them – dark and irregular, for example, in the Rothiemurchus Forest against a blue curtain of the Cairngorms. The Scotch Fir re-established in the south develops a flat top, this aboriginal subspecies develops a dome.

In the south, in Wales and England, it used to be thought that *Pinus sylvestris* had been wiped out. Nine thousand to seven thousand years ago it was the dominant tree, in the dryness of the Boreal period; which then gave way to a wet and warmer Atlantic climate, favouring alder, oak, elm, and lime (*c*. 5500–2000 B.C.). From about 2000 to 700 B.C. the climate improved again, before a slide into that cooler and wetter Sub-Atlantic phase, in which we are still involved, more or less. *Pinus sylvestris* all but disappeared, leaving its roots and stumps preserved underneath the peat. The peat also preserves pollen. Analysing the pollen tells the story, and it does suggest, by a continuity here and there, that a few descendants continue the old native stock. The Scots Pine is not so Scottish, after all.

However, in the eighteenth and nineteenth centuries, *Pinus sylvestris* was much planted – for example, in East Anglia, and across the sandy black heaths of Dorset, and around Bournemouth; and all over the country in small groups for picturesque emphasis and effect. It is one of the few conifers which 'fit' the English scene.

Visiting the aboriginal forests of Scotland, a thousand feet up, means the opportunity of seeing some of the rarer Highland plants which go with the pine – several Orchids and Wintergreens, and the little trailing, wiry *Linnaea borealis* with its pale pink corollas. This is the plant named after Linnaeus, of which Linnaeus himself wrote a famous sentence in his *Critica Botanica* (1737): '*Linnaea* was named by the celebrated Gronovius, and is a plant of Lapland, lowly, insignificant, disregarded, flowering only for a short while – named after Linnaeus who resembles it.'

The cones are known in Suffolk as DEAL APPLES or DEALIES.

II. Cupressaceae

1. Juniper. *Juniperus communis* L. 82, H 18

Local names. AITEN, Scot; AITNACH, Banff, Mor; BASTARD KILLER, Som; HORSE SAVING (i.e. savin), Cumb; MELMOT BERRIES, MELMONT BERRIES, Mor.

Juniperus gave the French word *genièvre*, and our 'geneva', or 'gin' for short, since proper Holland gin has its flavour from juniper berries. 'An admirable *solar* shrub', Culpepper calls *Juniperus communis*, and though, from downland and limestone hill and heath to Highland pine forest, it grew somewhat out of sight and out of mind, the herbalists considered Juniper 'scarce to be paralel'd for his Vertues'; among them the virtues of being 'a most admirable Counter-poyson' and a great 'resister of the Pestilence'. Juniper wood was also burnt indoors to give rooms a sweet smell (49, 141). The berries were swallowed to procure abortion (so the name of Bastard Killer); and also in childbirth. Yet the Juniper has been altogether more celebrated in other countries, not only as an apothecary's shop to itself, but as a powerful shrub, apotropaic against devils, elves, witches, etc. Oil of Juniper (*British Pharmaceutical Codex*, 1949) is still valued and sold as a carminative.

III. Taxaceae

1. Yew. *Taxus baccata* L. 56, H 20

Local names. HAMPSHIRE WEED, Hants; IFE (cf. French *if*), Suff; PALM, Dev, Kent, S Eng, Ire; PALM-TREE, Kent; VEW, Ches, Derb, Lancs, Yks.

The berries are elegantly known as SNOTTY GOGS, Suss; SNOTTER GALLS, Wilts, Berks; SNODER GILLS, Hants; SNOTTLE BERRIES, Yks; SNAT BERRIES, N'hants.

Yew is the third of the trio of native conifers – a coneless conifer. Various properties combined to suggest power and peculiarity in yew trees: they live to a great age, the timber is hard ('a post of yew outlives a post of iron'), the leaves are poisonous, and the berries are red. So the Yew was a protective, offensively defensive tree, one of the best to plant by your house, and the very best (though other timbers were as elastic) to make into bows. In several parts of England – Wiltshire, for example, and Dorset – protective yews were planted alongside farmhouse and cottage – normally, it is true, on the side of the prevailing wind, but that meant giving a double shelter, practical and, by our thinking, magical (though magic was a very practical affair), deferred, yet also immediate, because the protection would begin at once; it would begin the moment the small

1 Yew *Taxus baccata*

yew was set, and long before it had become large enough to keep off the wind.

> The smoke was blue, above the yew,
> The yew beside your house in sight.

The Yew beside the house, near to the gables and the chimneys, not only protected but looked comforting. It tied the house to the landscape. And as William Barnes suggested in those Dorset lines, the blue smoke contrasts pleasantly with the greenish black of the Yew, which is always darker than the night, and always sets off the brilliance of the planets, the stars, or the rising moon. Originally the Yew was either the protective 'lord of the home', the deity, or the dwelling-place of this protective deity.*

The Scandinavian god Ull, warrior god of bows and ski-ing, lived in yew dales. Yggdrasil, the world-tree, on which Odin hanged himself, was possibly a yew (193). Poisonous, then, and strong, fighting on the side of those who employed it, the Yew not only made bows, above all weapons, but it made spear shafts as well; and in Ireland, where it has been a sacred tree, yew timber was anciently cut into croziers and shrines (69, 137). Alongside the church, it was no doubt powerful against elves or spirits of the returning dead; and later on it furnished supplies of 'Palm' for the processions on Palm Sunday (see also *Salix caprea*). The churchyard Yews engendered the tallest tales about archery. But they cannot have been planted for the bowyer, since he always fashioned his bow from the trunk; and since a well-grown, straight trunk, free from knots, gives only timber enough for a bare half-dozen weapons. Anyhow, the mediaeval bowyer liked yew wood from overseas. The English timber was too brittle and too knotty. Yew, it should be added, is the material of the most ancient wooden weapons yet discovered: two early palaeolithic spears, one found at Clacton, one in Saxony with the skeleton of a forest elephant (K. P. Oakley, *Man the Toolmaker*, 1952).

Through the centuries we have been two-minded about this tree, we have felt its excellence and its funereal side, its durability and its malignancy. An Irish poem, or a fragment of an Irish poem, written about A.D. 800 to 1000, finds the Yew delightful:

> There is here above the brotherhood
> A bright tall glossy yew;
> The melodious bell sends out a clear keen note
> In St Columba's church.

An English passion carol finds it a tree of evil. The seven fair maids who are searching for sweet Jesus Christ, 'all under the leaves of life', are directed to the town where they will find him 'nailed to a big yew tree' (105, 55). In Elizabethan poetry, the Yew is persistently the grim, double-fatal, funereal tree. By then (though Gerard pooh-poohed it in his *Herbal* in 1597) the story in Dioscorides was well known – that sitting or sleeping under the shade of a yew was often fatal. Robert Turner, in his *Botonologia: The British Physician*, 1664, explained that the Yew was set in

* cf. the protective function of hawthorns in the Isle of Man.

churchyards, not, 'as some superstitious Monks have imagined', because it could drive away devils, but because it 'attracts and imbibes putrefaction and gross oleaginous vapours exhaled out of the Graves by the setting Sun, and sometimes drawn into those Meteors called *Ignes Fatui*'. Beliefs of that kind only darkened the dark side. Even Wordsworth always made his yews sombre – the four yews of Borrowdale, for example,

> beneath whose sable roof
> Of boughs, as if for festal purpose, decked
> With unrejoicing berries – ghostly shapes
> May meet at noontide; Fear and trembling Hope,
> Silence and Foresight; Death the Skeleton
> And Time the Shadow. –

although a yew with a full crop of 'unrejoicing' coral berries against a blue sky is one of the most tropical of English sights.

Wild Yew prefers the good drainage of chalk and limestone; it prefers on the whole a clean and cheerful scenery. Yews forming a black wood for jays, goblins, and ecological textbooks may be strange enough – the yew woods on Mickleham Downs in Surrey, for example, on Hambledon Hill in Dorset, or, best of all, in Kingley Vale near Chichester, in Sussex. Yet notice the excellent form of a single, full-skirted yew, the contrast of the reddish bark, in early or late sunshine, with the blackness of the yew itself, or the contrast of the yew's black, tufted density with the lighter and looser foliage of other shrubs and trees. By a happy chance, Yew and White Beam are both constituents of scrub on the chalk of Surrey, Sussex, Hampshire, etc. – black tuft and solemnity offset by a tossing of silver-white.

Yews need to be seen, as well, growing out of rock, darkening, to borrow from Wordsworth's line, 'the silver bosom of the crag', emerging from limestone and flattening themselves against the grey wall, as in the Cheddar Gorge in Somerset or on the wild walls of Gordale Scar in the West Riding. The seed is poisonous, like the foliage; the red aril surrounding the seeds is *not* poisonous.

For a story of the blended power of Yew and Rowan (which was also planted in churchyards), see *Sorbus aucuparia*.

IV. Ranunculaceae

1. Marsh Marigold. *Caltha palustris* L. 112, H 40

Local names. BACHELOR'S BUTTONS, Dor, Som; BEE'S REST, BIG BUTTERCUP, BILLY BUTTONS, Som; BLOGDA, BLUDDA, BLUGGA, Shet; BOBBY'S BUTTONS, Som; BOG DAISY, Yks; BOOTS, Shrop, Ches; BULL-BUTTERCUP, Ess; BULL-CUP, Som; BULLDOGS, Dev; BULL-FLOWER, Dev, Som; BULL-RUSHES, Dev, Som, Wilts; BULL'S EYES (cf. German *Ochsenauge*), Dev, Dor, Som; BUTTERBLEB, Yks; BUTTERCUP, Dev, Som; BUTTER-FLOWER (cf. German *Butterblume*), Wilts.

CARLICUPS, Som; CHIRMS, N'hants; CLAUT, Wilts; COW-LILY, USA; COW-CRANES, N'hants; COWSLOPS, USA; CRAZY, Som, Wilts, Glos, Bucks, Lancs; CRAZY BETS, CRAZY BETSEY, CRAZY BETTY, Dor, Wilts; CRAZY LILIES, Dor; CROW-CRANES, Oxf; CROW-FLOWER, CUP AND SAUCERS, DALE-CUP, DOWNSCWOBS, Som; DRUNKARD, Dev, Wilts; FIDDLE, Banff; FIRE O' GOLD, Bucks.

GILCUP, GIPSY'S MONEY, Som; GILTY-CUP, Wilts; GOLDEN-CUP, Som; GOLDEN KNOB, Som, Berks; GOLDEN KINGCUP, Som; GOLDICUP, Corn; GOLDILOCKS, Som; GOLLAND, Lancs, N'thum, Caith; GOWAN, Cumb, N'thum; GRANDFATHER'S BUTTONS, Som; HALCUP, Hants; HORSE-BLOB, Surr, N'hants, Worc, Leic, Notts; HORSE-HOOVES, Shet; HORSE-BUTTER-CUP, Dev, Som; JOHN GEORGES, Bucks; JOHNNY CRANES, N'hants; KING'S COB, Berks, Herts; KINGCUP, Dor, Som, Kent, E Ang, War, Yks, N'thum, Lanark, Ire.

LIVERS, Dor; MARE-BLOBS, Som, Glos, N'hants, War, Derb; MARIGOLD, Yks; MARSH-LILIES, Som; MARYBOUT, Lancs; MARYBUDS, Dor, War; MARY'S GOLD, Som; MAY-BLOB, Som, Wilts, N'hants, War, Leic, Rut, Yks; MAY-BLUBS, Wilts; MAY-BUBBLES, Som, Wilts; MAY-FLOWER, Wilts, Worc, Shrop, Ches, Lincs, Cumb, Ire; MEADOWBOUT, Shrop, Ches, Lancs; MEADOWBRIGHT, N'hants; MOLL-BLOB (cf. German *Mollesblume* – '*Molle*', cow), N'hants, Worc; MOLLY-BLOB, N'hants; MONKEY-BELLS, Som.

OLD MAN'S BUTTONS, POLICEMAN'S BUTTONS, Som; PUBLICANS (Publicans and Sinners are *Caltha palustris* and *Ranunculus* species growing alongside each other), Oxf, Yks; SOLDIER'S BUTTONS, Som; WATER BABIES, Som; WATER BLEBS, Lincs; WATER BLOBS, Oxf, N'hants, War, Leic, Notts, Rut, Derb, Yks; WATER BLUBBERS, Glos; WATER BUTTERCUP, Dev, Surr, Oxf, Yks; WATER GEORDIES, Som; WATER GOGGLES, Oxf; WATER GOLLAND, or GOWLAND, Yks, N'thum, N Eng, S Scot; WATER GOWAN, Cumb; WATER

LILY, Wilts; WILDFIRE, Kirk; YELLOW BLOBS, Leic; YELLOW BOOTS,* Ches; YELLOW CRAZIES, Wilts; YELLOW GOWLAN, Cumb, N'thum, N Eng, N Scot.

Caltha palustris is an ancient and nearly universal plant of northern latitudes, which was growing in England before the Ice Age (157) and must have forced itself from the earliest times on human consciousness in Great Britain. Shining, sun-like flowers opening while the year is still cold and

2 Marsh Marigold *Caltha palustris*

colourless, and lasting into May; flowers which illuminate grey moors, black woodland, or the black mud by the roots of alder; and which also spread into damp meadows. In Iceland the farmsteads on drier green knolls rising out of the bog are surrounded by flaming Marsh Marigolds, when there are still snow showers, and when hardly another flower has blown. Settlers in the British Isles must have found them wherever they went. They must have been impressed by the way they lit up the kind of

* See Richard Barnfield's *The Affectionate Shepherd*, 1594:

The yellow boots
That grows by rivers and by shallow brooks.

Barnfield grew up in Shropshire.

water-hole inhabited by boggarts (they would fit well into the dismal scenery of Beowulf's encounter with Grendel).

Gerard describes the plant vividly in his *Herbal* (1597): 'Marsh Marigold hath great broad leaves, somewhat round, smooth, of a gallant greene colour, slightly indented or purlde about the edges, among which rise up thicke fat stalkes, likewise greene; whereupon do growe goodly yellow flowers, glittering like gold.' Golden-flowered, yet why a 'marigold', properly the name (from 'Mary' and 'gold', according to the dictionaries) of the garden composite, *Calendula officinalis*? By confusion. The Anglo-Saxons knew a plant called *meargealla* or *mersc meargealla*, 'marsh meargealla', which by a guess has been identified with *Gentiana pneumonanthe* – an improbable guess, since Marsh Gentian is a rare plant the Anglo-Saxons would not have noticed, and since *meargealla* occurs in names of a few places, Marlborough, for one, where the Marsh Gentian could never have grown. *Mersc meargealla*, or a later form of it, could very easily have sounded like Marsh Marigold, as if the plant were a wild relative of the garden Marigold which grew in marshes. Indeed *meargealla* is from *mearh*, 'horse', and *gealla*, 'gall', 'swelling', or 'blister', and in several counties (see above) equivalent names survive for the Marsh Marigold – Mare-blob, Horse-blob. 'Blob' is a dialect form for 'bleb' – 'blister'. The Marsh Marigold was called a *meargealla*, a Horse-blob or Mare-blob, either because the tight round buds suggested a round swelling, or else because the flowers resemble large buttercups, and buttercup roots were used for raising blisters in counter-irritation. So Marlborough is likely to mean the 'Marsh Marigold Barrow', from the tall barrow or mound which rises by the banks of the Kennet, and from the Marsh Marigolds, which blaze in all the Kennet water-meadows.

Not always capable of distinguishing a dahlia from a daisy, etymologists have argued that Marlborough meant either the 'Gentian barrow' or the 'barrow of Maerla', a personal name which they had to invent (see Eilert Ekwall, *Studies in English Place-names*, 1936, and *The Place-names of Wiltshire*, 1939). A similar confusion between the golden flower of the marsh and the golden flower of the garden, or at least their resemblance, led to the application of the name *Caltha palustris* in the sixteenth century, since the *caltha* mentioned by Pliny was considered to be *Calendula officinalis*.

Many of its common names *Caltha palustris* shares with the various meadow species of *Ranunculus*, but one to notice most of all is May-flower. In Ireland, where customs much more general at one time still endure, the May-flower keeps its importance on May Day. On this fertility festival, witches and fairies are about and are unusually active, as on the summer festival of St John. So the May-flower is one of the many apotropaic plants used to avert their influence; May Day bunches are picked and hung over the doors, and the fertility of the cattle is protected. More of the notable defensive plants of May Day are the Rowan and the May or Hawthorn (q.v.).

Its show, its prominence even in man-made surroundings and beliefs of this kind, rather than use in medicine, have made *Caltha palustris* so much of a favourite. Gerard wrote that 'touching the faculties of these plants we have nothing to saie, either out of other men's writings or our owne experience'. But then it is not a Near Eastern or southern species, and so never attracted the attention of the older civilizations who handed on a knowledge of plants to mediaeval and Renaissance Europe. Anglo-Saxon leeches included *mersc meargealla* in a medicine for curing an eruptive rash, paper can be stained yellow with a dye made of Marsh Marigold petals boiled with alum, and in the United States, where the Marsh Marigold (native in the New World) has been called Cowslip (i.e. a plant growing where cows have dropped their dung. See *Primula veris*), it is eaten as a vegetable in the spring (88, 167, 131).

2. Globe Flower, Locker Gowlan. *Trollius europaeus* L. 56, H 3

Local names. BULL-JUMPLING, Kinrs; BUTTER-BASKET, BUTTER-BUMP, Yks; CABBAGE DAISY, Scot; GOLDEN BALLS, Ches, Lancs; GOLDILOCKS, West; GOLLAND, Cumb; GOWAN, N'thum; LAPPER GOWAN, Scot; LOCKER GOWLAN, West, Cumb, N'thum, Scot; LOCKIN GOWAN, West, Cumb, N'thum, Scot; LOCKIN-MA-GOWAN, Cumb; LOCKYER GOLDENS, Yks; MAY-BLOB, Leic; STOCKS, N'thum; WITCHES' GOWAN, Scot.

In Great Britain, as a mountainy species belonging to the Scottish range-type, the Globe Flower has never been constantly under the human eye, so the history of it is thin. It was unnoticed, it seems, until the intensification of plant studies in the sixteenth century. A 'gowan' or 'gowlan' is a yellow flower, a 'locken' or 'locker' gowan is a closed in, or locked in, yellow flower – apt and accurate, if not very lyrical.

> The lockety gowan an' the bonny bird-een
> Are the fairest flowers that ever were seen.

Trollius means much the same. The German *Trollblume*, latinized in the sixteenth century into *Trollius* or *Flos Trollius*, appears to be a contraction of *die rolle Blume*, so-called from the rolled in, or closed in, petals. It was *not* the flower of the trolls; and indeed Globe Flowers, in a squad or platoon in a limestone gorge in the West Riding, shaking in the wind, fresh, clean, butter-yellow, are not at all suggestive of evil, though the plant is sharp and poisonous.

Gerard first recorded Trollius as a British plant: 'This kinde of Crowfoote groweth in most places of Yorkshire and Lancashire, and other those

3 Globe Flower *Trollius europaeus*

bordering shires of the North countrey . . . but not founde wilde in these southerly or westerly parts of Englande, that I could ever understand'. Not quite accurate, because it does, in fact, come down through a damp westerly band of neighbouring counties, from Derbyshire – thinning out all the while – until it reaches Glamorgan and Monmouth.

Lyte in 1578 and Gerard in 1597 both called it Troll Flower, from the Latin. Gerard also used 'Locker Goulans' and 'Lockroun Goulans', 'Globe Crowfoote', and 'Globe Flower', the book name which has endured, although perhaps we ought to prefer Locker Gowlan.

'The country people of Westmorland, Scotland, and Sweden consider this as a sort of festival flower, going in parties to gather it, for the decoration of their doors and apartments, as well as their persons' (Smith, *English Flora*, 1829).

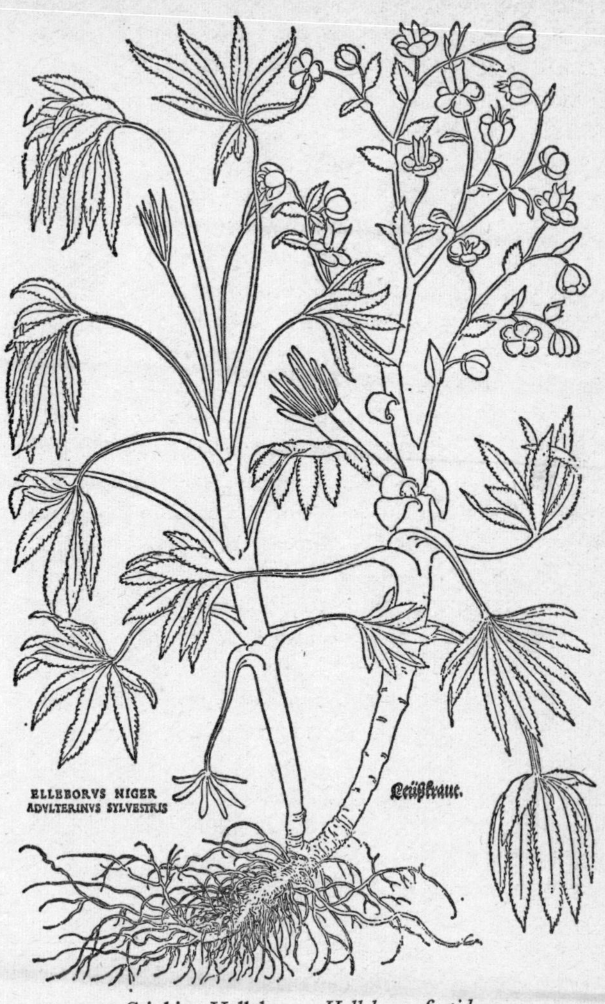

4 Stinking Hellebore *Helleborus foetidus*

3. Stinking Hellebore. *Helleborus foetidus* L. 32

Local names. BEAR'S FOOT, Dor, Som, Wilts, War, Worc, Yks; HE-BARFOOT, War; SETTER, Norf; SETTER-GRASS, Yks; SETTERWORT, Hants, I o W, Ess, Yks, N Eng.

4. Green Hellebore. *Helleborus viridis* L. ssp. *occidentalis* (Reuter) Clapham 45

Local names. BEAR'S FOOT, Glos; BOAR'S FOOT, Bucks; SHE-BARFOOT (in distinction to *H. foetidus*), War; FELLON-GRASS, West; GREEN LILY, Wilts.

For their rarity and their form, both kinds of Hellebore are excellent plants to discover, on a loose chalky or limestone slope, in a faint gloom under hazels overgrown with Old Man's Beard, or near yew trees (*H. ofetidus*), or in a moist wood on chalk or limestone (*H. viridis*); or even as relics of cultivation from an abandoned garden. Moreover, they both flower early in the year, between February and April. Often their dark leaves and sagging, livid green flowers rise up from a circle of snow. It is no good looking for them (as natives) very far north or in Ireland. *Stinking* goes up to Lancashire, *Green* to Westmorland. Gilbert White admired both species at Selborne, the *Stinking* because it 'continues a great branching plant the winter through', the *Green*, the smaller of the two, because it broke out in February, 'flowering almost as soon as it appears above ground'. A fact he did not know, and would have enjoyed, is that snails have a share in distributing the seeds of *H. foetidus*. The seed has an oil-body or 'elaiosome' which the snail eats. In the process, the seed sticks to the snail's slime and gets itself transported (159).

Both these hellebores are drastically, dangerously, and poisonously cathartic, and they were grown by English housewives partly against boils and spots (so the Westmorland name of Fellon-grass), partly against worms in children – a violent remedy, as Gilbert White recognized, by which the children were sometimes killed. 'Where it killed not the patient,' it was said, 'it would certainly kill the worms; but the worst of it is, it will sometimes kill both.' The doctor's kind was the Christmas Rose, *Helleborus niger*, which we still grow for its winter flowers. But the doctors also used the native Green Hellebore. In the eighteenth century a great quantity of the roots came up from the country every year to Guy's Hospital (50), no doubt making a fairly scarce plant scarcer still.

Also, by ancient prescription, the hellebores were needed for cattle. When a cow wheezed or coughed, an issue was made through its dewlap with a seton or thread, and a length of hellebore root was inserted to irritate – or counter-irritate – the flesh and keep it running. Both *Helleborus foetidus* and *H. viridis* (which in French is *herbe à setons*) were used in this way, so explaining 'Setter', 'Setter-grass', and 'Setterwort'. Setterwort and Oxheal were Gerard's designations for *H. foetidus*.

Either for men or cattle or both, the Green Hellebore was taken over to New England, where it is naturalized – a link across the Atlantic to Greek medicine, and to the mythical physician Melampus, who noticed the effect of Hellebore (*H. orientalis*) on goats, and then used it to purge a divine lunacy out of the daughters of the Kings of Argos.

5. Monkshood. *Aconitum anglicum* Stapf. 15

The wild Monkshood is one of the few kinds of plant peculiar to the British Isles, liking shaded or half-shaded ground along brooks and streams, and growing only in one or two of the south-western counties – in Gloucestershire, for example, and Monmouthshire and Glamorgan.

Aconites give some of the most virulent of all poisons, a fact which was known from early times in Europe and the Near East and through India. So the garden Monkshood, *Aconitum napellus*, which is very near the wild species, may have been introduced more for poison – to everyone his own supply – than for beauty. It was familiar by Gerard's time. In his *Herbal* of 1597, he commented on its 'very fair and goodly blew flowers, in shape like an helmet, which are so beautifull that a man woulde think they were of some excellent vertue, but *non est semper fides habenda fronti*' – appearances are not to be trusted, and if you eat Monkshood it kills you. The warning is repeated again and again. A hundred and fifty years later, Miller, in his *Gardeners Dictionary*, stated that Monkshood was in 'almost all old gardens', not to be put in the way of children 'lest they should prejudice themselves therewith'. Another two hundred years have gone by, and it is still in gardens, a reminder of ancient arrow poisons, murders, and fatal prejudicings. Old houses vanish, and the Monkshood will persist.

Local names for this garden plant are all of them charmingly innocent, relating mostly to the odd form of the flowers, and especially to the fluttering, dove-like nectaries. Some of them are: DOVES-IN-THE-ARK, Dor; BIRDS OF PARADISE, Som; NOAH'S ARK, Glam; LADY LAVINIA'S DOVE CARRIAGE, Som; VENUS' CHARIOT DRAWN BY DOVES (cf. French *char de Venus*), Dev, Berks, Ess; OLD WOMAN'S NIGHTCAP, Yks (and many variants of the last name from all over Britain). Monkshood is paralleled by the French *capuche de moine*, and equivalent names in German, Danish, Swedish, etc.

6. Herb Christopher. *Actaea spicata* L. 7

Herb Christopher is not unlike the *Anemone hybrida* of gardens, though it carries its small white flowers in a raceme, and these are followed by glittering green berries, which change to a glittering ebony. It is one of the rarer plants of Great Britain, occurring in a narrow band of scattered limestone localities across Yorkshire and Lancashire. Around Ingleborough, this plant fills hole after hole between the grey chunks of limestone or in the grey clefts of the limestone pavement.

In Norway, it is *troldbaer*, the troll's berry, since the berries are poisonous – 'deadly', wrote Gerard, 'and remediless' – a poison which acts upon the heart (134). Other names are *Hexenkraut*, 'witch's plant', *Teufelsbeer*, 'devil's berry', and *raisin du diable*. By the fourteenth century this evil creature had already been converted into *herba Christofori*, the herb of St Christopher. Perhaps it suggested St Christopher the ferryman by carrying its flowers, so to say, in a shoulder raceme, as the saint carried the infant Christ over the river. Another St Christopher's Herb is *Osmunda regalis*, the Royal Fern, which so conspicuously 'carries' its fertile leaves. Herb and Saint are certainly well matched. Before he was baptized and renamed, Christopher was the giant Reprobus, the rejected one, the troll, the ogre with the face of a dog who devoured men. As if the killing plant of the witches, wizards, trolls, and the devil had been baptized and ordered to kill no more, the rootlets of Herb Christopher reveal in section a cross or a star. St Christopher was also the patron saint of wizards.

7. Wood Anemone. *Anemone nemorosa* L. 109, H 40

Local names. The commonest names for a nearly universal plant, missing only from the Shetlands, the Orkneys, the Outer Hebrides, and some of the smaller islands, are: NEMONY, Dev, Som, Glos, Kent, Lincs, Lancs (or ENEMY, Dev, Wilts, Som, Lincs; and EMONY, Dev, Som, Lincs), and WIND-FLOWER, Dev, Som, Hants, Glos, Bucks, N'hants, War, Ches, Leic, Notts, Yks, Peeb; and probably in other counties.

Also, BOW BELLS, Worc; BREAD AND CHEESE AND CIDER, Dor; CANDLEMAS CAPS (Candlemas is 2 February, the Purification of the Blessed Virgin; see the names after the VIRGIN, below), Som; CHIMNEY SMOCKS, Som; CUCKOO-FLOWER, Som, Wilts, Bucks, Ches, Yks; CUCKOO, Wilts; CUCKOO-SPIT, Worc; DARN-GRASS (for giving cattle a disease called the Darn), Scot; DROPS OF SNOW, Suss; EASTER-FLOWER (cf. French *pâquette* and *fleur de vendredi saint*), Dev, Dor; EVENING TWILIGHT, Som.

FAIRIES' WINDFLOWER, Dor; GRANNY'S NIGHTCAP, Som, Wilts, War; GRANNY THREAD THE NEEDLE, Som; JACK O' LANTERN, Dor; LADY'S SHIMMY (i.e. chemise), LADY'S MILKCANS, Som; LADY'S NIGHTCAP, Glos, Heref; LADY'S PETTICOAT, Wilts; LADY'S PURSE, Dor.

MILKMAIDS, Som; MOLL O' THE WOODS, Dor, War; MOON-FLOWER, Worc; NANCY, Dor; NEDCULLION, NE Ire, Donegal; SHAME-FACED MAIDEN, Wilts; SHOES AND SLIPPERS, SILVER BELLS, Som; SMELL FOXES, Som, Hants; SMELL SMOCK, Herts; SNAKES AND ADDERS, Som; SNAKE'S EYES, Dor; SNAKE

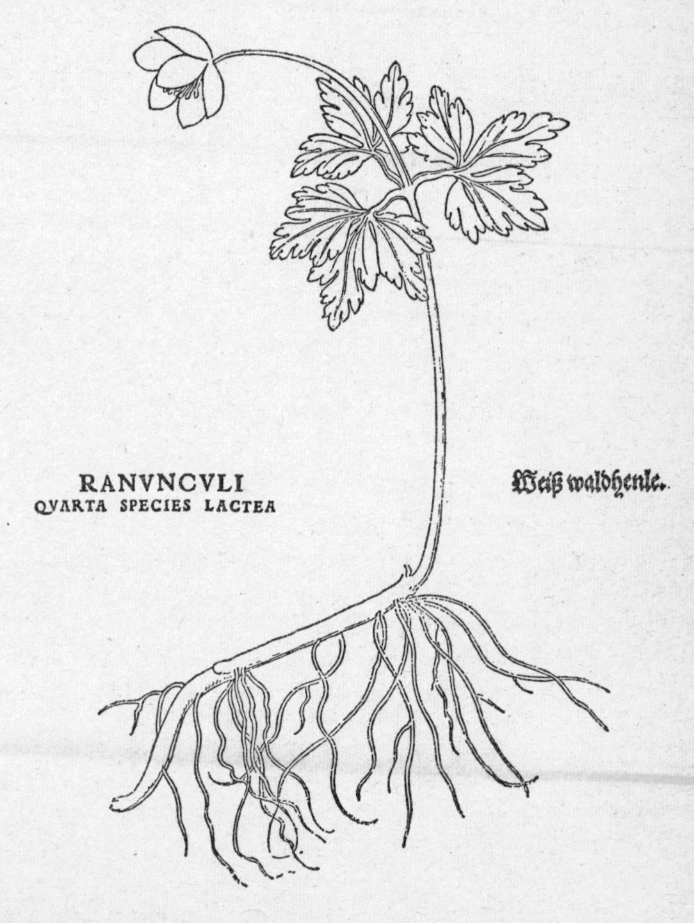

5 Wood Anemone *Anemone nemorosa*

FLOWER, Dor, Som, Lincs; SOLDIERS, Bucks; SOLDIER'S BUTTONS, Som; WHITE SOLDIERS, Bucks; STAR OF BETHLEHEM, Som, Donegal; WILD JESSAMINE, Dumf; WIND PLANT, Lincs; WOOLLY HEADS, Som.

Anemone and Windflower are names borrowed from the famous *Anemone coronaria* of Greek legend. They seem to fit our species well; the flowers do hang and nod and shake in the wind, although to the Greeks, Anemone, or Daughter of the Wind, probably signified something altogether different. Pliny's feeble explanation in the *Natural History* is that anemones do not open until the wind blows.

Other English names for *Anemone nemorosa* are not all so innocent. The plant has been linked with girls and their chemises and smocks (cf. the various German names containing *Hemd*, chemise) and with the wanton habits of the cuckoo, and with snakes – a connotation discussed under *Cardamine pratensis* (q.v.) and *Stellaria holostea*, which have a number of identical names. Matters were evidently corrected by transferring these white yet not untarnished flowers to the Virgin. One element (cf. 'Smell Smock', 'Smell Foxes') was the sharp, rather unpleasant, if faint, yet intriguing smell of *Anemone nemorosa*.

8. Pasque Flower. *Anemone pulsatilla* L. 16

Local names. BLUE EMONY, Rut; COVENTRY BELLS, Cam; DANES' BLOOD, Herts, Cam, Ess, Norf; DANES' FLOWER, Cam.

Pasque Flower has a fair claim to being the most dramatically and exotically beautiful of all English plants. 'Pasque Flower', French *passefleur*, Dutch *paaschbloem*, German *Osterblume*, since it blooms at Pasch, or Eastertide. Gerard wrote the best description of the flowers, which open out of the grey, chalky herbage: 'The first of these Passe flowers hath many small leaves finely cut or jagged, like those of carrots; among which rise up naked stalkes, rough and hairie; whereupon do growe beautifull flowers bell fashion, of a bright delaid purple: in the bottom whereof groweth a tuft of yellow thrums [stamens] and in the middle of the thrums thrusteth foorth a small purple pointell: when the whole flower is past there succeedeth an head or knoppe, compact of many graie hairie lockes, and in the solid parts of the knops lieth the seede flat and hoarie, every seede having his own small haire hanging at it.'

Gardeners have removed the Pasque Flower to the border – it was too much of a temptation. Farmers have ploughed up many of its old stations. In his *Wild Flowers of Chalk and Limestone*, J. E. Lousley has published a

6 Pasque Flower *Anemone pulsatilla*

distribution map showing where it still grows and where it is now extinct. It is on the decrease, yet there must still be several hundred localities for it across the chalk and limestone – colonies, for instance, by the dozen along the Cotswolds (see Riddelsdell, Hedley, and Price's *Flora of Gloucestershire*, 1948).

Probably it was between Cambridge and Suffolk that the Pasque Flower first became known as Danes' Blood or Danes' Flower. Unusual beauty suggested unusual origin. The flowers appeared on the great earthworks of the Devil's Dyke and Fleam Dyke. The dykes were associated with the Danes, so these flowers must have grown from the blood of the Danes, on the analogy of the Dwarf Elder, or Danewort (q.v.).

In support of two famous lines in his *Omar Khayyám*:

> I sometimes think that never blows so red
> The Rose as where some buried Caesar bled –

Edward Fitzgerald, who was an East Anglian, added a note about flowers, Danes, and Danes' blood, and the Pasque Flower on the side of Fleam Dyke.

9. Old Man's Beard. *Clematis vitalba* L. 53

Local names. BEDWIND, Dor, Wilts; BEDWINE, Dor, Wilts, Hants, Glos, I o W, Berks; BELLYWIND, Hants; BETHWIND, Glos; WITHYWINE, Som, Wilts, Glos; WILLOWWIND, Glos; BLIND-MAN'S-BUFF, Som; BULLBINE, Hants, Herts.

BOY'S BACCA, Hants, Suss; GIPSY'S BACCA, Som; SMOKING CANE, Dev, Som, Herts; TOM BACCA, Suss; SHEPHERD'S DELIGHT, POOR MAN'S FRIEND, Som.

BUSHY BEARD, DADDY'S BEARD, Som; DADDY'S WHISKERS, Som, Wilts; GRANDFATHER'S WHISKERS, Corn, Som; GRANDFY'S BEARD, Som; GREY-BEARD, Wilts, Hants; OLD MAN'S BEARD, Dev, Dor, Som, Wilts, Hants, Glos, Suss, Kent, Bucks, Oxf, Herts, Hunts, N'hants, War; OLD MAN'S WOOZARD, Bucks; OLD MAN, Som, Suss; FATHER TIME, Som.

CLIMBERS, Kent; DEVIL'S GUTS, Dor, Som, Wilts; HAG-ROPE (cf. German *Hexenseil*, *Hexenstrang*, witches' rope), Som; HALF-WOOD, Glos; HEDGE FEATHERS, Yks; HONESTY, Wilts, Glos, Berks, Oxf, E Ang, Heref, Worc, War, Lancs, Yks.

LADY'S BOWER, Som, Glos; VIRGIN'S BOWER, Som, War; MAIDEN'S HAIR, Bucks; SKIPPING ROPES, Wilts; SNOW IN HARVEST, Som, N'hants; TUZZY-MUZZY, Glos.

Mathias de l'Obel called this hedge-climber Viorna, which, rightly or wrongly, Gerard interpreted as '*viorna quasi vias ornans*' 'of decking and adorning waies and hedges'; and since we travel along ways and hedges, he coined the name Travellers' Joy. Fifty years before, William Turner had coined Hedge-Vine and Downi-Vine (183). Probably names enough existed already. 'Bethwind' and 'Withywine' have an old look about them. But then Turner came from Northumberland and Gerard from Cheshire, whereas *Clematis vitalba* grows mainly in the south of England.

It is Old Man's Beard (cf. German *Altermannsbart*) from the long

7 Old Man's Beard *Clematis vitalba*

feathery styles. But observe that the Old Man, as so frequently in English plant names, may also be the Devil. This is a devil's twister, devil's guts, which can twist and choke trees to death, and turn a south-country copse into an Amazonian forest. On the one hand, the plant of the Devil and witches; on the other, the plant of God and the Virgin: *Herrgottsbart*, *barbe au bon dieu*, *Muttergotteshaar*, *chevaux de la Bonne-Dame*, *berceau de la Vierge*, beside the English 'Maiden's Hair', 'Lady's Bower', 'Virgin's Bower'.

Why should this plant be so very widely named Honesty? Parkinson wrote, in his *Theatrum Botanicum* of 1640, 'In English, of most country people where it groweth, Honestie; and the Gentlewomen call it Love'. John Aubrey, a few years later, gave the name in his *Natural History of Wiltshire* as Maiden's Honesty. It is Boy's Bacca because boys (as well as gipsies – Gipsy's Bacca – and shepherds – Shepherd's Delight – and the poor generally – Poor Man's Friend) smoked cigar lengths of the dry stems, which draw well and do not burst into flame. These cigars or cigarettes were smoked elsewhere – cf. the names *Rauchholz*, 'smoke wood', in German, *fumailles*, *bois à fumer* in French, and *smookhout* in Dutch.

10–12. Buttercup. *Ranunculus* species of the meadow

The three kinds, *R. acris*, *R. repens*, and *R. bulbosus*, are not very much marked from one another in their common names, which are legion, as Nicholas Culpepper wrote: 'Many are the Names this furious biting Herb hath obtained, almost enough to make up a Welch-mans Pedegree, if he fetch it no further than *John* of *Gaunt*, or *William* the Conqueror' (49). But why Buttercup? As long as there have been meadows and cows to feed in the meadows, there have been Buttercups (as Culpepper also remarked: 'They grow common everywhere. Unless you run your Head into a Hedge, you cannot but see them as you walk'), and it was natural to link the yellow of butter and cream with the rich yellow of the petals or of the pollen, which will colour the hooves of a grazing cow. The connection was magically emphasized. On May Day the Irish still rub the flowers on the cow's udder (137), which may once have been a general practice; and in the Highlands *Ranunculus ficaria* (q.v.) had its own function in the byre.

The name Buttercup seems to have come late into general use – rare (at least in print) before the middle of the eighteenth century. 'Butter-flower', little heard now in the country, was used by Gerard in 1597 for *R. acris* and *R. repens*, along with the exactly equivalent names in German, and with Kingkob, Gold cups, Gold knops, and Crowfoot. In 1538 William Turner had distinguished *R. bulbosus* (apparently) as a Crowfote, Kyngcuppe, or Golland (182). Crowfoot was the general name with the early botanists. Every child now learns in the village school that a Buttercup is a Buttercup, but that has yet to kill other names such as Crazy, Gilcup, Goldcup; and Turner's 'Kingcup', which is still used more widely for Buttercups than for Marsh Marigold.

Local names. General names covering the three species are: BACHELOR'S

BUTTONS, BUTTER AND CHEESE, Som; BUTTERBUMP, Yks; BUTTERCHURN, War; BUTTERCREESE, BUTTER-DAISY, Bucks; BUTTER FLOWERS, Som, Herts, Derb; BUTTER ROSE, Dev.

CALTROPS, Som; COWSLIP, Dev; CRAZY, Dev, Som, Wilts, Hants, Glos, Berks, Bucks, War, Worc, Lancs; CRAZY BET, Wilts; CRAZY WEED, Bucks; CROWFOOT, Som, Hants, Glos, Suss, Ches, Yks, Cumb, Lanark; CROWTOE, Dev; KRAA-TAE, Shet; CUCKOO-BUDS (used by Shakespeare in *Love's Labour's Lost*, v. ii. 885), Som, N'hants, Worc; DALECUP, DELLCUP, Som; DEWCUP, Dor; DILLCUP, Dor, Hants; FAIRIES' BASINS, Dev.

GILCUP (meaning Gilt or Gilded Cup), Dev, Dor, Som, Wilts, Hants; GILTED-CUP, GILDCUP, Som; GILTYCUP, Dev, Dor, Som; GOLD BALLS, GOLD CRAP, Som; GOLDCUP, Corn, Dev, Som, Wilts, Hants, Suss, Kent; GOLDENCUP, Dev, Som; GOLD KNOP, Glos; GOLDEN KNOP, Oxf; GOLD-WEED, GOLDY, Som; GOLLAND, Yks, N'thum, Berw; GLENNIES, Wilts.

KINGCUP, Corn, Dev, Som, Suss, Bucks, Ess, Norf, Cam, N'hants, Ches, Yks, Cumb; KING'S CLOVER, KING'S COB, Suff, Norf, Cumb; LAWYER-WEED, MARYBUDS, MAYBUDS, OLD MAN'S BUTTONS, Som; PAIGLES, Suff; SITSICKER, S Scot; SOLDIER BUTTONS, Som; YELLOW CAUL, I o W; YELLOW CREAMS, Som; YELLOW CREES, Bucks, Herts; YELLOW CUP, Wilts, Hants, I of W, Bucks; YELLOW GOLLAN, N'thum, S Scot; TEACUPS, Som.

10. Meadow Buttercup. *Ranunculus acris* L. 112, H 40

These names have been recorded for *R. acris*, though most of them are no doubt applied to other species as well:

BASSINET, Som (Lyte in 1578 gave Brave Bassinet for Marsh Marigold. *Bassinet* is a general name for Buttercups in Normandy); BLISTER CUP, Lincs; BLISTER PLANT, Som; CLOVEWORT, N'hants; CROWFLOWER, Staffs, Rut, Midlands; GOWAN, Wigt; LADY'S SLIPPER, Som.

11. Creeping Buttercup. *Ranunculus repens* L. 112, H 40

Distinct enough in its creeping habit to have some distinctive names:

BUR-CROWFOOT, Yks; CAT'S CLAWS, Lancs; CRAWFEET, Yks, Scot; CRAZY-MORE, Wilts; CREEPING CRAZY, Glos; CROWTOE, Donegal; DELTY-CUP, Wilts; DEVIL'S GUTS, N'thum; LANTERN LEAVES, Som, W Eng; MANY FEET, Yks; MEG MANY FEET, Cumb; MEG WI' MANY TEAZ, Cumb, Lakes; OLD WIFE'S THREADS, Yks; RAM'S CLAWS, Dor, Som, Wilts; SITFAST, Dumf, Ire; TANGLE-GRASS, TETHER TOAD or TOAD TETHER, Yks.

12. Bulbous Buttercup. *Ranunculus bulbosus* L. 110, H 40

Local names. CROWBELLS, War; CROWPICKEL, N'hants; CROWPIGHTEL, Beds, N'hants; EGGS AND BACON, Ches; FAIRGRASS, Rox; KING'S NOBS, MAIDEN IN THE MEADOW, Som.

For their 'furious biting' qualities, these and other kinds of Ranunculus have had their use in medicine and country medicine, especially for producing blisters. The blistering substance is anemenol. In the sixteenth century, bubonic plague cried out for specifics. One of these was the blistering root of the Crowfoots, 'which being stamped with salt is good for those that have a plague sore . . . by means whereof, the poison and malignitie of the disease is drawen from the inward partes . . . into those partes of lesse account. For it exulcerateth and presently raiseth a blister to what part of the bodie soever it is applied' (Gerard, 1597). From the ancient *Herbarium* of Apuleius Platonicus Gerard also quotes one passage which probably explains the name Crazy and its variations, such as Crazymore (which means Crazy root). If the root 'be hanged in a linnen cloth about the necke of him that is lunatike in the waine of the moone, when the signe shall be in the first degree of *Taurus* or *Scorpio*', according to Apuleius, 'then he shall foorthwith be cured'.

A pleasanter note by Culpepper says of the flowers that 'Virgins in Ancient time did use to make powder of them to strew Bride-Beds' (49).

13. Corn Buttercup. *Ranunculus arvensis* L. 71

Local names. CLENCH, N'hants; COGWEED, Yks; CROWCLAWS, Hants, Suss, Ess; CROWPECK, Wilts; DEVIL'S CLAWS, Hants, I o W; DEVIL-ON-ALL-SIDES, Yks; DEVIL-ON-BOTH-SIDES, Bucks, War, Dur; DEVIL'S COACHWHEEL, Hants; DEVIL'S CURRYCOMB, Shrop.

EGGS AND BACON, Ches; GYE, Ess, Suff, Norf, Lincs; HARDINE, HARD-IRON, Midlands, Derb, Leic; HELLWEED, Yks; HORSE-GOLD, Herts, N'hants; HUNGERWEED, Glos, Norf; JACKWEED, Oxf; JACK O' BOTH SIDES, Leic, Notts; JACK O' TWO SIDES, Shrop; SCRATCH-BUR, Beds.

A nasty weed of cornfields. The achenes, much larger than those of the meadow Buttercups, are covered with spines, real claws of the Devil. When the fruit develops, the achenes stick out all round, Devil-on-both-sides, or more exactly, Devil-on-all-sides, a Devil's coachwheel made up of Devil's currycombs. Gerard knew the plant, and noticed the 'clusters of rough and sharpe pointed seedes'.

14. Lesser Spearwort. *Ranunculus flammula* L. ssp. *flammula*. 112, H 40

Local names. BUTTERPLATE, Berw; COWGRASS, N Ire; CROWFEET, Som; GOOSETONGUE, Carm, Scot; WATER BUTTERCUP, Wilts; SNAKE'S TONGUE, Berw; WILFIRE, Scot; YELLOW CRANE, N'hants.

Another *Ranunculus* used for blistering, a practice which survived late in the Highlands and the Islands. It was pounded in a mortar and then neatly applied in limpet shells (128). Gerard also called the plant 'Banewoort' 'bicause it is dangerous and deadly for sheepe; and that if they feede of the same it inflameth their livers, freeteth and blistereth their guts and entrailes' – i.e. it caused liver rot, an accusation which Fitzherbert had made against this 'sperewort' in his *Boke of husbandrie*, 1523.

15. Water Crowfoot. *Ranunculus fluitans* Lam., *R. aquatilis* L., etc.

Local names. BACON AND EGGS, Som; COW-WEED, Hants; EELWARE, N'thum; EEL-WEED, Donegal; PICKEREL-WEED (i.e. young pike weed), E Ang; RAIT, W Eng, Som, I o W, Glos, War, Heref, Worc, Shrop, Leic, Midlands, N Eng, Scot; RAWHEADS, Shrop; WATER LILY, W Eng, Wilts, Donegal; WATER NEMENY, Som, Wilts; WHITE CROWFOOT, Shrop.

A short list of names. The buttercups in his cow ground the farmer observed, the flowers snowing the stream at the bottom of his cow ground he did not observe – unless the stream was blocked. Then he had to cut and clear the 'Rait', which is more the tangle than the plant itself, with its white blossoms under a blue sky. When people began to look outside themselves, Wordsworth made them aware of the next plant in this book, the Lesser Celandine: half a century later William Barnes wrote excellently on this Crowfoot of his Dorset streams. The poem is not too familiar:

> O small-feäced flow'r that now dost bloom
> To stud wi' white the shallor Frome,
> An' leäve the clote to spread his flow'r
> On darksome pools o' stwoneless Stour

When sof'ly-rizèn aïrs do cool
The water in the sheenèn pool,
Thy beds o' snow-white buds do gleam
So feäir upon the sky-blue stream,
As whitest clouds, a-hangèn high
Avore the blueness o' the sky. . . .

An' oh! as long's thy buds would gleam
Above the softly-slidèn stream,
While sparklèn zummer-brooks do run
Below the lofty-climèn zun,
I only wish that thou could'st staÿ
Vor noo man's harm, an' all men's jaÿ.
But no, the water man 'ull weäde
Thy water wi' his deadly bleäde,
To slay thee even in thy bloom,
Fair small-feäced flower o' the Frome

The 'clote' in the third line is the Yellow Water-lily.

16. Lesser Celandine. *Ranunculus ficaria* L. 112, H 40

Local names. BRIGHTEYE, Dev; BUTTER, BUTTER AND CHEESE, BUTTERCHOPS, CHEESECUPS, Som; CRAIN, N'hants; CRAZY, Wilts, Glos; CRAZY BET, Wilts; CRAZY CUP, Som; CREAM AND BUTTER, Dev; CROWPIGHTLE, Beds; CUPS, Glos.

DILLCUP, Wilts; FOALFOOT, Ayr; FOGWORT, Dor; FOXWORT, FROG'S FOOT, GENTLEMAN'S CAP AND FRILLS, Som; GILCUP, Som, Wilts; GOLDEN CUPS, Dev, Som; GOLDEN STARS, Som; GOLDEN GUINEAS, N'hants; GOLDY KNOB, Oxf; KINGCUP, Dev; KING'S EVIL, Som.

LEGWORT, POWERWORT, Som; SPRING MESSENGER, Dor; STARFLOWER, STARLIGHT, Som.

William Turner in 1548 made the first mention of *Ranunculus ficaria* as a British plant: 'Figwurt groweth under the shaddowes of ashe trees.' It was Figwort, or Pilewort, because the root tubers were the signature for the 'fig' or piles: 'The later age use the rootes and graines for the piles, which being often bathed with the juice mixed with wine, or with the sick man's urine, are drawne togither and dried up, and the paine quite taken away' (Gerard). Like the larger Figwort, *Scrophularia nodosa* (q.v.), it also cured 'kernels by the ears and throat, called the King's Evil'. – 'Let good people make much of it for these uses', wrote Culpepper; 'with this I cured my own Daughter of the King's-Evil, broke the Sore, drew out a quarter of a pint of Corruption, cured it without any scar at all, and in one week's time' (49).

More pleasantly, the root tubers suggested the teats of a cow, while the yellow petals suggested butter. So in the Highlands and Islands, the herb was important as *lus an torranain*, the 'herb of St Torranan', or St Ternan, the missionary of Uist. It was gathered with Gaelic charms:

> I will pluck the figwort
> With the fulness of sea and land,
> At the flow, not the ebb of the tide,
> By thine hand, gentle Mary . . .

And to produce a high tide of cream in the cow's udder, the root, the 'cluster of four bulbs like the four teats of the cow', was now hung in the byre or on the cow fetter (31a).

As *Chelidonium minus*, translated into Lesser Celandine, this plant was linked to the Greater Celandine, *Chelidonium majus* (q.v.), though the two are unrelated. *Chelidonium majus* was the larger 'swallow herb'; *Ranunculus ficaria* was the smaller 'swallow herb' (Greek *chelidon*, a swallow) 'bycause that it beginneth to springe and to flowre at the comming of the swallows, and withereth' – though it withers long before – 'at their return' (Lyte, 1578). A third Scrophularia, from the signature of the glands of King's Evil in its tubers, was *Sedum telephium*.

In the Faeroes, the Lesser Celandine is an introduced plant, surviving in churchyards, in which it was planted as a medicinal herb (139).

17. Pheasant's Eye. *Adonis annua* L. 11

Local names. JACK IN THE GREEN, Wilts; LOVE LIES BLEEDING, Glos.

'The red flower of Adonis groweth wilde in the west parts of Englande among their corne,' Gerard observed in 1597, and he was wise enough to get seeds from the West of England and grow Red Maythes or Red Chamomile or Rose-a-ruby in his garden, 'for the beautie of the flowers sake'. In the corn, this naturalized foreigner has always been scarce or local, though it became a garden favourite for the scarlet petals and the dark centres. According to the *Flora Londinensis*, it was cried around the streets of eighteenth-century London as 'red Marocco'.

The pheasant has a red eye. So has the Devil, and in Germany this plant is not Pheasant's Eye, but *Teufelsauge*, 'devil's eye' – cf. the Welsh *llygad y bwgan*, 'sprite's eye', for the Poppy, *Papaver rhoeas*.

18. Columbine. *Aquilegia vulgaris* L. 66, H 31

Wild blue Columbines are so pleasant to meet in a damp hedge in Devon or Cornwall or Somerset that one does not immediately think of gardens, or of the newest Columbines in the Chelsea Flower Show. The wild ones look wild, each flower, as Gerard remarked, 'with five little hollowe hornes, as it were hanging foorth, with small leaves standing upright of the shape of little birds'.

Here are some of the local names for Columbine (from *columba*, a dove or a pigeon, on account of the dove-like flowers), though most of them must have been prompted by plants in cultivation:

BABY'S SHOES, Som; BACHELOR'S BUTTONS, CAINS AND ABELS, Wilts; BONNETS, Som, War; BOOTS AND SHOES, Corn, Som; DOLLY'S BONNETS, DOLLY'S SHOES, DOVES AT THE FOUNTAIN, DOVES IN THE ARK, DOVES ROUND A DISH, Som; FOLLY'S FLOWER, Dor, Som; FOOL'S CAP, Yks.

GRANDMOTHER'S BONNET, Dor, Som; GRANNY HOODS, Yks; GRANNY-JUMP-OUT-OF-BED, Wilts; GRANNY'S BONNET, Corn; GRANNY'S NIGHTCAP, Dev, Dor, Wilts, Som, Glos; HEN AND CHICKENS, Norf; LADY'S PETTICOAT, Som; LADY'S SHOES, Ess, Norf, Cam; LADY'S SLIPPERS, Corn, Wilts; NIGHTCAPS, Som, Wilts; NOAH'S ARK, OLD LADY'S BONNET, OLD MAID'S BASKET, OLD WOMAN'S BONNET, OLD WOMAN'S NIGHTCAP, RAGS AND TATTERS, SHOE AND STOCKINGS – all names recorded in Som; SKULLCAPS, Corn; SNAPDRAGON, Dev; SOLDIER'S BUTTONS, Som, Wilts; STOCKING AND SHOE, Corn; THIMBLES, Som; TWO FACES UNDER A HAT, Suss; WIDOW'S WEEDS, Wilts.

An old name which has been much discussed is Culverkeys – 'azure Culverkeys' in Walton's *Compleat Angler*, and Calverkeys in John Aubrey's *Natural History of Wiltshire*. In each place, wild Columbine was probably meant. Though it may be used no longer, Culverkeys for Columbine has been recorded in Dorset and Somerset.

Columbine seems altogether a mild and gentle plant, yet it was believed in the Middle Ages to be the food of lions. Rub Columbine on your hands, and you had the courage of a lion. The possible source of this is that doves were the sacred birds of the love goddess Aphrodite and her Phoenician counterpart Astarte, who was also associated with lions.

V. Paeoniaceae

1. Peony. *Paeonia mascula* (L.) Mill. 1

This Peony of southern Europe has lived on the island of Steep Holm, in the Severn estuary, for a hundred and fifty years at the least. It was discovered in 1803, soon catching the fancy of botanists, antiquaries, and poets. Here was a plant new to the British flora, which lived in this unique, isolated station. Steep Holm, 'abrupt and high' (258 feet high), wrote William Lisle Bowles,

> And desolate, and cold, and bleak, uplifts
> Its barren brow – barren, but on its steep
> One native flower is seen, the peony;
> One flower, which smiles in sunshine or in storm,
> There sits companionless, but yet not sad.

He went on as you might expect –

> – so Virtue, a fair flower
> Blooms on the rock of Care, and, though unseen,
> So smiles in cold seclusion.

A moral and romantic account (16). But was the Peony a *native* flower? In the Middle Ages monks had lived for a while on Steep Holm, and monks had physic gardens, and Peony came of a medicinal family, and there were other plants on the island, where the Peony did not sit 'companionless' by any means. William Turner in the sixteenth century knew that Alexanders, one of the oldest of pot-herbs, grew on the desert of Steep Holm. It is still there today, making a tall forest of shabby vegetation. In 1668, John Ray knew that the plants of Steep Holm included *Allium ampeloprasum*, one of the rarest of garlics. In 1775, visitors found the Caper Spurge, which was the Catepeuce of mediaeval physic. Later on, Coriander and Red Valerian were added to the list ('Flora of Steep Holme', R. P. Murray, *Journal of Botany*, vol. 29, 1891).

As for the monks, the occupants were a small community, probably of Austin Canons – the Brethren of St Michael of Steep Holm – who lived there for a number of unworldly years in the thirteenth century, and then gave up and went back to the mainland before the century was out. But if the monks had left the Peony, as well as Alexanders, Caper Spurge,

Coriander, etc., how was so obvious a species overlooked by every botanical visitor between the sixteenth century and 1803? Were they blind? Perhaps they were, or perhaps they always came too late. Or perhaps there was no Peony before the beginning of the nineteenth century. Perhaps there was no Peony until after a chance seed had fallen from the pocket of some gardener who came on a visit. Or did some romantic mystifier plant *Paeonia mascula* on the island deliberately, as a hoax, to catch all the botanists of England? No one can ever tell how the Peony got to Steep Holm, though the case for the Brethren of St Michael ought not to be dismissed too easily.

Do not imagine that the Peony colours the whole limestone mass of Steep Holm. It grows only in two places on the cliff, and there have never been a great many plants.

VI. Berberidaceae

1. Barberry. *Berberis vulgaris* L. 104

Local names. BERBER, Scot; GUILD or GUILD TREE, Selk; JAUNDERS BERRY, Som; JAUNDICE or JAUNDERS TREE, Corn, Som; PIPPERIDGE or PIPRAGE, Hants, S Eng, Herts, E Ang, Lincs, N Eng; WOODSOUR, WOODSORE (cf. German *Sauerdorn*, and the old botanists' Latin *spina acida*), Oxf.

Barberry in the hedges often survives from a mediaeval garden, as at Kington St Michael in Wiltshire, where it is a relic of the Benedictine Nuns, or on the site of Godstow Nunnery, near Oxford. It was a useful shrub. The red fruits make a delicious jam, or jelly. They were candied and eaten as sweets, they were used in a fever drink, punch was flavoured with them, and with the bark leather could be tanned or dyed. Above all, Barberry was a shrub with a yellow bark for a yellow disease. William Coles, in his *Adam in Eden* (1657), wrote that the yellow inner bark of the branches and the roots bore the signature of the yellow jaundice; and in Ireland herb doctors improved matters by mixing yellow sulphur with the yellow bark and administering them to the yellow patient in stout (100).

Pipperidge is an odd name of uncertain derivation. William Turner used it in 1538. Earlier still the plant had been called Berbarya or Barboranne, from the mediaeval Latin *barbaris*, *berberis*, the meaning of which is uncertain, though perhaps it is connected with *barb*, in reference to the tripartite spines of the Barberry which resemble arrow-heads.

VII. Nymphaeaceae

1. White Water-lily. *Nymphaea alba* L. 98, H 37

Local names. BOBBINS, Bucks; CAMBIE-LEAF, N Scot; CAN-DOCK, Som, War, Notts; CAN-LEAF, War, Leic; FLATTERDOCK, FLOATING DOCK, Ches; LADY OF THE LAKE, Som; LOUGH LILY, Donegal; SWAN AMONGST THE FLOWERS, Dor, Wilts; WATERBELLS, N Eng; WATERBLOB, N'hants, Yks.

2. Yellow Water-lily. *Nuphar lutea* (L.) Sm. 96, H 38

Local names. BLOBS or WATERBLOBS, Dor, Wilts, N'hants, Yks; BOBBINS, Scot; BRANDY BOTTLE, Wilts, Som, Berks, Suss, Norf; BULL'S EYES, Dor, Som; BUTTER-CHURN, Mor; BUTTERPUMPS, Som; CAMBIE-LEAF, N Scot; CAN-DOCK, Som, War; CHURN, Oxf; CLOTE, Dev, Dor, Som; CRAZY or CRAZY BET, Som; FLATTERDOCK, FLOATING DOCK, Ches; LILY-CAN, Fife, Perth; PATTY-PANS, Worc; QUEEN OF THE RIVER, SOLDIER'S BUTTONS, WATER CAN, WATER CUPS, Som; WATER GOLLAND, Yks.

Thirteenth-century sculptors often carved the Yellow Water-lily on roof bosses, etc. There are examples in Bristol Cathedral, in Westminster Abbey, and in the Angel Choir at Lincoln, where the sculptor gave a ripple and a liquid sinuosity to the leaves (33).

Both kinds of Water-lily have had some curious applications. The rhizomes were steeped in tar and applied against baldness. The black rhizomes of the White Water-lily in red wine were given for leucorrhoea, and over-warm Elizabethans ate the seeds and the powdered rhizomes in broth and with meat to induce chastity.

Yellow Water-lily is called Brandy Bottle because the flowers are credited with the smell of stale dregs of brandy. More exact would be the smell of stale dregs of a sweet white wine. 'Can-dock' is a name which needs explanation. Look at a capsule of the Yellow Water-lily when the sepals and petals have fallen away: each capsule is a green-glazed carafe, a can – an old kind of can before the word began to suggest cans only of metal, and of a different shape. So a Can-dock implies a plant with leaves like a dock and capsules like carafes. In Germany, *Kanne* is still a pot or a jug or a tankard.

Both species were called Nenuphar by Elizabethan apothecaries, and by *The Grete Herball* of 1526. This is a fancy word, which came by way of Mediaeval Latin from *nilotpala*, the Sanskrit for the Blue Lotus of India (*Nymphaea stellata*).

VIII. Papaveraceae

1. Field Poppy. *Papaver rhoeas* L. 104, H 39

Local names. BLIND EYES, N'hants; BLIND MAN, Wilts; BLINDY BUFFS, Yks; BULL'S EYES, BUTTERFLY LADIES, Som; CANKER, Suff, Norf; CANKER-ROSE, E Ang; CHEESEBOWLS, Som; COCKENO, N'thum, Berw; COCK-ROSE, Yks, Scot; COCK'S COMB, Berw; COCK'S HEAD, Scot; COLLINHOOD, Rox, Loth; COP-ROSE, Som, E Ang, Yks, N'thum, Ire; CORNFLOWER, Dev; CORN-POPPY, Corn; CORN-ROSE, Dor, Som; CUP-ROSE, Yks; CUSK, War.

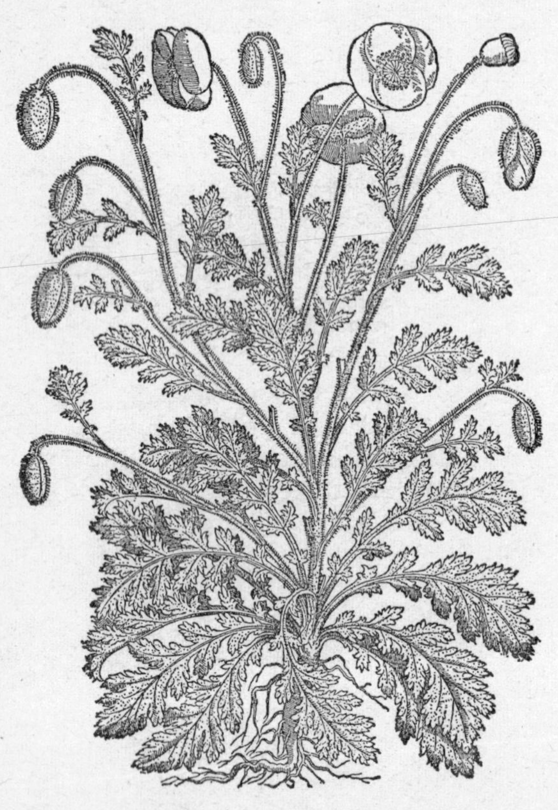

8 Field Poppy *Papaver rhoeas*

DEVIL'S TONGUE, Corn; EARACHES, Derb, Notts; FIREFLOUT, Som, N'thum; GOLLYWOGS, Som; GYE, Suff; HEADACHE, Som, Suss, E Ang,

N'hants, Ches, Derb, Leic, Notts, Rut, Lincs, Yks, Cumb, Ire; HOGWEED (cf. Cornish *rosen-mōgh*, 'pig rose'), E Ang.

LIGHTNINGS, N'thum; OLD WOMAN'S PETTICOAT, PARADISE LILY, PEPPER BOXES, Som; POISON POPPY, Bucks; POPPET, War; POPPLE, Glos, Kent, Oxf, N'hants, Notts; REDCAP, REDCUP, RED DOLLY, RED HUNTSMAN, RED NAP, Som; RED PETTICOAT, Kent; RED RAGS, Dor; RED SOLDIERS, Som; RED-WEED, Dor, Som, Wilts, Hants, I o W, Suss, Kent, Berks, Bucks, Herts, Norf, E Ang; SLEEPYHEAD, Som; SOLDIERS, Wilts, Norf, N'hants; THUNDER-BALL, War; THUNDERBOLT, Dev, W Eng, Shrop, Ches; THUNDERCUP, Berw; THUNDERFLOWER, Wilts, Berw; WARTFLOWER, Corn; WILD MAWS, Derb.

Through the Latin *papaver*, our English word poppy may have come from the Sumerian *pa pa*. At any rate, although cornfields 'garnished and overspread' with poppies are no longer so common a sight (except where the poppies are favoured by lighter soil), for thousands of years corn and poppy and civilization have gone together. If farmers think of the poppy as a weed, in earlier ages it was rather the corn's inseparable companion. Assyrians respectfully named it 'Daughter of the Field', in Greece it was a flower of Aphrodite as goddess of vegetation, Romans looked on it as sacred to their corn goddess, Ceres, who taught men to sow and reap and thresh and winnow. On her yellow hair Ceres had a crown of wheat, in one hand she held a lighted torch, in the other a bunch of corn and poppies. Fierce sheets of poppy seem to have been regarded as a life blood sprouting from the soil along with the grain of nourishment. Poppies after a battle seemed a natural consequence in more than one European war, our own story and symbol of the Flanders poppy showing that life and the blood-red flower of the cornfields are not altogether divorced, even now: out of blood comes blood (see *Sambucus ebulus*).

So the poppy was under protection. It has been recorded in Ireland that women had a dread of touching poppies. English people (see the names above) have had the notion that smelling poppies induces a headache, and that staring at them too long makes you blind. However, the poppy also cured headaches. According to the *Agnus Castus* of the late fourteenth century, 'If a man hawe ony peyne aboutyn his eyne or if a man hawe a mygreyn tak this herbe and stamp it and nedle it with olye de olywe and anoynt ther-with the forhed and it schal amende the syth and slake the peyne and distroye the mygreyn' (20). The belief has lingered among English children that poppies must not be picked for fear of a thunder-storm – so the names Thunderbolt, Thundercup, Thunderflower, Lightnings (160). So in the neighbourhood of Liège, the poppy was the *fleur di tônit*, the thunder-flower. Picking it brought thunder; but poppies placed among the timbers under the roof warded off the lightning (165). For other thunder plants, see *Sempervivum tectorum* and *Melandrium album*. In Welsh, too, the poppy is *llygad y bwgan*, 'sprite's eye', 'bucka's eye', 'goblin's eye'.

The Field Poppy was used in medicine, nevertheless; though it lacked the properties of *Papaver somniferum*, which is the source not only of opium, but of poppy-oil for cooks and artists. The Opium Poppy was known to neolithic lake villagers in Switzerland, and was perhaps cultivated by them.

2. Yellow Horned Poppy. *Glaucium flavum* Crantz. 51, H 16

Local names. BRUISEROOT, Dor, Hants, S Eng; GOLD WATCHES, Dor; HORN POPPY, Dev; INDIAN POPPY, Som; SQUAT, SQUATMORE, Dor, Hants, S Eng.

William Turner in 1548 called this fine plant the Horned Poppy or Yellow Poppy, Gerard added Sea Poppy in 1597, but it was too local, and on the 'sandes and bankes of the sea' it grew too much away from men to have had any general impact. John Aubrey, though, wrote to tell Ray the naturalist, in 1677, that 'by the salt pits at Lymington, Hampshire, grows a plant called Squatmore' – which is the Yellow Horned Poppy – 'of wonderful effect for bruises'. A squat is a bruise, and Squatmore is 'bruise root'.

In 1698 an odd story was reported to the Royal Society, that 'a certain person made a pye of the roots of this plant, supposing them to be roots of the Eryngo [Sea Holly, q.v.], of which he had before eaten pyes which were very pleasant, and eating it while it was hot, became delirious, and having voided a stool in a white chamber pot, fancied it to be gold, breaking the pot in pieces, and desiring what he imagined as gold might be preserved as such. Also his man and maid servant eating of the same pye, fancied of what they saw to be gold' (144). The dangers to life were considerable in the days of herbal medicine.

3. Welsh Poppy. *Meconopsis cambrica* (L.) Vig. 21, H 13

This Atlantic plant of Wales and the West (Cheddar Gorge, for instance), of Ireland, Western France and Northern Spain, has a right to its name. It was first found and recognized in North Wales, by the Flemish botanist Mathias de l'Obel (1538–1616), probably when he was visiting Wales in company with his friend and patron Lord Zouche, who was made Lord President of Wales in 1602 (151). From de l'Obel's papers John Parkinson, in his *Theatrum Botanicum*, 1640, was able to give a list of habitats in North Wales for this delicate yellow-flowered poppy, the only species of its genus outside South-western Asia and the Himalayas.

4. Greater Celandine. *Chelidonium majus* L. 96, H 40

Local names. DEVIL'S MILK, Yks; JACOB'S LADDER, Shrop; KILL-WART, Dev; ST JOHN'S WORT, Dev; SALADINE, Ches, Yks, Cumb; SOLLENDINE, Ire; SWALLOW WORT, Som; WARTFLOWER, Dev; WART PLANT, Som; WARTWEED or WRETWEED, E Ang; WARTWORT (cf. *Warzenkraut*, in German), Som, Wilts, Glos; WITCH'S FLOWER (cf. German *Hexenkraut*), Som; YELLOW SPIT, Hants.

Break a stem of Celandine and a drop of orange latex comes out, a drop of that 'yellow spit' which is still applied to warts. It is acrid, it makes the skin go slightly red, but it is not a very effective remedy. When Dioscorides described *chelidonion*, he wrote nothing of warts, but he told a story which appealed to every herbalist. The name *Chelidonion*, from the Greek word for a swallow, was given to the plant either, he explained, because it flowers at the coming of the swallows or because mother swallows, so it was said, employed it to give sight back to their blinded nestlings. It was an eye herb. 'The juice' – the same orange latex – 'is good to sharpen the sight', wrote Gerard, following the Master, 'for it cleanseth and consumeth awaie slimie things that cleave about the ball of the eye'. Sir Hugh Plat, in his *Delights for Ladies* (1602), mentioned the excellence of May-dew from Fennel and Celandine in the treatment of sore eyes. The juice of Celandine took away cloudy spots on the eyeball which were called 'kennings'; and as 'Kenning-wort', Celandine was taken to New England by the West Countrymen (111). It is now a naturalized plant in the Eastern states; and the Americans call it Swallow Wort, which was one of the names printed both by Lyte and Gerard.

In Great Britain, the status of Celandine is not very different. It was introduced. It was here a thousand years ago. To take mistiness away from the eyes, Anglo-Saxons used it with Fennel, Wormwood, and honey, abating the sharpness with mother's milk (42). Once in the garden, it is now outside the fence; it is now left to its own devices on the bank, or beside the ditch, or in a corner of waste land, where it ranks with such abandoned plants as Danewort and Soapwort and Evergreen Alkanet.

IX. Fumariaceae

1. Fumitory. *Fumaria officinalis* L. 110, H 33

Local names. BABE IN THE CRADLE, Som; BIRDS ON THE BUSH, Dor; DICKY BIRDS, FEVETORY, Wilts; FAMINTERRY, I o M; GOD'S FINGERS AND THUMBS, Dor; HEMITORY, Kent; JAM TARTS, LADY'S LOCKETS, Som; LADY'S SHOE, Som, Wilts; SNAPDRAGON, N'hants; WAX DOLLS, Som, Kent, Herts, War, Yks, N'thum.

Pliny, and Dioscorides before him, wrote of a plant named *kapnos*, which was identified with Fumitory. Kapnos is the Greek for smoke. The juice, according to Dioscorides, made you weep when it was put into your eyes. Smoke makes you weep, so *kapnos* or smoke the plant was called. In the Middle Ages, Fumitory was *fumus terrae*, 'smoke of the earth', and *The Grete Herball* of 1526, quoting from a twelfth-century herbal, explains that Fumitory was 'engendered of a coarse fumosity rising from the earth' – as though the plant were generated spontaneously.

Does this pretty weed have a smoky appearance, as its pale, blue-green leaves sprawl delicately over the ground? Perhaps. William Coles thought so. Fumitory, he observed in his *Adam in Eden* (1657), once more repeating older authority, is of 'a whitish blew colour as smoak is', and 'appeareth to those that behold it at a distance as if the ground were all of a smoak'. Perhaps the smell also helped. Pull up a root, and it certainly does smell gaseous – 'remarkably like the fumes of nitric acid', Britten and Holland say in their *Dictionary of Plant-Names*. A smell of fumosity, a look of fumosity, and an effect of fumosity (for the juice of *Fumaria officinalis* does also make you weep).

Fumitory is well known to everybody who turns the ground:

> Fumiter is erbe, I say,
> That springyth in April and in May
> In feld, in town, in yerd and gate
> There land is fat and good in state

– as the poet of the Stockholm Medical Manuscript (*c.* 1400) wrote with a fair accuracy. Most of the recorded names, such as the charming Wax Dolls (from the flowers which have the waxy texture of the Victorian doll), or the Wiltshire and Somerset names after the Virgin, suggest affection and appreciation. In East Anglia – see Cockayne's *Leechdoms*, iii, pp. xxxii and 320 – a coarser and not inappropriate name survived out of Middle English (and perhaps survives even now).

X. Cruciferae

1. Sea Cabbage. *Brassica oleracea* L. 15

Sea Cabbage is the ancestor of the garden cabbages. Growing mainly on cliffs in Great Britain, often where it can be seen more easily than reached, it has rarity and habitat as recommendations, as well as a boldness of character lost in the tame rows of the allotment. William Turner first observed it four centuries ago, when it grew abundantly, as it does still, on the cliffs around Dover. His name for it was Sea Cole.

Cabbages, though, were anciently domesticated from Mediterranean stock, and not from British plants, which cling only to a few scattered localities. Greeks and Romans – see Pliny's long account in the *Natural History* – praised both the wild cabbage and the varieties which they grew, less as a food than as a concentrated apothecary's shop. Cabbage for medicine was often mixed with other herbs. It was good for headaches, gout, rheumatism, deafness, poor sight, boils, bad dreams, impetigo, jaundice, rupture, baldness, and much else; and since there was antipathy between cabbage and vine, it prevented drunkenness.

At Dover, the Sea Cabbage was cut and sold in the market, though the leaves are bitter and need much washing and two boilings.

2. Charlock. *Sinapis arvensis* L. 112, H 40

Local names. BAZZOCKS or BRAZZOCKS, Yks; BREAD AND MARMALADE, Som; CADLOCK, Som, Wilts, Glos, Suss, Kent, Surr, Herts, N'hants, War, Heref, Shrop, Staffs, Ches, Derb, Leic, Notts, Lincs, Lancs, Yks; CALVES-FEET, Glos, War; CARLOCK, Glos, Oxf, Beds, N'hants, Heref, Worc, Shrop, Notts, Rut, Yks; FIELD KALE, Cumb; HARLOCK, Ess; HEADRIDGE, Pemb; KELK, Wilts, Suss, Kent, Surr; KINKLE, Kent.

POPPLE, Cumb; PRASSHA BWEE, Donegal; RUNCH, Yks, N'thum, N Eng, Scot; RUNCH-BALLS, Yks, Cumb, Rox; RUNCHIK or RUNCHIE, Scot, Ork, Shet; SCALDRICKS, Scot; SKEDLOCK, Lancs; SKELLOCKS, Scot; SKILLOCKS, SKILLOGS, Donegal; WARLOCK, I o W; WILD KALE, Kirk, Wigt, Lanark, N Ire; WILD MUSTARD, Cumb, Scot; WILD TURNIP, Yks; YELLOW, N'thum;

YELLOW-FLOWER, Ches; YELLOWTOP, N'thum; YELLOW WEED, N Eng, Scot; ZENRY, Som.

'Kedlokes hath a leafe lyke rapes, and beareth a yelowe floure, and is an yll wede, and groweth in al maner corne,' wrote Fitzherbert in *The Boke of husbandrie* (1523). Certainly a vicious pest for the farmer, living from generation to generation by an abundance of seeds. Bury the seed under grass and it can remain viable for at least eleven years, possibly even for fifty years or more (G. E. Fogg, in *Journal of Ecology*, 38, p. 415, 1950). Plough the old grass, and up comes the Charlock, like a vegetable rat. However, it seems that selective weedkillers will murder it, eliminate it, or greatly reduce it, at last. It is a plant with a rat's individuality and lack of charm, although young wheatfields yellow with Charlock are among the undeniable pleasures of landscape.

Charlock's record is not entirely evil. It was one of those border plants which men were not quite sure whether to accept or reject. Forcing itself into notice in the crops, it was once boiled and eaten. Caleb Threlkeld wrote in 1727: 'It is called about the Streets of Dublin before the Flowers blow, by the name of *Corn-cail*, and used for boiled sallet.' On the island fringes, this vegetable use endured still later, as on Colonsay in the Hebrides (177, 128).

William Turner in 1548 used the names Carlock or Wyld Cole. The Pembrokeshire name 'Headridge' may have come from the Flemings who settled in Pembrokeshire about 1107 (cf. the German *Hederich*); many of them have left their personal names behind on farm and parish.

3. Wild Radish. *Raphanus raphanistrum* L. 111, H 37

Another rattish and world-conquering farmer's weed, which shares names and likeness with Charlock – CADLOCK for *Raphanus raphanistrum* has been recorded in Glos, War, Staffs, and Yks; RUNCH in Cumb, Berw, Rox, Ayr, Mor; WHITE RUNCH in Yks; RUNCHIK and RUNSHIE in Shet; RUMP in Oxf; HORSE RADISH in Dor; and WHITE CHARLOCK in Berks and Ess. It is not the ancestor of the garden Radish, which is an ancient vegetable introduced and living only in cultivation, or as a chance escape.

Donegal names (see *Sinapis arvensis*) also include SKELLOG and SKELLIGS.

4. Sea Kale. *Crambe maritima* L. 41, H 8

Sea Kale is rather a fine wild plant, smooth surfaced, with curly, crisp, blue-green foliage and green-clawed white flowers. Living on cliffs and sand, 'upon the bayche and brimmes of the sea, where there is no earth to be seene, but sande and rowling pebble stones', its seeds float around with the tide. But Sea Kale is uncommon. Collection must have reduced it greatly, at any rate on the south coast.

Before cultivation began, Sea Kale had probably been an English wild food for hundreds of years. It was brought to market in the south coast towns, much as Somerset people still bring their bunches of the Bath Asparagus, *Ornithogalum pyrenaicum* (q.v.), into the greengrocer's shops at Bath. In Hampshire, on the long sands by Calshot Castle, the country people watched for the young shoots to appear, heaped sand around them to blanch out the bitterness, and collected bundles to dispose of in Southampton (21). Miller, in his *Gardeners Dictionary* of 1731, remarked that the wild plant was found in great plenty on the Sussex coast, 'where the Inhabitants gather it in the Spring to eat, preferring it to any of the Cabbage kind'. By that time the 'Sea-colewort or Cabbage' had reached the garden. Miller gives instructions for growing it. But it was a Hampshire man, William Curtis (1746–99), the botanist and the author of the great *Flora Londinensis*, who popularized Sea Kale by writing a pamphlet of *Directions for the culture of the Crambe maritima or Sea Kale, for the use of the Table.*

So here is one of the few vegetables which we ourselves have developed from the native flora, and which has no previous history in the Mediterranean or the Near East (Sea Kale does not occur on the Mediterranean coasts). After it had reached Covent Garden Market and the gardens and the dinner tables of the polite, the French considered it. A French gardening writer of 1807 described the *chou marin d'Angleterre* – 'But he appears to have tried to use the broad green leaves, instead of the blanched shoots. Disgusted with his preparation, he denies the merits of sea-kale; and resigns the plant, with a sneer, to colder climates' (124). However the greatest of French chefs Antonin Carême thought highly of Sea Kale. In his *L'Art de la Cuisine française* (1833–35) he gives instructions for cooking '*sickell*' and serving it on a folded napkin with buttered toast and fresh butter, or with fried breadcrumbs and lemon juice, and a bechamel sauce.

Sea Colewort in the sixteenth century, Sea Cabbage, Sea Kale – maritime plants are seldom rich in names, and that is the full list, except for

Strand Cabbage in Donegal. Sixteenth-century botanists knew Sea Kale as the *brassica Anglica*, or English kale.

5. Dittander. *Lepidium latifolium* L. 26, H 5

9 Dittander *Lepidium latifolium*

Gerard grouped the Horse Radish (not used, it seems, as an English condiment in his day) with Dittander, which was 'extreme hot' and 'burning and bitter' in the taste of its roots and leaves. Dittander was grown in gardens, and it was employed in sauce until pepper and Horse

Radish drove it out. In medicine it was used awhile longer. As a native, it is not common, growing in salt marshes along the coast, a tall striking perennial with long smooth leaves of a blue green and panicles of small crowded white flowers. It might be worth putting back in the garden as a curiosity but only in a corner, since it behaves like Horse Radish: every scrap of root will grow, and the roots 'will spread and shoot up at a great distance, so as to over-run the ground' (Philip Miller).

William Turner wrote in 1548 that 'Dittany', for this species, was a 'false name'; properly it belonged to the handsome *Dictamnus albus* of flower gardens, which was to be introduced later in the sixteenth century. The Germans, he added, called it *Pfefferkraut*; and 'Pepperwort' it became. The name 'Dittany' refused to disappear, though gradually it was replaced by the form 'Dittander'.

6. Wart Cress. *Coronopus squamatus* (Forsk.) Aschers 86, H 29

Local names. HOG GRASS, War; SOW GRASS, Yks; STAR OF THE EARTH, Suss; SWINE'S CRESS, War.

Henry Lyte first recorded *Coronopus squamatus* as an English plant in his translation of the herbal of Dodoens, in 1578, mentioning that in some places in England it was called 'Swine's Cress'. 'Wart Cress' obviously from the small round, pitted, clustred siliculae, which contain the seeds.

7. Woad. *Isatis tinctoria* L. 9

Local names. WAD, E Ang, N'hants, Yks, Lakes; WOAD, Dor, Wilts, E Ang, N'hants, Lincs, Yks, West, Cumb, Scot; WOOD, Dor, Wilts, Yks, N Eng, Kirk, Wigt.

Woad is a now scarce relic of ancient cropping which lasted from Anglo-Saxon times, or Celtic times, until the nineteen-thirties, when the two remaining woad-mills in Lincolnshire – the only two left in the world – were closed. As an alien, introduced for dye, Woad does not take easily to an independent English life. There is one place, though, to see it in a 'wild' mass, in May and June – on the Mythe, a red marl cliff above the Severn, just outside Tewkesbury. The cliff turns golden with the blossom, and has done so ever since 1818. Here on the Mythe, the plant has sunshine and freedom from interference, and remains as a monument of the Gloucestershire cloth trade (158).

The literature of Woad does not suggest a beauty. Plant by individual plant, it may look cruciferous and dull, yet grown as a mass, making a clear, stately cloud of yellow, and giving out a smell of honey, it goes well in a garden. Moreover, it can be raised without difficulty, and it ripens an abundance of dark fruit to continue its biennial life.

How long a history has Woad in the British Isles? Many place-names and the print left by a fruit of Woad on the clay of an Anglo-Saxon pot from the Midlands carry it back more than a thousand years (40, 107). Caesar's report in *De Bello Gallico* that British fighting men painted themselves with Woad may be true, as well as Pliny's statement that British women stained their bodies with Woad and paraded in certain religious ceremonies naked and coloured like Ethiopians. Otherwise there seems no Celtic evidence except in the name Glastonbury, or rather in the presumption of an original Celtic name 'very likely something like *Glastonia*, a derivative of Old Celtic *glasto-*, Gaulish *glastum* "woad". The meaning would be "place where woad grew"' (66). Centuries on, Woad brought money to the mediaeval Abbey of Glastonbury.

Coming from Assyria into the Mediterranean world, probably the use and cultivation of Woad spread northwards – though rather late, since a fair degree of civilization is required for the elaborate preparation of the dye. The leaves are green, without much hint of blue. They have to be pulped, moulded into balls, dried, powdered, wetted again, and fermented, or 'couched', to bring out the colour by oxidization, and then dried a second time. If the *isatis* which Dioscorides wrote about is Woad (as it seems to be), the British may have used the blue dye not only to make themselves more terrible in war, but to help the wounds they might get. The leaves, if not the dye, according to Dioscorides, closed up bloody wounds and stopped the violent flow of blood.

Woad culture and Woad milling were never popular. Giving four or five crops a year and up to a ton of leaves per acre, the plant exhausted the soil; the juice from the crushing mills fouled the streams (there are complaints of the 'wood-gore' in mediaeval records), and the couching of the Woad-balls filled the air with the vilest stench (95, 103). The business was 'profitable to some few, and hurtfull to many'. Gradually Woad was driven out of use by the tropical Indigo; and Indigo, or Anil, itself, in the end, gave way to synthetic dyes.

Woad was used not only for blues, but 'for the foundation of many colours, especially all sad colours' (130).

8. Shepherd's Purse. *Capsella bursa-pastoris* (L.) Medic. 112 H, 40

Local names. BAD MAN'S OATMEAL (i.e. the devil's oatmeal), Dur; BLIND-WEED, Yks; CASEWEED, CASEWORT, Som; CROW-PECKS, Wilts; FAT HEN, Glos; GENTLEMAN'S PURSE, GUNS, HEN AND CHICKENS, Som; LADY'S PURSES, E Ang, Berw; MONEY-BAGS, Som; MOTHER'S HEART, Glos, Mddx, Lancs, N Eng, Scot.

NAUGHTY MAN'S PLAYTHING (i.e. the devil's plaything), War; OLD WOMAN'S BONNET, Som; PEPPER AND SALT, Mddx; PICK-YOUR-MOTHER'S-HEART-OUT, War; PICKPOCKET TO LONDON, Yks; PICKPOCKET, Dev, Dor, Som, Wilts, Suss, Kent, Bucks, Ess, Norf, N'hants, Worc, Ches, Notts; PICKPURSE, POOR MAN'S PARMACETTY, POOR MAN'S PURSE, Som; POVERTY PURSE, Lincs; PURSEFLOWER, Som; RIFLE THE LADIES' PURSES, Banff.

ST JAMES'S WORT, Som; SHEPHERD'S PEDLAR, Wilts; SHEPHERD'S POCKET, Bucks; SHEPHERD'S POUCH, Som, Herts; SHEPHERD'S SCRIP, SNAKE FLOWER, Som; STONY-IN-THE-WALL, Lincs; TACKER WEED, Som.

Here is a mean little plant of verges, gateways, gardens, and fields, which has hit the fancy for one thing – its seed cases or siliculae (hence the names Casewood, and *Capsella*, which is a little case or box). Pick off a ripe case, and surprisingly it breaks in two, the seeds pour out. These cases are shaped like hearts, but hearts which break. So 'Mother's Heart', and the sinister, no doubt ancient game, German as well as English, in which one child makes another child pick one of the cases, and then, when it breaks, tells him he has broken his mother's heart. A cruel little game of this kind was, or is still, played with the quickly falling petals of *Veronica chamaedrys* (q.v.).

However, to Europeans in general, this oddly behaving silicula also suggested a purse, with the seeds in it for the money – one of the purses like a pouch, or bag, which were carried from the waist on strings. To get the likeness, you must reverse the silicula, pedicel and narrow end upwards, wide end downwards. Then look at hanging purses in sixteenth-century paintings by Bruegel, such as his 'Dance of the Peasants'.

A small purse on a poor, small, miserable weed, so usually it belongs to a poor man – a shepherd by European consent. Thereby our own Shepherd's Purse, with its variations, 'Shepherdes pouch', as William Turner wrote in his *Herbal*, 'of the likeness that the seed vessel hath unto a shepherdes pouch or skrippe', and the German *Hirtentäschen* and the French *malette de berger*, and so on in many languages from Icelandic to Russian. But purses are stolen – by cutpurses, who cut the strings, as the thief cuts them from behind the old man in Bruegel's 'Perfidy of the World'. Thieves (not the reason usually advanced, that this plant beggars the farmer) explain why it was called Pickpocket, with similar names in Dutch and German. So, with Shepherd's Purse, the other German and English game of the accusation of theft: the child invites another child to pick off a seed case, and then, as the seeds or coins pour out, he chants derisively:

> Pick pocket to London,
> You'll never go to London.

or:

> Pick pocket, penny nail,
> Put the rogue in the jail.

Shepherd's Purse and emigrants from the Old World have travelled about together. Along with a few more plants, such as Sorrel and Sweet Vernal Grass, it still survives in Greenland (138), around the farm sites where the Norsemen settled in A.D. 985 or 986.

9. Wild Candytuft. *Iberis amara* L. 22

A charmer, but common only in a few districts, on the Chilterns, for example, or in upland Berkshire at the edge of the long chalky fields. It is the ancestor of many races of garden Candytuft.

The name was transferred from the foreign *Iberis umbellata*. In the sixteenth century Edward, Lord Zouche, a nobleman fond of flowers and of poets (including Ben Jonson and William Browne), collected seeds of *Iberis umbellata* and gave them to Gerard. The plant was known as *Thlaspi Candiae*, since it grew, among other places, in Candia or Crete. At first 'Candie Mustard' or 'Candie Thlaspi', the name before long was affectionately and happily turned into 'Candytuft'.

10. Scurvy-grass. *Cochlearia officinalis* L. 87, H 25

Officinalis is the epithet of the *officina*, the shop, the apothecary's shop, and so of the plants used in medicine. The juice of Scurvy-grass gives the antiscorbutic vitamin C, and it may have been an old folk-remedy around the coast before the long voyages of the sixteenth century, which brought the plant into a wider recognition. For those who would lie 'lamentable sicke of the scurvies' in brine and ballast, the dried herb or a bottled distillation was taken aboard.

Gerard in the *Herbal* gives an extra long piece to the virtues of Scurvy-grass in curing sailors of 'this filthie, lothsome, heavie and dull disease', which is a formidable and recurrent theme in Hakluyt's *Voyages*, and in which 'the gums are loosed, swolne, and exulcerate; the mouth greevously stinking; the thighes and legs are withall verie often full of blewe spots, not much unlike those that come of bruses: the face and the rest of the bodie is often times of a pale colour; and the feete are swolne, as in the dropsie'. But not only sailors were attacked. Deficiencies of diet made scurvy a mainland disease which scourged the English for centuries after Gerard's time. Scurvy-grass medicines, in which the unpleasant taste of the herb was disguised with spices and saffron, were much taken in the sixteen-hundreds, when Scurvy-grass was cried around London:

> Hay'ny Wood to cleave,
> Will you buy any scurvy Grasse?
> Will you buy any Glasses,
> Ripe St Thomas onions?

Anthony à Wood observed how in the sixteen-fifties there was a fashion for a Scurvy-grass drink in the mornings, like our own glass of orange juice (142, 39). Early in the nineteenth century, Scurvy-grass sandwiches were still eaten, and antiscorbutic 'spring juices' were concocted from Scurvy-grass, Watercress, and Seville oranges (124). It was the pleasanter Watercress and lime juice which put the older remedy out of fashion.

Cochlearia officinalis was commonly grown in the physic corner of the garden. *Cochlearia anglica*, with its narrower, less rounded leaves, was not so easily cultivated, and herb-women collected it for the apothecaries from the Thames marshland. When Scurvy-grass became known, botanists and herbalists greatly argued whether it was or was not the *Britannica* or *herba Britannica* of ancient doctors, including Dioscorides. The legionaries of Germanicus Caesar, in his campaign across the Rhine, were taught the use of it by the Frisians and were rescued from a scurvy-like disease (146). New medicines went better in the early days if there could be found for them the faintest shadow of a classical authority.

There seem to have been few other names for Scurvy-grass ('Screeby' in Scotland, 'Scrooby Grass' in Yorkshire and the North). Spoonwort was a book name, from the German *Löffelkraut*, since the root leaves have more or less the shape of a spoon. *Löffelkraut* was latinized in the sixteenth century into *cochlearia* (Latin *coclear*, a spoon).

11. Horse Radish. *Armoracia rusticana* Gaertn., Mey. & Scherb. 89, H 6

Horse Radish will spread its shabby leaves in waste ground and along road verges, but always as an escape from the garden. It has been introduced, after making its way across Europe from the Near East; and with us it does not often ripen its seeds. William Turner and Gerard write of it in the sixteenth century as though it were unfamiliar. Turner, in his *Herbal*, in 1568, remarked that in Friesland it was employed to make sauce for boiled meat. Gerard recommended it in preference to mustard. Yet either for sauce or medicine, it must already have been growing in England for a long while, because Gerard knew it here and there outside the garden and Turner in his young days had seen it apparently wild in Northumberland. Also the name by which Turner knew it, Redco (Redcole, Redcoll, Rotcoll), is first on record towards the end of the fifteenth century.

Yet why *Horse* Radish? The German name is *Meerrettich*, 'sea radish', and *meer* seems to have been taken for *mähre*, an old horse, as if for the rankness and toughness of the roots. It was Gerard who first gave 'Horse Radish' to print. Lyte had called it Mountain Radish or Rayfort, which is the French *raifort*, 'strong root'. Another French name is *moutarde des moines*, 'monks' mustard'.

12. Watercress. *Nasturtium officinale* R.Br. 112, H 40

Local names. BILDERS, E Ang, Ire; BROOKLIME, Bucks; CARPENTER'S CHIPS, Glos; CREESE, KERSE (OE *cerse*, cress), Som; RIB, E Ang; TANG-TONGUES, Yks; TONGUE-GRASS, Ire; WATER-GRASS, N Ire; WELL-GIRSE, Scot; WELL-KARSE, WELL-KERSE (OE *wielle-cerse*, spring or stream cress), N'thum, Scot.

Boiled, or eaten raw, for centuries; and to the eye (if you forget the untidy watercress farms of the Midlands or the south), delicious in its viridian tufts and pillows along a clear stream, or when it is topped with the white flowers. Watercress is often mentioned with delight in early nature poems of the Irish (105), in the twelfth century:

> Well of Tráigh Dhá Bhan
> Lovely is your pure-topped cress,

or:

> Blackthorn, little thorny one,
> Black little sloe-bush;

Watercress, little green-topped one
On the brink of the blackbirds' well.

The Anglo-Saxons distinguished it as *ea-cerse*, watercress, or (see above) as *wielle-cerse*. Though the victory was late, it conquered all classes rather more quickly than the local and less familiar Sea Kale. Philip Miller gave directions for propagating watercress in beds in the early editions of his *Gardeners Dictionary*, and spoke of its increasing popularity as a salad and against scurvy. Commercial cultivation began about 1808, in Kent, soon spreading around London. Cultivators from the beginning knew of three types of Watercress, which had to wait a long while for botanical recognition. In fact, they were *Nasturtium officinale* and *N. microphyllum*, which are about equally common, and much alike, except that the leaves of *microphyllum* usually turn brown in the winter (101), and a hybrid between these two. Watercress farmers grow both *officinale* and the hybrid, which is also brown-leaved; though it is perhaps reasonable unreason to prefer buying the green to the brown. Either way, few plants are richer in vitamin C.

Pliny interpreted *nasturtium* as 'nose-twister', *quod nasum torqueat*, but the proper derivation seems to be from a Greek compound *mnastorgion*, 'that longs for wet soil'.

The place-name Ribbesford (Worcs) probably signifies the ford where watercress grew (cf. the East Anglian name rib, above), although OE *ribbe* is usually glossed as ribwort, i.e. plantain, or as hound's-tongue (66).

13. Stock. *Matthiola incana* (L.) R. Br. 9

The classic home for the Stock or Gilliflower as a 'wild' plant (though the odds are against it as a native) is on the cliffs of the Isle of Wight, where this southern species has taken refuge in the warmth like an old lady. Arnold Bromfield, author of the *Flora Vectensis*, used to employ one of the island cliffsmen and samphire gatherers to collect it for him from the chalk declivities. According to him, the flowers, 'of a fine purplish pink varying to violet-blue or lilac on the same plant', have a clove perfume which is even more delicious than the perfume of garden Stocks. Though it is perennial on the cliffs, *Matthiola incana*, or one variety of it, gave rise to the annual Ten Weeks Stock in all its colours.

14. Sea Stock. *Matthiola sinuata* (L.) R. Br. 8, H 2

By contrast with the last species, the Sea Stock ranks as a rare native, though a southerner at the limit of its range, confined to dunes or cliffs in two English counties, Devon and Kent, two Welsh ones across from Devon, Pembrokeshire and Glamorgan, and two Irish ones, Co. Clare and Wexford. First discovered in Wales (but in Merionethshire) early in the seventeenth century, sweetly perfumed again, with flowers pale purple, you can have an idea of the situation of the Sea Stock from an early poem by Andrew Young:

> Save for frail shade of jackdaw's flight
> No night was there,
> But blue-skied summer and a cliff so white
> It stood like frozen air;
> Foot slipped on damp
> Chalk where the limpets cramp.
>
> Like a soul strayed in Paradise,
> Dazed by deep light
> I held my hand to shade my blinded eyes
> Till I saw how the white
> Hardness of chalk
> Was purpled by sea-stock.

15. Gold of Pleasure. *Camelina sativa* (L.) Crantz. 62, H 9

A tallish, yellow-flowered, yellow-seeded crucifer, with a poetic name which goes back at least to the sixteenth century. With us, Gold of Pleasure is only a casual, an unnaturalized weed; yet a species with a very old human history. Giving a fibre which could be made into brooms, and oil from the seeds which was edible and useful for light, it was long grown in continental Europe. Perhaps it brought itself to human notice first of all as a weed in crops of flax, with which it often flourishes. There is proof, in fact, that it was cultivated in Iron-Age Denmark and neolithic Hungary (40).

Perhaps 'Gold of Pleasure' requires no explanation. Such a plant, rather elegant, with leaves clasping the stem, displaying its yellow flowers in the flax field, was surprising and pleasurable. German names are *Leindotter* and *Saatdotter* – 'flax yolk' and 'crop yolk'.

16. Cuckoo Flower, Lady's Smock. *Cardamine pratensis* L. 112, H 40

Local names. APPLE PIE, Yks; BIRD'S EYE, Shrop, Yks, Cumb; BOG-FLOWER, Yks; BONNY BIRD EYE, Cumb; BREAD AND MILK (cf. French *pain-au-lait*), Glam; CARSONS (growing on *carse*, low, rich, damp land), SW Scot.

CUCKOO, Dev, Som, Glos; CUCKOO BREAD, Dev, Som; CUCKOO'S EYES, Glos; CUCKOO FLOWER (cf. German *Kuckucksblume*), Corn, Dev, Dor, Som, Wilts, Hants, Glos, Kent, Surr, Suff, War, Ches, Leic, Lincs, Yks, N'thum, Ire; CUCKOO'S SHOES AND STOCKINGS, Som, S Wales (where whiter flowers are the stockings, the more lilac ones the shoes); SHOES AND STOCKINGS, Bucks; WATER CUCKOO (distinguishing it from Dry Cuckoo, *Saxifraga granulata*), Wilts; CUCKOO-PINT, Wilts, Suss, Leic; CUCKOO-PINTLE (see *Arum maculatum*), Suss, Leic; CUCKOO SPICE, Yks; CUCKOO SPIT, N Eng; HEADACHE, Glos, Cumb.

LADY'S CLOAK, Som; LADY'S FLOCK, Notts; LADY'S GLOVE, N'hants; LADY'S MANTLE, Som; LADY'S MILK SILE or MILK SILE (a *sile* is a strainer), Yks; LADY'S PRIDE, Som; LADY'S SMOCK, Corn, Dev, Som, Wilts, Hants, Glos, Suss, Surr, Herts, Suff, N'hants, War, Worc, Shrop, Ches, Derb, Leic, Lincs, Lancs, Yks, Rad, Dumf, Banff, Ork; MY LADY'S SMOCK, Dev.

LAYLOCKS (i.e. lilacs), Dev, Wilts; LAMB'S LAKENS (i.e. toys), Cumb; LONESOME LADY, Dev; LUCY LOCKET, Derb, Rut; MAY-BLOB, N'hants; MAYFLOWER (cf. German *Maiblume*), Som, Hants, Ches, Lancs, Yks; MEADOW FLOWER, Cumb; MEADOW PINK, Dev; MEADOW KERSES (i.e. cresses), Dumf; MILKIES, MILK GIRLS, Dev; MILKING MAIDS, Som; MILK MAIDS, Dev, Som, Wilts, Mddx, Ess, Suff, Yks, Donegal; MILKY MAIDENS, Dev (for milk names, cf. German *Milchblume*); MOLL-BLOBS (i.e. 'cow blobs'), N'hants; NAKED LADIES, Som.

NIGHTINGALE FLOWER, Hants; PAIGLE, Suff; PICK-FOLLY, N'hants; PIGEON'S EYE, Yks; PIG'S EYES, Ess; PINK, N'thum; SMELL SMOCK, Glos, Bucks, Kent; SMICK-SMOCK, Hants, Glos, Oxf; SPINK or BOG SPINK, N'thum; WATER LILY, Norf (cf. Faeroese *Vatnlilja*).

A spring flower associated with milkmaids and their smocks, the cuckoo, and the Virgin. As Gerard wrote, *Cardamine pratensis* comes out 'for the most part in Aprill and May, when the Cuckoo begins to sing her pleasant note without stammering'. But this is less than half the story. The plant was the Cuckoo-pint and the Cuckoo-pintle (see *Orchis mascula* and *Arum maculatum*), it was the Cuckoo Flower and the Cuckoo Plant in other countries, Holland, France, Germany, etc., and the English 'smock' names appear to have come from the OE *lustmoce*. The 'smick', more usual as 'smicket', of the name 'smick-smock' was another word for smock, and 'smickering', to 'smicker', were words of amorous looks and purposes. 'Smock' was used coarsely, especially in the sixteenth and seventeenth centuries, as we use 'skirt', or 'piece of skirt', etc. A plant, clearly, which needed christianizing and handing over to the Virgin. The smock of the Virgin Mary – to which the name no doubt refers – was one of the relics which St Helena was supposed to have found in the cave at Bethlehem: 'And whan seynt Elene was come in to this derk place, sche founde the same heighe that crist was leyde in yn the manger, and the clothes that oure lord ihesu crist was wounde yn, and our lady smok – and all thes thyngys oure lady had forgete behynde her whan sche gede oute of that place in to Egipt.' The English translation of John of Hildesheim's account of the Three Kings written in the second half of the fourteenth century, from which that comes, goes on to explain that the relics were lodged in St Sophia in Byzantium, and later taken by Charlemagne to Aix-la-Chapelle – 'and ther oure lady smok and othir Relikes be worschepped of the cristen pepil yit in to this daye'. (Horstmann, *The Three Kings of Cologne*, Early English Text Society, 1886.)

The plant had other undesirable associations. In parts of France, *Cardamine pratensis* was a plant *not* to be included in May Day garlands. In Vienne, it was considered the favourite flower of adders. Those who picked it would be bitten before the year was out (165). In Germany, it was a *Donnerblume*, a *Gewitterblume*, a thunder and storm flower (in France also a *fleur de tonnerre*), not to be picked, not to be brought into the house, for fear of a storm and of the house being struck by lightning (129). In some parts of England too (see the name Pick-folly), picking *Cardamine pratensis* was unlucky. Having this character, it seems that the plant was little used in medicine. Gerard wrote that there was no 'proofe or authoritie' of its virtue, though for a while – but this was after the sixteenth century – it was employed against epilepsy. It was also used in love-divination – in a tinker-tailor game, picking off the leaflets (pick-folly = ? pick-foil).

Gerard, as late as 1597, was the first English botanist to record *Cardamine pratensis*. 'In English,' he wrote, 'Cuckowe flowers: in Northfolke, Caunterburie bels: at the Namptwich in Cheshire where I had my beginning, Ladie Smockes, which hath given me cause to christen it after my countrie fashion.' Shakespeare, who seems to have gone to Gerard's *Herbal* for much of his knowledge of plants, probably lifted the famous 'lady-smocks all silver white' of the song in *Love's Labour's Lost* (in the quarto of 1598, 'newly corrected and augmented') straight out of Gerard's volume, which was published the year before. Both poet and botanist call the flowers white, when they are more usually lilac (118). Still, that does not mean that Shakespeare was unfamiliar with the Lady's Smock before he explored the *Herbal*; and the song, with the coupling birds, the maidens bleaching their smocks, and the cuckoo mocking married men, suggests that Shakespeare knew all about its less delicate associations. In a fifteenth- or sixteenth-century Irish poem, *Cardamine pratensis* and the young girls go together (105):

> Tender cress and cuckoo-flower:
> And curly-haired, fair-headed maids,
> Sweet was the sound of their singing.

17. Jack-by-the-hedge. *Alliaria petiolata* (Bieb.) Cavara & Grande. 101, H 37

Local names. BEGGARMAN'S OATMEAL, Leic; HEDGE GARLICK, Cumb; JACK-BY-THE-HEDGE, Dev, Suss, Ches, Yks, Merion; JACK-BY-THE-HEDGESIDE, Dor; JACK-IN-THE-BUSH, Glos, Heref; JACK-IN-THE-HEDGE, Dev, I o W, Herts, Leic, Lincs; JACK-OF-THE-HEDGE, Ches; JACK-RUN-ALONG-BY-THE HEDGE, Wilts.

LADY'S NEEDLEWORK, LAMB'S PUMMY, Som; PENNY-HEDGE or PENNY-IN-THE-HEDGE, Norf, Worc; PICKPOCKET, Dev; POOR MAN'S MUSTARD, Lincs; SAUCE ALONE, Som.

In a brilliant sunshine, in May, one is always freshly struck by platoons of this familiar plant, at starched attention, the starch-white flowers above the new green leaves and against the green bank. Dr Prior, in his *Popular Names of British Plants*, outdid himself in pedantry when he came to Jack-by-the-hedge – from '*Jack* or *Jakes*, latrina, alluding to its offensive smell'. A garlic smell isn't necessarily offensive, and the plant does not smell at all unless you crush it or sniff the root. Is it nothing but a Jack, like a Jackstraw, or a Jackanapes, or a Jack in the Green, or rather any Jill's Jack, so named with a half-depreciatory affection.

The species was 'Jack-of-the-Hedge' to William Turner in 1538, and 'Sauce Alone' – since, like many of the wild crucifers, it was used for a condiment – a spring sauce, according to Turner's *Herbal*, a sauce with salt fish according to Gerard.

Loudon still found it worth mentioning in his *Encyclopaedia of Gardening*, boiled (and eaten with boiled mutton), or for salad, or in sauces.

18. Winter Cress. *Barbarea vulgaris* R. Br. 107, H 40

Local names. CASABULLY (Old Cornish *cas-beler*), Corn; LAND CRESS, Som, I o W; ST BARBARA'S HERB, Som; YELLOW CRESS, Yks; YELLOW ROCKET, Som.

Winter Cress is tall and good looking enough to have caught the eye and mind, and to have been developed into a garden flower. Why, in many European languages, was it given to St Barbara, a favourite saint of the Middle Ages? Perhaps because her festival is 4 December, and this herb is green all winter long, as Gerard mentioned, and was 'thought to be equall with Cresses of the garden' in the winter months. Or because it was tall like the tower (often Barbara's symbol on badges, etc.) in which this virgin saint was shut up by her father?

Turner in the *Names of Herbes* (1548) put the species under 'newe founde Herbes, whereof is no mention in any olde auncient wryter'. Since it had leaves like a Rocket and was good for a wound, he gave it the name Wound Rocket, which was never adopted. Winter Cress was borrowed from the German, or translated from the botanical Latin of the sixteenth century, *nasturtium hibernum*. Lyte also translated the *Herba Sanctae Barbarae* into Herbe St Barbe.

XI. Resedaceae

1. Weld, Dyer's Rocket. *Reseda luteola* L. 98, H 40

Local names. BASE ROCKET, DYER'S WEED, Som; GREENWEED, Kent, Suss; YELLOW, N'hants; YELLOW ROCKET, N Eng, Scot; YELLOW WEED, Wilts, Donegal.

Weld, in spite of the rather anaemic, greeny-yellow flowers, has a fine growth on the thin soil of the Cotswolds, on the chalk, or on the ledges of an abandoned limestone quarry. It is one of the most ancient and one of the best of dyer's plants. Found in Swiss neolithic settlements (40), it endured in use after the development of chemical dyeing, giving specially brilliant, pure, and fast yellows. A native, unlike Woad; but the wild supply was too small. Weld was imported from France, and crops were grown, especially in Kent, around Canterbury, in Essex, and in Yorkshire. Farmers in the eighteenth century followed the odd practice of

10 Weld/Dyer's Rocket *Reseda luteola*

sowing Weld with the corn – with barley or oats. It is not uncommon to see plants of Weld among the patchy corn of some limestone field. Before the days of cultivation, perhaps Weld had always been collected from the corn, so that an association between the two crops persisted in the mind.

Here again, with a plant of commerce, you would not expect many different names, which could only have confused collector, farmer, market man, and dyer. Weld – *gaude* in France and *Wau*, earlier *Waud*, in German – must be one form of the ancient general name in northern Europe. The explanation of the few other names is clear, except perhaps for 'Base Rocket', a name, according to Parkinson in the *Theatrum Botanicum* of 1640, which distinguished this Rocket-like plant, 'a base and wild herbe', from the garden Rocket or Rocket Gentle, *Eruca sativa*. So Dyer's Greenweed (*Genista tinctoria*) was Base Broom to Gerard – a 'base kinde of Broom'.

Curiously the first English record of the name Weld comes late in one of the minor poems of Chaucer, a moving lament for the past age of innocence:

Allas, allas! now may men wepe and crye!
For in our dayes nis but covetyse
And doublenesse, and tresoun and envye,
Poysoun, manslauhtre and mordre in sondry wyse.

Before tyrants, dictators, wars, evil, and perfidy, and greed, and luxury

A blisful lyf, a paisible and a swete
Ledden the peples in the former age.

But then there were no ships, no mills, no merchants, no farmers, no 'litesteres', which is to say, dyers:

No mader, welde, or wood [woad] no litestere
Ne knew; the flees was of his former hewe.

Madder (not our own wild plant, but *Rubia tinctorum*) and Weld and Woad were the three staples of the litestere's craft which naturally came into Chaucer's head.

XII. Violaceae

1. Sweet Violet. *Viola odorata* L. 87, H 39

Perfume raised the Violet out of its wild obscurity more than two thousand years ago; and its ubiquity around the Mediterranean and throughout Europe helped it to become eminent, or pre-eminent, among the flowers of European affection. Scent suggested sex, so the violet was a flower of Aphrodite and also of her son Priapus, the deity of gardens and generation. *Priapeion* was one of the Greek names for the Violet. All this was well understood by the designer of the seventh and final tapestry of the Hunt of the Unicorn, which was woven probably for the marriage of Francis I of France in 1514. In this last exquisite scene (the tapestries hang in the Cloisters, in New York), the captured unicorn lies within a fence, tethered, as a symbol of consummation, to the pomegranate tree of fertility. Round the white unicorn grow various plants of sex: Bistort, Lords-and-Ladies, Early Purple Orchis, Bluebells – and *Viola odorata*.

A flower so deeply and finely scented must also have its virtues in physic, and since classical doctors held it to be cool, bland, and soothing,

Oil of Violets and Syrup of Violets, etc., were made from the petals. Our own candied violets are a relic of this long medical history; and our bottles of violet scent recall Priapus and Aphrodite.

Gerard, in his *Herbal*, like Shakespeare in his plays, expresses the sixteenth-century affection for 'the black or purple violets, or March Violets of the Garden'. They are eminent, 'not only bicause the minde conceiveth a certaine pleasure and recreation by smelling and handling of these most odoriferous flowers, but also for that very many by these Violets receive ornament and comely grace: for there be made of them Garlands for the heade, nosegaies and poesies, which are delightfull to looke upon and pleasant to smell to, speaking nothing of their appropriate vertues; yea Gardens themselves receive by these the greatest ornament of all, chiefest beautie, and most gallant grace; and the recreation of the minde which is taken heere by, cannot be but verie good and honest: for they admonish and stir up a man to that which is comely and honest'.

2. Dog Violet. *Viola riviniana* Rchb. 112, H 40

Local names. BLUE MICE, Som; BLUE VIOLET, Dev, Ches; CUCKOO'S SHOE, Shrop; CUCKOO'S STOCKINGS, Caith; HEDGE VIOLET, Dev; HORSE VIOLET, Dev, Som, Ess, War; HYPOCRITES, Som; PIG VIOLET, Ches; SHOES AND STOCKINGS, Pem; SNAKE VIOLET, Dor; SUMMER VIOLET, War.

'Dog Violets', and these other names for the violets without scent, cover several species: *Viola riviniana*, the very close *V. reichenbachiana* Jord. (62, H 26, the Wood Violet), and *V. canina* ssp. *canina* (106, H 38), etc.

Gerard called these scentless violets outside the garden 'Dog Violets, or wilde Violets', apparently translating the botanists' *viola canina*. So Dog Violet is really a book name, though Dog, like Horse, is a common English prefix for distinguishing an inferior species from its superior relative.

Before 'Violet' came into currency from the French, perhaps Violets were known by cuckoo names – at any rate in the Celtic countries (cf. Cuckoo's Shoe, Cuckoo's Stockings). In Irish they are *salchuach*, 'cuckoo's heel', from the spur of the flower.

3. Heartsease, Wild Pansy. *Viola tricolor* L. 112, H 36 *Viola arvensis* Murr. 112, H 40

11 Heartsease, Wild Pansy *Viola tricolor*

Local names. BIDDY'S EYES, BIRD'S EYE, CALL-ME-TO-YOU, Som; CAT'S FACE, Som, Suss; COACH-HORSE, EYEBRIGHT, Som; GENTLEMAN-TAILOR, Dor; HEART PANSY, HORSE VIOLET, Dev; JACK-BEHIND-THE-GARDEN-GATE, Suff.

KISS-AND-LOOK-UP, Som; KISS-AT-THE-GARDEN-GATE, Suff; KISS-BEHIND-THE-GARDEN-GATE, War; KISS-ME, Suss, Lincs; KISS-ME-BEHIND-THE-GARDEN-GATE, KISS-ME-LOVE, KISS-ME-LOVE-AT-THE-GARDEN-GATE, Dev; KISS-ME-OVER-THE-GARDEN-GATE, Norf; KISS-ME-QUICK, Dev, Som; LEAP-UP-AND-KISS-ME, Som, Hants, Suss; MEET-HER-IN-THE-ENTRY-KISS-HER-IN-THE-BUTTERY, Lincs; KITTY-RUN-THE-STREET, Som, Kent.

LARK'S EYE, Som; LOVE-A-LI-DO, Wilts; LOVE-AND-IDLE, Dor, Som, Wilts, Hants, Glos, Berks, Oxf; LOVE-AND-IDLENESS, War; LOVE-IN-IDLENESS, Glos, Oxf, N'hants, War, Lanark; LOVE-IN-VAIN, Som; LOVE-TRUE, N'hants; LOVER'S THOUGHTS, Som; MONKEY'S FACE, Suss.

PINK-EYED JOHN or PINK O' MY JOHN, Beds, N'hants, War, Leic, Lincs; PUSSY-FACE, Som; SHASAGH-NA-CRIODH, Donegal; STEPMOTHERS, Som,

Yks; THREE-FACES-UNDER-A-HOOD, N'hants; THREE-FACES-UNDER-ONE-HOOD, Som; TITTLE-MY-FANCY, E Ang; TRINITY VIOLET, Yks.

'It groweth ofte among the corne,' Turner wrote of the Wild Pansy in 1548. In or out of the garden, it was and it has remained a favourite plant, from the 'beautie and braverie' of the colours, yellow, blue, pink, or all three upon the same flower – 'which colours', according to Gerard, 'are so excellently and orderly placed, that they bring great delectation to the beholders', though the flowers have no scent. The contrast of colours was striking and unusual enough to set the mind working upon the Wild Pansy – which grows everywhere – in the matter of names and belief.

How the love names arose is not exactly clear. William Turner knew the Wild Pansy in 1548 as Two Faces in a Hood, as if the two side petals were kissing within a hood of the upper petals and the lower petals, so the kiss gave ease of heart, 'Heartesease'; but it was also called Three-faces-in-one-hood, or under one hood, the large lower petal being perhaps the girl, and the petals on either side her lovers, of whom one must be left out – Love-in-Idleness, Love-in-Vain. However, here was a gay flower associated too much with wantonness; so it was also christianized: by way of its three colours in unity, it was called *Trinitaria*, *Trinitatis herba*, the herb of the Blessed Trinity. 'Three faces in one hodde . . . herba Trinitatis,' wrote William Bullein in his *Booke of Simples*, 1562, 'but I read in an old Monkish written Herball, where in the auethour writeth, that this herb did signify the holy Trinitie: and therefore was called the Herbe of the trinity, and thus he made his allegorie. This flower is but one in which said he, are three sondry colours, and yet [though it is scentless] but one sweete savour . . .' Bullein, as a good protestant, found the religious name no more proper than the names which were wanton and venerous: The majesty of God 'may not with reverence, be compared or lykened, by any alligory, to any base, vayne, venerous flower'. The *Trinitatis herba* – insult on insult – 'may rather be called, thre faces in a Monk's hodde'.

Turner, in 1548, recorded the names Pansy and Two Faces in a Hood, Lyte in 1576 added Heartsease and Love-in-Idleness, and as a venerous flower, Shakespeare, in *A Midsummer Night's Dream* (II. i. 168), made Oberon squeeze the juice of Love-in-Idleness into Titania's eye so that she would fall in love with the ass-headed Bottom, when she awoke. Gerard, in 1597, gave, as well as Three Faces in a Hood, the name Cull Me to You. In France, *Viola tricolor* was known as *pensée* (thought), in the late Middle Ages, and, so derived, Pansy was in use in English by 1500. 'Heartesease' was applied in the sixteenth century both to Pansy and

Wallflower. The name Stepmothers, paralleled by the German *Stieffmütterchen*, is perhaps explained by the large rich lower petal, the Stepmother, the two side petals her own children, and the two upper, half-hidden petals the neglected step-children.

4. Mountain Pansy. *Viola lutea* Huds. 66, H 7

Local names. SHEPHERD'S PANSY, N'thum; YELLOW VIOLET, Cumb.

The older garden pansies were cultivated forms of the variable *Viola tricolor*. Our modern pansies of the garden are said to have arisen as hybrids between *Viola tricolor* and a subspecies of *V. lutea*. In upland Yorkshire or on the Lakeland mountains the Mountain Pansy looks up deliciously from the hard grass. Gerard, giving the first record of it as an English species, complained of the difficulty of growing it: 'The Yellow Violet is by nature one of the wilde Violets, for it groweth seldome any where but upon most high and craggie mountaines, from whence it hath been divers times brought into the garden, but it can hardly be brought to culture, or growne in the garden without great industrie. And by the relation of a Gentleman often remembered, called Master Thomas Hesketh, who found it growing upon the hils in Lancashire, neer unto a village called Latham, and though he brought them into his garden, yet they withered and pined away.'

The Mountain Pansy one may suppose to have been the flower of Wordsworth's *Intimations of Immortality*, which he began to write in the Lakes in 1802. A field, a tree, speak of something that has gone, and

> The Pansy at my feet
> Doth the same tale repeat.
> Whither is fled the visionary gleam?
> Where is it now, the glory and the dream?

XIII. Polygalaceae

1. Milkwort. *Polygala vulgaris* L. 106, H 40

Local names. CROSS-FLOWER, Som; FAIRY SOAP, Donegal; FOUR SISTERS (from the different colours in the flower), Ire; GANG-FLOWER, N Eng; JACK-AND-THE-BEANSTALK, MOTHER MARY'S MILK, PROCESSION FLOWER,

Som; ROBIN'S EYE, Hants; ROGATION-FLOWER, Wilts, N Eng; SHEPHERD'S THYME, Wilts; WAXWORKS, Som.

Cross-flower, Gang-flower, Rogation-flower, Procession-flower, are probably all book names by origin. In other countries, Milkwort was picked and carried in the processions in Rogation week, when the bounds were beaten and the crops blessed. Jacobus de Voragine, in *The Golden Legend* (he died in 1298), says that in the Rogations we implore the help of the saints, asking that God should 'preserve and multiply the fruits of the earth which are beginning to bud': in the procession the Cross is carried (therefore 'Cross-flower') and bells are rung 'that the demons may flee in terror'. These processions became a junketing which was frowned upon by the reformers. However, they go back to the Roman rites of *Ambarvalia*, so Milkwort was named *Ambarvalis flos*: 'In English we may cal it Crosse flower, Gang flower, Rogation flower,' wrote Gerard; but in England its May blossoming must have been too late usually for Rogation.

'Milkwort' also comes out of the book tradition, since the species was taken to be the *polugalon*, or 'much milk', of Dioscorides; of which he wrote that 'it is believed to make milk more abundant'. He does not say whose milk, but the herbalists took it to be the milk of nursing mothers, not of cows in a Milkwort pasture or on a Milkwort hillside. They prescribed it accordingly.

In Wales, it was named *llysiau Crist*, 'Christ's herb', but no doubt from the same continental tradition of the Milkwort garlands and the junketing before Ascension Day. In Donegal, they believe fairies make a lather from the root and leaves, hence the name Fairy Soap.

XIV. Hypericaceae

1. Tutsan. *Hypericum androsaemum* L. 82, H 40

Local names. AMBER, Kent; BIBLE FLOWER, Corn; BIBLE LEAF, Corn, Som; BOOK LEAF, Dor, Som; DEVIL'S BERRIES, Corn; PARK LEAVES, Som; SWEET AMBER, Suss; SWEET LEAF, Dev; TIPSEN, Dev, Bucks; TIPSY LEAVES, Som; TITSUM, Dev, Som, Suss; TITZEN, Corn; TOUCH-AND-HEAL, Bucks; TOUCH LEAF, Wales; TOUCHEN LEAF, Hants, Wales; TUTSAN, Bucks, Suff, S Eng; TREACLE LEAF, Cumb.

Most of the names come from the south-west, and in the high earth hedges of Devon and Cornwall, Tutsan is a common plant, however local it may be in many other counties. There you cannot miss it along the roads, either in golden flower or in black fruit. The fruits may shine like Devil's Berries, they may look nearly sinister enough, as Hebridean islanders believed, to induce madness, but Tutsan is a plant of good reputation. The leaves were applied to grazed legs and to wounds. In Normandy, *Hypericum androsaemum* is *toute-saine*, 'all wholesome', which presumably gave our 'Tutsan' and its variations. In England, and also on the Channel coast of France, the dry leaves, for their scent (likened to ambergris, so the names 'Amber' and 'Sweet Amber') and for good luck, were put between the pages of prayer books and the Bible (160, 165). The smell, in fact, is sharp and aromatic, resembling the smell of phlox flowers in the sun, or of silver paper which has been wrapped around tobacco. 'Park' in Park Leaves (by which the plant was well known to the herbalists) is a contraction and corruption of *Hypericum*. 'Treacle Leaf' is from the older meaning of 'treacle' as a sovereign remedy.

Tutsan was ascribed all these virtues because mediaeval herbalists took it to be the Agnus Castus described by Pliny – *Vitex agnus-castus* L. The leaves of the Agnus Castus, dark green and grey underneath, crush into fragrance. The Greeks associated it with Artemis and chastity, perhaps for the onomastic reason that 'hagnos' in Greek means chaste, and so *agnos* the shrub must be chaste as well. Pliny wrote of matrons who wished to be chaste, putting leaves of Agnus Castus underneath them in their beds. Coming at the head of the alphabet, Agnus Castus in its confused identity gave its name to the mediaeval herbal compiled in the late fourteenth century: 'Agnus castus is an herbe that men clepyn totsane or parkeleuys . . . the vertue of this herbe is this that he wylle gladly kepe men and women chast. . . . Also it dystroyeth and drywyth awey the fowle lust of lecherye if men drynke it' (20). By Gerard's time Agnus Castus and *Vitex agnus-castus* had been correctly correlated, and the shrub introduced to English physic gardens.

2. St John's Wort. *Hypericum perforatum* L. 105, H 38

Local names. AMBER (see *Hypericum androsaemum*), USA; BALM OF THE WARRIOR'S WOUND, Som; CAMMOCK, Hants; JOHN'S WORT, Som; PENNY JOHN, Norf; ROSIN ROSE, Yks, USA; ST JOHN'S WORT, Som, Ches, Yks; TOUCH-AND-HEAL (cf. *Hypericum androsaemum*), N Ire.

Magically, in white magic rather than black, here is one of the most famous of European plants, one of the chief herbs of St John the Baptist. The vigil of St John comes on 23 June – his day is 24 June. On the evening of the 23rd, fires were lit, more for their smoke than for sparkling, crackling flames. The smoke was purifying. It strengthened the magic of plants already magical, it strengthened against the powers of evil all those who jumped across the fire. The herbs of St John were picked on the morning of the 23rd before sunrise, while they were still wet with dew – itself a magical and strengthening substance. When the *feux de joie*, as they were called in France, were lit and sanctified in the evening, the flowers were smoked. They were now better for medicine, or for immediate use in protecting stables, cow-stalls, horses, animals, or men, against elves, devils and demons, witchcraft and all evils. So the French phrase, *avoir toutes les herbes de la St-Jean*, to be ready for anything. As well as *Hypericum perforatum*, the herbs of St John included *Artemesia vulgaris*, *Plantago major*, *Chrysanthemum segetum*, *Sambucus ebulus*, *Achillea millefolium*, *Hedera helix*, *Verbena officinalis*, and *Sedum telephium*.

Most books on folklore write of St John's Eve and the fires and ceremonies as a christianized survival of fire-worship or sun-worship. But these ideas have been criticized, especially by the French anthropologist, Arnold van Gennep (188). St John's Day, 24 June, is not, in fact, the summer solstice, shortest night and longest day, when the sun climbs to its zenith – which is 21 June; and there is no true evidence of a European cult either of the sun or fire. Moreover, there is no evidence to connect the picking of the plants with the sun. In general, the sun's rays were harmful, the rays of the moon advantageous, so the plants had at this festival, and at all times, to be picked *before* sunrise. Still, van Gennep thinks that in pre-Christian times there may have been a ceremonial cycle from 23 June to 30 June, in which the magic ceremonies, the fires, etc., were used to protect the harvest, the farm animals, and the people; in which 'fear of storms and fires, so frequent in the summer, could have been the creative psychical element'. The summer bonfires would then have risen out of the analogical reasoning of like for like.

Christianity adopted the festival, with St John's Day at the beginning and St Peter's Day, 29 June, at the end. St John, the Forerunner of Christ, was well fitted to the season. He was born (Luke i. 36) six months before Christ. He was light – 'a burning and shining light', as Christ called him (John v. 35). Possibly *Hypericum perforatum*, among the most common though not the most flaming of its genus, also fitted St John above other midsummer herbs, because of its yellow sun-flowers and a red juice which could be likened to the blood of St John at his beheading. Yet no doubt it had been an elf-chaser long before; long associated with a dangerous season, which excited the fairies and the spirits of the dead, and in which it was necessary to be alert and armed. As elves turned into the devils and demons of Christianity, *Hypericum perforatum* kept its place as a *fuga daemonum*, a devil-chaser, the *chasse-diable*, as it is named in France. Many things, besides the smoking of St John's Wort and other plants, show the strengthening and purifying virtue of the plants of St John. In some French towns living cats, as animals of witch and devil, were burnt on top of the bonfire (188).

Perforatum – that, too, is a key word. More peculiar than its flowers or general habit are the glandular dots of *Hypericum perforatum*, which look like perforations when you hold the leaves up to the light. Perhaps it was these glandular dots which first drew attention to the plant. Later on the dots were a 'signature' of perforations or wounds, reinforcing the signature of the red juice. So St John's Wort was vulnerary as well as magical.

So very few English names, in such scattered districts, have been recorded for *Hypericum perforatum* that it may have become known in England as a magical and holy herb only rather late and through mediaeval learning. In the *Great Life* of St Hugh, the Burgundian bishop of Lincoln, the story is told of a woman who rid herself, after the bishop's prayers, of a demon lover (112). Another supernatural being had revealed the remedy, a plant, which she was told to thrust inside her dress and scatter over the house. This had been too much for the demon. Afterwards she showed the plant to a monk of Canterbury and asked the name. The monk replied that it was called *hypericon* by the Greeks and *herba Sancti Johannis* by the English (112). This life of St Hugh was written early in the thirteenth century, and St John's Wort now gathered fame as the centuries went by. It was linked in ballads and tales of other demon lovers with Vervain (*Verbena officinalis*), the demon lover saying

> If thou hope to be lemman mine
> Lay aside the St Johns grass, and the Vervine.

'Herba Joannis. This is called Saynt Johannes worte,' says Banckes's *Herball* of 1525: 'The vertue of it is thus. If it be putte in a mannes house there shall come no wycked spryte therein.'

St John's Wort became familiar in Ireland and Wales, and in the Highands and Islands. The Welsh, from its power and its opposite leaves, knew it as *Ysgol Grist* or *Ysgol Fair*, Christ's, or Mary's, Ladder; and they are said to have hung it on their houses on St John's Eve. In Ireland it was (and is?) given to children on the same night to keep sickness away (137). The Irish and the Highlanders and Islanders also named it after their own St Colum Cille and after the Virgin – in Irish *lus Cholmcille*, *Allus Muire*, Mary's Sweat (at the birth of Jesus, and because of the glandular dots?), and *lus na Maighdine Muire*, Virgin Mary's Herb. St Colum Cille, or Columba, specially venerated St John the Baptist, and taught the use of his plant, so it was said, and carried it on his own body. In Gaelic some of the names mean 'jewel of Columba', 'charm of Columba', 'hail of Columba', 'glory of Columba'. One name in particular is *achlasan Chaluimchille*, 'armpit package of Columba': St John's Wort was worn by men and women under the left armpit, though it was effective only when they had found it by accident. Alexander Carmichael wrote that it was 'one of the few plants still cherished by the people to ward away second-sight, enchantment, witchcraft, evil eye, and death, and to ensure peace and plenty in the house, increase and prosperity in the fold, and growth and fruition in the field' (31a). He gives several Gaelic poems on picking St John's Wort, and its power:

> Arm-pit package of Columba, kindly
> Unsought by me, unlooked for!
> I shall not be lifted away in my sleep
> And I shall not be thrust upon iron . . .
>
> Better the reward of it under my arm
> Than a crowd of calving kine;
> Better the reward of its virtues
> Than a herd of white cattle. (31b)

Reginald Scot, in *The discoverie of Witchcraft*, 1584 – to add black magic to white – gives instructions for raising the ghost of a hanged man, with the aid of a hazel wand tipped with an owl's head and a bundle of St John's Wort. John Aubrey tells how the house of a London friend of Henry Lawes, the musician, was haunted and cleansed by a doctor, who put St John's Wort under the householder's pillow (5).

A few names have been recorded for other species which resemble *Hypericum perforatum*. For *Hypericum hirsutum*, 'Devil's Bane' in Somerset; and 'Thousand Holes' (cf. the French *mille-pertuis*, the *Perforata* and *Millies Perforatum* of the herbalists, for the true St John's Wort) in Yorkshire; for *Hypericum dubium* or *H. tetrapterum*, 'Golden Rod' in Devon and Somerset, and 'St Peter's Wort' in the Orkneys – St Peter's, since the

12 St John's Wort *Hypericum perforatum*

magic cycle and religious cycle ended on 29 June with St Peter's Day. St Peter's Wort or Hardhay was given by Gerard for the kinds which are easily distinguished by the square stem from *H. perforatum*. Americans have dealt St John's Wort the last blow. Introduced and a vile nuisance in the U.S.A., the *fuga daemonum*, the *Christi Kreuzblut*, 'Christ's Cross-blood', has become a prosaic demon itself, under the name of Klamath Weed (from the Klamath Indians or the Klamath River in California and Oregon).

As though this sun plant contained an extra dose of the sun, it has been found that fluorescent substances in *Hypericum perforatum* can make the nerve endings in white skinned cattle or sheep sensitive to strong sunlight, causing severe inflammation and blistering (134).

XV. Cistaceae

1. Rockrose. *Helianthemum chamaecistus* Mill. 93, H 1

Local names. GOAT'S FOOT, Som; SOL FLOWER, Mor; SOLDIER'S BUTTONS, Som; SUN DAISY, Lincs.

Before the pleasure-garden developed, the Rockrose was a neglected and overlooked plant. Pretty golden or sulphur-yellow flowers were not enough to mark it in any special way. Also, it is patchily distributed, preferring chalk and limestone. Gerard had no name to give it except 'English Yellow Dwarffe Cistus'. John Ray called it Sunflower. Then, as various species entered the garden in the eighteenth century, they were given the book name of Rockrose, which has stuck. For the wild plant, if not for many of its garden forms with white or rose flowers, the Scottish name 'Sol Flower' would be more apt, or else an Englishing of the French *herbe d'or*, Herb of Gold.

XVI. Tamaricaceae

1. Tamarisk. *Tamarix anglica* Webb and *Tamarix gallica* L.

Local names. BRUMMEL, CYPRESS, Corn.

Tamarix anglica comes from the south-west of Europe, and the very near *Tamarix gallica* from southern Europe. It seems as if the Tamarisk was first introduced in Tudor times. William Turner wrote in 1548: 'I dyd never see thys tree in Englande, but ofte in high Germany, and in Italy.' Edmund Grindal, who was in exile in Germany during Queen Mary's reign, is supposed to have brought back with him the first Tamarisk, which he planted (he was then Bishop of London) in his garden at Fulham Palace. He had 'found by experience that it was a soveraigne remedie against the great and indurate passion of the Spleene' (28). This use goes back to Dioscorides, and Gerard makes much of it in the *Herbal*. He also grew Tamarisk in his garden.

Tamarisk must have been planted widely for medicinal reasons. Later on, Miller in his *Gardeners Dictionary* declared scornfully that Tamarisk was only recommended by its oddness. The branches straggled and could not be trained and the leaves were thin on the branches. One may think differently about the Tamarisk naked and waving its racemes of pink blossom in a Cornish sea wind, or gnarled and tall on St Agnes, in the Isles of Scilly, as a windbreak. Cornishmen and Scillonians have used its willowy branches for making lobster-pots.

XVII. Caryophyllaceae

1. Bladder Campion. *Silene cucubalus* Wibel 106, H 40

Local names. ADDER-AND-SNAKE PLANT, Dev; BILLY BUSTERS, Som; BIRDS' EGGS, Shrop; BLADDER BOTTLE, BLADDERS OF LARD, Som; BLADDERWEED, BLETHERWEED, Dor; CLAPWEED, Herts; COCKLE, War; CORN-POP, Wilts; COWBELL, Som; COW-CRACKER, Dumf; COWMACK, Scot; COW-RATTLE, Bucks.

FAT BELLIES, KISS-ME-QUICK, Som; POP GUNS, Dor; POPPERS, Som, Wilts; POPPY, Wilts; RATTLE-BAGS, Corn, Dev; RATTLEWEED, Wilts; SHACKLE BACKLE, SNAGGS, Som; SNAPPERS, Suss, Kent; SPATTLING POPPY, Cumb; THUNDERBOLT, Kent.

WHITE BOTTLE, Som, Cam; WHITE COCKLE, Berw; WHITE COCK ROBIN, WHITE HOOD, Som; WHITE MINTDROPS, N'thum; WHITE RIDING HOOD, Som; WHITE ROBIN HOOD, Som, Wilts.

Bladder Campion shares a few names with Ragged Robin, Red Campion, and White Campion, though it was noticed and named chiefly for its bladder-shaped calyx, which will pop before the flowers expand, and for its rattling seeds. Spattling Poppy was a name coined by Gerard, 'in respect of that kinde of frothie spattle, or spume, which we call Cuckow Spittle, that more aboundeth in the bosomes of the leaves of these plants, then in any other one plant that is knowne'; yet this was due to a guess by one of his predecessors, who identified Bladder Campion with a 'foam poppy' mentioned by Dioscorides.

Of Bladder Campion and other Campions, Gerard also says they are good against venomous bites; his Spattling Poppy will even prevent damage or hurt by any venomous beast if you merely hold it in your hand. Though snakes have given the Bladder Campion only one of its names, compare the snake associations of Red Campion and White Campion, and of the Greater Stitchwort.

The explanation that Bladder Campion was called 'Thunderbolt' because children made miniature explosions by popping the calyx (19) seems thin; for the more probable explanation, see *Papaver rhoeas* and *Melandrium album*. In French Bladder Campion is also *herbe du tonnerre*. A likeness between cow bells and the inflated calyx has suggested two names. The Scottish name 'Cowmack' rose out of a belief that Bladder Campion helped to make cows want the bull.

The young shoots when they are boiled, it was long ago claimed, make a vegetable so delicious that it ought to be improved for the garden (25).

2. Sea Campion. *Silene maritima* With. 82, H 25

Local names. BUGGIE-FLOWER, Shet; DEAD MAN'S GRIEF, N'thum; THIMBLES E Ang; WHITE SNAPJACKS, Som; WITCHES' THIMBLES, N'thum.

Names which probably apply also to the preceding species – except for Buggie-flower, since *S. cucubalus* does not occur in Shetland.

3. Red Campion. *Melandrium rubrum* (Weig.) Garcke. 112, H 40

Local names. ADDER'S FLOWER, Som, Herts; BACHELOR'S BUTTONS (often the garden variety with double flowers), Dev, Som, Suss, Kent, Ess, Suff, N'hants, War, Worc, Lancs, Yks, Cumb; BILLY BUTTONS, Ess, War; BIRD'S EYE, Dev, Som; BRID EEN, Ches; BULL RATTLE, Bucks; BULL'S EYE, Som; CANCER, Scot; CUCKOO, Dev, Som, Notts; CUCKOO-FLOWER, Dev, Som, N'hants, Leic; CUCKOOPINT, N'hants; DEVIL'S FLOWER, Lancs; DOLLY WINTER, Corn; DRUNKARDS, Som.

FLEABITES, Corn; GIPSY-FLOWER, Som; GRAMFER-GREYGLES and RED GRAMFER-GREYGLES, Dor; GRANFER-GRIGGLES, Dor; GRANFER JAN, Dor, Wilts; JACK-BY-THE-HEDGE, Suss; JACK-IN-THE-HEDGE, Som, Suss; JACK-IN-THE-LANTERN, Dor; JAN GRANFER, Som, Wilts; JOHNNY WOODS, KETTLE SMOCKS, Som; LOUSY BEDS, Cumb; MARY JANES, Som; MINTDROP and RED MINTDROP, N'thum; MOTHER-DEE, MOTHER-DIE, Cumb; PLUM-PUDDING, Suff; POOR JANE, PUDDINGS, Som; RED BIRD'S EYE, Rad; RED BUTCHER, Glos, Ches; RED CATCHFLY, Berks; RED JACK, Ches; RED RIDING HOOD, Dev, Dor, Som.

BOB ROBIN, Corn; COCK ROBIN, Corn, Dev; POOR ROBIN, Corn, Dev, Som; RED ROBIN, Dev, Dor, Som, Wilts; RED ROBIN HOOD, Wilts; ROBIN FLOWER, Dev; ROBIN HOOD, Dev, Dor, Som; ROBIN-I-TH'HEDGE, Yks; ROBIN-RUN-IN-THE-HEDGE, Dor; ROBIN REDBREAST, Corn, Dev; ROBIN'S EYE, Dev, Som; ROBIN'S FLOWER, Corn, Dev; ROUND ROBIN, Dev, Kent; WAKE ROBIN, Yks.

ROBROYS, Som; ROSE CAMPION (from the *Lychnis coronaria* of gardens), Glos, Yks; SARAH JANES, Som; SCALDED APPLES, Shrop; SOLDIERS and RED SOLDIERS, Ches, N Eng; SOLDIER'S BUTTONS, Yks; LOUSY SOLDIER'S BUTTONS, Lancs; SWEET WILLIE, Shet; WATER POPPIES, Lincs.

A plant of snake (in Wales a local name is *blodau'r neidr*, snake's flower), of devil, goblin (Robin Goodfellow and Jack a Lantern – see *Geranium robertianum*), and of death, if it is picked (for Mother-dee, see *Melandrium album*). In May, you often find Red Campion, Bluebells, and Early Purple Orchis flowering together in the same patch of wood, and there is a curious relationship between the three species in some of the names they share. In Devon and Somerset, all three are sometimes called Cuckoo-flowers. In Dorset, Wiltshire, and Somerset, the Bluebell is Gregle or Griggle; in Dorset and Somerset, Granfer-griggle is given to the Early Purple Orchis, in Dorset to the Red Campion; and in both Dorset and Somerset the Bluebell is sometimes named Granfer-griggle. In Somerset all three plants are Adder's Flowers, in Gloucestershire both the Bluebell and Orchis are Bloody Man's Finger. *Orchis mascula* in several counties is named Robin (see the Robin names above). Possibly a kinship was imagined between these plants of the same habitat and altogether different appearance, on a sexual and supernatural basis no longer absolutely clear. In Manx the Red Campion is called *blaa-ferish*, 'fairies' flower'.

In the second, revised edition of Gerard's *Herbal* (1633), the name Bachelor's Buttons (which belongs to other species as well) is explained by a resemblance between the Campion flowers – the double ones of the garden – and the 'jagged cloath buttons anciently worne in this kingdome'; and the name was conferred by 'our gentlewomen and other lovers of floures in those times'. This is only half of an explanation. If one may trust Robert Greene, in *A Quip for an Upstart Courtier*, 1592, wanton maids wore bachelor's buttons beneath their aprons – to entice the love of the owner. See also *Knautia arvensis* for the method of love divination practised by Belgian girls, and *Centaurea nigra*. 'Campion' itself is a book name of the sixteenth century, probably first applied to the cultivated Rose Campion (*Lychnis coronaria*), meaning 'champion', i.e. of the garden.

4. White Campion. *Melandrium album* (Mill.) Garcke. 107, G 26

Local names. BACHELOR'S BUTTONS, Glos; BILLY BUTTONS, Som, Glos, War; BUTCHERS, BUTCHER'S COW, Glos; COCKLE, Norf, Rut; COW RATTLE, Bucks; CUCKOO FLOWER, Som; EVENING CLOSE, Dor; GRANDMOTHER'S NIGHTCAP, Suss; GRANNY'S NIGHTCAP, Bucks.

MILK-FLOWER, Wilts; MILKMAIDS, Som; MOTHER-DEE, MOTHER-DIE, Cumb; PLUM-PUDDING, Suff; POOR JANE, Som; SNAKE'S FLOWER, Mddx, Oxf; SUMMER SAUCERS, Som; THUNDERBOLT, Rut; THUNDER-FLOWER, Cumb; WHITE RIDING HOOD, Dev; WHITE ROBIN, Dor; WHITE ROBIN HOOD, Som.

White Campion grows on disturbed or intermittently disturbed soil, often making a fair show of white blossom, or 'Summer saucers', along a main road. So it is a plant perhaps of ancient introduction, which entered the country with neolithic farmers. It goes about with agriculture, and remains of it have been found in Great Britain, in Scotland, and Wiltshire (and in Switzerland), on neolithic and Bronze Age sites (see *Journal of Ecology*, 35, p. 271, 1947).

Moths flutter from white flower to flower during the night, and do the business of pollination. But this, too, is a sinister plant, as though under protection; called 'Mother-dee', because picking the flowers brought death to your mother, whereas picking *Melandrium rubrum* (though it was also Mother-dee) killed your father. A North German name for White Campion is *Todtenblume*, 'death-flower'. Luxemburg beliefs about *Melandrium rubrum*, the Red Campion, may explain why the White Campion is called 'Thunder-flower' and 'Thunderbolt'. In Luxemburg the Red Campion is *fleur de tounouar* (*tonnerre*, thunder), and children believe they will be killed by lightning if they pick it (160). Other English thunder flowers are *Lychnis flos-cuculi*, the *Stellaria holostea* of adders, fairies, and the devil, *Silene cucubalus*, *Papaver rhoeas*, *Ajuga reptans*, *Sempervivum tectorum* (q.v.), and possibly *Veronica chamaedrys*. For Bachelor's Buttons, see *Melandrium rubrum*; and for the Robin names see *Geranium robertianum*.

5. Ragged Robin. *Lychnis flos-cuculi* L. 112, H 40

Local names. BACHELOR'S BUTTONS, Dev, Som, Suss; BILLY BUTTONS, War; COCK ROBIN, Som; COCK'S CAIM, Lanark; CUCKOO, Dev; CUCKOO-FLOWER, Dev, Som, Berks, Suff, War; DRUNKARDS, GIPSY-FLOWER, Som; INDY and INDIAN PINK, Glos.

MEADOW PINK, Dev, Som; MEADOW SPINK, Scot; POLLY BAKER, Som; RAG-A-TAG, Shet; RAGGED JACK, Som, Suss, Kent, Ess; RAGGED ROBIN, Dev, Som; RAGGED WILLIE, Shet; RED ROBIN, Som; ROBIN HOOD, Dev, Dor, Som, Dur; ROUGH ROBIN, Cumb; SHAGGY JACKS, Dev, Som; THUNDER-FLOWER, Yks; WILD WILLIAMS, Berks.

Flowers of a good colour, 'finely and curiously snipped in the edges', early blossoming, universal distribution, and a liking for damp meadows – in spite of all these advantages, Ragged Robin has few associations. 'These are not used either in medicine or in nourishment: but they serve for garlands and crowns, and to decke up gardens.' Gerard had no more to say, and Ragged Robin has to rest upon its charm. Yet, by the names, the Cuckoo names and the Robin names, it may have shared curious associations with Red Campion, Herb Robert (q.v.), and other species.

The pleasant name of Wild Williams was known to Gerard, and must have marked *Lychnis flos-cuculi* from the Sweet Williams of the Tudor garden (cf. Indian Pink and Meadow Pink). For Bachelor's Buttons and Billy Buttons, see Red Campion, and for Thunder-flower, see *Papaver rhoeas* and *Melandrium album*.

6. Corn Cockle. *Agrostemma githago* L. 104, H 22

Local names. CAT'S EARS, Dor; COCKEREL, Suff; COCKLEFORD, Glos; COCKLES, Som; COKEWEED, Scot; CORNCOCKLE, Dor, Glos, Ess, Yks; CORNFLOWER, Ches; CORN PINK, N'hants; CROWN OF THE FIELD, Som; GYE, Suff, Lancs; HARDHEADS, N'thum; LITTLE AND PRETTY, Som; PINK, N Ire; PAPPLE, PAWPLE, POPILLE, Scot; POPPLE, E Ang, Lincs, Yks, Cumb, N'thum, Scot, N Ire; POPPY, Ches; PUCK NEEDLES, Suss; ROBIN HOOD, Dor.

Older floras describe Corn Cockle as common among the crops of corn. The *Flora of the British Isles* in 1952 calls it 'common, but decreasing'. Decreasing is true, common now seems an exaggeration – for most parts of the country. Thus in North Wiltshire I have not found one of its bright flat eyes staring out of the wheat in the last ten summers. It is going the way of the Bluebottle (q.v.), after troubling the farmer for hundreds of years. The seed corn is better cleaned, so this ancient weed from southern Europe must now vanish. 'Cuckole hath a longe small lefe, and wyl beare fyve or vi floures of purple colour, as brode as a grote,' Fitzherbert wrote in his farmer's manual, *The Boke of husbandrie*, in 1523. Gerard wryly remarked that 'what hurt it doth among corne, the spoil unto bread, as well in colour, taste, and unholsomnes, is better known than desired'.

The Grete Herball of 1526 gives for one of its names *herba indica*, the Indian Herb, which fits an invader from the warmth. Popple is an old name going back to the Middle Ages. The Somerset name 'Crown of the Field' translates the French name for Corn Cockle, *couronne des blés*. The Puck Needles of Sussex (Puck the goblin) must refer to the long teeth of the calyx-tube which spread out between the bright petals. In Germany it has been likened by name to the Devil's eye, in Flanders and France to God's eye, or Christ's eye.

7. Deptford Pink. *Dianthus armeria* L. 54

All our wild Pinks are unfortunately too rare. The Deptford Pink is one of the least rare, though when you find a colony of it at last, it looks a little scant and thin, and not quite the plant you hoped for, for all its little spirts of rose colour. The name we owe to Gerard: 'There is a little wilde creeping Pinke, which groweth in our pastures neere about London, and other places, but especially in the great field next to Detford, by the path side as you go from Redriffe to Greenewich, which hath many small tender leaves, shorter than any of the other wilde Pinkes; set upon little tender stalks, which lie flat upon the ground, taking holde of the same in sundrie places, whereby it greatly encreaseth; whereupon doth growe little reddish flowers.' This was taken to be *Dianthus armeria*, though from the description the species must be the Maiden Pink.

8. Cheddar Pink. *Dianthus gratianopolitanus* Vill. 1

If you look up at the grey walls of the Cheddar Gorge in the Mendips, you will see more of Ivy and of Pretty Betsy or Red Valerian (*Kentranthus ruber*) than anything else. But there was a time when the Cheddar Pink gave colour to the rock, and had not to be searched for through binoculars. The Cheddar Pink or Cleeve Pink – Cliff Pink, that is to say – was discovered early in the eighteenth century by a Wiltshire botanist, Samuel Brewer. It must have been a bright moment in a sad life, since Brewer failed as a cloth manufacturer at Trowbridge, lost a fortune, and quarrelled with his family, and became head gardener to the Duke of Beaufort. But he had other triumphs. At Holyhead, soon after finding the Cheddar Pink, he also discovered a Rockrose of extreme rarity – now *Helianthemum guttatum* ssp. *breweri* – which was named after him.

Delicate on the grim rock, with flowers larger than those of our other native Pinks, blush-coloured and sweetly scented, and with bluey-green leaves, the Cheddar Pink was as famous as Cheddar cheese. Villagers traded roots to those who began to visit Cheddar and the caves after a road had been made through the Gorge in 1801. The Pink was in every guide-book, and Murray's *Handbook for Travellers in Wiltshire, Dorsetshire, and Somersetshire* of 1882 declared that the excursionists had nearly made it extinct. They never succeeded. The Cheddar Pink remains in its one British station – out of reach. Still, if you want this pink for a garden, it is easy enough to buy stock from the flower firms.

9. Maiden Pink. *Dianthus deltoides* L. 58

Rare again (though less so than *Dianthus armeria*), and for the most part preferring the chalk. Though Gerard was describing a different species of the sixteenth-century garden, the name must come from his *Caryophyllus virgineus* or Maidenly Pink – 'this Virginlike Pinke', as he calls it, whose flowers 'are of a blush colour, whereof it took his name'. Since the Wild Pinks are aristocrats of rarity or beauty, or both, it should be mentioned that Dianthus means the 'Flower of Zeus'.

10. Soapwort. *Saponaria officinalis* L. 71, H 29

Local names. BOUNCING BETT, Dor; BRUISEWORT, Som; FAREWELL SUMMER, Mon; GILL-RUN-BY-THE-STREET, Suss, Kent; HEDGE PINK, Hants.

Crush a handful or two of Soapwort leaves, and bring them to the boil in a saucepan of water. Strain off the liquid, and you will find it makes an appreciable lather. It will give you a dry, comfortless, slightly stinging, but not altogether ineffectual wash. The lather does not last very long.

Soapwort is a sad name for an attractive plant. The great but unimaginative William Turner invented Sopewort and Skowrwurt (183), because the herbalists called it *Saponaria* and *Herba Fullonum*, the 'fullers' herb'. (In French, it is still *herbe à foulon.*) The English names in *The Grete Herball* of 1526 were Burit, Herbe Phylyp (? St Philip), Saponary, Fuller's grasse, and Crowsope. Another of its mediaeval names was Foam Dock. Possibly soapwort was used by the early mediaeval fullers for soaping cloth before it went under the stamps of the mill, and no doubt it was one of the ancient washing plants before the invention or the general employment of soap. Washing plants are still used by the Arabs, and Soapwort is still cultivated for washing woollens in Syria. It was so used in France in the nineteenth century. In the Swiss Alps, sheep were washed with a mixture of the leaves and roots and water before they were shorn, and linens were washed in Soapwort juice and ashes. Moreover, Soapwort may have been the plant *mastakal* known to the Assyrians (30). It hangs on pertinaciously as an escape, and it might be interesting to discover if it grows at all by the site of mediaeval fulling-mills.

The thirteenth-century Italian monk Rufinus, a descriptive botanist of marked ability, described the properties of Soapwort, and found it valuable for what we might call a disinfectant wash (176). The centuries go by and Gerard talks of Soapwort only as an ornament of gardens, but it also was laid on cuts (and bruises?), according to Culpepper, and extolled as 'an absolute cure in the French-Pox' (49). The settlers thought it valuable enough to take with them to New England. There water made soapy with the leaves and roots was applied to the horrible rash caused by Poison Ivy (162). Soapwort is naturalized in the U.S.A., and Americans still call it by the west country name of Bouncing Bett (also Lady-by-the-Gate). It may be no less of an introduced and naturalized species in Great Britain. Gerard thought it wild 'neere to rivers and running brookes in sunnie places', and if it looks native at all, it is only beside streams in Cornwall, Devon, and Wales.

11. Chickweed. *Stellaria media* (L.) Vill. 112, H 40

Local names. ARVA or ARVI, Shet; BIRD'S EYE, Som; CHICK WITTLES (i.e. victuals), Suff; CHICKEN'S MEAT, E Ang; CHICKENWEED, Norf, Ches, Lincs, Yks, Cumb, Scot; CHICKENWORT, Scot; SCHICKENWIR, Shet; CHICKNYWEED, Dev; CLUCKENWEED, CLUCKWEED, CLUKENWORT, N'thum; MISCHIEVOUS JACK, Som; MURREN, Yks; SKIRT BUTTONS, Dor; TONGUE-GRASS, Ire; WHITE BIRD'S EYE, Bucks. 'Winter-green' has also been recorded.

Growing on the open soil of gardens and fields, and as Gerard remarked, on old walls and in gutters of houses, Chickweed is everywhere, winter-green and happy in the coldest months. Coles, in his *Adam in Eden* (1657), explained Chickweed by saying that chickens and birds loved to pick the seed of it, which they do. Gerard wrote that 'little birdes in cages' are refreshed with it, 'when they loath their meate'. Turner in 1538 called it Chykwede or (see above) Chykenwede. German names are *Hühnerdarm*, 'hen's guts', and *Vogelkraut*, 'bird plant'. French names include *mouron* – hence 'Murren' above – *mouron des oiseaux*, or *morsgeline*, which preserves the *Morsus gallinae*, or 'hen's bite' of mediaeval botany.

12. Greater Stitchwort. *Stellaria holostea* L. 111, H 40

Local names. ADDER'S MEAT, Corn, Som; ADDER'S SPIT, Corn; ALL BONES, Som; BAALAM'S SMITE, Suff; BACHELOR'S BUTTONS, Som, Bucks, Suff; BILLY BUTTONS, Som, War; BILLY WHITE'S BUTTONS, War; BIRD'S EYE, Worc, Derb, Yks, Dur; BRANDY SNAP, Suss; BREAK JACK, Dor; CUCKOO'S MEAT, CUCKOO'S VICTUALS, Bucks; CUCKOO FLOWER, I o W.

DEAD MAN'S BONES, Berw; DEVIL'S CORN, Shrop; DEVIL'S EYES, Dor, Denb; DEVIL'S FLOWER, DEVIL'S NIGHTCAP, DEVIL'S SKIRT BUTTONS, Som; EASTER BELL, Dev; EASTER-FLOWER, Suss; EYEBRIGHT, Som; GRANNY'S NIGHTCAP, Dor; HAGWORM-FLOWER (i.e. adder flower), Yks; HEADACHE, Cumb; JACK-IN-THE-BOX, Dor; JACK-IN-THE-LANTERN, JACK SNAPS, LADY'S BUTTONS, LADY'S CHEMISE, Som; LADY'S LINT (i.e. flax), Dev, Som; LADY'S NEEDLEWORK, Som; LADY'S SMOCK, Corn, Som; LADY'S WHITE PETTICOATS, Corn, Heref; LITTLE JOHN, Som. The Manx name is LIEEN-FERISH, 'fairy flax'.

MAY-FLOWER, Cumb; MAY GRASS, Shrop; MILK-CANS, Wilts, Ches; MILKMAIDS, Som; MILK-MAIDENS, Dev, I o W, Surr, Yks; MILK-PANS, Ches; MILLER'S STAR, Som, Suss, Yks; MOON-FLOWER, Worc; MOONWORT, Yks; MORNING STARS, Dor; MOTHER SHIMBLE'S SNICK-NEEDLES, MOTHER'S THIMBLE, Wilts; NANCY, also PRETTY NANCY and SWEET NANCE, Som; NIGHTINGALE, Wilts; OLD LAD'S CORN (i.e. devil's corn, as above), Shrop; OLD MAN'S SHIRT (i.e. devil's shirt), Corn; ONE O'CLOCK, Dev; PICK-POCKET, Dev, Som, Kent; PISGIE or PIXIE, Corn, Dev, Som; PISGIE-FLOWER, Corn; POP-GUNS, Som; POP JACK, Som; POPPERS and POPPY, Wilts.

SAILOR-BUTTONS, Hants; SATIN FLOWER, Corn; SCURVYGRASS, Worc; SHEPHERD'S WEATHERGLASS, Lancs; SHIMMIES (i.e. chemise), SHIMMIES AND SHIRTS, Som; SHIRT BUTTONS, Som, Wilts, Hants, Kent, Yks, Ess, Norf, Camb, Yks; SMOCK-FROCKS, Dev, Bucks; SMOCKS, Bucks; SNAKE-FLOWER, Som, Wilts, Hants, Worc, Notts, Lincs; SNAKE GRASS, Hants; SNAKEWEED, Som.

SNAPCRACKERS, Ess; SNAPJACKS, Dev, Dor, Som, Wilts; SNAPPER-FLOWERS, Suss; SNAPPERS, Dev, Suss; SNAPS, Wilts; SNAPSTALKS, Ches; SNAPWORT, Kent; SNOW, SNOWFLAKE, Suss; SNOW ON THE MOUNTAIN, Wilts; STAR-FLOWER, Suss, Lincs; STAR GRASS, Yks; STAR OF BETHLEHEM,

Dev, Som, N Eng, N Ire; STAR OF THE WOOD, Dev, Som; STARWORT, Dev, Som, Glos, Scot; STEPMOTHERS, SWEETHEARTS, Som.

THUNDERBOLTS, Dor; THUNDER-FLOWER, Cumb; TWINKLE STAR, Som; WATCHES, Dor; WEDDING FLOWERS, Glos; WHITE BIRD'S EYE, Rad; WHITE BOBBY'S EYE, Hants; WHITE-FLOWER, WHITE-FLOWERED GRASS, Wilts; WHITE SUNDAY, Dev; WILD PINK, Bucks.

Stitchwort was clearly a plant under protection, belonging to the devil, piskies, the Jack a Lantern (the lantern-carrying elf or goblin of the *ignis fatuus*), and to snakes; and then, to banish evil, belonging to the Virgin, and associated with Whitsunday and the Star of Bethlehem. Cornish children feared picking the flowers; if they did, the adder would bite them, or they would be piskey-led (often in the spring, adders come out to sun themselves on the earthen hedges of the west country, which are whitened with Stitchwort). And it was another thunder flower, the picking of which presumably provoked thunder and lightning (see *Papaver rhoeas* and *Melandrium album*). Add also the rather dubious associations by name with cuckoos, smocks, milkmaids, and bachelor's buttons (see *Cardamine pratensis*, *Melandrium rubrum*, and *Geranium robertianum*).

The name Stitchwort seems to have dropped from the vernacular. Bullein used Stitchwort and Stitchgrasse in his *Booke of Simples* (1562), and Turner gave Stitchwort in the part of his *Herbal* published in the same year. The name is an old one, which goes back at least to the thirteenth century. It must have been the wort for a stitch in the side, or any sudden pricking or pain ('They are woont to drinke it in wine with the powder of Acornes, against the paine in the side, stitches, and such-like' – Gerard). And pains of this kind, in the magical beliefs of the Anglo-Saxons and the Celtic peoples, were likely to be 'elf-shot', to be caused by those elves to whom the plant belonged. Many of the names come from the easy breaking of the stems, and some reflect both this and the name *Holosteum* given to the plant by the French botanist, Jean Ruel (1474–1537). Ruel identified it, wrongly, with the *holosteon*, the 'whole bone' or 'all bone' of Dioscorides, which was employed against fractures.

The Somerset name of Eyebright was transferred by the first New England settlers to the allied American species *Stellaria longifolia* (111). With the English names after the Virgin, cf. the French *collerette de la Vierge*, 'Our Lady's little collar'.

13. Pearlwort. *Sagina procumbens* L.   112, H 40

Local names. BEADS, Wilts; BIRD'S EYE, Suss; LITTLE CHICKWEED, Som; POVERTY, Norf.

Known as *mòthan* or *mòlus*, in the Highlands and Islands, Pearlwort has special powers, attributed to the blessing of it by Christ, St Bride, and St Columba; also it was the first plant Christ stepped on when he came to earth, or when he had risen from the dead. It was gathered to an incantation in the Gaelic, part of which runs

> While I shall keep the pearlwort
>   Without wile shall be my lips,
> Without guile shall be mine eye,
>   Without hurt shall be mine hand,
> Without pain shall be my heart,
>   Without heaviness shall be my death.

Over the lintel, it kept fairies from entering the house and spiriting away the inhabitants. Under the right knee of a woman in childbed, it soothed her mind and protected her child and herself from the fairies. It protected cows, calves, and milk, placed in the fore-hooves of the bull, or in a bag with iron under the milk vessels. If a cow had eaten Pearlwort, the protection extended to the calf, the milk, and all who drank the milk. Girls drank the juice, or wetted their lips with it, to attract lovers. If they had a piece in their mouths when they were kissed, the man was bound for ever (31b). Pearlwort is a botanist's book name.

14. Corn Spurrey. *Spergula arvensis* L.   112, H 40

Local names. BEGGARWEED, Beds; BOTTLE BRUSH, Yks; CAT'S HAIR, Corn; COWQUAKES, E Eng; DEVIL'S FLOWER, Corn; DODDER, Ches, Cumb; DOTHERS, Berw; GRANYAGH, Ire; GUANO WEED, Corn; MELDI, Shet; MOUNTAIN FLAX, Dor, Shrop, Yks; PICKPOCKET, Ches; PICKPURSE, E Ang, Lincs; SANDWEED, Norf; TAILOR'S NEEDLE, Corn; TOADFLAX, TOAD'S BRASS, Ches; YARR, Lancs, Cumb, N'thum, Scot, Donegal.

Spurrey will crowd into fields on a light soil, and look up from the ground, very neat and fetching with its white petals and its leaves set with such regularity round the stem, like green spokes of a wheel. It has only declined into a weed. Still grown as fodder in continental Europe, and grown with us in the sixteenth century (as 'Francke', or a fattening herb, from an obsolete word for a fattening pen or sty), Spurrey once gave food for men. There is evidence that it was a 'utility plant' in Denmark as far back as the pre-Roman Iron Age; in Great Britain spurrey seeds have been found with oats of the Roman period and with prehistoric flax (107). They were part of the last meal of the Tollund Man (see *Polygonum convolvulus*); and in Shetland in historic times they were ground into meal – hence the Shetland name of Meldi, the meal plant (166). Our word Spurrey comes from the Dutch.

XVIII. Portulacaceae

1. Spring Beauty. *Claytonia perfoliata* Willd. 57

America has its weeds from Europe. The Spring Beauty (as the Americans call it) is a *quid pro quo*, a smooth, delicate, peculiar little plant with perfoliate stems, which was not observed in Great Britain until 1852. It grows from the Isles of Scilly up to Scotland, but a preference for sand and a dislike of heavy soils prevent it from being a plague to most gardeners. The small bulb fields in the Isles of Scilly are sometimes white with its pretty flowers. It came late to us, no doubt because it is a plant of farthest America, on the Pacific. Other American names for it are Winter Purslane, Indian Lettuce and Miners' Lettuce. Tender and succulent, you can boil it and eat it like spinach.

XIX. Ficoidaceae

1. Hottentot Fig. *Carpobrotus edulis* (L.) N.E. Br. Introduced

Better known by its old name of *Mesembryanthemum*, whence the Cornish name of Sally-my-handsome. In the warmth of Cornwall or Devon, this garden plant from South Africa is quite at home near the sea, escaping, or crawling in thick, fleshy mats of green up the cliffs and over rocks and banks. It looks well on the anchovy sauce cliffs about Dawlish (where you can see colonies of Sally-my-handsome from the train). The flowers are silky, and when they are rose-red, and not a frail yellow, they splendidly suggest the warmth, sun, and colour of other climates. The fruits – the Hottentot Figs – can be eaten. The taste is described as 'pleasantly acid'.

XX. Chenopodiaceae

1. Good King Henry. *Chenopodium bonus-henricus* L. 101, H 39

Local names. ALL GOOD, Som, Hants; FAT HEN, Berks, Kent, Surr, Suff, Ches; GOOD KING HENRY or HARRY, Cam, N'thum; GOOD NEIGHBOURHOOD, Wilts; JOHNNY O' NEALE, Shrop; MARCARAM, Yks; MARGERY, Lincs;

13 Good King Henry *Chenopodium bonus-henricus*

MERCURY, Lincs, Yks, West, Cumb, N Eng; MIDDEN MYLIES, Selk; ROMAN PLANT, Lancs; SHOEMAKER'S HEELS, Shrop, Rad; SMEAR DOCKEN (i.e. fat or grease dock, cf. German *Schmerbel*, *Schmeerwurz*, etc.), Scot; SMIDDY LEAVES, Berw; WILD SPINACH, Som, Hants, I o W.

Since the young shoots and flowering tops boiled, and eaten with butter, are neither very pleasant nor unpleasant, this plant hardly lives up to its names as an old pot-herb – the French *toute-bonne* and the English All Good (which are translations of the *tota bona* of the herbalists), or Fat Hen ('in handling it is fat and oleous'), or Good King Henry. This last name comes to us – by way of the Tudor herbalists – from the German *Guter Heinrich*, 'Good Henry'. Dioscorides had described a plant known as the little herb of Hermes – the god Mercury. This *mercurialis* of the Romans was identified with the medicinal herb and garden weed, *Mercurialis annua*, which faintly resembles *Chenopodium bonus-henricus*. But as well as this useful Mercury, there was also the more sombre, more decidedly poisonous *Mercurialis perennis* of the woods. Since this was called the Bad Henry, *Böser Heinrich*, the edible and virtuous Chenopodium needed distinction as the 'Good Henry'. The King in Good *King* Henry is our interpolation. Like the Robin in many English plant-names, this Heinrich of German plant-names may have been an elf – in this case a substitute for Hermes or Mercury. The fourteenth-century *Agnus Castus* calls the plant not only 'Mercurie' but 'Papwourt' (20).

Gerard observed that Good King Henry grew 'in untilled places, and among rubbish neere common waies, olde walls, and by hedges in fields'. It had been introduced, and then it escaped and persisted; though – especially in the North – it was cultivated long after Gerard's day. In more than one place, near the ruins of Llanthony Priory in South Wales, for example, colonies of Good King Henry can be found on mediaeval sites. Unexpectedly, it is an attractive plant, in the contrast between its leaves and the tall flowering tops, which are spires of a greenish yellow.

2. Fat Hen. *Chenopodium album* L. 112, H 40

Local names. ALL GOOD, Hants; BACON WEED, Dor; CONFETTI, Som; DIRTWEED, Som, Lincs; DIRTY DICK, Wilts, Ches; DIRTY JACK, DIRTY JOHN, Ches; DOCK FLOWER, Som; DUNGWEED, Glos; FAT HEN, Som, Suss, Surr, Suff, E Ang, Beds, N'hants, Heref, Ches, Leic, Notts, Lincs, Yks, Cumb, N'thum; LAMB'S QUARTERS, Som, I o W, N Eng.

MELDWEED (cf. German *Melde*, usually for *Atriplex* species, but also for species of *Chenopodium*), Scot; MELGS, Mor; MIDDEN MYLIES, N Scot; MUCKHILL WEED, War; MUCKWEED, Glos, Suff, Norf, Yks, N'thum; MUTTON CHOPS, Dor; MUTTON TOPS, Dor, Som; MYLES, Berw; PIG WEED (cf. German *Säumelde*, 'sow meld'), Som, Hants; RAG JACK, Ches; WILD SPINACH, Midlands. For muck and dung names, cf. German *Mistmelde*, 'dung meld'. It is also MILDS, or MILES, in N Ire.

Easy to see how Fat Hen became one of men's food plants. It is a species, as so many of the names indicate, and as you can see any day in farmyard or field, of the muck heap. Ostentatiously it would have grown wherever men threw out their rubbish on to a midden, and its remains have been identified from neolithic villages in Switzerland (40). It will grow also in the fields; the seeds, in which there is fat and albumen, were once eaten as a supplementary food – they have been identified from the Bronze Age in Sussex and in Scotland (98, 107), and they formed part of the last meal of the Tollund Man between 400 B.C. and A.D. 400 (see *Polygonum convolvulus*). The plant can also be eaten for a green vegetable, and it can be used for a red or golden red dye. This species (though perhaps *Chenopodium bonus-henricus* as well, also *Atriplex patula*) seems to have been the *melde* of the Anglo-Saxons, a food plant which gave rise to place-names – Melbourn in Cambridgeshire, which was *Meldeburna* in 970, the stream where *melde* grew, and Milden in Suffolk, anciently *Meldinges* and meaning probably a '*melde* place' (67).

Curtis, in his *Flora Londinensis*, in the last years of the eighteenth century, mentioned the gathering of Fat Hen or Lamb's Quarters from dunghills and gardens, warning readers not to pick *Solanum nigrum* by mistake. Eating Fat Hen continued a long while in Ireland and in the western Highlands, where it was called *praiseach fiadhain*, 'wild pottage'. The leaves were boiled, pounded, and mixed with butter (29, 177, 128). It was the introduction and increasing cultivation of spinach, an allied plant from south-west Asia, that at last put an end to the eating of this species with meat and bacon. Fat Hen in Normandy is called *grasse-poulette*.

3. Beet. *Beta vulgaris* L. 52, H 25

The wild species around the coasts is *Beta vulgaris* ssp. *maritima* (L.) Thell, drab and unattractive, but the founder of a great family. More than 2,000 years of cultivation, selection, and accidental crossing on the wind between wild and tame, have produced all the useful forms we eat, or feed to cattle, or turn into sugar. In the Isle of Wight the Beet was called Wild

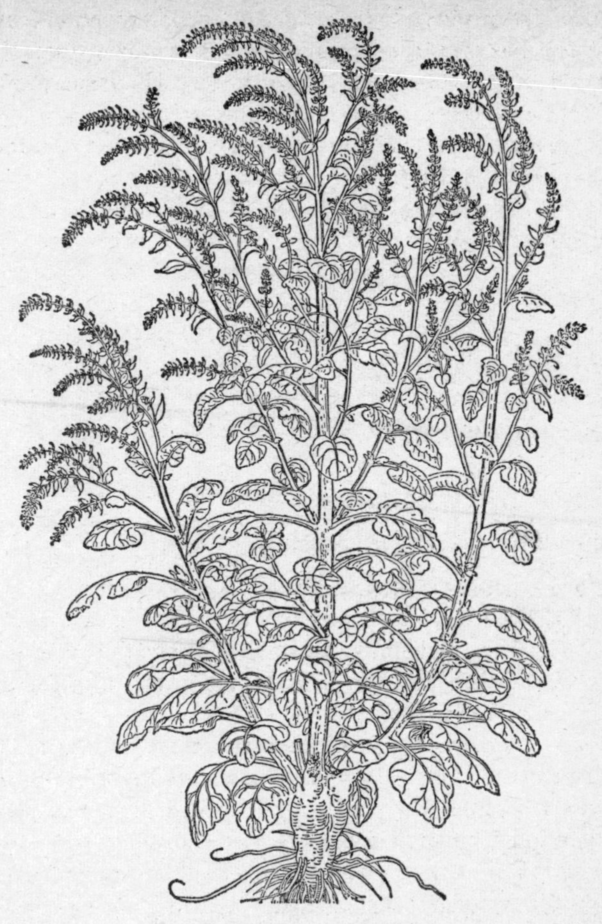

14 Beet *Beta vulgaris*

Spinach or Sea Spinach, and the leaves were eaten by cottagers with pork or bacon (21). So an old practice survived. But long before, Persian gardeners had begun to cultivate the Beet for its leaves, and especially for the long sturdy roots which anchored it into the sands of the eastern Mediterranean and the Caspian. Through centuries of diversification and improvement, the root thickened into the crimson Beetroots of the table, and the bulky root of the Mangold and the Sugar Beet.

William Turner in 1548 knew Beets only in the garden. Gerard was also ignorant of the wild plant (which was not recorded till 1629), but he was very pleased by the scarlet Beetroot, 'full of a perfect purple juice tending to rednesse'. The seeds had been brought to him from overseas 'by that courteous merchant, master *Lete*'; and in 1596 he grew a plant some twelve feet high. The leaves, he added, of this 'great red Beete or Romaine Beete' were excellent with oil, vinegar, and pepper – 'but what might be made of the red and beautifull roote (which is to be preferred before the leaves, as well in beautie as in goodnesse) I referre unto the curious and cunning cooke, who no doubt when he hath had the view thereof, and is assured that it is both good and holsome, will make thereof many and divers dishes both faire and good'. Mangolds were still to arrive out of Germany; and also the Sugar Beet, by way of Germany (where the sugar in beetroot was noticed in 1747) and France, to transform the look of so much of the agricultural countryside of northern Europe and East Anglia.

A fantastic change, from the wild species in the sand to the scarlet slices in vinegar.

4. Common Orach. *Atriplex patula* L. 109, H 39

A poor man's pot-herb, like *Chenopodium bonus-henricus* or *Chenopodium album*, and close enough to both plants to share the names of Fat Hen (E Ang, Yks, Dur, N'thum) and Lamb's Quarters (Som). Also named Lamb's Quarter or *ceathrama caothrach* in Ireland. The odd-seeming name of Orach belongs rather to the garden species, *Atriplex hortensis*, which comes from Asia, and which can be grown for spinach, or in several forms for colour – particularly *Atriplex hortensis cupreata* with violet stems and red leaves. Orach is derived through French from *atriplex*, which in turn is from the Greek *atraphaxis* (Dioscorides).

5. Glasswort. *Salicornia stricta* Dum. 75, H 25

Crisp and sparely and neatly adapted to its exposed life in the salt marshes, Glasswort reminds one of a miniature succulent out of the desert. Turner in his *Herbal* (1568) called it Glaswede for want of an English name, since the ashes of the plant were used in making glass. They provided an impure carbonate of soda for mixing with the sand. Foreign glass-workers in the sixteenth century may have taught the English that it was valuable in this way, like other seaside plants and like the kinds of Kelp which began to be collected and calcined for the same purpose. Gerard's names for *Salicornia* included Crab Grass, Frog Grasse, and the now familiar Glasswort.

Glasswort also makes as good a pickle as Samphire, and it has been called for that reason Samphire (E Ang, Lincs, N Eng), Semper (N'thum), Sampion (Ches), and Pickle Plant (Cumb). Early in the last century, it was occasionally sold in the markets (124), and it is still collected for a pickle in the eastern counties.

XXI. Tiliaceae

1. Lime Tree. *Tilia × vulgaris* Hayne

The common Lime Tree appears to be a hybrid between the Large-leaved Lime, *Tilia platyphylla* Scop., which is possibly native, and the Small-leaved Lime, *Tilia cordata* Mill., which is certainly native. No one is sure when, or by whom, it was introduced. The Romans have been given the credit, or it may not have been planted till the seventeenth century (68).

Lime flowers and their warm scent of honey have made the tree's reputation. Lime walks were planted, with the branches trained inward like a roof, or 'brave sommer houses and banketting arbors', as Gerard wrote, were contrived under or within the boughs; and the Lime was a tree of love – as in the mediaeval lyric by Walther von der Vogelweide (who died about 1230):

> Under the lime tree on the daisied ground
> Two that I know of made this bed.
> There you may see heaped and scattered round
> Grass and blossoms broken and shed
> All in a thicket down in the dale;
> Tandaradei – sweetly sang the nightingale . . .

Lime tea deliciously yields this scent and savour of honey, so long as you pick the flowers and dry them yourself, and do not use shop packets, which often make a tea tasting of newspaper. It is also possible to ferment a wine out of the flowers. Light and soft – 'whitish, plaine, and without knots, yea very soft and gentle in the cutting or handling' (Gerard) – the timber has been used for carving and sculpture from very early times. It has been made into sounding boards and keys in musical instruments – made even into aircraft. In the second world war, much of the native *Tilia cordata* and the perhaps native *T. platyphylla* along the limestone gorges of the Wye was felled to provide plywood for Mosquito planes (76).

If ever you have to cut a Lime – not for logs, since Lime wood smells rather unpleasant in the fire – it ought to be worth experimenting with the fibre of the inner bark, which is 'white, moyst, and tough, serving very well for ropes, trases, and halters' (Gerard). Ropes of lime bark used to be woven in Devon and Cornwall and in Lincolnshire (125).

A small plant may have a hundred local names. Since trees give timber, and timber is sold and is an essential of life, the names of one species do not vary a great deal. Turner, in the second part of his *Herbal* (1562), wrote of the 'Lind tre'. Lyte called it Linden or Linden tree. 'Line' was common in the sixteenth century. 'Lin' survived in Yorkshire, 'Line' in Lincolnshire, 'Lind' in Scotland. 'Whitewood' has been recorded in Worcestershire and 'Pry' was an old Essex name. Linnaeus – Carl Linné – owed his family name, very aptly for a botanist, to the tall Lime, or Linden, which guarded the family home.

XXII. Malvaceae

1. Musk Mallow. *Malva moschata* L. 92, H 34

Thrusting its pink flowers (sometimes they are white) and its delicately cut leaves out of the grass along a road, the Musk Mallow is among the prettiest of all English plants – pretty as *Sidalcea* – and it does well, and looks well, in gardens. Musky it is. You do not notice the smell out of doors, but take the flowers into a warm room, and the musk soon becomes obvious.

2. Common Mallow. *Malva sylvestris* L. 102, H 40 3. Dwarf Mallow. *Malva neglecta* Waler. 90, H 25

Local names. BILLY BUTTONS, Som; BREAD AND CHEESE, Dor, Som; BREAD AND CHEESE AND CIDER, Som; BUTTER AND CHEESE, Dev, Som; CHEESE-CAKE FLOWERS, Yks; CHEESE FLOWER, Som, Wilts, Suss; CHUCKY CHEESE, Som; CUSTARD CHEESES, Lincs; FAIRY CHEESES, Som, Yks; FLIBBERTY GIBBET, Som; FRENCH MALLOW, Corn; GOOD NIGHT AT NOON, Som; HORSE BUTTON, Donegal; LADY'S CHEESE, Dor; LOAVES OF BREAD, Dor, Som.

MALLACE, Dev, Som, Hants, I o W, Bucks; MALLOW-HOCK, Som; MARSH-MALLICE (by confusion with the name of *Althaea officinalis*), Dev, Som, Shrop, Lakes, N'thum; MAWS, Notts, N'thum, Scot; OLD MAN'S BREAD AND CHEESE, Som; PANCAKE PLANT, Som, Lincs; RAGS AND TATTERS, Dor, Som; ROUND DOCK, Som; TRUCKLES OF CHEESE, Som.

These two Mallows are very much a species of waste and wayside; but rather than the gay flowers, it was the disk of nutlets which caught the fancy, the 'knap or round button, like unto a flat cake' (Gerard), and like a cheese. Children still eat these disks or 'cheeses', as they are known from Cornwall to the Border (cf. the name *fromages* in France). Crisp and slimy, they taste not unlike monkey-nuts.

Like the Marsh Mallow and the Tree Mallow, the Common Mallow is

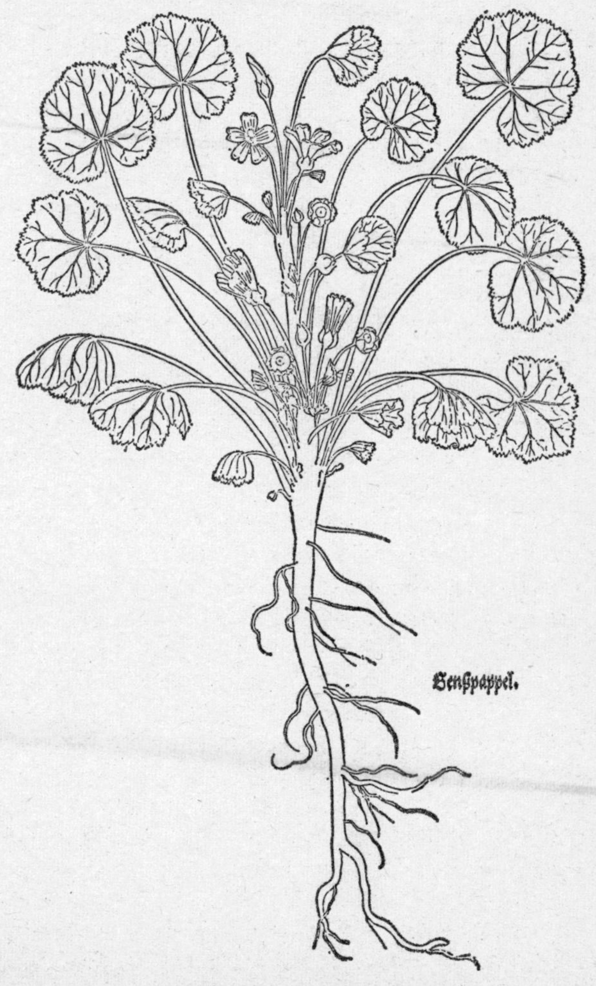

15 Dwarf Mallow *Malva sylvestris*

soft and full of mucilage. So for centuries it has been used for poulticing, and the roots have been dug up to make a soothing ointment.

4. Tree Mallow. *Lavatera arborea* L. 23, H 8

Tree Mallow is tall as a garden hollyhock, and in Devon, Cornwall, and Wales it has often been transplanted into cottage gardens to supply poultices and ointment. Where hens scratch, and old lobster-pots and broken bedsteads and broken oars and the skeleton of a boat and dry wisps of seaweed lie about, there you are likely to find the ragged handsome sticks of this plant. But it is a true native, a maritime species able to withstand the full scythe of Atlantic gales and spray. Out on the rim of the Isles of Scilly, among the granite boulders of the little desolate island of Roseveare, a miniature forest of Tree Mallow is whitened with the droppings of gulls, and shaken, but not shattered by every storm. Here the Tree Mallow can claim to be the last plant between England and North America.

5. Marsh Mallow. *Althaea officinalis* L. 33, H 15

Local names. MARSH MALLICE, I o W; MARSH MALLOW, Suff, Yks; MESHMELLICE, Dev; WIMOTE, Shrop.

Go to a shop and buy marshmallows, and they are made of starch, syrup, gelatine, and sugar. Once they were made of the roots of *Althaea officinalis*, 'thicke, tough, white within' and containing 'a clammie and slimie juice' (Gerard). Laying this mucilaginous juice in water 'will thicken it', Culpepper noted (before giving two close pages to the virtues of Mallows and Marsh Mallow), 'as if it were jelly'. For the London apothecaries, herbwomen picked the leaves and dug the roots of Marsh Mallow on the salt marshes of the Thames estuary (where it still grows); and every wife included it at one time in her physic garden for poultices and ointment and sweetmeats. 'Belly, Stone, Reins, Kidneys, Bladder, Coughs, Shortness of Breath, Wheesing, Excoriation of the Guts, Ruptures, Cramp, Convulsions, the King's Evil, Kernels, Chin-cough [i.e. Whooping Cough], Wounds, Bruises, Falls, Blows, Muscles, Morphew, Sun-burning' is Culpepper's marginal list for the application of Marsh Mallow in one form or another (49).

In itself, pink-flowered, and velvet (velvet leaves, velvet stem), tall, though not as tall as the *Lavatera arborea*, it is an exquisite surprise in muddy, salty, desolate, smelly surroundings.

XXIII. Linaceae

1. Purging Flax. *Linum catharticum* L. 112, H 40

Local names. FAIRY FLAX, N Eng, N Ire; FAIRY LINT, N'thum; LAVEROCK'S LINT, Lanark; MILL MOUNTAIN, Som, Hants; MOUNTAIN FLAX, Shrop, Ches, Derb, Yks, Cumb; PURGING FLAX, Ches.

This common little annual of chalk downs, especially, and limestone, and moorland and heath, with its white flowers and stems like fine wire, looks at first glance nearer to a Stitchwort than a Flax. It has not the attraction of the cultivated Flax or the blue-petalled wild Flaxes, the rare *Linum anglicum* or the more abundant *Linum bienne*. Like the Greater Stitchwort (*Stellaria holostea*) which has the west country name of Lady's Lint – Lady's Flax, that is to say – it was connected with fairies. In the Highlands and in Ireland it is the Fairy Woman's Flax – *líon na mban sídhe* in Irish. Whatever the reason, probably it is not so much because the plant is small and fairies were small. Elves and fairies in general were no smaller than humans; they have dwindled in size as the belief in them has dwindled, writers (Shakespeare and others of his time, particularly) making them sentimentally elfin and minute.

Purging Flax purges: Thomas Johnson in the second edition of Gerard's *Herbal* (1633) describes the way in which this plant, known as Mil-mountaine in Hampshire, was to be bruised and gently cooked in a pipkin of white wine, and taken for its somewhat fierce effect. The laverock, in 'Laverock's Lint', is the lark.

XXIV. Geraniaceae

1. Meadow Cranesbill, Blue Basins. *Geranium pratense* L. 104, H 1

Local names. BLUE BASINS, BLUE BUTTONS, BLUE WARRIOR, Som; FLOWER OF DUNLUCE, Antrim; GIPSY, Wilts; GRANNY'S BONNETS, Som; LOVING ANDREWS, Wilts.

Geranium pratense is local, but where it is common, in north Wiltshire, for instance, or in Co. Antrim, round about Dunluce Castle, the purple-blue flowers are a colour element of the scene along the roads, on banks and in fields, the leaves turning red and tawny in the autumn.

16 Meadow Cranesbill *Geranium pratense*

In his version of Dodoens, Henry Lyte translated a German name for it, *Gottesgnade*, as Grace of God. This Grace of God, or one of the Somerset or Wiltshire names, ought to be firmly used. 'Meadow Cranesbill' is a dull book name. Loving Andrews is charming, Blue Basins is apt; and the plant (see above) haunts the verges like an encampment of gipsies.

2. Wood Cranesbill, King's Hood. *Geranium sylvaticum* L. 56, H 1

A northern species, climbing high into the mountains, and resembling *Geranium pratense*, though the flowers are smaller and darker. Less bookish names for it than Wood Cranesbill are King's Hood and Mountain Flower (N'thum, Berw, Rox).

3. Painted Lady. *Geranium versicolor* L. 29, H 6

Local names. PAINTED LADY, Dev; QUEEN ANNE'S NEEDLEWORK, N'hants.

A fortunate introduction from Italy into English gardens in the seventeenth century, which was noticed as an escape more than a hundred years ago. Plenty of it grows damply and happily out of the earthen hedges of Devon and Cornwall, the leaves sometimes blotched with brown, the flowers pale lilac and scrawled with violet lines.

4. Dusky Cranesbill. *Geranium phaeum* L. 73, H 11

Also an introduced species, given the garden name in Lancashire and Yorkshire of Mournful Widow. Its flowers come as near as can be from purple to black. Lurking on the edge of shrubberies and drives, it will often escape and maintain itself on banks outside the garden.

5. Bloody Cranesbill. *Geranium sanguineum* L. 64, H 11

Bloody Cranesbill is the most striking of all the native geraniums, the one it is altogether most cheerful and refreshing to meet on sand dunes, and still more on wild limestone rocks – in Derbyshire, or on the cliffs (and sands) of Gower, on Great Orme's Head in North Wales, on the sides of the Bristol Gorge, on the limestone pavements of Burren in Co. Clare (which every plant-hunter should visit before he dies), or the limestone pavements around Ingleborough in the West Riding. Only 'bloody' is the wrong description: the wide flowers are a glorious magenta, admirably set off by the grey rock.

6. Dove's Foot, Dove's Foot Cranesbill. *Geranium molle* L. 112, H 40

Local names. CULVERFOOT, DOLLY SOLDIERS, JAM TARTS, Som; MOTHER OF MILLIONS, Yks; STARLIGHTS, Bucks.

The name the herbalists gave to Dove's Foot (it should be that alone, without the Cranesbill) was *Pes Columbinus*, presumably from the softness of the plant and the shape of the leaf. For a geranium, it is rather undistinguished. It was used, like Herb Robert, as a vulnerary, 'a very gentle, though Martial plant', as Culpepper described it (49). Gerard found Dove's Foot miraculous against ruptures, if it was powdered and drunk in red wine or old claret. He added credulously that if the ruptures were in old persons, the herb should be fortified with the powder of nine red slugs, dried in an oven.

7. Shining Cranesbill. *Geranium lucidum* L. 98, H 34

Local names. BABY CAKES, Som; BACHELOR'S BUTTONS, Lancs; JOHN'S FLOWER, Som; ROBIN, Dev.

A small, lucid, appealing, widely spreading annual, with shining stems, and tiny but vivid pink flowers, each set above a puffy, angled calyx. The seeds are worth collecting and scattering in moist, half-shaded parts of the garden.

8. Herb Robert. *Geranium robertianum* L. 112, H 40

Local names. ADDER'S TONGUE, Ess; ANGELS, Dor; BABY'S PINAFORE, Dev; BACHELOR'S BUTTONS, Dev, Suss; LITTLE BACHELOR'S BUTTONS, Suss; BILLY BUTTON, Bucks; BIRD'S EYE, Corn, Dev, Bucks, N'thum; BISCUIT, Dev; BISCUIT FLOWER, Som; BLOODY MARY, Yks; BLOODWORT, Som, Cumb; BOBBIES, Som; CANDLESTICKS (from the erect beak), Dev, Dor; CAT'S EYE, Dor, Hants; CHATTERBOXES, Dor; CHINESE LANTERN, Som; CRY BABY, Dev, Som; CUCKOO'S EYE, Bucks, Kent; CUCKOO'S MEAT, CUCKOO'S VICTUALS, Bucks.

DEATH COME QUICKLY, Cumb; DOG'S TOE, Donegal; DOLL'S SHOES, Som; DOLLY'S APRON, Dev, Som; DOLLY'S NIGHTCAP, DOLLY'S PINAFORE, DOLLY'S SHOES, Dev; DRAGON'S BLOOD, Dor, Som, Hants, Shrop; DRUNKARDS, Som; FELLON GRASS and FELLONWORT, Yks; GARDEN GATES, Bucks; GIPSIES, GIPSY FLOWER, GRANNY'S NEEDLES and GRANNY-THREAD-THE-NEEDLE, HEADACHE, Som; HEDGE LOVERS, Dev; HEN AND CHICKENS, HOP-O'-MY-THUMB, Som.

JACK-BY-THE-HEDGE, Som; JACK FLOWER, JACK HORNER, Dor; JAM TARTS, JENNY FLOWER, Som; JENNY HOOD, Dev, Som; JOE STANLEY, Dor; JOHN HOOD, Som; KISS ME, KISS-ME-LOVE, Dev; KISS-ME-LOVE-AT-THE-GARDEN-GATE, Bucks; KISS-ME-QUICK, Dev, Wilts; KNIFE AND FORK, Bucks; KNIVES AND FORKS, Som; LADY JANES, LITTLE JACK, Dev; LITTLE JAN, Dev, Som;

LONDON-PINK, WILD PINK, Glos; MARY JANES, MOTHER-THREAD-MY-NEEDLE, Som; NIGHTINGALE, Bucks; PINK BIRD'S EYE, Bucks; PINK PINAFORE, Dor; POOR JANE, Som; POOR ROBERT, Dev, Som; PRINT PINAFORES, Dor.

REDBREASTS, Som, N'hants; REDSHANK, Yks; REDWEED, Ches; ROBERT, ROBERT'S BILL, Dev; BOB ROBERT, Dev; RED BIRD'S EYE, Oxf; RED BOBBY'S

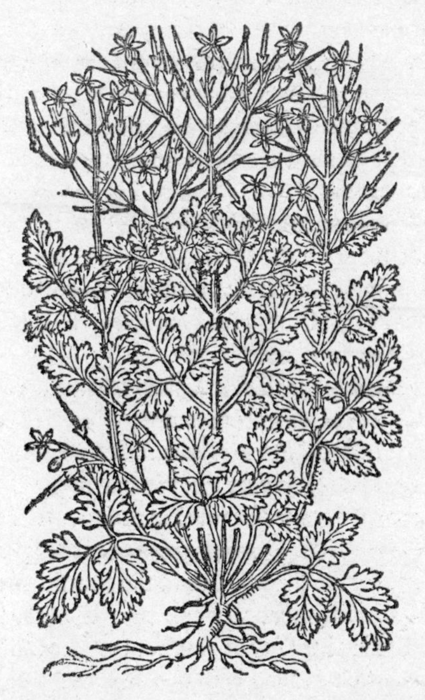

17 Herb Robert *Geranium robertianum*

EYE, Wilts, Hants; ROBIN, Dev, Wilts; LITTLE RED ROBIN, Som; LITTLE ROBIN, Dev; POOR ROBIN, Dev, Som; RAGGED ROBIN, Bucks; RED ROBIN, Som, Glos; ROUND ROBIN, Dev, Kent; ROBIN HOOD, Dev, Som; LITTLE JOHN ROBIN HOOD, Dor; ROBIN I' TH' HEDGE, Yks; ROBIN REDBREAST, N'hants; ROBIN REDSHANKS, Yks; ROBIN'S EYE, Corn, Dev, Som, Wilts, Suff; SMALL ROBIN'S EYE, Glos; ROBIN'S FLOWER, Dev, Som; RUBWORT, Ches.

SAILOR'S KNOT, Bucks; SNAKE FLOWER, SNAPJACK, Som; SOLDIERS, Dev; SOLDIER'S BUTTONS, Bucks; SPARROW-BIRDS, Dev, Som; SQUINTER-PIP, Shrop; STARS, STINKER BOBS, STINK FLOWERS, Som; STINKING BOB, Suss,

Bucks, Herts, Cumb; STINKING JENNY, Som; STINKING ROBERT, Donegal; STINKING ROGER, Lancs; STORKS, Dor; STORK'S BILL, Som.

WREN, Corn; WREN FLOWER, Som; WREN'S FLOWER, Dev; JENNY WREN, Som.

Herb Robert is very familiar: it lives with man, much as the Robin flips into his garden and to his back door – 'Herbe Robert groweth upon old wals, as well those made of bricke and stone, as those of mudde or earth: it groweth likewise among rubbish, in the bodies of trees that are cut downe, and in moist and shadowie ditch banks' (80). But neither familiarity nor charm of appearance – though it has no charm in the smell, which Gerard exaggerated as 'a most lothsome stinking smell' – are enough to explain so many names and the attention which they argue. Other plants are as obvious and common, and have hardly an English name other than those concocted for them by the botanist. True that the stems of Herb Robert are often shiningly red, which made it into a plant for the staying of blood, as the herbals and folk medicine show. In Ireland, too, it was given to cattle suffering from red water.

Yet why Herb Robert – the *Herba Roberti* of mediaeval Latin, turned into the 'herbe Robert' of thirteenth-century English? Not merely because it was given to St Robert – *Herba Sancti Ruperti* – probably St Robert of Salzburg. In England at least it has never been a sanctified flower. Here and abroad I suspect it was the plant of the house goblin, the German *Knecht Ruprecht* (in German the name is *Ruprechtskraut*), our own Robin Goodfellow, the French Lutin, the Scottish Brownie, and the Black Piet of Holland; and perhaps the plant as well of the house-haunting Robin Redbreast. Robin Goodfellow was not the pleasantest of creatures, unless he was treated well and fed with cream. Knecht Ruprecht, dressed in red usually, and Black Piet, are still rather terrifying Christmas figures in Germany and Holland, coming with Santa Claus when he brings the presents, but having a stick for children who have not been good. The red-breasted Robin, too, in northern Europe, England, Wales, Ireland, Scotland, has been a dangerous bird, who brought illness, death, or bad luck when he flew indoors (a belief which is not done with, by any means). Kill a robin, and your cow's milk turns to blood. Destroy his nest, and lightning will strike your roof, crockery will break in your kitchen, your hands will tremble for ever, or you will be paralysed. Treat the robin kindly and he will bring you good luck, and protect your house from lightning. The bird and the plant, then, may have belonged to Robin Goodfellow; the plant may have been the bird's vegetable counterpart.

The clues are perhaps the stench of Herb Robert, the colour, the beaks, and the way it keeps human company round the house. The names show a link with adders, cuckoos, and death (and also with the wren, who was the robin's wife, and another bird both sacred and frightening).

The name Robin, a diminutive of Robert by way of French, seems innocent in its attachment to flowers, but most of the Robin flowers appear to have been linked to goblin, robin, and evil. *Melandrium dioicum*, *Arum maculatum*, *Orchis mascula*, *Calystegia sepium*, *Solanum dulcamara*, and *Stellaria holostea* (of which one name is White Bobby's Eye, cf. Red Bobby's Eye, above) – all these share Robin names with Herb Robert; and by other names as well they are all, or several of them, associated with the following: snakes, death, the devil, fairies, sex, and cuckoos (most flowers with cuckoo names have a link with sex). See also Ragged Robin, and White Campion. In Cumberland it was unlucky – since they were goblin plants? – to pick either the Mother-Dee (= die), which was Red Campion, or the Death-come-quickly, which was Herb Robert. In Cornwall and Devon it was unlucky to pick the adder's plant, and pisky's plant, the plant of smocks, and elsewhere of cuckoos, *Stellaria holostea*. So, too, bushes of Hawthorn, as fairy plants, must not be disturbed. Herb Robert's connection with sex is at least suggested by several names, including Cuckoo's Eye and Cuckoo's Meat, and by a particular virtue ascribed to it by the herbalists, including Gerard and Culpepper, who makes it 'an Herb of *Venus*, for all it hath a man's name' (49).

As for the names linking Herb Robert and Robin Hood, it is an old suggestion that Robin Hood grew out of elf into outlaw, and was tied in some way with the mischievous Robin Goodfellow, who haunted house and woodland. To complete the household family of plant, bird, and goblin, the woodlouse was called Robin Goodfellow's louse, as though woodlice crawled off his body. Other qualities in common between Herb Robert and Robin Goodfellow are possibly colour, hairiness, and candlestick beaks. The Cornish pisky, often equivalent to Robin Goodfellow, wore a red cap (37). Sixteenth-century mentions of Robin Goodfellow make him a hairy goblin, sometimes wearing a red suit, and red-featured. He usually carried a candlestick; and he appears to have been the same being as Kit with the Canstick (114). Notice, too, that Herb Robert was called after the goblin Hop-o'-my-thumb, in Somerset; for which goblin, see *Lotus corniculatus*.

9. Musk Storksbill, Muscovy. *Erodium moschatum* (L.) Ait. 36, H 20

Gerard knew this local seaside plant as Musked Cranesbill or Musked Storksbill, Ground Needle, and Pink Needle; and wrote that it was planted in gardens 'for the sweet smell that the whole plant is possessed with'. Banks in his *Flora of Plymouth*, two hundred and thirty-three years later, said that it was still grown as a garden annual and was always put among the nosegays on market-stalls. The Devonshire names for it were Muscovy or Pick-needle. In my copy of Gerard's *Herbal*, 'Moscovie' is written against the woodcut in a sixteenth- or seventeenth-century hand. A 'pink-needle' and a 'pick-needle' are synonymous – needles (from the long beaks of the carpel) for pinking seams and edges of cloth.

In the U.S.A., where *Erodium moschatum* is a naturalized weed, it goes by the pleasant names of White-stem Filaree (from the Spanish *alfiler*, a pin) and Musk-clover.

10. Storksbill. *Erodium cicutarium* L. 111, H 25

This is the plant which William Turner in 1548 called 'Cranes byl' or 'Pynke nedle' (see the preceding species). Few other names for it have been recorded except Five-leaved Grass (Hants), which is usually *Potentilla reptans*. Also an introduced weed in America, it is alfilaria, or Red-stem Filaree, since the plant is less covered with white hairs than *Erodium moschatum*. In Colonsay, in the Hebrides, the root was used for a dye (128).

XXV. Oxalidaceae

1. Wood-sorrel. *Oxalis acetosella* L. 112, H 40

Local names. ALLELUIAH, Dor, Som, Wales; HALLELUJAH, Som; BIRD'S BREAD-AND-CHEESE, Dev, Cumb; BREAD-AND-CHEESE, Dev, Som, Ches, Lancs; BREAD-AND-CHEESE-AND-CIDER, Som; CHEESE-AND-BREAD, Lakes; BREAD-AND-MILK, BUTTER-AND-CHEESE, BUTTER-AND-EGGS, Som.

CUCKOO-BREAD (cf. Cornish *bara-an-gok*, 'cuckoo bread'), Dev, Dor, Som; CUCKOO'S BREAD-AND-CHEESE, Som, Glos, Worc, Shrop, Lancs, Cumb; CUCKOO-CHEESE, Dev; CUCKOO-CHEESE-AND-BREAD, Cumb; CUCKOO'S CLOVER, N Ire; GOWK'S CLOVER, N'thum; CUCKOO-FLOWER, Wilts, Kent, Bucks, Notts, Yks; CUCKOO'S MEAT or CUCKOO-MEAT, Som,

Glos, Suss, Bucks, War, Shrop, Ches, Lancs, Yks, N'thum, SW Scot (in the north Gowk's Meat); CUCKOO-SORREL, Suss, Worc, S Scot, N Ire; CUCKOO'S SOUR, Shrop; CUCKOO'S VICTUALS, Dor, Glos, Berks.

FAIRY BELLS, Dev; FOX'S MEAT, Corn; GOD ALMIGHTY'S BREAD-AND-CHEESE, GOOD LUCK, Som; GREEN SAUCE, Corn, Dev, Glos, War, Ches, Derb, Leic, Notts, Lincs, Lancs, Yks; GREEN SNOB, War; GREEN SORREL, Bucks; HARE'S MEAT, Corn, Som; HEARTS, N'thum; KING-FINGER, Bucks; LADY CAKES, Dumf; LADY'S CLOVER, Perth, Scot; LAVEROCKS, THREE-LEAVED LAVEROCKS, Yks.

RABBIT'S FOOD, Lancs; RABBIT'S MEAT, Corn, Dev, Som; SALT CELLAR, Dor; SHAMROCK (with *Trifolium repens* and *T. dubium*, q.v.), Ire; WILD SHAMROCK, Som; SHEEP-SOORAG, Caith; SHEEP'S SORREL, Dor, E Ang, Ire; SLEEPING BEAUTY, Dor; SLEEPING CLOVER, Dor, Oxf; SOOKIE-SOURACH, Inv; SORREL, Ches; SOUR CLOVER, Berw; SOUR DOCK, Dev; SOUR GRASS, Dev, Yks; SOUR SAB, Corn, Dev; SOUR SALLY, Som; SOUR SUDS, Dev; SOUR TREFOIL, STUBWORT, Som; SOUROCK, Donegal; WHITSUN FLOWER, Dor; WILD CLOVER, Lincs; WOODSOUR, Yks.

Generations of children have bitten the sharp, pleasant taste out of the flowers of Wood-sorrel. Several of its names are shared with Sorrel (*Rumex acetosa*) and the leaves were eaten in a green sauce. John Gardener's early fifteenth-century poem, *The Feate of Gardening*, makes it clear that the 'Wodesour', as he calls it, was actually cultivated for sauce (79), and John Evelyn, in some manuscript notes on gardening in the Evelyn Library, at Christ Church, Oxford, includes 'Wood-Sorrell' among the 'Plants for the Kitchen-Garden'. (He wrote the manuscript between 1688 and 1706.) *The Grete Herball* of 1526 called the plant Cuckowes meate and Alleluya; William Turner's names were Allelua, Cuckowes meate, Wodsorel, and Woodsore; and in his *Herbal* in 1568, Turner explained that it was Alleluia 'because it appeareth about Easter when Alleluya is song agayn' – when, that is to say, Alleluia was sung once more in triumph on Easter Sunday and through Easter week after the psalms, the *Ite missa* and *Deo gratias*, etc. (see the Roman Missal). In mourning for the death of Christ, between Septuagesima and Holy Saturday, Alleluia was dropped and *Laus tibi, domine, Rex aeterne gloriae* was substituted:

> Whan alleluya is a lofte
> I go gaye and syt softe
> And than I am mery ofte
>   As any byrde on brere.

Whan laus tibi cometh to towne
Than me behoveth to knele downe
And ever to be in orisowne
  As it were a frere.

So the flowering of Alleluia or wood-sorrel was a sign of delight. Alleluia is the name for Wood-sorrel in other languages as well, in French, for instance, and Italian; a Christian name to offset associations in several countries with magic, the cuckoo, etc. Cuckoo's Meat, says Gerard, 'because either the Cuckoo feedeth thereon or by reason when it springeth forth and flowereth, the Cuckoo singeth most'. In France it is *pain de coucou*, Cuckoo's Bread, as well as Alleluia. Our 'Sorrel', too, is a borrowing from French, ultimately from the OF *sur*, 'sour'.

Wood-sorrel seems to have been a magic plant from the ternate leaves, not unlike the trifoliate leaves of Shamrock. The leaves, too, fold back, so the names of 'Sleeping Beauty' and 'Sleeping Clover'. In the Hebrides, the herbs were put into herb plasters against scrofula or King's Evil (128); and if this was so elsewhere, it may explain how Wood-sorrel acquired its Buckinghamshire name of King-finger. Gerard Manley Hopkins aptly described the new leaves of Wood-sorrel as having the sharp appearance of green lettering.

For the Somerset name Stubwort see the fourteenth-century *Agnus Castus*, which gives the names of Wood Sorrel as Alleluya, Wodesour, and Stopwourt (20).

XXVI. Balsaminaceae

1. Touch-me-not. *Impatiens noli-tangere* L. 8

The only native of the genus. In Cumberland called Old Woman's Purse, from the shape of the hanging flowers. As a garden plant, it has been called Quick-in-hand in Devonshire (a name used in Ray's *Synopsis*), and Biggoty Lady (i.e. Saucy Lady) in Somerset. The balsams have no patience, the ripe capsules bursting and ejaculating the seed – in the excruciating lines of Erasmus Darwin:

With fierce distracted eye Impatiens stands,
Swells her pale cheeks and brandishes her hands,
With rage and hate the astonished groves alarms
And hurls her infants from her frantic arms.

So the names – including Touch-me-not, which is borrowed from the words Christ spoke to Mary Magdalene after the resurrection, in the Vulgate, *noli me tangere* (John xx. 17).

Touch-me-not was discovered wild in this country on a botanizing journey made by George Bowles in 1632. He found it in Shropshire on the banks of the Camlad and among alder trees in the nearby parish of Church Stoke. In this same expedition, he found the Ivy-leaved Bellflower, *Wahlenbergia hederacea*, for the first time (81).

2. Orange Balsam. *Impatiens capensis* Meerburgh 21

A garden introduction from the eastern parts of North America, which had the distinction of being found in the wild for the first time – on the Wey in Surrey, in 1822 – by the philosopher John Stuart Mill. Americans call it Snapweed, Jewel-weed, Kicking Colt, and Celandine (because they used the juice against warts – see *Chelidonium majus*). They also used it against the rash and blisters caused by Poison Ivy.

3. Small Balsam. *Impatiens parviflora* DC. 36

From Siberia and Turkestan; but the meanest of the four balsams, first noticed as a weed in Battersea in 1851. Londoners can always find it, looking a little shabby on waste ground behind iron railings, or on the edge of shrubberies (e.g. on Hampstead Heath), or alongside paths.

4. Himalayan Balsam. *Impatiens glandulifera* Royle 48, H 12

In the western Himalaya this balsam, showiest and noblest of the four, grows up to ten feet. Introduced in 1839, it was cultivated at first as a greenhouse annual by gardeners who never imagined the career ahead of it. Experience proved it would grow in borders without any molly-coddling; and from the border it had escaped to stream banks by 1855. After nearly a hundred years its pink to purple flowers and thick red stems fill the edges of river after river. Push a boat among the stems and you come under rifle fire from the snapping capsules. Two extremes: you can see this Indian plant growing out of dirty water for mile after mile along a canal outside Leeds, or you can travel down the valley of the East Looe River in Cornwall alongside billows of *Impatiens glandulifera*, which hide the old canal and the banks of the river.

There in the Looe valley A. O. Hume found it first in 1900. 'I notice that it has been called "a cumbersome and weedy thing"; but growing in the soft warm south-west, with the base of its stem in the clear running stream, it is a magnificent plant, 5 to 7 feet or more in height, stalwart, with a stem from 1 to 2 inches in diameter just above the surface of the water, erect, symmetrical in shape, with numerous aggregations of blossom, the central mass as big as a man's head, and those terminating all the principal lateral branches, though smaller, still most striking – masses of bloom varying on different plants through a dozen lovely shades of colour from the very palest pink imaginable to the deepest claret, and with a profusion of large, elegant, dark green, lanceolate leaves, some of them fully 15 inches in length. Stunted specimens of this Balsam are common in Cornwall in orchard and cottage gardens; but in the upper Looe River the plant has become thoroughly naturalized, and I have never seen it quite as fine even in its native habitats' (*Journal of Botany*, 1901).

It will be curious to see how such a landscape plant gathers names. On record so far are Nuns, Jumping Jack, and Policeman's Helmet.

XXVII. Aceraceae

1. Sycamore. *Acer pseudoplatanus* L. 112, H 42

Local names. FADDY-TREE, Corn; MAPLE, Cumb; MAY, Corn; MAY-TREE, Corn; PEWEEP-TREE (i.e. whistle tree), Corn; PLANE, Yks, Cumb, Scot; SEGGY, Yks; SEGUMBER, Dev; SHARE, W Eng; TULIP-TREE, Dev, Wilts, Yks; WHISTLE-TREE, Corn; WHISTLE-WOOD, N Eng, N'thum, Scot.

The fruits are known variously as CHATS, Suff, Norf, Yks; HORSE-SHOES, Wilts; KNIVES-AND-FORKS, Kent; LOCKS-AND-KEYS, Ess, Norf, Camb, Lincs.

Sycamores are so common and familiar and so much fit the landscape that we accept them as native. Really *Acer pseudoplatanus* is a tree of European mountains from the Pyrenees to the Balkans. In Scotland it may have been introduced as early as the fifteenth century. William Turner mentions it in the second part of his Herbal (1562), but not as a species planted and familiar in England. Lyte sixteen years later says 'there is here and there a tree of it planted in England'. In 1597 Gerard calls it 'a stranger in England, only it groweth in walkes and places of pleasure of noble men, where it especially is planted for the shadowe sake, and under the name of Sycomore tree' – a wrong name, as Gerard knew (he preferred to call it the great Maple). Sycamore properly is an Egyptian and Syrian species of fig, *Ficus sycomorus*, but the word Sycamore was already used rather vaguely in fourteenth-century English.

It must soon have been found that Sycamore, as a mountain tree, stands up well to the wind, so it spread from noblemen's walks and pleasure grounds to farmsteads, including the loneliest, loftiest farms in Wales or the Lakes or the Pennines. Sycamores were much planted through the seventeenth and eighteenth centuries – in Cornwall, for instance, where Sycamore is almost a weed and where clergymen delighted to set it in their gardens and churchyards (Ray in 1690 wrote of it *in coemetariis et circa nobilium aedes*), and on their glebe (178). They paid no attention to John Evelyn's dislike of the Sycamore for its honey-dew leaves, 'which fall early, like those of the Ash, turn to mucilage and noxious insects, and putrefy with the first moisture of the season, so as they contaminate and marr our walks'. Evelyn wanted it 'banished from all curious gardens and avenues' (70). Wind-bent on the coast and around the farms and self-sown along the earthen hedges, Sycamore became so much a matter of course in Cornwall that branches were broken off when the Cornish went a-faddying on May Day, the children making peweeps or whistles of the twigs, which is easy to do when the sap is running: the bark then slides off unbroken after a few taps with a knife-handle (53). Three centuries have naturalized the Sycamore completely.

It is curious to find sycamore among the plants carved on St Frideswide's Shrine (A.D. 1282) in the Cathedral at Oxford. The sculptor must have been familiar with the tree elsewhere in Europe. On the shrine, though, the Green Man or May-Lord is also represented. Perhaps supernatural virtue had been transferred to the Sycamore from the Maple, which is so frequently carved in thirteenth-century churches and cathedrals – for instance, in the chapter-house at Southwell Minster along with the Hawthorn (q.v.), Ivy, Buttercup, and Oak, and half a dozen or more representations of the Green Man.

2. Maple. *Acer campestre* L. 68

Local names. CAT OAK, Yks; DOG OAK, Som, Notts, Yks; MAPLIN-TREE, Glos; OAK, Dev, Som; WHISTLE-WOOD, Scot; WHITTY BUSH, Shrop. The fruits are known as BOOTS-AND-SHOES, Som; HASKETTS, Dor; KETTY-KEYS, KIT-KEYS, or KITTY-KEYS, Yks; and SHACKLERS, Dev.

A well-grown Maple always imparts an idea of the tight, the close, and the strong. With its sturdy, branchy habit, its delicate leaves, and its close-furrowed bark, it also looks intensely and assuredly native, it *looks* that old inhabitant, which has given itself – as the OE *mapulder* – to so many place-names, Mapledurham in Hampshire, Maplestead in Essex, Mappowder in Dorset, Mappledore in Wiltshire, and so on.

Sycamore timber has its uses, Maple wood is harder and finer and belongs to the age of craftsmen. 'The timber is far superior to Beech,' Evelyn declared in 1664, 'for all uses of the turner, who seeks it for dishes, cups, trays, trenchers, etc., as the joiner for tables, inlaying, and for the delicateness of the grain, when the knurs and nodosities are rarely diapered, which does much advance its price. Our turners will work it so

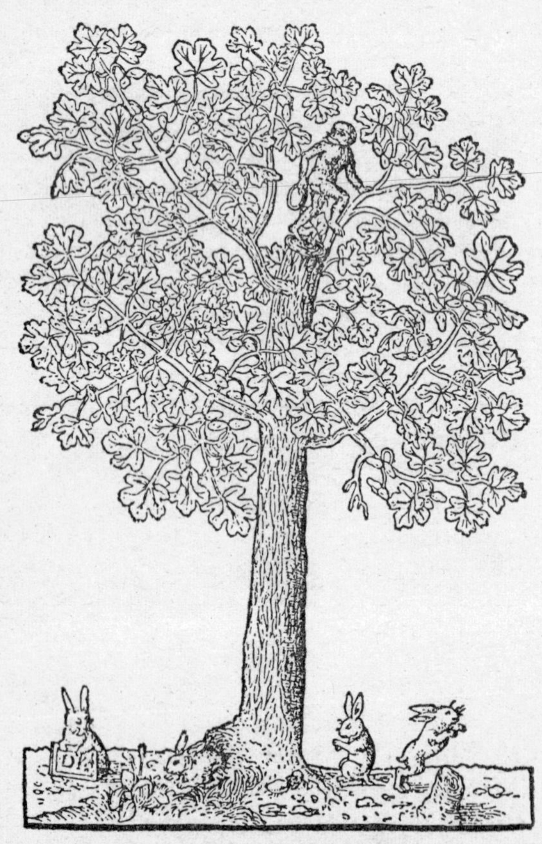

18 Maple *Acer campestre*

thin, that it is almost transparent.' Maple wood was anciently used for harps. A maple harp was excavated from the Saxon barrow at Taplow, in Berkshire. The main frame and the peg arm of the Saxon harp found in a sealskin bag with the Sutton Hoo ship treasure were also of Maple. For Maple in church carvings, see *Acer pseudoplatanus*, above.

XXVIII. Aquifoliaceae

1. Holly. *Ilex aquifolium* L. 107, H 40

Local names. AUNT MARY'S TREE, Corn; BERRY HOLLY, Som, Wilts; BERRY HOLM, Som; CHRISTMAS (applied to Holly branches cut for Christmas, but often to the tree as well, as May is used for Hawthorn and for the branches cut on May morning), Corn, Dev, I o W, Hants, Suss, Suff, Norf, Lincs; CHRISTMAS-TREE, Suff; CHRIST'S THORN (cf. *Kristtorn* in Norwegian), Yks; CROCODILE (i.e. low, hedge holly), Dev, Som.

HOLLIN, Shrop, Ches, Derb, Notts, Lincs, Lancs, Yks, Cumb, Lakes, N'thum, I o M, Scot; HOLLY, Kent, Yks, West, Cumb, N'thum, Pemb; HOLM, Corn, Dev, Dor, Som, Hants, Glos, Suss; HULVER, E Ang; KILLIN (Cornish *kelen*), Corn; PRICK-BUSH, Lincs; PRICK-HOLLIN, PRICK-HOLLY, Lincs, Yks; PRICKLY CHRISTMAS, Corn.

The prickly evergreen in the leafless wood, with red berries and leaves 'of a beautiful greene colour, smooth and glib', and the twining evergreen which crawls up the leafless trees:

> The Holly and the Ivy
> When they are both full grown
> Of all the trees that are in the wood
> The holly bears the crown . . .

As Henry VIII wrote in a song:

> A! the holy grouth grene
> With ive all alone,
> When flowerys can not be sene,
> And grene wode levys be gone.
> Grene growith the holy, so doth the ive;
> Thow wynter blastys blow never so hye,
> Grene growth the holy.

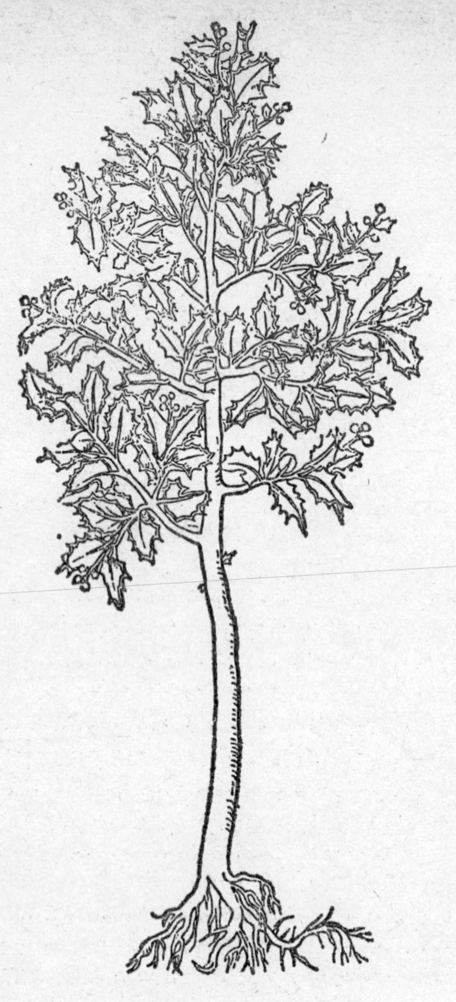

19 Holly *Ilex aquifolium*

Both Holly and Ivy were plants with power, and they were specially suitable for protection in the dead of the year, although they were effective at all times and in many ways. The red berries help, since red is a colour against evil (cf. the red-berried Mountain Ash, or Hawthorn, or *Solanum dulcamara*). In Austria today, if people are always staring at you in the bus or the streets, it is well to have something red about you against the evil eye. In London, in the fifteenth century, poles garlanded with Holly and

Ivy were set up alongside the Christmas sports (24); and John Parkinson wrote in his *Theatrum Botanicum* (1640) that 'the branches with berries, are used at Christ tide to decke our houses withall, but that they should defend the house from lightening, and keepe themselves from witchcraft, is a superstition of the Gentiles, learned from Pliny, saith Matthiolus.' Pliny does say that Holly planted by the house keeps away witchcraft (*Natural History*, XXIV. 71), but the power of Holly was a general belief independent of classical authority – so strong that it had to be accepted by Christianity. No doubt a function of Holly inside the house was to deal, not only with demons and witches, but with the house goblins, the Robin Goodfellow, the Brownie, the Hobthrust, etc., and keep them down in the Christmas season. When belief in the house goblins dwindled, the more or less pleasant side of their character was exaggerated. Yet in one Jacobean play a charm suggests that a little of Robin Goodfellow went a long way:

> Saint Francis and Saint Benedight
> Blesse this house from wicked wight,
> From the night-mare, and the goblin,
> That is hight Good-fellow Robin;
> Keep it from all evil spirits,
> Fairies, weezels, rats and ferrets,
>   From curfew time
>   To the next prime. (32)

Holly and Ivy would have subdued the house goblin precisely from Christmas Eve, when the decorations went up, to Candlemas Eve, when they were taken down.

Holly's apotropaic power was employed in other ways. John Aubrey related that a garland of Holly and Bittersweet (*Solanum dulcamara*) would cure a hag-ridden horse, if it was hung around its neck (8). Whooping-cough was cured, so people believed in the Forest of Bere in Hampshire, if children drank cow's milk out of a cup of holly wood. In Shropshire cups of ivy wood were made for the same purpose (37). Leaves from a She-holly without prickles were placed under pillows in the north of England for divination by dreams. There was an ambivalence about Holly, as about other magic plants (cf. Hawthorn). Though it was apotropaic, it was also in Ireland a 'gentle tree' liked by the fairies. In 1907 a farmer's house in Northern Ireland was troubled with the stones of a poltergeist. It was explained that the farmer had annoyed the fairies by sweeping his chimney with a holly branch (180).

Working the Holly into Christian beliefs was not difficult. It could still keep out demons from house or church. It was thorny and the berries were red – the Crown of Thorns and the drops of blood combined; and on the evil side it was held to be one of those trees from which the Cross was made. The symbolism is plain in *The Holly and the Ivy*, or in the St Day Carol taken down in Cornwall. The blossom of Holly is white as milk or white as the lily flower and can be linked to the birth of Christ. Red as any blood, the berries, with the prickles sharp as any thorn and the bark bitter as any gall, stand for the passion. Compare *The Holly and the Ivy* with three unchristianized mediaeval poems printed by E. K. Chambers and F. Sidgwick in their *Early English Lyrics* (1926). One is a poem warning men to do no ill to the Holly:

Who so ever ageinst holly do sing
He may wepe and handes wring.

The others are poems of contention for mastery between the two plants:

Then spak Holver 'I am frece* and joly;
I will have the maistry
  In londės where we go.'

In the third poem, Holly is the man's plant, Ivy the woman's, Holly has fine red berries, Ivy only black ones:

Holly berith beries,
  Beris rede enough;
The thristilcok, the popingay
  Daunce in every bough.
Welaway, sory Ivy!
  What fowlės hast thou,
But the sory howlet
    That singeth 'How how'?

Ivy bereth beris
  As blak as any sloe,
There commeth the woode colver,
  And fedeth her of tho;
She lifteth up her taill
  And she cakkės or she go;
She wold not for an hundred pound
  Serve Holly so.

* *Frece*, fresh, quick.

Holly with his mery men
  They can daunce in hall;
Ivy and her jentell women
  Can not daunce at all,
But like a meine of bullokes
  In a water fall,
Or on a hot somers day
  Whan they be mad all.

Holly and his mery men
  Sitt in cheires of gold;
Ivy and her jentell women
  Witt without in fold,
With a paire of kibėd*
  Helės caught with cold.
So wold I that every man had
  That with Ivy will hold!

The poem is explained by an unpleasant custom which survived at Tenby in South Wales early in the last century. On Boxing Day men and boys went 'holly-beating': they ran through the streets hitting girls on their bare arms with holly branches until they bled (173).

Holly, from the OE *holen*, *holegn*, is common in place-names. Killin, above, one of the names for Holly in Cornwall, is from the Cornish *kelen* (cf. Pencalenick, 'head of the Holly place').

* Chilblained.

XXIX. Celastraceae

1. Spindle-tree. *Euonymus europaeus* L. 79, H 40

Local names. BITCHWOOD, Worc; CAT-TREE, Bucks; CATTY-TREE, Shrop; CATWOOD, DEATH-ALDER, Bucks; DOG-TIMBER, Dev, Som; DOG-TREE, War; FOULRUSH, Bucks, N'hants; GATTERIDGE, Kent, E and S Eng; GATTER-TREE, Kent; IVY-FLOWER, Som; PEGWOOD, Dev; PINCUSHION SHRUB, Bucks; PRICKWOOD, Som, Suss, Cumb; SKEWERWOOD, Dev, Dor, Wilts, I o W, Glos, Berks, Yks; SKIVER, SKIVER-TIMBER, Som, Wilts; SKIVER-TREE, Dev; SKIVERWOOD, Dor, Som; SPINDLEWOOD, Glos; WITCHWOOD (from its pliant branches? See *Ulmus glabra*, Wych Elm, and *Sorbus aucuparia*), Suff.

20 Spindle Tree *Euonymus europaeus*

The fruits are called DOG-TOOTH BERRIES, Surr; HOT CROSS BUNS, Dor; LOUSE-BERRIES, PINCUSHIONS, Glos, War; and POPCORNS, Som.

Few things in the range of English plants are more surprising than the fruits of the Spindle-tree, the four golden oranges lapped brilliantly in the four pink lobes. But plants get named more often for use than beauty. 'Spindle-tree' was one of William Turner's prosaic inventions, or adaptations: 'I have sene this tree oft tymes in England and in most plentye between Ware and Barkwaye, yet al that I coulde never learne an English name for it, the Duche men call it in Netherlande Spilboome, that is

Spindel tree, because they used to make spindels out of it in that countrey, and me thynke it may be so wel named in English' (*Herbal*, 1568). Spindles, skewers (pricks and skivers), and pegs have given it most of its names; and gipsies still make skewers out of the hard, white, tough wood, which was also fashioned into viola bows, keys for the virginal, and toothpicks (70).

Gatteridge and Gatter-tree are names shared with the Dogwood (q.v.). They mean the 'goat bush', the 'goat tree', from the OE *gāt*. In the *Nonnes Preestes Tale* of Chaucer, the rooster Chauntecleer has a bad dream on the night before he is taken off by the fox. Pertelote, his favourite hen, thinks he is upset and needs a 'laxatyve' – an older word than you might think. She suggests some of the laxative plants out of the widow's garden – Spurge Laurel, Centaury, Caper Spurge, Hellebore, Ground Pine – or 'gaytres beryis'. The glossary says that Chaucer meant the grimly purging berries of Buckthorn; Gerard, who knew the *Nonnes Preestes Tale*, thought that he meant Dogwood berries. Dogwood was not used as a purge, and Buckthorn does not seem to have been called Gaiter or Gatter-tree. I think Chauntecleer's dose may have been a few 'berries' from the Spindle-tree: 'If three or fower of these fruits be given to a man,' wrote Gerard, 'they purge both by vomit and stoole.'

In Gerard, also, you can find why *Euonymus europaeus* became this goat tree, or Gatter-tree. He quotes Theophrastus. The *euonymus* mentioned in the great *History of Plants* (which, rightly or wrongly, the apothecaries and herbalists supposed to be our species) was fatal to goats – 'the leaves and fruite destroy Goates especially, unless they scoure as well upwards as downwards'. In fact, sheep and goats have been made ill by Spindle-tree leaves, though they cannot, all the same, be called violently poisonous (123). Not only goats. According to Theophrastus, 'this shrub is hurtfull to all things . . . the fruite heerof killeth'. Seeping out of the books into folk belief, it was that piece of information, no doubt, which made Buckinghamshire people call *Euonymus europaeus* the Death-Alder, unlucky, they imagined, in the house.

With names from goat and cat, dog and bitch, pegs and skewers and spindles, the beauties of *Euonymus* have had a raw deal. Add Louse-berries, since the fruits were baked, powdered, and sprinkled on the hair of small boys to kill their nits and lice (70). The Germans, Italians, and French make some amends by calling the fruit, and the species, *Pfaffenhütchen*, *Berretta da Prete*, and *Bonnet-de-prêtre* – 'priest's biretta', 'priest's bonnet'. The four ridges dividing a biretta correspond to the division between the four pink lobes of the capsule. And birettas are not always black. Cardinals wear them scarlet, if not Spindle-berry pink.

XXX. Buxaceae

1. Box. *Buxus sempervirens* L. 5

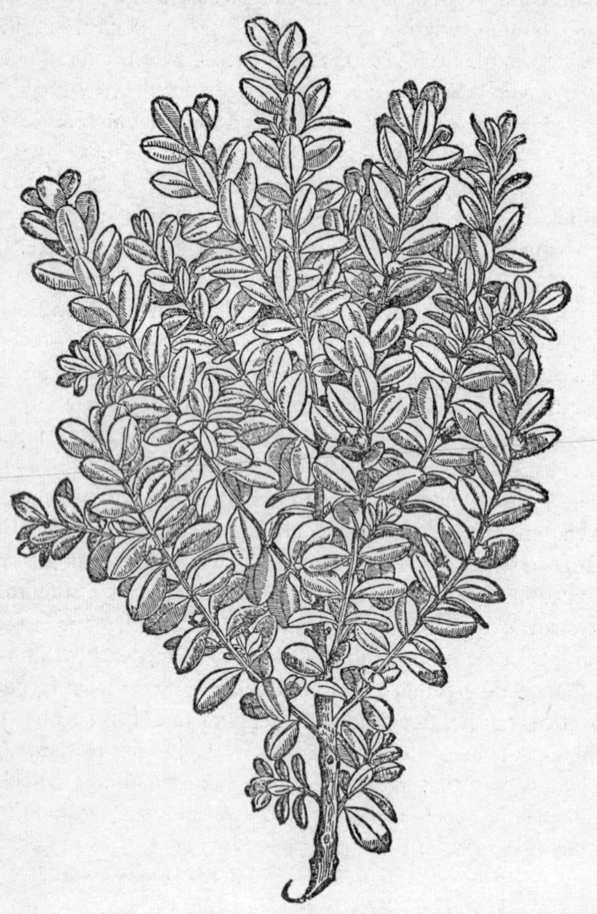

21 Box *Buxus sempervirens*

There are four counties where Box is considered to be wild: Gloucestershire, Surrey, Kent, and Buckinghamshire. The place-names Boxwell, the 'Box Stream' (Glos), Box Hill (Surrey), and Boxley, the 'Box *leāh*' or 'wood' (Kent), prove that it has been flourishing in three of its stations for more than a thousand years. According to John Aubrey, there had been trees at Box, just outside Bath. In his time, in the seventeenth century, they had 'worn out' (7). Several more names of places on the limy soil which fits the tree, suggest wild rather than cultivated Box – Boxted, the 'Box place', and Bexford in Suffolk, for instance, Boxgrove in Sussex, Box Hall in Hertfordshire, and Boxford, originally *Boxora*, the 'Box slope', in Berkshire. 'Box slope' precisely; for wild Box here and abroad is a plant of chalk and limestone escarpments. Gerard, in 1597, wrote of Box as a native which grew 'upon sundry waste and barren hills'. John Evelyn in his *Sylva* echoed him and prayed that we should 'furnish our cold and barren hills and declivities with this useful shrub'.

Living near by in Surrey, Evelyn was familiar with the ancient wood on Box Hill, where the trees – trees rather than shrubs – grow down the steep chalky face to the black waters of the River Mole. This wood, and the less familiar one at Boxwell on the Cotswold escarpment, belong to the select and queer company of yew woods on Mickleham Downs and elsewhere, the hornbeam portions of Epping Forest or the twisted oaken fantasia of Wistman's Wood on Dartmoor. At Boxwell there are thirty or forty acres on the steep side of a narrow valley (John Ray in 1695 was the first botanist to enter and examine this goblin covert). The trees grow close and form an evergreen roof overhead, the floor below is bare in the green light, except for scattered fragments of limestone and for moss, which also climbs a little way up the grey or yellow trunks. Many of the trees are twenty-five or thirty feet high, some higher still.

Box trees in the mass – to add another peculiarity – smell like foxes. Gerard called this smell loathsome and evil. It is grotesque, certainly, and strong. Evelyn wrote that the smell was driving the Box out of gardens, though he wanted the trees planted for the 'excellency of the wood'. We still know box-wood rulers, box-wood draughtsmen and chessmen. 'So hard, close, and ponderous as to sink like lead in water', indeed nearly twice as hard as oak and a timber which is sold by weight, not volume (15), it has been used in a hundred ways. Imported timber made, and still makes, the best block for wood engraving. Rolling-pins, nut-crackers, pestles, weaver's shuttles, even axle-trees, were fashioned of box wood. Gerard wrote that the roots were harder than the trunk and were cut into dagger hafts. John Aubrey said that London comb-makers purchased timber from Boxley in Kent and Box Hill (7). So the woods paid a good dividend – rather too good a one at Box Hill, where the trees (now guarded by the National Trust) were savagely exploited. An assault in 1815 gave timber worth nearly £10,000 (125). The Boxwell wood has been better treated. A narrow swathe used to be felled every year from top to bottom of the escarpment; and one widow of the family which still owns the wood, and still lives in the manor-house below, had the profits of the Box as her jointure.

Box trees take a long while to grow, and when they are young need to be saved from the unhealthy grip of Old Man's Beard. Greed in past centuries and ruthless felling must have destroyed many of the woods and turned the Wild Box into so rare a species.

XXXI. Rhamnaceae

1. Buckthorn. *Rhamnus cathartica* L. 61, H 23

The black, shining berries look innocent, but do not try eating them. They were employed as a mighty and muscular purge for men and cattle. When the *necessarium* or latrine pit of the Benedictine monks of St Albans was excavated in the nineteen-twenties Buckthorn seeds were recovered in quantity with fragments of old cloth which served the monks for lavatory paper (*Architect's Journal*, March 1937). Gerard, who used the names Buckthorn, Waythorn (which has been recorded in Shropshire), and Laxative Ram, described how the berries were meted out by number. 'Strong bodies' took between fifteen and twenty or more, but he advised breaking them and boiling them in flesh broth – 'for so they purge with lesser trouble and fewer gripings'. Lyte before him had written that they were only fit to be administered to 'young and lusty people of the country, which do set more store of their money than their lives'. For a long while, nevertheless, a doctor's Syrup of Buckthorn worked a more orderly yet fierce purge upon the constipated. Well into the nineteenth century the berries were collected for chemists from the bushes in Buckinghamshire, Oxfordshire, and Hertfordshire (73). The cascara that we know, originally *cáscara sagrada* in Spanish, 'holy bark', comes from *Rhamnus purshiana*, a native of the southern United States.

More gently the berries of *Rhamnus cathartica* were used in dyeing and for pigments. Unripe berries, dried, powdered, and boiled with alum, gave a saffron which was used in the seventeenth century for tinting leather (especially leather for gloves) and playing cards. The black, ripe berries, boiled with alum, afforded the painters' Sap Green in Gerard's time and for long afterwards; and a red was made from the last berries to cling to the bush (141, 7). The berries also give a Bladder Green, when mixed with lime-water and gum-arabic.

2. Alder Buckthorn. *Frangula alnus* Mill. 73, H 16

Local names. BLACK ALLER, Corn, Dev, Som, I o W; DOGWOOD, Corn, Hants, Suss, Kent; STINKING ROGER, Ches.

Gerard first recorded the Alder Buckthorn for the British Isles, finding it near Islington and Hornsey and on Hampstead Heath. The herbalists knew it as *Alnus nigra* – 'the leaves be like those of the Alder tree, or of the Cherrie tree, yet blacker, and a little rounder'. So the English 'Aller tree' or 'Black Aller Tree', of Gerard's *Herbal.* He observed that countrymen purged themselves and made themselves sick with the inner bark, 'not without great trouble and paine to the stomacke'. Therefore it is 'a medecine more fit for clownes, then for civill people, and rather for those that feede grossely, then for daintie people'. Evelyn warned his readers in the *Sylva* (1644) to dry the bark in the shade, and not use it green, and to let the decoction settle for two or three days before they drank it. As the Dogwood of gunpowder manufacturers, *Frangula alnus* gives the finest charcoal for small-arms powder – for a powder which has to burn with extra speed. For this it was grown at one time in Sussex and Kent (21).

XXXII. Papilionaceae

1. Dyer's Greenweed. *Genista tinctoria* L. 78

Local names. ALLELUIA, Shrop; BRUMMEL, Corn; DYER'S WEED, Cumb; GREENING-WEED, E Ang; GREENWEED, Glos, E Ang; SARRAT, West; SHE-BROOM, Ches, Yks; WOADMESH, Yks, N Eng, Kirk, Wigt; WOADWAX, Dor, Som, Wilts; WOADWAXEN, Som; WOADWEX, Dor, Wilts; WOADWISE, Kirk, Wigt; WOODAS, West; WOODWAX, Dor, Som, Wilts; WOODWAXEN, Glos; YELLOW (cf. German *Gilbe*), N'hants, Midlands.

Little noticed except when the yellow flowers are out, Dyer's Greenweed often grows with a cheerful, tough persistence in the tangle of old grass along roads and lanes; or sometimes, as you go by in the train, you glimpse a colony in possession of a railway embankment. In the Wessex counties it is still known by its ancient names, from the OE *wuduweaxe* and the ME *wodewex* (of which the dative plural was *wodewaxen*).

Wuduweaxe, since *G. tinctoria* is not a woodland plant, must have meant 'woody growth', of a plant neither herb nor tree (cf. German

Wildes Holz, *Holzgekraidich*). Some dictionaries give it as Woadwax, by an understandable confusion with Woad, the prince of English dyeing plants. Chiefly it was used to give a yellow as a basis for greens, a dipping with Woodwax first and Woad second; or else the blue wool was 'greened' with this Greenweed or Greening-weed (12) – which was valuable enough for the settlers to take to New England. John Josselyn, once more, in his *New-Englands Rarities* (1672) writes of it in its new surroundings by the old name, 'Wood-wax, wherewith they dye many pretty Colours'. It is now one of the naturalized aliens of eastern America – for instance, on dry hills in Massachusetts.

In Gloucestershire, as a great cloth county, poor women, poor 'wood-waxers', collected the wild crop – tough labour, since they had to pull it up by the roots. The dyers paid them 1*s*. 6*d*. a hundredweight. J. L. Knapp complains a little unkindly in his *Journal of a Naturalist* (1829) that the women tried to add to this poor return by watering the Woodwax and increasing the weight.

2. Petty Whin. *Genista anglica* L. 91

Local names. CARLIN-SPURS, Scot; CAT-WHIN, Yks, Cumb, Kirk, Wigt; GALLOWAY-WHIN, Scot; HEATHER-WHIN, Berw; MOOR-WHIN, Berw; MOSS-WHIN, Berw; NEEDLE-FURZE, NEEDLE-GREENWEED, Som; NEEDLE-WHIN, Som, Yks; STITCH-HYSSOP, Hants; WOADWAX, WOODWEX, and WOODWAX, Wilts.

Genista anglica is armed with spines; therefore 'Carlin-spurs' in Scotland, the spurs of a hag or a witch. The gay yellow flowers are smaller and the stems shorter than in *Genista tinctoria*, but both these plants of poor soil might be grown more often in gardens. Miller gave instructions for raising them in his great *Gardeners Dictionary*: *Genista anglica* afforded 'a very agreeable prospect' in May, though he advised supporting the weak young stems with stakes for two or three years until they could support themselves. Gerard found Petty Whin growing on Hampstead Heath – 'and divers other barren grounds, where in maner nothing else will grow'.

3. Gorse, Furze. *Ulex europaeus* L. 109, H 40

The three most general names are Gorse (OE *gorst*), Furze (OE *fyrs*), and Whin, which may originally have been a Scandinavian word. Gorse is more general in the Midlands, sometimes in the old form of Gorst; Furze (often Fuzz or Vuzz) is commonly used in the south-west and in Ireland,

and Whin is used more in eastern and northern counties, Scotland, and Ireland. There are other names:

FINGERS-AND-THUMBS, Wilts; FRENCH-FUZZ (in distinction to Dwarf Furze), Corn, Dev, Ire; FURRA, Norf; HAWTH or HOTH, Suss; HONEY-BOTTLE, Som; LING, Derb, N Eng; PINS-AND-NEEDLES, THUMBS-AND-FINGERS, Som.

If your liner puts in at Plymouth and lies off the breakwater, and you come on deck after a long while at sea, a warm exciting smell of gorse blossom blows across from the Cornish hills, an emotive welcome from one of the landscape plants of Great Britain. Before the coal trade developed, gorse was also of great value – a fuel for bakers, brickmakers, lime-burners, and farmers' wives. In Cornwall and Devon, for instance, every farm had its 'vuzz' brake. The Vuzz was cut and faggoted and used to fire the cloam ovens. Crush the spines, and Gorse made valuable feeding for the stock in winter time. The crushing was often done in cider mills. The Irish used wooden mauls and a block of stone (69). In the Isle of Man wooden mallets were made to rise and fall on the green gorse with water-wheels (125). In Ireland, if you wear a sprig of *aiteann mhuire*, Blessed Furze, you do not stumble (137), and bits of furze blossom are sometimes included in the 'Summer' (see Hawthorn) which the Irish put over their doorways on May Day to ward off witches and fairies. Furze was also used in this way by the Welsh in Anglesey (11).

On the whole, not one of the sinister plants. But do not forget the grim progress of the soul in the *Lyke Wake Dirge* which Yorkshire people sang over their dead:

When thou from hence away art past,
Every nighte and alle,
To Whinny-muir thou com'st at last;
And Christ receive thy saule.

If ever thou gavest hosen and shoon,
Every nighte and alle,
Sit thee down and put them on;
And Christ receive thy saule.

If hosen and shoon thou ne'er gav'st nane,
Every nighte and alle,
The whinnies shall prick thee to the bare bane;
And Christ receive thy saule . . .

– after which the whin-pricked soul made its way to the Brig o' Dread, from which it was liable to tumble to the fires of Hell, and to the fires of Purgatory, which burnt it to the bare bane.

In the meantime, while you are still on earth, a straight stem of whin, furze, or gorse will make you an excellent walking-stick, and the flowers will make a Gorse Wine, for which see a recipe given in *Farmhouse Fare* (1950).

4. Dwarf Furze. *Ulex gallii* Planch. 66, H 31

Local names. BED-FURZE, Hants; CAT-WHIN, Cumb; CORNISH FUZZ, TAM-FUZZ, Corn.

Perhaps, on second thoughts, it was the low prickly cushions of *Ulex gallii* that the Yorkshire soul had to cross without hosen and shoon. It is beastly stuff to walk through on a Cornish moor or a Cardiganshire hill-side, but it is wonderful to contemplate when the cushions are mixed with Heather or Bell-heather, and both flowers are out in July, August, and September, a mingling of late summer gold and late summer purple. Of the two dwarf species, *Ulex gallii* belongs more to the west and to Ireland, and will not tolerate lime in the soil; *Ulex minor* is absent from Ireland and belongs more to the home counties, the south-eastern counties, and East Anglia.

5. Broom. *Sarothamnus scoparius* (L.) Wimmer. 110, H 40

Local names. BANADLE, Wales (in North Welsh it is *banadlen*, in South Welsh *banhallen*); BANATHAL and BANNEL (Cornish *banathel*), Corn; BASOM, Corn, Dev, Som; GREEN BASOM, Som; BEESOM, Dev; BRUSHES, Dor; CAT'S PEAS, Corn; GOLDEN CHAIR, Dor, Som; GREEN-WOOD, LADY'S SLIPPER, Som; SCOBE, Donegal.

Broom, though the form varies in dialects, is nearly universal. The OE is *brōm*.

One of the great landscape plants, and since it breaks into flower in May and June, one of the plants of love and romance in European poetry. Everyone knows that the Plantagenets take their name from the broom, the *Planta genista*. A spray – in pod, curiously, and not in blossom – was the badge of Henry II. Plantagenet is supposed to have been the nickname of Henry's father, Geoffrey, count of Anjou, and certainly broom marks the landscape of Anjou (the modern department of Maine-et-Loire) as it marks so much of Brittany alongside.

Broom was a plant of many virtues, amorous, magical, scenic, medicinal, and practical. With broom, love and magic go together, since it was one of those plants used by witches and powerful against witches, liked by other-world beings, and useful to keep them away. Read the ballad of *Broomfield Hill* (36). The lady has a tryst to keep among the flowering broom, and fears that she will lose her maidenhead. But the witch-woman tells her the knight will be asleep, and adds a way of making him sleep still more soundly:

> Take ye the blossom of the broom,
>   The blossom it smells sweet,
> And strew it at your true-love's head
>   And likewise at his feet.

The blossoms will let him know, when he wakes up, that the tryst was kept, that 'his love was at command'.

Their magic power is made more clear in another version:

> But when ye gang to Broomfield Hills,
>   Walk nine times round and round;
> Down below a bonny burn bank,
>   Ye'll find your love sleeping sound –

The nine times' walking has put him to sleep:

> – Ye'll pull the bloom frae aff the broom,
>   Strew't at his head and feet,
> And aye the thicker that ye do strew
>   The sounder he will sleep.

When the fairies ride by in the famous ballad of *Tam Lin*, it is from a bush of broom that the Queen of the Fairies makes herself known. Janet has recovered Tam Lin from fairy capture:

> Out then spak the Queen o Fairies,
>   Out of a bush o broom:
> 'Them that has gotten young Tam Lin
>   Has gotten a stately groom.'

In another ballad which Lucy Broadwood contributed to the *Journal of the Folk-Song Society* (iii, 12) the lady has been seduced by an elfin knight, who will only marry her if she can answer his riddles. The ballad begins:

> There was a lady of the North country,
> *Lay the bent to the bonny broom.*

'Bent' in the border counties is used for the Heath Rush (*Juncus squarrosus*). Rushes, or plaited and patterned rushes, avert evil and the evil eye, so the line about broom and bent may have been incantatory against the wiles of the elfin lover. Going back earlier, to the thirteenth century, there is an odd little poem in which a wife consults a being in the broom:

> Tell me, being in the broom,
> Teach me what to do
> That my husband
> Love me true.
> When your tongue is still,
> You'll have your will.*

A powerful plant, the broom. If Geoffrey of Anjou wore it in his cap, it may not have been for sheer delight in broom flowers or the broom scenery of his dominions.

The magic of broom no doubt was felt to increase its power in medicine. Chiefly it was, and remains, a diuretic, the twigs – the *scoparii cacumina* – are still approved by doctors, and no doubt there are still people who make broom tea of twigs or flowers. 'That woorthie Prince of famous memorie *Henrie* the eight King of England,' Gerard wrote in a famous sentence, 'was woont to drinke the distilled water of Broome flowers against surfets, and diseases thereof arising.' People often believed in 'he' and 'she' plants of the same kind, suitable in medicine for men or women. The Cornish distinguished a He-Bannal – which made a diuretic drink for men – from a She-Bannal administered to women (53). To broom medicine must be added broom wine, for which *Farmhouse Fare* (1950) provides a recipe. Here too you have the plant which swept the floors, and has bequeathed its name to the universal instrument. The best of all sweeping plants are broom, birch, and heather.

Broom hates lime and does best on sand or gravel. That puts a limit to its distribution. Yet broom was once far more abundant. Broom place-

*Modernized. The original is in *Early Middle English Texts*, ed. Bruce Dickins and R. M. Wilson, 1952.

names can be counted by the score – familiar ones are Bromley in Kent, the 'broom clearing', or Brompton in Middlesex, the 'broom *tun*' or 'farmstead' (66). Before the coal merchant delivered coal, and factory brooms entered the shops, and when there was little to feed to the sheep in winter, broom was as desirable on a farm as gorse. John Aubrey says that sheep were preserved from the rot by broom. Farmers he knew had destroyed their broom, and the rot then attacked the sheep, 'so ever since they doe leave a border of broome about their grounds for their sheep to browze on, to keep them sound' (7). Prosaic uses; but not every wild plant has given a word to the language, if one can put it that way, or a name to a line of kings.

6. Restharrow. *Ononis repens* L. ssp. *repens*. 105, H 33

Local names. CAMMOCK, Corn, Dev, Dor, Som, Wilts, Hants, I o W, Suss, Bucks, Herts, Worc; CAT-WHIN, Som, Yks; CORNETS, Dor; DUMB CAMMOCK, Som; FIN and FINWEED, N'hants; FURZE, Wilts; GOOSEBERRY PIE, GROUND FURZE, Som; HARROW-REST, Lincs; HEN-GORSE, Ches, Midlands, N Eng; HORSE-BREATH, Som, Worc; LADY-WHIN, Scot; LAND-WHIN, E Ang; LEWTE, POVERTY, Som; RAMSONS and RAMSEY, Dev; RASSALS, Suff; RUSTBURN, Yks; SITFAST, Mor; SPANISH ROOT, Cumb; STAYPLOUGH, Som; STINKING TAM, N'thum; WHIN, N'hants; WILD LIQUORICE, Yks, Cumb, Dumf, Inv, Mor.

Cammock (OE *cammoc*) is the proper English name. Restharrow was adapted into book English from the mediaeval names *remora aratri*, 'plough hindrance', and *resta bovis*, 'ox stop', which also gave the French name *arrête-bœuf*. In his translation of Dodoens, Henry Lyte wrote: 'The roote is long and very limmer, spreading his branches both large and long under the earth, and doth ofttimes let, hinder, and staye, both the plough and oxen in toyling the ground, for they be so tough and limmer, that the share and coulter of the plough cannot easily divide, and cut them asunder' – which is the conventional explanation of the conventional book-name.

English farmers certainly disliked Restharrow for another reason – for tainting milk, butter, and cheese. Cammocky butter was a nuisance in Sussex, Hampshire, and the Isle of Wight. In the north children dug up the root and chewed it. Hence the names of Wild Liquorice and Spanish root, since liquorice was known as 'Spanish Liquorice' or 'Spanish Juice', or 'Spanish' for short.

7. Black Medick. *Medicago lupulina* L. 111, H 42

Local names. BLACK GRASS, Bucks; BLACK NONESUCH, Norf; BLACKSEED, Bucks; BLACK TREFOIL, Norf; DOG-CLOVER, Som; FINGERS-AND-THUMBS, Dor; HOP-CLOVER, Dev, Wilts; HOP-MEDICK, N'thum; LAMB'S TOES, Staffs; NATURAL GRASS, I o S; NONESUCH, Hants, Suss; SANFOIN, Bucks.

The pretty word Medick comes from *herba Medica*, i.e. the Median or Persian herb. Strabo wrote that the name was derived from the country of the Medes; and the description of the plant in Dioscorides shows that the Median herb, in fact, was the allied *Medicago sativa*, our familiar Lucerne, which was already a fodder plant in the older civilizations of the Mediterranean and the Near East.

This kind is Black Medick from the pods: when the plants begin to dry up, the ripe pods become odd curly objects as black as night.

8. Spotted Medick. *Medicago arabica* (L.) All. 55, H 6

Local names. COGWEED, Som; DEVIL'S CLOVER, SPOTTED CLOVER, SPOTTED GRASS and ST MAWES CLOVER (from the parish), Corn.

Spotted Clover and Spotted Grass – so-called from the blotched leaflets. The spiralling, spiny pods explain Cogweed, and no doubt Devil's Clover as well (cf. the devil names of *Ranunculus arvensis*, with its spiny achenes).

9. Melilot. *Melilotus altissima* Thuill. 90, H 10

Local names. HART'S CLOVER, Yks, N Eng; KING'S CLAVER, Scot; WILD GOLD CHAIN (i.e. wild laburnum), Som; WILD LABURNUM, Som.

Perhaps a naturalized alien, introduced by the herbalists of the sixteenth century. It is the plant William Turner in 1548 called *Lotus sylvestris*, which he fashioned into 'Wild Lote', adding that it was much grown in the gardens of East Friesland, and that it had been raised in the garden of Syon House (on the other side of the Thames opposite Kew Gardens), where Turner was physician to Edward Seymour, Duke of Somerset. Turner had been in East Friesland not long before, and perhaps he had brought the Melilot back with him (151). In medicine it was a chief ingredient of the famous Melilot Plaister, or poultice, frequently described in the herbals.

As it dries, Melilot gives off a delicious hay smell, since, like Woodruff or the Meadow-grass, *Anthoxanthum odoratum*, it contains coumarin. Not surprisingly, it was one of the French herbs of St John and of the St John's Eve ceremonies (188). It was 'King's Clover', from the herbalists' and apothecaries' name *Corona regis*, 'King's Crown': the species is kingly among clovers, and the racemes of yellow bloom suggested the golden spikes of a crown.

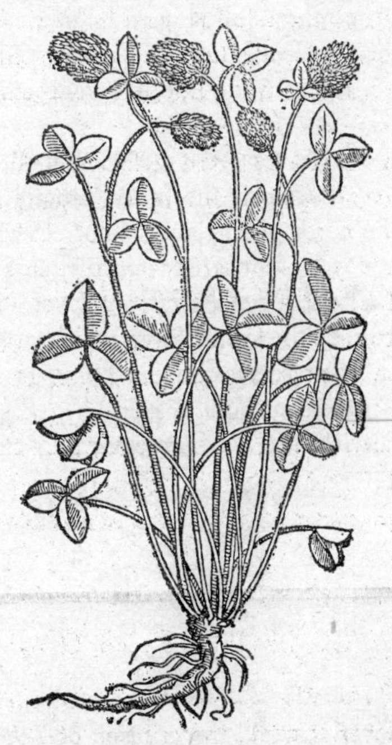

22 Red Clover *Trifolium pratense*

10. Red Clover. *Trifolium pratense* L. 112, H 40

Local names. BEE-BREAD, Kent; BROAD-CLOVER, I o W; BROAD-GRASS, Dor; CLAVER, Dev, Glos, Yks, Dur, Cumb, N'thum, N Eng, Ayr; CLOVER-ROSE, Dev; COW-CLOOS, Scot; COW-GRASS, Som, I o W, Cumb, N'thum, Rox; HONEYSTALKS, Som; HONEYSUCK, Som, Hants, Glos, N'hants, War; HONEYSUCKLE, Som, Wilts, Herts, Oxf, War, N'hants, Lincs, Yks, Donegal.

KING'S CROWN, Som; KNAP, Dor; LADY'S POSIES, Som; MARL-GRASS, Som, Wilt; PINKIES, Dor; PLYVENS, Scot; RED CUSHIONS, Som; SLEEPING

MAGGIE, N'thum; SOUKIES, N'thum, Scot; SOUKIE SOO, Scot; SUCKBOTTLES, SUCKERS, N'hants; SUCKLINGS, Suff, Norf, Shrop; SUGAR-BOSSES, Som; SUGAR-PLUMS, Bucks.

Red Clover and White Clover were broad distinguishing terms long before the Conquest, but more or less generically the Meadow clovers were taken to be grasses which happened to have flowers – flowers you could pick off one by one and suck for the nectar or honey, honeysuckles. Gerard calls *Trifolium pratense* the 'three-leafed grass', as well as Honeysuckles, Suckles, and Cocksheads. He mentions *Trifolium arvense*, the Hare's Foot, and realizes that there are many kinds, but leaves them to be distinguished by 'the curious'.

Three-in-one Clover leaves were lucky. Four-leaved clovers were luckier still. *Trifolium dubium* is the lucky, sanctified, and emblematic shamrock, but in the north of England, if not elsewhere, leaves of Red Clover were also employed as a charm against witches and evil (192); and the flowers of Red Clover – see *Farmhouse Fare*, 1950 – make a good country wine. Before 1600 the form Claver, which still survives, was more general than Clover. It preserves the OE *clāfre*, or *clǣfre*, which is in many Claver place-names, such as Clavering in Essex, Claverdon, the 'clover hill', in Warwickshire, or Claverton, the 'clover farmstead', in Cheshire (66).

11. Crimson Clover. *Trifolium incarnatum* L. 45

Local names. BLOODY TRIUMPH, Dor; NAPOLEON, I o W, Suff; RED FINGERS, SOLDIERS, Som.

The red of Red Clover is weak, the crimson of *Trifolium incarnatum* is decided and military. You can tell the species at a glance by the colour and the way the flower heads stand up like a muff on end, instead of resembling a cloth ball. Parkinson had the Crimson Clover in his garden in 1640. It hangs on as a relic of a crop which is now rare, an alien from the south of Europe. The name Napoleon is supposed to be a corruption of *Trifolium*, but it looks like a deliberate 'corruption', the bloody Frenchman of battles. Crimson Clover is commoner in France, where fields of it are magnificent by the summer roadsides.

12. Hare's-foot Clover. *Trifolium arvense* L. 102, H 16

The prettiest of the Clovers. The downy heads are exactly hare's feet or rabbit's feet, especially when they are brown and dry in the late summer. The only dialect name on record appears to be Dogs-and-Cats from Morayshire, but then *Trifolium arvense* is a 'useless' Clover. In the sixteenth century it was the *lagopus* or *pes leporis*, Greek and Latin for hare's foot; in French *pied-de-lièvre*. William Turner seems to have added Hare's-foot to the English language in 1548: 'Lagopus called also Logopyrus groweth much among the corne, it hath a rough toppe lyke Doune, and leaves lyke a Claver, wherfore it maye be named in Englishe rough Trifoly or harefote, the duche men cal it, Katzenklee [Cat's Clover], the French men Pede de leure. It dryeth manifestly.'

13. White Clover. *Trifolium repens* L. 112, H 40

Local names. BAA-LAMBS, BEE-BREAD, Som; BOBBY-ROSES, Corn; BROAD-GRASS, Dor; CLAVER, Dev, Glos, Yks, Dur, Cumb, Berw, Ayr; CURL-DODDY (i.e. curly head), Ork; DUTCH, Dor; HONEYSTALKS, War; HONEY-SUCKLE, Som, Wilts, Midlands, Beds; LAMB-SUCKLINGS, Yks, Cumb; MULL, Corn; PUSSY FOOT, Som; QUILLET, Corn; SHAMROCK, Ire; SHEEP'S GOWAN, Scot; SMOORA, SMARA (Icelandic *smári*), Shet; SUCKLINGS, Suff, Norf, Shrop; SUCKLERS, Berw; WHITE SOOKIES, Berw; THREE-LEAVED GRASS (see *Trifolium pratense*), Corn.

Honeystalks for the White Clover (given also to the red *Trifolium pratense*) demands the quotation from Shakespeare's *Titus Andronicus*, Act IV, Scene iv:

> I will enchant the old Andronicus
> With words more sweet, and yet more dangerous,
> Than baits to fish, or honeystalks to sheep.

Forget, too, that it is Dutch Clover, the clover whose cultivation we borrowed from the Dutch, and turn to the tenth-century Welsh tales of the *Mabinogion*, in which Olwen is described as a girl with breasts whiter than a white swan's breasts, and cheeks redder than the Foxglove, and hair more yellow than the flowers of the Broom. White clover plants would spring up in her footprints (Olwen means 'white footprint') wherever she walked.

14. Hop Trefoil. *Trifolium campestre* Schreb. 110, H 39

Local names. CRAID, Scot; HOP-CLOVER, Dev, Wilts, Glos, Yks, Berw; LOVE-AND-TANGLE, Dor; TOM THUMBS, Som.

Hop-clover, because the flower heads, changing from yellow to brown, are like miniature brown 'cones' of the Hop (*Humulus lupulus*).

15. Lesser Yellow Trefoil, Shamrock. *Trifolium dubium* Sibth. 112, H 40

At last the Irish shamrock – *seamrog*, 'little clover', 'cloverlet', a diminutive from *seamar*, clover. Irishmen with shamrock on St Patrick's Day are wearing what was once a charm against sorcery, witches, fairies – a charm christianized by the legend of St Patrick explaining the Trinity, the three-in-one, with the aid of a three-leaved clover. St Patrick's Day, 17 March, marked the beginning of spring and the end of winter, a time when evil was abroad, and protection was required. Shamrock is lucky too in the Highlands and Islands. The Shamrock of Luck and Blessing is a four-leaved or a five-leaved Clover:

> Thou Shamrock of promise on Mary's Day,
> Bounty and blessing thou art at all times,

says a Gaelic incantation to be uttered over the leaf. But it is no good if you search for it. It must be found accidentally. Scottish people also buried a soft substance coughed up by new-born foals in the belief that the Shamrock of Luck and Blessing would grow from it (31a).

There has always been argument, rather pointlessly, about the identity of shamrock, whether it was clover, and which clover, or whether it was wood-sorrel. It was clover first and foremost, and a small clover; though in a free world there was nothing to stop you picking the triple leaf of wood-sorrel and calling it shamrock, and using it as if it were.

There is a charming simile from the little clover in a love song which Peig Sayers sang to Professor Kenneth Jackson on Great Blasket, and which he recorded and translated (*Folklore*, 1933). The girl in the song had eyes 'greener than the dew-wet shamrock in the grass'.

16. Birdsfoot Trefoil. *Lotus corniculatus* L. 112, H 40

Local names. BIRD'S CLAWS, Dev; BIRD'S EYE, BIRD'S FOOT, Som; BOOTS-AND-SHOES, Corn, Dev, Som, Suss; BOXING GLOVES, BREAD-AND-CHEESE,

BUNNY RABBITS, BUNNY RABBITS' EARS, Som; BUTTER-AND-EGGS, Som, Glos, War; BUTTERED EGGS, Cumb; BUTTER-JAGS, N Eng.

CAMMOCK, Dev; CAT'S CLAW, Som, Bucks (i.q. CATCLUKE, Scot, and KATTAKLU, Shet); CATTEN-CLOVER, Scot; CAT'S LOVER, N'thum; CAT-PEA, Corn; CAT-PUDDISH, Cumb; CHEESE-CAKES, Som, Worc, Yks; CHEESE-CAKE GRASS, Yks, N Eng; CLAVER, N Ire; COCKLES, Som; CRAA'S FOOT, N'thum; CRAA-TAES, N'thum; CROWFEET, Som, Glos, Suff; CROW-TOES, Som, Scot; CUCKOO'S STOCKINGS, Som, Suss, Shrop.

DEAD-MAN'S FINGERS, Hants; DEVIL'S CLAWS, DEVIL'S FINGERS, Som; EGGS-AND-BACON, Som, Glos, Suss, N'hants, Rut; EGGS-AND-COLLOP (collop: a rasher of bacon), Yks; FELL BLOOM or BLOOM-FELL, Scot; FINGERS-AND-THUMBS, Dev, Dor, Som, Hants, Oxf, Herts, Rut; FINGERS-AND-TOES, Dev, Ess, Norf, Cam; FIVE FINGERS, Ess, Norf, Cam; GOD ALMIGHTY'S FLOWERS, Dev; GOD ALMIGHTY'S THUMB-AND-FINGER, Som, Hants; GRANDMOTHER'S SLIPPERS, Hants; GRANDMOTHER'S TOE-NAILS, Dev, Som; HEN-AND-CHICKENS, Som, Glos, Oxf; HONEYSUCKLE or GROUND HONEYSUCKLE, Ches; HOP-O'-MY-THUMB, Som; JACK-JUMP-ABOUT, N'hants; KING'S FINGERS, Bucks; KING'S FINGER GRASS, Mddx; KITTY-TWO-SHOES, Dor.

LADY'S BOOTS, Dev; LADY'S CUSHION, Dor, Wilts; LADY'S FINGERS, Som, Wilts, Hants, Suss, Herts, Hunts, N'hants, Lincs, Lancs, Yks; LADY'S FINGERS-AND-THUMBS, Som, Wilts; LADY'S FINGER GRASS, Herts; LADY'S GLOVE, Dor, Wilts, N'hants; LADY'S PINCUSHION, Dor; LADY'S SHOES, Hants; LADY'S SHOES-AND-STOCKINGS, Som, Kent, Bucks; LADY'S SLIPPER, Som, Wilts, Herts, Yks; LAMB'S FOOT, Lancs; LAMB SUCKLINGS, Yks, N Eng; LAMB'S TOES, N'hants, War, Leic, Notts, Lincs; LOVE ENTANGLED, Corn.

MILK-MAIDENS, Suss; OLD WOMAN'S TOE-NAILS, Dev; PATTENS-AND-CLOGS, Som, Glos, Suss; PEA-THATCHES (i.e. vetches), Som; PIG'S FOOT, Suff; PIG'S PETTITOES, Suss; ROSY MORN, Som; SHEEP-FOOT, Cumb; SHOE-AND-SOCK, Kent; SHOES-AND-STOCKINGS, Dev, Som, Hants, Glos, Suss, Bucks; STOCKINGS-AND-SHOES, Dev, Som; THUMBS-AND-FINGERS, Dor; TOM THUMB, Dev, Som, Wilts, Oxf, Yks; TOM THUMB'S FINGERS-AND-THUMBS, Som; TOM THUMB'S HONEYSUCKLES, Wilts; TOMMY TOTTLES, Yks; YELLOW CLOVER, Yks.

When a species has been endowed with more than seventy names, and when it is a small plant (however common) to which people could very well have been indifferent, one has to look for reasons. The blue flowers of the Tufted Vetch, *Vicia cracca*, stand out in grassland scenery as much as

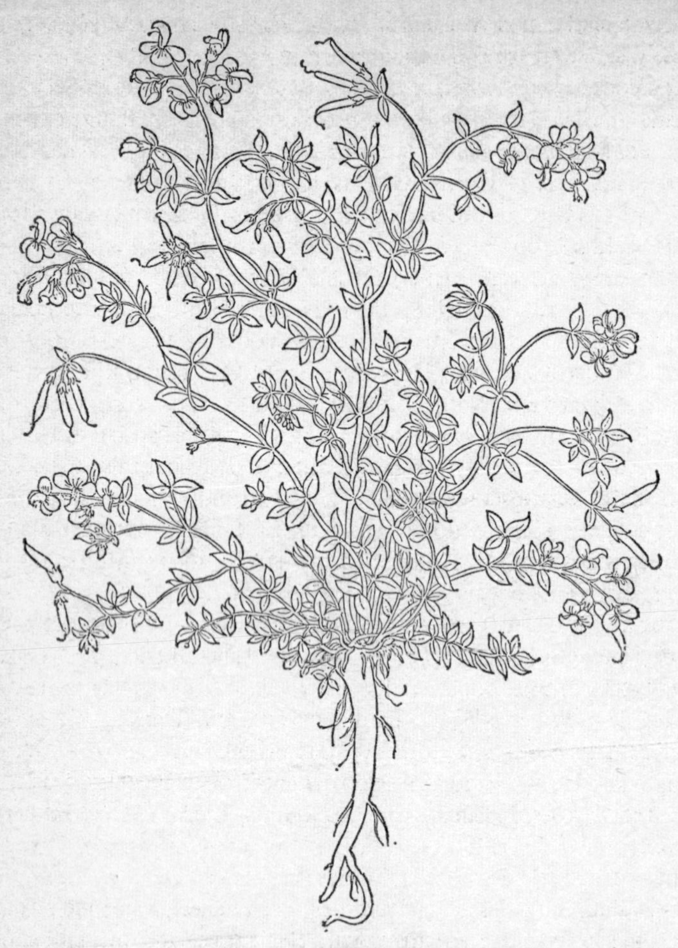

23 Birdsfoot Trefoil *Lotus corniculatus*

the yellow flowers of *Lotus corniculatus* – indeed a great deal more. Why, on the evidence of the names, a handful fixed to the Vetch and more than seventy to the Birdsfoot Trefoil, have people been aware more of one plant than the other? Some names are due to the flowers, some to the pods, which turn black and conspicuous and end in a claw. The flowers can be shoes (in French *petit sabot de la mariée*). The pods, usually between four and six to a head, can be bird's claws, animal's claws, toes and feet (in French also *pied du bon Dieu*), clawed fingers or clawed thumbs.

As so often, the names divide into neutral, evil, and good. As so often, one may assume that the evil associations were so strong that the plant needed conversion, and was made into a plant of God and the Virgin. On the one side the claws and the pods are bird's claws, especially of the evil bird, the crow. They are claws of the evil cat, claws and fingers of the Devil. More to the point, they are the fingers of one, or possibly more than one, goblin – of Tom Thumb. In five counties across England *Lotus corniculatus* has been known as Tom Thumb, in Somerset, too, as Hop-o'-my-Thumb,* in Northamptonshire as Jack-jump-about. One may infer that it was the plant of this particular goblin, and probably one which it was not wise to pick. Though most goblins and fairies were thought of as large, Tom Thumb or Hop-on-my-thumb came near the modern, literary concept of the miniature fairy.

When Tom Thumb came to print and paper in the seventeenth century, he was the merry, mad, supernaturally powerful Lilliputian, born to the ploughman's wife by intercession with Merlin, he was King Arthur's dwarf, and godchild of the Queen of the Fairies – transformed even more drastically than Robin Goodfellow. But no doubt he had been among the black and fearful company of bugs, of whom Reginald Scot gave a list (sprinkled with a few classical beings) in his *Discoverie of Witchcraft* (1548): 'In our childhood, our mother's maids have so terrified us with an ugly devil . . . and have so fraied us with bull-beggers, spirits, witches, urchens, elves, hags, fairies, satyrs, pans, fauns, sylens, kit with the can-sticke, tritons, centaures, dwarfes, giants, imps, calcars, conjurors, nymphs, changelings, Incubus, Robin good fellow, the spoorn, the mare, the man in the oke, the hell waine, the firedrake, the puckle, Tom thombe, hob-goblin, Tom tumbler boneles and such other bugs, that we are afraid of our own shadowes.'

Tom Thumb, Hop-o'-my-thumb, Jack-jump-about, was a restless, fidgety bug. If you contemplate Tom Thumb's plant, and push your fancy so far, you can discover the manner of bug our ancestors thought him to be, a bug with dry black fingers ending in a claw. Odd fingers to transfer to Our Lady. The flowers, as her shoes, boots, or slippers, were less inept. (See also *Lathyrus pratensis*.)

* In Somerset Robin Goodfellow's plant, *Geranium robertianum*, was also called after Hop-o'-my-Thumb.

17. Lady's Fingers, Kidney Vetch. *Anthyllis vulneraria* L. 112, H 40

Local names. BUTTER-FINGERS, Som; CAT'S CLAWS (cf. German *Katzenprankerl*, Swedish *cattklor*, etc.), Som, Mor; CRAE-NEBS (i.e. crow-beaks), N'thum; DOG'S PAISE (i.e. peas), Banff; DOUBLE LADY'S FINGERS-AND-THUMBS, DOUBLE PINCUSHION, Wilts; FINGERS-AND-THUMBS, Dor, Som; GOD ALMIGHTY'S FINGERS-AND-THUMBS, Dor; GRANFER GRIZZLE, HEN-AND-CHICKENS, Som.

LADY'S CUSHION, Wilts; LADY'S FINGERS, Dor, Hants, E Ang, Rut, Lincs, Yks, Lanark; LADY'S SLIPPERS (cf. German *Muttergottesschühlein*, etc.), War; LAMB'S FOOT, Som; LAMB'S TOE, Som, Rut; LUCK, Norf; PINCUSHION, Wilts; TWINS, Yks; YELLOW CROW'S FOOT, Bucks; YELLOW FINGERS-AND-THUMBS, Dor, Som.

More to look at, with more of a personality than *Lotus corniculatus*. The red and yellow flowers bunched together so comfortably can mark a whole cliff or slope. On this plant the pods are minute, they are not fingers, thumbs, or claws, though several of the claw, thumb, and finger names of *Lotus corniculatus* serve for *Anthyllis vulneraria* as well. Contrariwise, names one may think more apt for *Anthyllis vulneraria* have been transferred to the *Lotus*, such as Pincushion or Lamb's Foot. In Gaelic this plant is *meoir Mhuire*, Mary's Fingers, and *cas an uain*, lamb's foot. It was known throughout Europe as a wound herb, or vulnerary.

18. Sainfoin. *Onobrychis viciifolia* Scop. 37

Local names. BABY'S CRADLES, Dor; COCK'S HEAD, Som; EVERLASTING GRASS (it is perennial), Oxf; FRENCH GRASS, Som, Wilts, Hants; THATCH (i.e. vetch), Glos.

In the sunlight of June and July, Sainfoin flowers are exquisite, even startling. In bold, crowded racemes against a blue sky, the flowers have the pink of Quattrocento paintings, or of pink stripes upon white silk. Astrological herbalists placed Sainfoin under the dominion of Venus, which seems right.

Sainfoin has been cultivated for a long while, but though often naturalized, there is every reason for thinking it a native on limestone or chalky hills – the chalk of Wiltshire, for example, or Hampshire or Dorset. Gerard (who called it Cock's Head – 'Onobrychis sive *Caput Gallinateum*'

– and Red Fetchling, and Medicke Fitch) found it growing in Cambridgeshire and on a Bedfordshire hillside. The common name is French, borrowed by us in the seventeenth century. It means, not 'Holy Hay' or St Foyne, as the plant has been called, but simply 'wholesome hay', i.e. a dried crop good for cattle.

Identifying it on little evidence with the *Onobruchis* of Dioscorides, the herbalists followed him in recommending it for softening boils, opening the bladder, and procuring a sweat. 'It is a singular food for Cattel,' added Culpepper, 'to cause them to give great store of milk, and why then may it not do the lyke being boyled in ordinary drink of nurses?'

19. Hairy Tare. *Vicia hirsuta* (L.) S. F. Gray 111, H 37

Local names. BINDWEED, Herts; DILL, Glos; DOTHER, Ches; LINTELS, N'hants; MOUSE PEA, Donegal; TARE-FITCH, W Eng, Ches, Shrop; TARE, Mddx, Ess, Suff, Notts, Lincs, Ches, Yks, Scot; TARE-VETCH, Dor, I o W, S Eng; TARE-GRASS, Kent, Staffs; TINE, TINE-GRASS, Herts; TINE-TARE, Kent; WILD FITCH, Cumb.

Vicia hirsuta was the first of the Vetches the Englishman needed to know and recognize. It strangled his corn as an old weed of cultivation. He called it Vetch, from the Latin *vicia* by way of Norman French, and also Tare and Tine-tare. 'Tare' in Middle English meant first the seeds, apparently, then the plant. 'Tine' – the verb, at any rate, meant to suffer loss or deprivation, which the farmer indeed suffered from the Tine-tare clinging by its tendrils to his oats, his barley, or his wheat.

Farmers, when they listened to St Matthew xiii. 30 in the Authorized Version, would well understand the parable of the tares and the good seed – 'Let both grow together until the harvest: and in the time of harvest I will say to the reapers, Gather ye first the tares, and bind them in bundles to burn them: but gather the wheat into my barn', or the point would be well taken when John Redford wrote of vice and the tine in his Tudor morality play, *Wyt and Scyence*:

> This vice I lyken to a weede
> That husbondmen have named tyne
> The whych in corne doth roote or brede.

'The herbe,' said Gerard, allowing it no virtues at all, 'is better knowne then desired.'

20. Tufted Vetch. *Vicia cracca* L. 112, H 30

Local names. BLUE GIRSE, Shet; BLUE TAN-FITCH, Ches; CAT-PEAS, Scot; COW-VETCH, Glos; FINGERS-AND-THUMBS (cf. *Lotus corniculatus.* The pods of *Vicia cracca* also end in a nail or claw), Som; FITCHACKS, Aber; GOOSE-AND-GANDERS, Som; HUGGABACK, Cumb; MOUSE'S PEASE (used in the fifteenth century for *Vicia hirsuta*), N'thum, Mor; TARE-FITCH, W Eng, Ches, Shrop; TARE-GRASS, Kent, Staffs; TARE-VETCH, Dor, Som, I o W, S Eng; THATCH (i.e. vetch), Glos; TINE, TINE-GRASS, TINE-WEED, TWINE-GRASS, Herts; WILD FETCHES, WILD TARE, Scot; WILD VETCH, Oxf, Cumb (see above, Hairy Tare, *Vicia hirsuta*).

A finely etched and patterned species, bearing brilliant ladders of blossom, yet a plant lacking in attributes.

21. Common Vetch. *Vicia sativa* L. 112, H 40

Local names. FATCHES, Glos, War; FETCHES, Shrop; FITCHACKS, Aber, Mor; FITCHES, Suss, War, Shrop, N'hants, Ches, Yks, Cumb, Ire; GIPSY-PEAS, Som; LINTS, Derb, Lincs, Yks; TARE, Mddx, Ess, Suff, Ches, Notts, Lincs, Yks, Scot; TAR-VETCH, Som; TWADDGERS, Yks; VATCH, Glos; WILD FITCH, Cumb.

The virtuous Vetch of the family, introduced by the farmer for his cattle, and coming originally, it is thought, from western Asia. Miller, in his *Gardeners Dictionary*, says that in his day, early in the eighteenth century, *Vicia sativa* was cultivated in the fields 'for the seed, which is the common food of pigeons'. Grown everywhere for early forage, this ancient farmer's plant has naturalized itself throughout the British Isles.

Of the native Vetches, Bush Vetch, *Vicia sepium* L. (112, H 40), has been called Crow-peas, Cumb; Dill, Leic; Tare, Suff; Thetches, Dor, Som, Wilts, W Eng, Bucks, Herts; and Twaddgers, Yks; the local and handsome Bitter Vetch, *Vicia orobus* DC. (41, H 6), is Horse-pease in Cumberland; and the Wood Vetch, *Vicia sylvatica* L. (86, H 24), has been known in Somerset, Kent, and Northamptonshire as Culverkeys, a name also given to Cowslips, Bluebells, Columbine, and the keys of Ash.

22. Grass Vetchling, Crimson Shoe. *Lathyrus nissolia* L. 46

Grass Vetchling is a miserable book name, with no alternative except Shoes-and-Stockings (Suss). Why not Crimson Shoe? *Lathyrus nissolia* is a local species, always exciting to find, always surprising, when in the long grass you detect what appears to be another grass with a flower, or two flowers, of brilliant crimson attached to it. If you pick them, the flowers quickly shrivel.

As the books say, this Crimson Shoe is overlooked all too easily. But not always. Now and again there are good (extra moist?) years in which Crimson Shoe will colour a whole patch of rough grass along a south-country road.

23. Meadow Vetchling. *Lathyrus pratensis* L. 112, H 40

Local names. ANGLE-BERRY, N Eng; FINGERS-AND-THUMBS, Som; LADY'S FINGERS, Wilts, Yks; LADY'S SLIPPER, Som, Wilts; MOUSE PEA, Donegal; OLD GRANNY'S SLIPPER-SLOPPERS or SLIPPER-SLOPPERS, Dor; TOM THUMB, Suss, Berw; YELLOW TARE-TINE, Mddx; YELLOW TAR-FITCH, Ches; YELLOW THATCH, Wilts.

So far as it has been noticed at all, Meadow Vetchling has been taken both as a Vetch (for the last three names, see *Vicia cracca*) and as a plant like enough to *Lotus corniculatus* to share the names Tom Thumb, Lady's Fingers, and Lady's Slippers.

24. Bitter Vetch. *Lathyrus montanus* (L.) Bernh. 108, H 38

Local names. CAPEROILES, Scot; CORMEILLE, CORRA-MEILE, N Scot, Heb; CORNAMELIAGH, FAIRIES' CORN, Donegal; KNAPPERTY, N Ire; KNAPPERTS, NAPPLE, Scot; NIPPER-NUTS, Som; PEASLING, Yks; WOOD PEA, Som.

First recognized as an English species by William Turner, who wrote in 1548 that it was 'called in lowe duchelande Erde nut, and in Berglande Erdeklin', so he proposed to call it 'peaserthnut' in English, to distinguish it from the 'ernut' or 'earth nute' (*Conopodium majus*), the tubers of which he must have eaten as a boy in Northumberland.

In the Hebrides and the north of Scotland, and no doubt elsewhere, and long before Turner's day, the tuberous rhizomes were dug up and eaten. On Colonsay they ate them fresh and raw, or tied them up in bundles to dry under the thatch. They also used them for flavouring whisky (128).

Gerard knew all about this wild vegetable: 'The Nuts of this Pease being boiled and eaten, are hardlier digested then be either Turneps or Parsneps, yet do they nourish no lesse then the Parsneps: they are not so windie as they, they do more slowly passe thorowe the belly by reason of their binding qualitie, and being eaten rawe, they be yet harder of digestion, and do hardlier and slowlier descend.' These roots, 'thicke, long, like after a sort to acorns, but much greater, blacke without, graie within', he likened in taste to chestnuts.

XXXIII. Rosaceae

1. Meadowsweet. *Filipendula ulmaria* (L.) Maxim. 112, H 40

Local names. BITTERSWEET, Yks; BLACKIN-GIRSE, Shet; COURTSHIP-AND-MATRIMONY (from the difference in scent before and after crushing the plant), Cumb; GOAT'S BEARD, Dor, Dev; HAYRIFF (cf. *Asperula odorata*), Shrop; KISS-ME-QUICK, Som; LADY OF THE MEADOW, S Scot.

MAID OF THE MEAD, Ches; MAIDS OF THE MEADOW, Som; MAY OF THE MEADOW ('may', i.e. 'maiden'), War; MEADOW-MAID, Pemb; MEADOW-QUEEN, Renf, Perth; MEADOW-SOOT, Wilts; MEADUART, Scot; MEADWORT, Som.

NEW MOWN HAY, Som; QUEEN OF THE MEADOW, Dev, Dor, Som, War, Lincs, Yks, Dur, N'thum, N Eng, Scot, N Ire, Ork; QUEEN'S FEATHER, Som; SUMMER'S FAREWELL, Dev, Dor; SWEET HAY, Dor, Suss; TEA FLOWER, Som; WIREWEED (from the tough stems), Hants.

The milky foam of Meadowsweet flowers is more pleasant than the smell, one of those scents which are sweet but nauseating in too great a volume – hence the old apothecary's name of *Barba Caprae*, Goat's Beard. One can never be quite sure about these shop names of the sixteenth and earlier centuries. Many of them must be latinizations from the vernacular. Re-translated, they spread round from country to country.

As well as *Barba Caprae*, Meadowsweet was *Regina prati*, Queen of the Meadow, which, perhaps through shop and book, became one of the commoner English names – in French *Reine-des-prés*, in Italian *Regina de' prati*. Gerard calls the species Queene of the medowes, Medowsweete, and Medesweete. Turner in his *Herbal* in 1568 calls it Medewurte. Meadsweet goes back at least to the fifteenth century (cf. the German *Mädesüss*), Meadwort to the OE *medowyrt*; and it is clear that originally it was not the plant of the meadow or the sweetness of the meadow, but the plant used to flavour mead. The drink in OE was *meodu*; mead or meadow was *mæd*, *mædwe*, so a later confusion between the *meodu* plant and the plant of the damp meadows where it grew, the *Regina prati*, would not have been unlikely. The connection with mead is confirmed by the Swedish name *mjödört*. According to Culpepper, a leaf of Meadowsweet gave a fine relish to a cup of claret (49). In Ireland, where it was used for scouring milk churns and (as in Shetland) to give a black dye with copperas, Meadowsweet is pleasantly called *airgead luachra*, 'silver rushes'. In the Highlands and Islands it is *crios Chu-chulainn*, the belt of Cú Chulainn, the legendary warrior and hero of Ulster (29).

Meadowsweet had a number of medicinal virtues – for instance, against fluxes of blood and against malaria, which was once so common in England; and when floors were carpeted with rushes and plants, it was much in demand: 'The leaves and flowers far excell all other strowing herbes, for to decke up houses, to strowe in chambers, hals, and banketting houses in the sommer time; for the smell thereof maketh the hart merrie, delighteth the senses: neither doth it cause headach, or lothsomnesse to meate, as some other sweete smelling herbes do' (Gerard). Which smell did Gerard mean – the faintly sexy and sweet smell of the open flowers or (see the name Courtship-and-Matrimony above) the smell of the crushed plant which has the freshness of carbolic? In America a species allied to this *Regina prati* is known as Queen of the Prairies.

2. Blackberry, Bramble. *Rubus fruticosus* L. 112, H 40

'Bramble' is generally used in England, Scotland, and Ireland, though the forms vary according to the dialect, e.g. brimmel, brummel, bummel.

Local names. BRIER, Yks, N'thum, Donegal; COCK-BRUMBLE, Suff, Norf; DRISAG, Donegal; GATTER-TREE (cf. *Euonymus europaeus*), Rox; HAWK'S BILL BRAMBLE, E Ang; HE-BRIMMEL, Som; LADY'S GARTERS, Rox; LAWYERS, Suss, Worc; COUNTRY LAWYERS, Leic; YOE-BRIMMEL, Som; THIEF, Leic.

Names for the fruit include: BLACKBIDES, Kirk, Wigt; BLACK BLEGS, Yks; BLACK BOWOURS, Berw; BLACK BOYDS, W Scot; BLACK KITES, Cumb; BLACK SPICE, Yks; BLAGGS or BLEGGS, Yks; BRAMBLES, Perth; BRAMMEL KITES, Dur; BRUMMEL KITES, Cumb; BUMBLE KITES, Hants, Yks, Cumb, N'thum; BUMMEL-BERRIES, Cumb; BUMMELTY KITES, Cumb; DOCTOR'S MEDICINE, Som; GATTER-BERRY, Rox; GARTEN-BERRIES, Scot; LADY GARTEN-BERRIES, Rox; MOOCHES, Glos; MULBERRIES, Suff, Norf; MUSHES, Dev.

Blackberry seeds have been found in the stomach of a human being, probably neolithic, who was dug up from the submerged clay of the Essex coast near Walton-on-the-Naze (*Essex Naturalist*, xvi, 198–208, 1911), confirming what one would safely guess, that blackberries were eaten many thousands of years ago as they are today. We think of brambles as a thorny population, a tanglement not made up of individual plants or the many individual species loosely assembled under the name of *Rubus fruticosus*. Yet the bramble has given more than fruit for casual eating and jam and jelly and blackberry tart. It was *an druise beannaichte*, 'the blessed bramble', in the Highlands, the roots yielded an orange dye, the leaves were placed on burns and swellings, a remedy familiar to the sixteenth-century herbalists. Highlanders twined a bramble with Ivy and Rowan to ward off witches and evil spirits. It was with a blessed bramble, they said, that Christ switched the donkey on the way to Jerusalem, and drove the moneylenders out of the temple (128, 31a). In Cornwall, nine bramble leaves – compare the novena, the intercession on nine days, one after the other – were picked, and given the purification of spring water and then laid to swellings and inflammations with the charm:

> There came three angels out of the east,
> One brought fire and two brought frost.
> Out fire and in frost,
> In the name of the Father, Son and Holy Ghost

– which was repeated three times for each leaf. Diseases could be cured by passing the sufferer beneath an archway of bramble rooted at either end. Graves, too, in English churchyards were neatly tucked round with a plaiting either of brambles or osiers – no doubt to keep the dead from walking.

Bramble thorns pluck dags of wool out of the sheep as they go by, for which cottage wives were grateful in the days of spinning. So the pleasant story of the cormorant, the bat, and the bramble. The cormorant was once a merchant dealing in wool. He went shares with the bat and the bramble in a boat, to carry his wool overseas. The boat was wrecked, so the cormorant is always diving after the lost cargo, the bat owes money and hides from its creditors till dark, and the bramble makes up its losses by stealing wool from the sheep.*

A last item about blackberries. Late in the autumn, as every blackberry picker knows, the most rich and luscious berry may be a snare. A lazy fly drops off, the juice has started to ferment, some of the druplets are grey

* This story is quoted in *Plant-lore and Garden-craft of Shakespeare*, 1884, by H. M. Ellacombe, who took it from Charles Waterton.

with a mildew. On Michaelmas, it was believed in many parts of England, the Devil defiled all the blackberries by spitting or urinating on them. After Michaelmas it was unwise to pick any more. In Ireland this was done by the *púca*, but a few weeks later, on Hallowe'en (137).

3. Cloudberry. *Rubus chamaemorus* L. 42, H 1

Local names. AVERIN, EVRON, Banff, Mor; CLOUDBERRY, Lancs, Yks, N Eng; FINTOCK, Perth; KNOTBERRY, KNOUTBERRY, Lancs, Yks, Cumb, N'thum, Scot; NUB-BERRY, NUB, Dumf, Scot; NOOPS, Berw.

For the orange berries of *Rubus chamaemorus*, creeping among the heather and the cotton-grass, you have to go north, to the high mountains and peat bogs. The first botanist to find the plant was the Lancastrian, Thomas Penny (*c.* 1530–88), who knew the mountain country on the borders of Lancashire and the West Riding. He sent a description and drawing of it to Charles de l'Écluse of the Netherlands, saying that it grew plentifully on the summit of Ingleborough (151). Penny told l'Écluse that the English called it Knotberry because the fruit were like knots. Gerard put *Rubus chamaemorus* twice into his herbal, as though he was dealing with two different species. One description he lifted from de l'Écluse, the other (atrociously illustrated) he had from a Lancashire simpler, Thomas Hesketh, who told him the name was Cloudberry. 'This plant groweth naturally upon the tops of two high mountaines (among the mossie places), one in Yorkshire called Ingleborough, the other in Lancashire called Pendle, two of the highest mountaines in all England, where the cloudes are lower than the tops of the same all winter long, whereupon the people of the countrie have called them Cloud berries.'

Hesketh and Penny were both wrong. Both names mean 'hill berry' – 'cloud' from OE *clūd*, a rock or hill, 'knot' from ME *knot*, a hill. The 'nub' of Nub-berry also means 'knub', or 'knob', in the sense of a hill, no doubt.

In Wales, in the Berwyn mountains, the Cloudberry has been called *Mwyar Dogfan*, 'Dogfan's Berries', after St Dogfan, and *Mwyar Berwyn*, 'Berwyn Berries'. The Berwyn shepherds believed that a quart of the berries had been St Dogfan's stipend for a year at Llanrhaiadar. (See T. W. Hancock, *History of Llanrhaiadar-ym-Mochnant*, 1870, *Montgomeryshire Collections*, Vols 4 and 5.) In their *Mountain Flowers*, 1956, Raven and Walters seem to misinterpret this legend, the point of which is the scarcity of the berries – since Cloudberry is not a free fruiter.

4. Raspberry. *Rubus idaeus* L.   109, H 40

Local names. ARNBERRY, Yks; HINDBERRY, Staff, Lancs, Yks, Cumb, N'thum, Scot; RASP (properly the fruit), N'hants, Lincs, Yks, Scot, Ire; WOOD-RASP, Selk; SIVVEN (Gaelic *suibhean*), Scot.

Why are raspberries called raspberries? No one seems to know for certain. Hindberry is simple enough: it is the berry eaten in the woods by deer, by hinds, originally the wild raspberry; and this is the older name, the OE *hindberie*, *hindberig*. Gerard calls *Rubus idaeus* the Raspis bush, William Turner in 1548 wrote of Raspeses which grew wild in East Friesland and on the hills near Bonn: 'They growe also in certayne gardines of England'. Raspis, in fact, is not recorded in English until about 1532. Perhaps it came into use only after *Rubus idaeus* had left the woods for the garden – or after it had come into English gardens out of France. The English had been familiar since the fifteenth century with a dark red, sweet French wine which they called respyce or raspis – in French it had been *raspeit* or *raspei*. Possibly they connected the colour of the wine with the luscious,

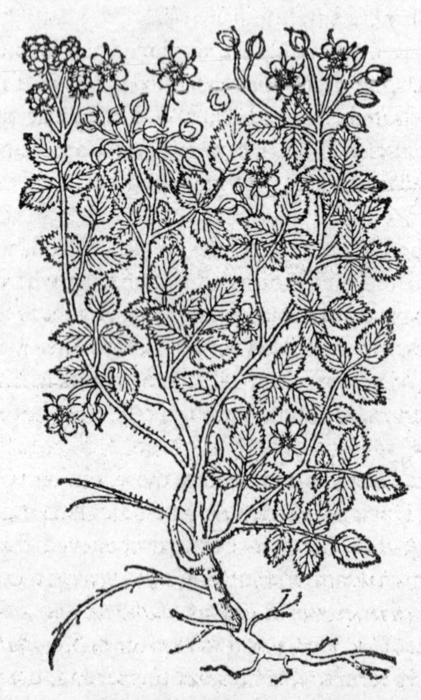

24 Raspberry *Rubus idaeus*

dark colour of the crushed berries of *Rubus idaeus*, which were thereupon christened Raspis: or perhaps home-made wine from the berries had the appearance of the raspis wine.

Easy to understand why *Rubus idaeus* was cultivated (wherever the cultivation began) in preference to other species. For instance, the Cloudberry and the Stone Bramble and the Dewberry all have only a few druplets. Raspberry (wild) and Blackberry have many druplets. Blackberries are easy to come by, and no one at first wanted this tangling, rampageous, sharp-thorned weed in a garden. Raspberry prickles do no great damage, and the plants are not difficult to control. So we have developed the many varieties of this most delicious of all soft fruits.

Of the other *Rubi*, the Stone Bramble, *Rubus saxatilis* L. (70, H 31), whose scarlet druplets shine like cabochon garnets in damp stony woods, has few names – Bungleberry in Cumberland, Bunchberry in Yorkshire and Cumberland, Stoneberry in Donegal, and Roebuck in Cumberland and Scotland. The fruits of *Rubus caesius* L. (86, H 29), the Dewberry or Blue Bramble of the low lands, can be uncommonly good eating if you strike a colony where the fruits have rather more druplets than usual. The very look of the fruits is appetizing, each druplet covered with a bloom as if the dew had settled on it and never dried under the August sun. Every year I visit a clump of Dewberry on a chalk escarpment in north Wiltshire: in a good year the fruits are full and plentiful, another year they will be mean, scanty, and not worth the picking or eating, so one may understand the Wiltshire name of 'Token Blackberry'.

5. Silverweed. *Potentilla anserina* L. 112, H 40

Local names. BLITHRAN (Irish *brioslan*), Ire; BREAD-AND-BUTTER, Som; BREAD-AND-CHEESE, Som; DOG-TANSY, Scot; FAIR-DAYS, N'thum, Berw; FAIR-GRASS, Rox; FERN-BUTTERCUP, Wilts; FISH-BONES, GOLDEN FLOWER, GOLDEN SOVEREIGNS, Som; GOOSE-GRASS, Som, Hants, Glos, Hunts, Lincs, Yks, Berw, Dumf; GOOSE-TANSY, Norf, N'hants, Lincs, Cumb.

MASH-CORNS (the roots), Ire; MIDSUMMER-SILVER, Surr; MOOR-GRASS (cf. Faeroese *múrgras*), Scot; MOSS-CORNS, Selk, Mor; MOSS-CROP, Scot; SILVER FEATHER, Oxf; SILVER FERN, Som, Wilts; SILVER GRASS, Wilts; SILVER LEAVES, Som; SILVERWEED, Suff, Yks, Berw, Ire; SWINE'S BEADS, Ork; SWINE'S MURRIKS (*murrik*, i.e. root), Shet; TANSY, Yks, Cumb, N'thum; WILD TANSY, Som, Berks, Bucks, N'hants, Cumb; TRAVELLER'S EASE, War.

A *potentilla* is a little powerful one, and *Potentilla anserina* has been more, in its human day, than one of the pretty plants along roads and paths and in gateways. The roots were a marginal food or a famine food in the upland zone of Great Britain. In the Highlands and Islands, according to Alexander Carmichael (*Carmina Gadelica*, vol. iv, 1941), the *brisgein* was actually cultivated for its root before the introduction of the potato. The roots were boiled or roasted or eaten raw, and were dried and ground into meal for bread and porridge. The Irish ate them as well; and John Ray noted in his *Catalogus Plantarum Angliae* (1670) that children around Settle in the West Riding dug them and ate them; which was no doubt the practice elsewhere, to judge from the Somerset names of Bread-and-cheese or Bread-and-butter. Ray said they tasted like parsnip. Another of the Gaelic names was *an seachdamh aran*, 'the seventh bread'.

Herbalists gave *Potentilla anserina* against ulcers in the mouth, the stone, inward wounds, and wounds of the privy parts (80). The little powerful one still flourishes though you tread upon it, so perhaps sympathetic magic suggested placing the leaves under the feet – 'it is certain that your carriers wear the leaves in their shoes, which keep them cool and prevent a too immoderate sweating of the Feet which causes a soreness in them' (56). Pleasant to think of the carriers taking off their shoes and picking the silver leaves by the roadside. The magic *Potentilla* above the rest was *Potentilla reptans*, the Cinquefoil, but the apotropaic power may have spilled into this species, too, if John Aubrey (who was no botanist) meant Silverweed when he wrote of Midsummer Silver, 'a little Herb, which continues all the Year of a bright Ash colour'. He noticed that the people of Lingfield made garlands of Midsummer Silver and 'crowded the church and their own Houses with them' (*Natural History of Surrey*, 1718).

It is natural enough to name *Potentilla anserina* after the leaves, which are silver with a felting of long fine hairs; yet 'Silverweed' was probably coined by Henry Lyte in 1578 to match the Dutch *zilverkruid* or the German *Silberkraut*. The older name was Wild Tansy (used by Turner in 1549) since the leaves are like those of *Tanacetum vulgare* in miniature. Gerard makes a curious statement about Silverweed: 'The later Herbarists do call it *Argentina*, of the silver drops that are to be seene in the distilled water thereof when it is put into a glasse, which you shall easely see rowling and tumbling up and down in the bottom'.

6. Marsh Cinquefoil. *Potentilla palustris* (L.) Scop. 107, H 40

Really the most distinguished *Potentilla*, hanging its sultry purple flowers in the tangled vegetation of a bog. But living in out-of-the-way places, it has gathered few names to itself – in the Isle of Man 'Bog-strawberry', in Scotland 'Cowberry', in Donegal 'Bog-berry'.

Gerard was the first to record *Potentilla palustris* for Great Britain and to give it the obvious name of Marsh Cinquefoil. It 'groweth in a marrish ground adjoining to the land called Bourne pondes, halfe a mile from Colchester; from whence I brought some plants for my garden, where they flourish and prosper well'.

7. Barren Strawberry. *Potentilla sterilis* (L.) Garcke. 109, H 40

Local names. CRAISEY (cf. *Ranunculus* species), Wor; LAZY-BONES, Dor; STORYTELLERS, Som; STRAWBERRY PLANT, Dev.

The small and charming liar of the spring, posing as a Strawberry. 'This wilde Strawberrie hath leaves spred upon the grounde ... of a russet greene colour: among which rise up slender stems bearing such flowers as the common Strawberris do, but lesser, which do wither away, leaving behinde a barren or chaffie head, in shape like a Strawberrie, but of no woorth or value.' Gerard's russet green (i.e. grey-green) should really be blue-green. This shade of green distinguishes the Barren Strawberry from the proper wild Strawberry, *Fragaria vesca*, which has leaves of a pure, bright green.

8. Tormentil. *Potentilla erecta* (L.) Raüsch. 112, H 40

Local names. BLOOD-ROOT (cf. Norwegian *blodrod*), N'thum, Scot; EARTBARTH (i.e. earth-bark), Shet; EWE-DAISY, N'thum, Berw; FIVE-FINGERS, Suff; FIVE-FINGER-BLOSSOM, Suff; FIVE-FINGER-GRASS, I o W, Glos; FIVE-LEAVED-GRASS, Bucks, War, Worc, Notts, Lincs; FLESH-AND-BLOOD, Berw; SHEEP'S KNAPPERTY (cf. *Lathyrus montanus*), N Ire; SHEPHERD'S KNOT, N'thum, Scot; SNAKE'S HEAD, STAR, STARFLOWER, Wilts; TORMENTING ROOT, N Ire.

Tormentil is an intriguing word concealing all too prosy a significance. The woody astringent roots were boiled in milk, and the milk was then given to calves and children to bind their loose insides. *Tormentilla* is a mediaeval diminutive, from the Latin *tormentum*. *Tormina* in Latin means colic, which is a torture. So Tormentil, with the two meanings apparently in one word, is the plant for the torture or torment of colic. Tormentil medicine survived a long while in upland or outlying parts of Great

Britain, in Northumberland, for instance, or the Hebrides. The Shetland name Eart-barth (above) explains another ancient use: the roots not only gave a red dye, they could be used in tanning as a substitute for the tanner's 'bark' (which was usually oak bark). In the Western Isles the fishermen tanned their nets with Tormentil roots (166, 128).

Gerard has the odd note that the powdered roots cure diarrhoea ('the laske'), and a bloody flux, especially if they are given 'in the water of a smithes forge, or rather the water wherein hot steele hath been often quenched of purpose'. The Irish have preserved a respect for blacksmiths, and their power to cure diseases and to avert evil. Wearing a smith's apron helps a woman in childbirth; and forge water, among other virtues, will remove warts (137). The Five Finger names seem to have been transferred from *Potentilla reptans*, which is the true Cinquefoil, with leaves divided into five leaflets instead of three, as in *Potentilla erecta*.

9. Creeping Cinquefoil. *Potentilla reptans* L. 109, H 40

Local names. CREEPING JENNY, Som; FIVE-FINGER-BLOSSOMS, Suff; FIVE FINGERS, Som, Suss, Ess; FIVE-FINGER-GRASS, Som, I o W, Glos; GOLDEN BLOSSOM, Dev; ST ANTHONY'S TURNIP, Berks.

Another of the plants of European magic or of supernatural power which was centred in the five leaflets of each palmate leaf, the five fingers. Dioscorides and Theophrastus wrote of the plant Pentaphyllon (*pente*, five; *phyllon*, a leaf) which seems to have been this *Potentilla reptans*. Dioscorides gives the names by which his Pentaphyllon or Pentadactylon (Five-finger) was known to the Gauls (*pempedula*), the Dacians, and the Egyptians, followed by a long list of its virtues in medicine. It was good against malaria, the leaves of four shoots against quartan malaria, of three against the tertian, of one against the quotidian, a piece of medical magic which Gerard denounced as 'most vain and foolish'. In OE *Potentilla reptans* was *fifleaf*, in Dutch it is *vijfvingerkruid*, in German *Fünffingerkraut*, in French *quintefeuille*, in Irish *cúig mhear Mhuire*, 'five fingers of Mary'.

I suspect the English knew of it more from classical and literary sources than from any ancient traditional use of *Potentilla reptans* by themselves. Reginald Scot, in *The discoverie of Witchcraft* (1584), scornfully refers to foreign papists who 'hang in their entries an herb called Pentaphyllon, Cinquefoil', together with hawthorn gathered on May Day, in order 'to be delivered from witches'; and Bacon, in his *Sylva Sylvarum* (1627),

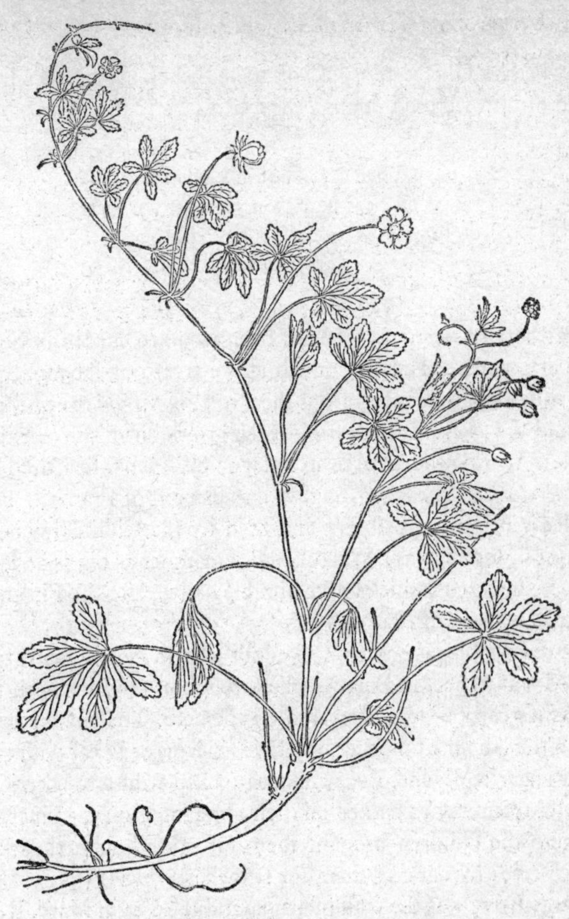

25 Creeping Cinquefoil *Potentilla reptans*

refers to the ointment witches were supposed to employ, reputedly made of 'the fat of children digged out of their graves; of the juices of smallage, wolfbane, and cinquefoil, mingled with the meal of fine wheat'.

Potentilla reptans, the five fingers clearly carved, is frequent among the stone foliage of the chapter-house at Southwell Minster, together with other apotropaic plants, including ivy and hawthorn (see *Crataegus monogyna*); but the thirteenth-century craftsmen, or the master-mason of the chapter-house, may well have been French or have had French experience (143).

10. Strawberry. *Fragaria vesca* L. 112, H 40

The strawberries Gloucester spoke of to the Bishop of Ely when he wished to get rid of him were no doubt *Fragaria vesca*:

> When I was last in Holborn,
> I saw good strawberries in your garden there;
> I do beseech you send for some of them.
> (*King Richard III*, III, iv. 33–5.)

Wild strawberries were transplanted from wood to garden in September or February, and then treated much as we treat our fat modern strawberries (which are descended from the *Fragaria virginiana* of the eastern U.S.A. and the *Fragaria chiloensis*, which grows down the maritime edge of the New World from Alaska to Patagonia). Even when the Hautbois, *Fragaria elatior*, had reached English gardens out of France, followed by the scarlet Virginian strawberry and then by the Chili Strawberry with fruit 'generally as large as a Walnut, and sometimes as big as an Hen-egg', people were inclined to stick to the humble native, 'for the Firmness of its Fruit, and Delicacy of Flavour' (130).

Not unreasonable, as anyone will admit who lives in a good strawberry district – for instance, in Surrey, where rough ground between the hawthorn bushes can be strawberry-red by the acre. The small strawberries are worth the labour of picking, whether for jam, or to eat with sugar and cream or sugar and wine. Poor children picked wild strawberries for the market. John Aubrey remarked on their abundance on the limestone near Bath where the children brought them for sale, though they killed the young ash trees by barking them for strawberry boxes (7).

The strawberry was the fruit of Venus and the Virgin Mary. Red strawberries litter the centre panel of that strange painting by Hieronymus Bosch which hangs in the Prado and is known as *The Garden of Earthly Delights*. Bosch (who died in 1516) depicts naked men and naked girls weaving in a medley of enigmatic and gently erotic actions. At one point a naked man offers a naked girl a huge ripe strawberry double the size of his own head. The allegory has been interpreted as an essay in paradisaical eroticism, painted for a sect of mystical Adamites (77).

Strawberry is one of the puzzle words – in OE *streawberige*, the berry to do with straw. It has been said that the runners look like straws, which they do not; that the plants are strewn over the ground, which they are, though it does not appear reason enough for the name. The straw could refer to manuring the cultivated plants with straw dung. More likely *strēaw* in this name had its OE meaning of chaff, indicating the chaff-like achenes all over a strawberry.

11. Herb Bennet, Wood Avens. *Geum urbanum* L. 112, H 40

Local names. BLESSED HERB, COLEWORT, GOLD STAR, Som; RAM'S FOOT, Dev; YELLOW STRAWBERRY, Som.

Knowledge of this common species seems to have come to us from France in the Middle Ages. Herb Bennet and Avens are both names from the Old French – Herb Bennet (in modern French *benoît des villes*, or *herbe de Saint-Benoît*) from *herbe beneite*, not the herb of St Benedict, but the 'blessed herb', the mediaeval Latin *herba benedicta*. It was marked less by appearance than by the sweet, spicy smell of the roots, the clove smell, which gave it the old shop name of *Caryophyllata* and the German name of *Nelkenwurz*, 'clove root'. Fragrance repels evil, so the *Ortus Sanitatis*, published at Mainz in 1491, says of Herb Bennet that if the root is in the house the devil is powerless and flees from it, 'so it is blessed above all other herbs'.

It was grown as a pot herb in sixteenth-century gardens (Gerard calls it 'the common garden Avens'), and was boiled in pottage or broth or decocted in wine against an upset stomach, wind, stitch, and 'the biting of venomous beasts'. The roots repelled moths, so it was thought, as well as the devil – 'The rootes taken up in Autumne and dried, do keep garments from being eaten with Mothes, and make them to have an excellent good odour' (81).

12. Water Avens. *Geum rivale* L. 104, H 35

Local names. BILL BUTTON, BILLY'S BUTTON, Wilts; COCK-AND-HENS, N'thum; EGYPTIAN, Wilts; FAIRIES' BATH, GRANNY'S BONNET, Dor; GRANNY'S CAP, GRANNY'S NIGHTCAP, OLD WOMAN'S BONNET, Wilts; LONDON-BASKET, Yks; SOLDIER'S BUTTONS, Wilts.

An Egyptian, which is to say a gipsy, among the flowers of the damp roadsides and runnels. The colour of the petals is not easy to describe. Bentham and Hooker called it a 'dull purplish colour with a tint of orange'. The *Flora of the British Isles*, with more precision, describes the petals as 'dull orange-pink', and adds that the calyx is purple. These drooping, swarthy Egyptian blossoms give the plant a character, a personality strong like that of the purple-flowered Marsh Cinquefoil or the Fritillary.

A puzzle that so striking a species, southern as well as northern, should have been overlooked by our early botanists – by William Turner in his native county of Northumberland, for example. It was recorded first by the London apothecary and botanist, Thomas Johnson, in his revised edition of Gerard's *Herbal*, which he published in 1633 – 'the red floured mountaine Avens was found growing in Wales by my much honoured friend, Mr. Thomas Glynn, who sent some plants thereof to our Herbarists in whose gardens it thriveth exceedingly'.

13. Mountain Avens. *Dryas octopetala* L. 19, H 10

For the full enjoyment of this lovely and with us rare arctic species, there is no better place than the bizarre district of Burren in Co. Clare. *Dryas* toughly carpets the limestone pavements and terraces, hangs its broad, white, golden-centred flowers and dark green leaves over the ledges, and creeps down nearly to the Atlantic – acre upon acre of rarity among other rare plants, in a natural rock garden. The time to go is the second half of May, when the Irish clouds are scarce (perhaps) and you can stand among the *Dryas*, a blue sky overhead, a blue Atlantic underneath.

It grows further north on the opposite side of Galway Bay, and there it was first discovered late in the seventeenth century by the cleric and botanist, Richard Heaton, who sent the news to William How. How printed it in his *Phytologia Britannica* in 1650, noting another of the charms of *Dryas* – 'it makes a pretty show in the winter with his rough heads like Viorna' (which is Old Man's Beard). Like Old Man's Beard and like the Pasque Flower, *Dryas* develops long, silky, wriggling awns from each carpel, which are nearly as conspicuous as the flowers.

No plant so much needs an apt English name. In Co. Clare they have called it Wild Betony from the close resemblance of Betony leaves to *Dryas* leaves. But this is a poor name, like Mountain Avens. Bentham and Hooker preferred to anglicize the generic name. They called it White Dryas, which is better than the other names, though in its origin it is a little far-fetched. A leaf of *Dryas* also resembles an oak leaf, so it was named after the dryad, the wood-nymph of the oaks. Really it requires a name suggestive of gold, whiteness, and open sunlight.

14. Agrimony. *Agrimonia eupatoria* L. 108, H 40

Local names. AARON'S ROD, Som; CHURCH-STEEPLES, Som, Suss; CLOT-BUR, COCKLE-BUR, Som; FAIRY'S WAND, GOLDEN ROD, Dor; LEMONADE, LEMON-FLOWER, Som; MONEY-IN-BOTH-POCKETS, Dor; RAT'S TAILS, Wilts, Dur; SALT-AND-PEPPER, Corn; SWEETHEARTS (from the clinging receptacles of the fruit, cf. *Galium aparine*); TEA-PLANT, Som. In Hants the fruits have been called Harvest-lice.

Agrimony has been valued for many centuries, backed by the authority of the ancients, including Dioscorides and Pliny. Dioscorides recommends

26 Agrimony *Agrimonia eupatoria*

his *Eupatorion* against snake-bite, dysentery, and upsets of the liver, so the Anglo-Saxons made a snake-bite salve out of Agrimony, Bistort, and Plantain. The doctor was told to confine the poison by ringing the bite with Agrimony, which was also one of the fifty-seven herbs in the Anglo-Saxon 'Holy Salve' against goblins, evil, and poisons (42, 88). The 'subtile and fine partes' of Agrimony became general knowledge, and for its astringent qualities it is one of the plants still popular in cottage medicine. In the *Gardeners Dictionary*, in the eighteenth century, Miller stated that Agrimony, though 'common in the Hedges in many Parts of *England*', should not 'be wanting in a Garden', adding that some people preferred the sweet-smelling Agrimony (*Agrimonia odorata*) for 'pectoral decoctions' and 'a pleasant sort of tea'. The Somerset names of Lemonade and Lemonade Flower refer to Agrimony wine, made (for colds) with lemons, oranges, ginger, sugar, and Agrimony gathered in flower.

For the name Aaron's Rod (also given to *Solidago virgaurea* and *Verbascum thapsus*) turn to Numbers xvii, and the story of the rods laid up in the tabernacle, among which 'the rod of Aaron for the house of Levi was budded, and brought forth buds, and bloomed blossoms, and yielded almonds'. An apt name for a plant with rod-like racemes, and perhaps a name which offset less desirable associations indicated in the Dorset name of Fairy's Rod. It may have been used by the powers of evil and against them, as in the Holy Salve of Anglo-Saxon medicine-men. That other Aaron's Rod, *Verbascum thapsus*, was certainly believed to have supernatural power. In France Agrimony was sanctified as the *herbe de Saint-Guillaume*.

Agrimony itself, *agrimonie* in OE, *aigremoine* in French, and *Odermennig* in Germany, comes from Pliny's *agrimonia*. A native OE name, *garclife*, was superseded.

15. Lady's Mantle. *Alchemilla vulgaris* L. 109, H 40

Local names. BEAR'S FOOT, Hants, N'thum; DEW-CUP, Scot; DUCK'S FOOT, Dur, Berw; ELF-SHOT, Kirk, Wigt; LAMB'S FOOT, Lancs; LION'S FOOT, Som.

'In the night it closeth it selfe together lyke a purse, and in the morning it is found ful of dewe' – so William Turner described *Alchemilla vulgaris* in his *Herbal*, in 1568. John Parkinson improved upon Turner, when he explained the name Syndaw which Turner seems to have adopted into English from the German *Sinnau*, which means 'ever dew'; in Low

German the plant is called *Immertau*, 'because the hollow crumplings and the edges also of the leaves, will containe the dew in droppes like pearles, that falleth in the night' (*Theatrum Botanicum*, 1640). Dew was a powerful substance, and *Alchemilla vulgaris* a powerful, magic-working plant, with a reputation independent of the classical authors, such as Pliny. In his *Names of Herbes* (1548) Turner included it among 'newe founde Herbes wherof is no mention in any olde auncient wryter'. The German

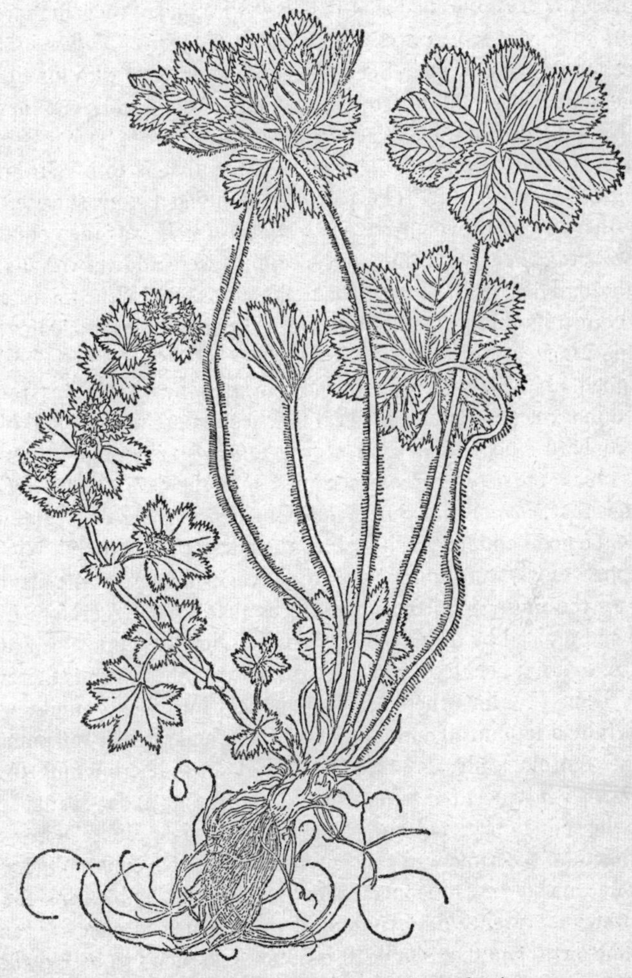

27 Lady's Mantle *Alchemilla vulgaris*

Hieronymus Bock had devised the name *Alchemilla* early in the sixteenth century. It was a coinage like *Potentilla*, the 'little powerful one', meaning, one may suppose, the 'little alchemical one', the 'little magical one'.

A plant of old northern magic, from the pearls of dew and perhaps also from the division of the leaf into nine serrated lobes (for which it was called Nine Hooks), *Alchemilla vulgaris* must have been new-found only for the botanists. In the Highlands (the Gaelic name is *copan an druichd*, 'dew cup') and in Ireland it was used to cure elf-shotten animals. Your cow fell sick – the fairies had shot the beast with one of their flint arrow-heads. The business was well described in the *Irish Naturalist* (xiii, 219–20, 1904), from a case in Sligo. The elf-doctor came with his elf-bag, which contained several flint arrow-heads, a silver coin, and some coppers. When he made sure of the cow's illness – that it had been 'struck' – a pail of 'three-mearing water' was fetched – water, that is to say, from the meeting of three parishes. The pail needed filling before sunrise. The juice of herbs, one of which was *Alchemilla vulgaris*, was then squeezed into the water, the silver coin and the coppers were added, with a single arrow-head. The cow was made to take three sips, or else the mixture was bottled down its throat in three doses. All the rest was sprinkled over its back and ears.

Stripped of the elf-doctor's magic, *Alchemilla vulgaris* was recommended in sixteenth-century medical practice against wounds and bleeding. Leonhard Fuchs, in his *De historia stirpium*, mentioned a curious virtue which the later herbalists copied into their volumes: Gerard's version is that *Alchemilla* 'keepeth down maidens paps or dugs, and when they be too great and flaggie, it maketh them lesser and harder'. Nicholas Culpepper, too, is worth quoting for the developed usage of the plant: 'It is one of the most singular wound-Herbs that is, and therefore highly prized and praised by the *Germans*, who use it in all Wounds inward and outward, to drink the Decoction thereof, and wash the Wounds therewith, or dip Tents (i.e. lint) therein, and put them into the Wounds, which wonderfully drieth up all humidity of the sores, and abateth Inflammation therein.' Astrologically it was a plant of Venus. Hieronymus Braunschweig had instructed that one should 'plucke or gader planta leonis before the sonne uprysynge and stampe that and dystylle it'. Then 'who so drynketh of the same water incontynent he shal have lust to the worke of generacyon' (17). Culpepper wrote that women should drink the distilled water for twenty days running if they wished to conceive, and to retain the birth; and they should also sit now and then in a bath made of a decoction of the herb (49). Originally it must have averted evil and evil beings at this dangerous moment in a woman's life. Aphrodisiac also for cows, a French name for it is *herbe à la vache* (160).

So powerful a plant needed a Christian aura. The Germans called it *Frauenmantel* or *Marienmantel* (from the leaf), which gave William Turner 'Our Ladies Mantel'. In Welsh it became *mantel Fair*, in Irish *falaing Muire*, in Norwegian *Marikaabe* (all 'Mary's Mantle'). In Irish the plant is also *bratog Muire*, 'Mary's Rag', and *dearna Muire*, 'Mary's Palm', i.e. the palm of her hand, again from the shape of the leaves, which was responsible for the old shop name of *Pes leonis*, 'Lion's foot'.

16. Parsley Piert. *Aphanes arvensis* L. 112, H 40

Local names. BOWEL-HIVE-GRASS, Berw; BREAKSTONE-PARSLEY, Staff; COLICWORT, Heref; PARSLEY-BREAKSTONE, Suff, War, Scot, Ork; PARSLEY-VLIX, Dor.

Mathias de l'Obel, the great Flemish botanist who settled and died in England, found this common little plant at Bristol, and described it in his *Stirpium adversaria nova* (1570) as the *Percepier Anglorum*, the 'percepier' of the English. A *perce-pierre* is a plant which pierces or thrusts its way through stony ground, and so, by sympathy, it is a plant to be employed against stone in the kidneys or bladder. De l'Obel observed that *Aphanes arvensis* was hardly known to the herbalists, although it was much used by the poor women of Bristol. No doubt it was used all over the country, and not only against the stone. It is Bowel-hive-grass in the north, since it was held to cure bowel-hive or inflammation of the bowels in children (cf. the Herefordshire name of Colicwort). On Colonsay, in the Hebrides, it was eaten either raw or pickled (128); and in his *English Physitian Enlarged*, Culpepper had remarked that it was a good salad herb and that 'the Gentry' ought 'to pickle it up as they pickle up Samphire for their use all the Winter'. The implication is that the gentry ought to learn the practice from the common people.

Presumably *Aphanes arvensis* was known as Parsley-breakstone before de l'Obel's botanical discovery of the plant: the multitude of its cut leaves give it some faint resemblance to a miniature of the garden parsley, which was also prescribed for the stone.* Thomas Johnson, in his revision of Gerard's *Herbal* (1633), says that the herb-women who came into market

* So was the rock-growing Samphire, which Culpepper mentioned. *Perce-pierre* is one of the French names for Samphire. Parsley was the *petroselinon* of Dioscorides, the '*selinon* of stony places'.

at Cheapside called it Parsley Breakstone. Parsley Piert, though, as Parkinson stated in *Theatrum Botanicum* seven years later, appears to be a corruption of the *percepier* which de l'Obel had applied to *Aphanes arvensis*, a mix-up of *perce* and the parsley of Parsley-breakstone. A hundred years before de l'Opel wrote of *Aphanes arvensis*, William Worcestre (who came from Bristol) made a note in his *Itineraries* (1478–80) that the plant called in English 'Perspeere' grew close to the ground in Cotswold country and was also to be found on the walls of Dover Castle. He had been told how to use it against the stone, powdered, with roots of garden parsley and ivy berries (*William Worcestre*, *Itineraries*, ed. J. H. Harvey, 1969, pp. 368–9).

17. Great Burnet. *Sanguisorba officinalis* L. 70, H 4

Local names. DRUM-STICKS, Som, Glos; MAIDEN'S HEADS, Yks; PARASOL, Wilts; RED HEADS, Yks; RED KNOBS, Notts.

Like the Salad Burnet (*Poterium sanguisorba*), the Great Burnet was a vulnerary and a stauncher of blood, perhaps from the heads of dark crimson or mahogany flowers, which make the plant so much worth finding in some old meadow of heavy clay. 'Burnet' is 'brunette', dark brown, Old French *burnette* or *brunette*.

Gerard collected *Sanguisorba officinalis* from a part of London long covered with bricks and road metal – 'myselfe have founde it growing upon the side of a cawsey which crosseth the one half of a fielde, whereof the one part is earable grounde, and the other part medowe, lying between Paddington and Lysson greene, near unto London'.

18. Salad Burnet. *Poterium sanguisorba* L. 76, H 21

Local names. DRUM-STICKS, OLD MAN'S PEPPER, Som; POOR MAN'S PEPPER, Dor; RED KNOBS, Notts.

Though not so often grown in the gardens, Salad Burnet still holds its reputation – the cucumbery leaves in salad or in summer drinks. In salads 'it is thought to make the hart merry and glad, as also being put in wine, to which it yeeldeth a certaine grace in the drinking'. Gerard distinguished the two Burnets by calling this species the Garden Burnet, and *Sanguisorba officinalis* the Wild Burnet. Both were good against wounds, since they were identified with a vulnerary plant mentioned by Dioscorides.

19. Wild Rose, Dog Rose. *Rosa canina* L., etc.

Local names for the fruit include: PUCKIES, Aber, Ire; BUCKIE-BERRIES, N Ire; CANKERS, Dor, Ess, Norf, Cam; CANKER-BERRIES, Kent; CAT-CHOOPS, Cumb; CAT-JUGS, Yks, Dur; CHOOPS (cf. Norwegian *kjupa*), Yks, Dur, Cumb, Rox, Dumf, Ayr; DOG-BERRIES, Hants, Yks; DOG-CHOOPS, Yks; DOG-HIPS, Scot; DOG-HIPPENS, Aber; DOG-JOBS, Yks; DOG-JUMPS, Yks; HAGISSES (cf. *Crataegus monogyna*), Hants; HAWS (cf. *Crataegus*), Dor; HAWPS, N Scot; HEDGE-SPEAKS, Glos; HEDGY-PEDGIES, Wilts; HIPS (OE *heope*), general from Wilts to Scot, in various forms, Hep, Epp, etc.; HIPPANS, Mor; DOGS HIPPANS, Aber; HIPSONS, Oxf; HUGGANS, Yks; ITCHING-BERRIES, Lancs; NIPPERNAILS, NIPS (cf. Norwegian *nypen*), Ches; PIG'S NOSES, Dev; PIXIE PEARS (cf. *Crataegus monogyna*), Dev, Hants; RED BERRIES, Yks; SOLDIERS, Kent.

The seeds, which children put down each other's necks to produce itching, are BUCKIE-LICE, S Scot, Ire; COW-ITCHES, Ches; TICKLERS and TICKLING TOMMIES, Dev.

The plant-names include: BRIAR (which is a prickly shrub, particularly a wild rose; and mainly northern – 'breer' frequently), N'hants, Worc, Shrop, Ches, Derb, Notts, Rut, Lincs, Lancs, Yks, Dur, West, Cumb, N'thum, Scot, Ire; BRIAR-ROSE, Dur, N'thum, Berw, Rox; BRID-BRIAR (i.e. bird-briar), Ches; BRIMMLE, Shrop; BUCK-BREER, N Ire; BUCKY, N Ire; CANKER, Dev, Dor, Som, Ess, Norf, Lincs, Cumb; CANKER-ROSE, Dev, Som, Kent; CAT-ROSE, Ches; CAT-WHIN, Yks, N Eng; CHOOP-ROSE, CHOOP-TREE, Cumb; COCK-BRAMBLE, Suff.

DIKE-ROSE (i.e. hedge-rose), Cumb; DOG-BREER, Yks; DOG'S BRIAR, Hants; HIP-BRIAR, Shrop; HIP-ROSE, Glos; HIP-TREE, Glos, N'thum; HORSE-BRAMBLE, E Ang; HUMACK, Som; KLONGER or KLUNGER (Old Norse *klung*), Shet; LAWYERS, Surr, War; NEDDY-GRINNEL, Worc; PIG-ROSE, Corn; PIG'S ROSE, Dev; ROE-BRIAR, Som; YOE-BRIMLE (i.e. ewe bramble), Dev, Som.

Roses have been so anciently cultivated that the wild roses were overshadowed or outshone. 'I had rather be a canker in a hedge than a rose in his grace,' John the Bastard remarked of his brother, the prince, in *Much Ado about Nothing* (I. iii), and Gerard, after dealing at length with the garden rose, which 'doth deserve the chiefest and most principall place among all flowers whatsoever, being not only esteemed for his beautie, vertues, and his fragrant and odoriferous smell; but also bicause it is the honor and ornament of our English Scepter', disposes quickly of the wild kinds – 'The faculties of these wilde Roses are referred to the manured Rose, but not used in Phisicke where the other may be had.'

Roses in the garden were equally there for physic and delight. At the time of the tulip mania, Thomas Fuller put a speech into the mouth of the rose at a solemn rendezvous of flowers and herbs. The Rose complained that he was being ousted by the 'toolip', 'neglected and contemned, and conceived beneath the honour of noble hands, and fit only to grow in the Gardens of Yeomen'. He had the precedency of all flowers under the patent of colours and scent, but he relied also on his virtues – 'Yea, when dead, I am more sovereign than living. What cordials are made of my syrups! How many corrupted Lungs (those fans of Nature), sore wasted with consumption . . . are with conserves made of my stamped leaves restored to their former soundness againe?' – whereas the toolip was no more than 'a well complexioned stink, an ill savour wrapt up in pleasant colours; as for the use thereof in Physick, no Physitian hath honoured it yet with the mention, nor with a Greek or Latin name, so inconsiderable hath it hitherto been accompted' (*Antheologia, or the Speech of Flowers*, 1660).

Garden roses were used against a hundred ills, from St Anthony's Fire to the French pox; but medicine had recourse to the wild roses for one thing – the reddish-yellow bedeguar or Robin's pincushion, the gall made by the gall wasp *Rhodites rosae*. These 'briar balls' were sold by the apothecaries. They were powdered and a decoction of the powder was taken to break the stone, as a diuretic, and for colic. Each briar ball contains the larva of *Rhodites rosae* and the larvae of various predators and parasites. These were no deterrent but an additional medicine: 'In the middle of the Balls are often found certain white worms, which being dried and made into powder, and some of it drunk, is found by Experience of many, to kill and drive forth the Worms of the Belly' (49). English country people have also hung bedeguars round their necks as an amulet against whooping-cough (72).

The opprobrious 'Dog-rose' first occurs in Gerard's *Herbal* to distinguish wild from garden roses. By way of the mediaeval Latin *Rosa canina*, it goes back to Pliny's *cynorrodon*, with the root of which a dog-bitten soldier of the Praetorian Guard cured himself of hydrophobia. And Pliny took *cynorrodon* from Theophrastus.

20. Burnet Rose. *Rosa spinosissima* L. 96, H 39

Local names. BURROW-ROSE, Pemb; BRID-ROSE (i.e. bird-rose), Ches; CANT-ROBIN, Fife; CAT-HEP, Berw, N Eng; CAT-ROSE, Yks; CAT-WHIN, Yks, N'thum; FOX-ROSE, War; SOLDIER'S BUTTONS, Kirk; ST DAVID'S ROSE, Pemb.

The rose with leaves like the Salad Burnet – 'leaves', wrote Gerard, 'consisting of divers small ones, set upon a middle rib like those of Burnet, which is called in Latine *Pimpinella*, whereupon it was called *Rosa Pimpinella*, the Burnet Rose'. Gerard, who gives the first English record of the Burnet Rose, points also to the fruit, the only ones among the British wild roses which are purple-black instead of scarlet or red.

The one name which makes a true distinction is Burrow-rose, i.e. the rose of Pembrokeshire sand dunes or burrows. *Rosa spinosissima* is characteristic of dunes. As St David's Rose, it is the emblem of the See of St David's.

21. Blackthorn. *Prunus spinosa* L. 108, H 40

Among the fruit names, Sloe is from the OE *slā(h)*, with Slon or Slan (Cornwall to the Midlands) as modern singulars from the ME plural *slon* and the OE plural *slan*. Slaa in the north of England is from the equivalent word in Old Scandinavian. Also, BULLENS, Shrop; HEG-PEGS, Glos; HEDGE-PICKS, Hants; HEDGE-SPEAKS, Wilts; SLAGS, Oxf; SNAGS, Dor, Som, Wilts, Hants; WINTER KECKSIES, I o W; WINTER PICKS, Suss.

Among the plant names: BLACKTHORN (general), BLACKHAW, Ire; BUCKTHORN, Lincs; BULLEN, Shrop; BULLISON, Wilts; BULLISTER (i.e. bullace tree), Cumb, Scot, Ire; EGG-PEG-BUSH, Glos; PIG-IN-THE-HEDGE, Hants; SCROGG, Notts, N Eng; SLAATHORN and SLAA-TREE, Yks, Cumb; SLACENBUSH, N'hants; SLON-BUSH, N'hants, Leic; SLON-TREE, Corn, Som, Leic; SNAG-BUSH, Dor, Som, Wilts, Hants.

The Whitethorn (q.v.) has gathered to itself a host of beliefs, because it blossoms in May, in the critical month. The Blackthorn blossoms bravely and too early. Yet it has also been something of a magical tree, both dangerous and powerful against other dangers. The blossoms, like Whitethorn blossom, are unlucky in the house, and unlucky in the buttonhole; but the shoots (again like Whitethorn, or Rowan) are powerful for walking-sticks and for the knotty shillelaghs of Ireland. The wry sloes go into the bottle for sloe gin, or are made into sloe wine. Blackthorn leaves have been made into tea, the blossoms and the juice of the sloes have been used in country medicine.

The herbalists gave the cue. 'The juice of Sloes do stop the belly, the lashe and bloody flixe, the inordinate course of womens termes, and all other issues of blood in man or woman, and may very well be used in stead of Acatia, which is a thornie tree growing in Egypt, very hard to be gotten, and of a deere price, and therefore the better for wantons; albeit our Plums of this countrie are equall unto it in vertues' (80). But the Blackthorn has a greater virtue. It is probably the ancestor, or one of the two ancestors, of all the luscious plums of the garden, including the Greengage. 'Our European plum *Pyrus domestica* has probably arisen on many different occasions in prehistoric and historic times from hybridization between diploid and tetraploid species, followed by chromosome duplication to give the fertile hexaploid *P. domestica*.' The diploid species, English and Russian workers believe, is *Prunus cerasifera*, the Cherry-plum, from Asia Minor, Persia, and Russian Central Asia, and the tetraploid species, *Prunus spinosa* (48).

22. Bullace. *Prunus domestica* ssp. *insititia* (L.) Poiret. 73, H 34

Bullace is either the tree or the fruit, probably from the Old French *beloce*. Other fruit names are BULLUMS, Corn; CHRISTIANS, Som; CRICKSEYS, Ess, Norf, Cam; CRYSTALS, Corn, Dev; KESLINGS, Dev; SLATHS, Lancs.

Tree or shrub names include (besides Bullace): BULLISON, Wilts; BULLUM-TREE, Corn; CREX, Cumb; SCAD and SCAD-TREE, Suss, Kent, Lincs; WILD DAMSON, Yks; WILD PLUM, Yks.

In its day the Bullace has been a cultivated plum, which we had, no doubt, by way of France in the Middle Ages. As better plums have been raised, the Bullace more and more has been abandoned, clinging to orchard hedges, old gardens, undisturbed corners, naturalizing itself in neglect. But country people still recognize the Bullace, the fruits are still collected; and Bullace wine is made now and again. Bullace wine, Damson wine, and Sloe wine, kept long enough, are all said to resemble port in taste as well as in colour.

23. Gean, Wild Cherry. *Prunus avium* L. 105, H 33

Local names. CRAB-CHERRY, Bucks; GASKIN, Som, Suss, Kent; GEAN, Suss, Dur, Cumb, N'thum, Scot; HAWKBERRY, Stir; MAZZARD, Corn, Dev, Dor, Som, Wilts, Glos, Kent, Worc, Ire; BRANDY-MAZZARD, Dev; MERRY, Dor, Som, Wilts, Hants, I o W, Berks, Bucks, Oxf, Hunts, Shrop, Ches,

Derb, Lancs, West; MERRY-TREE, Dor, Hants, Suss, Shrop, Ches, Shet, N Eng.

The fruits are Mazzards and Merries, and give their name to the tree.

A native tree, for which we use more than one French name. The north country Gean is *guigne* in modern French (a sweet cherry). Gaskin is Gascoine, from the Gascoine Cherry-tree of Elizabethan orchards. Merry is the older Merise of the sixteenth century, from the French *merise* (a wild cherry). Cherry itself comes to us from the Old French *cherise*, which goes back to the Greek *kerasos*, which goes back a stage further to the Assyrian *karsu*; and so back towards the origin of the cherries in cultivation, since they descend from the wild *Prunus avium*.

If you number the counties, it is clear that Gean has pushed itself too boldly into English books. We ought to call *Prunus avium* either Mazzard, a name of obscure origin, or Merry. Mazzard is too good a name to be occluded. Wild Cherries, or Mazzards, as John Evelyn wrote, 'thrive into stately trees, beautified with blossoms of a surprising whiteness, greatly relieving the sedulous bees'.

A tree of virtuous blossom, virtuous timber, and rather less virtuous fruit, except for the purposes of cherry brandy. The wild fruit used to be sold around the streets of London, on the branch. Gerard was not in favour of them – 'the best and principall Cherries be those that are somewhat sower: those little sweete ones, which be wilde and soonest ripe, be the woorst' – but for a special reason: 'They conteine bad juice, they very soone putrifie, and do ingender ill bloud, by reason whereof they do not onely breede woormes in the belly, but troublesome agues, and often pestilent fevers: and therefore in well governed common wealthes it is carefully provided, that they should not be sold in the markets in the plague time.'

Cherry-tree gum, which children suck and wish they could find a use for, is an old medicine which Dioscorides recommended for coughs, and a good complexion, a good appetite, and keen sight. Dissolved in wine, European herbalists believed that the gum helped to break up and expel the stone.

24. Bird-cherry. *Prunus padus* L. 69, H 32

Local names. BLACK DOGWOOD, Surr; BLACK MERRY (black from the fruit), Hants; HACKBERRY, West, Cumb, Berw, Rox, Dumf, Perth; HACKWOOD, West, Cumb, N'thum; HAGBERRY, Hants, Lancs, Yks, Dur, West, Cumb,

N'thum, Scot; HAWKBERRY, Stir; HECKBERRY, Yks, Cumb; HEGBERRY, Cumb; MAZAR-TREE, Dev; MAZZARD, Lincs. (For Mazar, Mazzard, Merry, see *Prunus avium.*) The fruit in the north are HACKERS, HACKS, HAGBERRIES; in Hants they are HOG-BERRIES.

Bird-cherry belongs mainly to the north, coming down to Gloucester and South Wales. To southern eyes (though it has been planted here and there in the south), it is the prettiest of cherries, hanging its racemes of clear white blossom in unfamiliar surroundings – the limestone canyons of Ribbledale, for instance, or the Peak, or of that peculiar district of gorges and waterfalls behind Neath in South Wales. A useless little tree, with black and wry fruit, rejoicing, as it should, in a name from the Old Scandinavian. The Old Scandinavian *heggr* gives Hag(berry), Hack-(berry), etc., in the north, and *hegg* in Norwegian.

25. Hawthorn, Whitethorn, May-tree. *Crataegus monogyna* Jacq. 112, H 40

The OE for the species is *hagathorn*, *haegthorn*. The name is similar in German, Dutch, Swedish, Danish, and Norwegian, and is usually taken to mean the thorn which produces the haw or fruit, OE *haga*.

Local names. AGLET-TREE, Corn; AZZY-TREE, Bucks; BREAD-AND-CHEESE TREE, Som; CUCKOO'S BREAD-AND-CHEESE TREE, Suss, Leic; MAY BREAD-AND-CHEESE BUSH, Som, Hants, E Ang; MAY BREAD-AND-CHEESE TREE, Norf, Lincs; CHEESE-AND-BREAD TREE, Yks (all from the habit, still common among children, of nibbling the Bread and Cheese, or young leaves).

HAG, Lancs; HAG-BUSH, Yks; HAGTHORN, Dev, Som, Yks; HAG-TREE, Dev, Som; HAW, Dumf, Selk; HAW-BUSH, Dumf; HAW-TREE, Glos, N'thum, Dumf, Loth, Lanark, Ayr; HEG-PEG BUSH, Glos; HIPPERTY-HAW TREE, Shrop; HOLY INNOCENTS, Wilts; MAHAW, Ire.

MAY (chiefly the flowers, or when flowering), Corn, Dev, Som, Wilts, Hants, Glos, Suss, Ess, Herts, Suff, Camb, War, Yks; MAY-BUSH, Hants, Norf; MAY-FLOWER, Som, Cumb; MAY-TREE, Lincs; MOON-FLOWER, Som; PEGGALL-BUSH, Wilts; QUICK (especially when used for hedges), Som, Norf, Shrop, Ches, Lincs, Ire; QUICKTHORN, Lancs, Yks; SCROG, SCROG-BUSH, Ire; SHIGGY, SKAYUG, SKEEOG, SGEACH (name for isolated 'fairy thorns', from Irish *sceach*, bush, bramble), Ire; WHITETHORN (in contrast to Blackthorn, *Prunus spinosa*), Corn, Dev, War, Ches, Yks.

The fruit names include AGALD, Wilts; AGARVE, Suss; AGASSE, Hants,

Suss; AGGLE, Dev; AGLET or EGLET, Corn, Dev, Som; AGOG, Berks; BIRD'S EEGLE, Ches; BIRD'S MEAT, Som; CHAW, Mor; CUCKOO'S BEADS, Shrop; HAG, Dev, Som, Hants, I o W, Suss, Kent, Berks, Bucks, Herts, Ches, Derb, Lancs, Yks; HAGGIL, Hants; HALVE, Dev; HARVE, HARSY, Ess; HAW, especially in the north and Ire (also CAT-HAW, N'hants, Lincs, Yks, West, Cumb; HAW-BERRY, Ches, Scot; HAW-GAW, Surr; HIP-HAW, Oxf, Lakes; BUTTER-HAW, Oxf, Norf; HIPPERTY-HAW, Oxf, Shrop; HOG-HAW, Hants, S Eng; PIG-HAW, Som, Wilts, Hants, E Eng).

HAZLE, Dev; HEETHEN-BERRY, Ches; HEG-PEG, Glos; HOGAIL, I o W; HOGARVE, Suss, Surr; HOGAZEL, Suss; HOG-BERRY, Hants; HOGGAN, Corn;

28 Hawthorn *Crataegus monogyna*

HOG-GOSSE, Suss; HOG-HAGHES, Hants; MAY-FRUIT, Yks; PIGALL, Som, Wilts, Hants, West; PIG-BERRY, Wilts (these pig names rise from seeing 'pig' in 'hog', from the OE *haga*); PIXY-PEAR (cf. *Rosa canina*), Dor, Som.

The flowers look white, clean, and delicious over the landscape, and smell delicious (if you do not get too strong a whiff). In the autumn no one can miss the red hags or haws shining after rain, glistening along the road in the car headlights. No doubt because of the scent, the thorns, the white flowers, and because it blossoms in May, and is conspicuous and common everywhere (on soils with or without lime), the Hawthorn became above most plants of the far west of Europe a supernatural tree. It has been a lover's tree in poetry, and for everyone it still symbolizes the change from spring to summer. It is only one of the plants of May Day, which opens the most significant month for vegetation, it is only one of the plants collected when 'the May' was brought into villages and towns, yet Hawthorn ends as the chief of them all, as the May or May-tree. Gerard, after giving the English names as Hawthorn and Whitethorn, adds that Londoners call it May-bush.

Irish customs of May Day, or *lá bealtaine*, have remained intact long enough to give an idea of the old significance of the festival in other parts of Great Britain. It is a festival of vegetation and farming, precisely of the bringing in of summer, though we remember chiefly the May Queen, the May Lord, the Maypole, and the dancing. By Irish belief the sun rose extra early on May Day. Through the Eve, and on the day itself (indeed, through the whole month), fairies and witches were active, excited no less than humans by the new summer. Milk, butter, and all 'profit', or farm produce, were liable to be stolen or bewitched. Protection was called for by plants with a powerful *mana* such as Rowan-tree or Quicken (*Sorbus aucuparia*) and that other May-flower, the Marsh Marigold (*Caltha palustris*). Collectively, the Irish call the plants which they bring in 'Summer', or *an Samhradh*, as we in England call them the May (137). In Ireland, also in the Highlands and Islands, the Rowan-tree was the surest of protectors on May Eve and May Day; whereas in England, and in France, the emphatic plant was the Hawthorn, which was among the plants put around the Maypole when it was carried in from the woods (170). In fact, reverence for the Hawthorn is part of our French, not of our Teutonic, inheritance.

Hawthorn spoke of sex and fertility, which needed protection so much at this critical time. Robert Herrick, in his poem *Corinna Going a-Maying* (from *Hesperides*, published in 1648), mentions no other plant – only

Hawthorn or Whitethorn. 'Come, my Corinna, come' (she was late out of bed on May morning) –

and coming, mark
How each field turns a street, each street a park
Made green, and trimm'd with trees: see how
Devotion gives each house a bough,
Or branch: each porch, each door, ere this,
An ark, a tabernacle is,

Made up of whitethorn neatly enterwove. . . .

The interweaving was important, since the magic of magic plants was always increased by weaving them into shapes, coronets, crosses, etc. The magic of the Hawthorn and the other May Day plants had already been increased during the night by the dew, which was always a magic liquid (cf. dew on the plants of St John, pp. 85, 168-9, 207).

May and May Day being propitious for engagements, though not for marriage itself, Herrick goes on:

And some have wept, and woo'd, and plighted troth,
And chose their priest, ere we can cast off sloth:
Many a green gown has been given;
Many a kiss both odd and even;
Many a glance too has been sent
From out the eye, Love's firmament:
Many a jest told of the keys betraying
This night, and locks picked, yet w'are not a-maying.

In France the Hawthorn and the other branches of May Day were set outside the windows of every young girl. The stale, sweet scent from the trimethylamine the flowers contain, makes them suggestive of sex,* and on most May Days, before we at last changed the calendar in 1732 and adopted the New Style, the Hawthorn was already in full blossom. May Day now comes thirteen days earlier. South of London, the flowers do not open, on the average, till 10 May – in the Midlands till 13 May (178).

May Day was much hated by the more extreme Puritans of the sixteenth and seventeenth centuries. At the end of the thirteenth century and early in the fourteenth, it looks as if the church made some attempt to

* Later there was a belief in some places in England that Hawthorn flowers preserved the stench of London during the plague. Trimethylamine is an ingredient of the smell of putrefaction.

sanctify it rather than stand in opposition. Though it is denied that the foliage carved in churches and cathedrals had any symbolical content, the churchmen must have known all about Hawthorn and all that it stood for in the May Day ceremonies. Yet Hawthorn entered the churches, exquisitely portrayed, for example, in the chapter-house of Southwell Minster, with other May Day or magical plants – Oak, Ivy, Cinquefoil, Buttercup, and Maple, and, moreover, with heads wreathed in Hawthorn foliage (143). These heads, decorated sometimes with Hawthorn, more frequently with Oak, are not uncommon on capitals, bosses, corbels. Lady Raglan ('The Green Man in Church Architecture', in *Folklore*, L. i, 1939) believes these oak and hawthorn men are carvings of the May Lord, or May King, otherwise the Green Man (she points to the Green Man as a name for inns), Jack-in-the-Green, and Robin Hood; and she mentions payment in the churchwardens' accounts at Reading for the 'gathering of Robin Hood' in 1499.

Like other species, Hawthorn was supernaturally powerful at all times and against a wide range of evil, not merely against the malice of fairy and witch. In France (and in England as well) it was powerful against lightning – in Touraine, especially if it was picked fasting on May Morning; in the Basses-Pyrénées and Charente, if branches were picked in June as one of the herbs of St John; in Shropshire, if it was cut on Holy Thursday. Oaths were sworn by the tree – at any rate, a large oath in the border ballad of *Glasgerion*; and its magical, apotropaic power – as well as the thorns – helped to make it a favourite plant for protective hedges; and a favourite, too, for walking-sticks (36, 137, 165).

There was the usual ambivalence in the magic of Hawthorn: it helped and it hurt. Thousands of people in England still believe it unlucky to bring the blossom indoors. An Irish poem of the twelfth century distinguishes Hawthorn from Blackthorn, Rowan, Holly, Birch, and others, as a sinister tree:

'Alder, you are not spiteful, lovely is your colour, you are not like the prickly hawthorn where you are in the gully' (106).

And everyone knows, if only from William Allingham's 'Up the airy mountain, down the rushy glen', that in Ireland lone hawthorns belong to fairies and cannot safely be disturbed:

> By the craggy hillside
> Through the mosses bare,
> They have planted thorn-trees
> For pleasure here and there.

Is any man so daring
  As dig them up in spite,
He shall find their sharpest thorns
  In his bed at night.

Worse still, after disturbing a 'Lone Thorn', a 'Fairy Thorn', or a 'Lone Bush', you would become ill or die. The Irish fairies meet at the thorns, or live under them, or indeed live inside them. Cut the lone thorn and it may bleed or scream. If you so much as put your clothes out to dry on a lone thorn, you may be interfering with washing spread across it already by the fairies (137). In the Isle of Man thorn trees were planted as guardians of the cottage (cf. the protection given by yew trees) – *Isle of Man Natural History and Antiquarian Society Proceedings*, V. 4, 1954. Ireland also has its sacred thorns above holy wells, on which rag offerings are left. There is an Irish belief, too, that if you eat the haws, they will give you jaundice.

An attempt has been made to disinter an English cult of such lone thorn trees, mainly because thorn trees are often mentioned in old charters as boundary marks and because occasionally they marked the meeting place of the hundred (47). The evidence is thin, though farmers in the Cleveland district of Yorkshire did hang the afterbirth of their animals on particular Hawthorns.

Supernatural ownership of the Hawthorn could be transferred respectably, and the supernatural power for good or evil could be explained, by etiological stories:

Under a thorn
Our Saviour was born –

So, by the English rhyme, the power and the ownership are made Christian, and the Hawthorn's efficacy against lightning is justified. Or in Gascony, where the Whitethorn or *aubépine* keeps thunder and lightning from the house, the rhyme goes:

La Vierge Marie
S'est endormie
Sous un aubépine
Depuis le soir
Jusqu'au matin. (192, 165.)

About A.D. 400, Marcellus Empiricus, in his *De Medicamentis*, had written of the 'salutaris herba, id est *Spina alba*, qua Christus coronatus est' (129); and this belief that Christ's crown was made of Whitethorn (*alba spina*) could explain wickedness or virtue in the plant, according to need. It was an Irish, as well as an English, belief (137). Thus according to Sir John Mandeville, the Jews set the crown on Christ's head 'so faste and sore that the blood ran down be many places of his visage, and of his necke, and of his schuldres. And therefore hathe the White Thorn many vertues; for he that berethe a braunche on him thereoffe, no thondre, ne no maner of tempest may dere him; ne in the hows that it is inne may non evylle gost entre ne come unto the place that it is inne'.

The fame of the Glastonbury Thorn, which is both legend and no legend, was partly tied to Hawthorn as the material of the Crown of Thorns. The Glastonbury Thorn is *Crataegus monogyna* var. *praecox*, putting out leaves and flowers in winter and again in May. There was such a tree at Glastonbury, and also a walnut which bore no leaves until the Feast of St Barnabas – indeed, according to *The Lyfe of Joseph of Arimathia*, written about 1502, not one Hawthorn, but three, on Weary-all Hill, which the pilgrims had to climb. First the Walnut, then the Glastonbury Thorn:

> Thre hawthornes also, that groweth in Werale,
> Do burge and bere grene leaves at Christmas
> As freshe as other in May.

The original story was no doubt that the Glastonbury Thorn or Thorns blossomed on Christmas Day because Hawthorn was the material of the Crown of Thorns. The second and final story was woven into the Glastonbury legend of Joseph of Arimathea – he had stuck his dry hawthorn stick into the hill, where at once it 'grew, and constantly budded and blowed upon Christmas Day' (4, 47). This story came into print for the first time in 1722, when Thomas Hearne published the *History and Antiquities of Glastonbury*. In the sixteenth century visitors were shown a single Glastonbury Thorn with two trunks. One was felled towards the end of the century, the other (by the soldiers, according to John Aubrey, in his *Natural History of Wiltshire*) during the Civil Wars. However, it had been widely propagated by cuttings. No doubt the celebrity of the Thorn conveniently offset the May Day paganism and any other dubious associations of this supernatural species.

26. Medlar. *Mespilus germanica* L. 18

Have you ever met anyone who enjoys eating a ripe and rotten medlar? More than a hundred years ago, when J. C. Loudon compiled his *Arboretum et Fruticetum Britannicum*, he wrote that the medlar might be very agreeable to some palates, 'though it is, as Du Hamel observes, more *un fruit de fantaisie*, than one of utility'. Each medlar tree in the garden, each thorny and tough medlar shrub gone wild in the hedges, endures as a vegetable antique from the civilizations of south-western Asia, one of the most curious plants ever tamed into service. Long cultivated in Great

29 Medlar *Mespilus germanica*

Britain, Gerard already knew the Medlar as a wild tree, more or less, found 'oftentimes in hedges among briars and brambles'.

Examine the fruit, brownish-green, 'bearing a crown as it were on the top, which were five green Leaves [the calyx], and being rubbed off or fallen away, the Head of the Fruit is seen to be somewhat hollow' (49). In Aelfric's Glossary of about A.D. 1000 the Medlar is given its old, blunt English name from an all too obvious likeness, and one which suggested various uses in medicine – '*mespila*, openaers'. The name served for Chaucer. Osewold the Reeve, before he begins his tale, talks of his old age, of faring 'as dooth an open-ers' –

> We olde men, I drede, so fare we;
> Til we be roten, can we not be rype.

The name served for *The Grete Herball* (1526), and for the blunt William Turner, but Gerard is more polite and mentions only 'Medlar', which came into English in the Middle Ages from the French, and which goes back through the Latin to the Greek *mespile* or *mesmilon*. Shakespeare fashions some bawdy fun for Mercutio out of medlar and 'an open et cetera', in *Romeo and Juliet* (II. i. 33–40). Sir John Harington, the Elizabethan courtier who invented the water-closet, refers obliquely to the old name in *The Englishman's Doctor*, 1608, his witty translation of the *Regimen Sanitatis Salernitanum*, the mediaeval treatise on health:

> Eate Medlers, if you have a loosenesse gotten,
> They bind, and yet your urine they augment.
> They have one name more fit to be forgotten.

But it is still not forgotten, it still endures in the country (Corn, Som, Glos, Norf, War, Worc, Lincs, Ches) and in Scotland as How-doup or Hose-doup ('how' is hollow). Medlar jelly is worth making and medlar trees are worth growing for the big flowers and the intricate, gnarled, eccentric shape of the tree.

In France the Medlar was a powerful tree: it protected houses from enchantment, and put witches and sorcerers to flight (160).

27. Rowan, Mountain Ash. *Sorbus aucuparia* L. 111, H 40

In the north, known by the Scandinavian names of Rowan or Roddin, also as Quicken, which has the forms Wicken or Wiggen. In the south and the west, a Quickbeam and a Whitty (but cf. *Sorbus aria*): 'The tre whiche we call in the North Countre a quicken tre or a rown tre, and in the South countre a quikbeme' (William Turner, *Herbal*, 1568).

Local names. CARE, Corn, Dev; CARE-TREE (Cornish *kerdhyn*, cf. Irish *caorthann*), Corn, Dev; CAYER, Pemb; CHITCHAT, Wilts; QUICKBEAM, Dev, Som, Hants, Suss, Ire; QUICKBEAM-TREE, Som, Hants, Suss; QUICKEN, Kent, War, Shrop, Ches, Derb, Lincs, Lancs, Yks, N Eng, Scot; QUICKEN-TREE, Glos, War, Heref, Shrop, Derb, Lincs, Yks, Dur, N'thum, N Eng, Ire; QUICKEN-WOOD, Lincs, Yks; WICKEN and WICKEN-TREE, Shrop, Ches, Derb, Lincs, Lancs, Yks, N'thum, N Eng; TWICKBAND, Hants; TWICK-

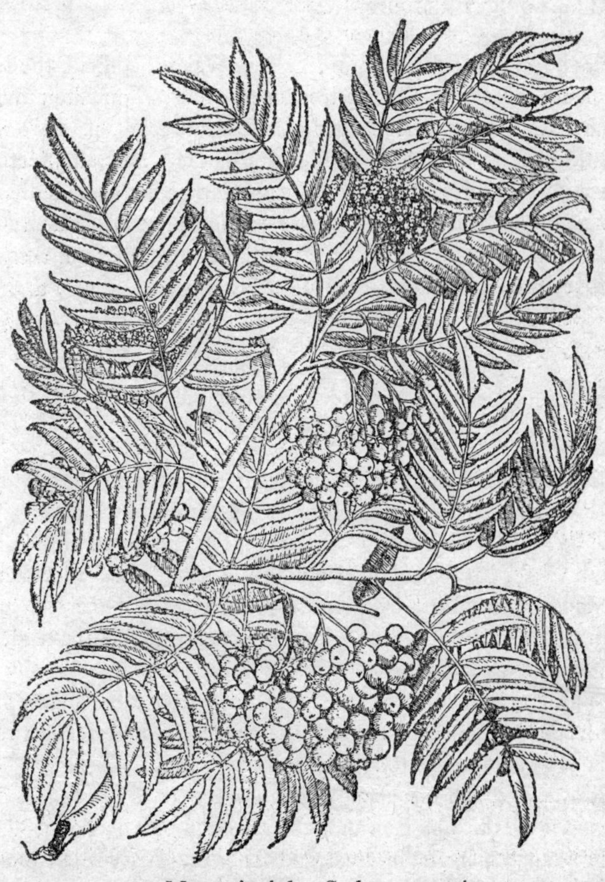

30 Mountain Ash *Sorbus aucuparia*

BINE, Dev; WHITTY, Shrop, Rad; WHITTY-TREE, Heref, Worc, Shrop, W Eng; WHITTEN-TREE, Shrop, Ire; WICKEY, Shrop, Ches, Derb, Dur; WIGGEN, Dev, Som, Ches, Derb, Lancs, Yks, West, Cumb, Wales; WIGGY, Dur; WITHEN, Lancs; WITHY, Heref, Shrop.

RODDIN, Scot; RODDIN-TREE, Scot; ROWAN and ROWAN-TREE, Kent, Lincs, Lancs, Yks, Dur, West, Cumb, N'thum, N Eng, Wales, Scot, Ire; SAP-TREE, Yks; SIP-SAP, Lancs, Yks; SHEPHERD'S FRIEND, Dor; WHISTLE-WOOD, Yks; WITCHEN, N'hants, Worc; WITCHBEAM, Dev; WITCHHAZEL, Yks; WITCH-WICKEN, Lincs; WITCHWOOD, Yks, Dur, Cumb, N'thum.

Names for the berries include CARES, Corn, Dev; COCK-DRUNKS, HEN-DRUNKS, Cumb; DOG-BERRIES, Ches, Cumb, Lakes; POISON-BERRIES, Som, Yks, N'thum; QUICKEN-BERRIES, Ire.

Wherever the Rowan flourishes, in England, Wales, Scotland, the Isle of Man, Ireland, it has been a prime tree for protection. Points to notice are the red berries – there is no better colour against evil, its universality (though the Rowan avoids clay and lime) in the Celtic and Germanic areas, even to Iceland; and that it will grow at a higher altitude than any other British tree. In fruit the Rowan is wonderfully conspicuous against a blue sky or below dark clouds, on a tor in Devonshire or a mountain in Wales or Ireland. Rowans are beautiful:

> Glen of the rowan trees with scarlet berries,
> With fruit praised by every flock of birds,
> A slumbrous paradise for every badger
> In their quiet burrows with their young

says an Irish poem of the fourteenth century, or later, supposed to have been written by Deirdre (105). More than beautiful, each Rowan was a powerful being. That is clear in the story of Diarmuid and Grania in the Irish Finn Cycle. On their flight from Finn, Grania's husband, they stay in the wood of Dubhros by permission of the fairy guardian of the Quicken-tree, who is thick-boned, large-nosed, crooked in the teeth, with one red eye in a black face. By day he sits at the foot of the tree, by night he sleeps in the branches. The tree had sprung from a berry dropped by the Tuatha de Danaan. Grania asks for some of the wonderful berries, and to get them Diarmuid has to kill the fairy guardian.

The Rowan is much respected among the few trees of Iceland, where there was a proverb 'The Rowan is the salvation of Thor.' This derived from the strange tale of the great god Thor and his visit to the giant Geirröth, in the land of the giants, as told in early Icelandic poetry and the

Prose Edda. On his journey Thor had to cross the river Vimur, dividing the world of men from the world of giants. It was hugely swollen by the menstrual blood and urine of the giantesses. Thor helped himself to safety on the far bank by catching hold of a Rowan. There is evidence to suggest that the Rowan was Thor's wife: the berries were sacred to the Ravdna, wife of the parallel thunder-god of the Lapps (193).

Break a piece off this being, and his power is in your hand for a thousand purposes. On May Day, for example, when fairies and witches are abroad. In Ireland pieces of Rowan are nailed over the doors, tied around the churn and placed in the milk pail to prevent the butter and milk from being stolen on this day of days. The fire is likely to be enchanted; Rowan is hung about the hearth (137). So also in Yorkshire, where 2 May was Rowan-tree Day or Rowan-tree Witch Day, and the houses were hung about with Rowan twigs; in the Isle of Man; in the Highlands and Islands, where barns, byres, stables, sheepfolds, and lambcots, and the milk in the boynes or tubs all had this necessary plant of May Day, and where special cakes were made over a Rowan fire (31a); in the north of England, where churn-staffs were made of Rowan, in the Lakes where the cream was stirred with a Rowan stick, and in Hereford, and Cornwall and Wales (a Welsh name for *Sorbus aucuparia* is *cas gan gythraul*, 'devil's hate'). In Wales the children of Narberth in Pembrokeshire dressed the Priest's Well on May Day with Rowan ('Cayer') and Cowslips – 'in order to keep "the witch" away from those families who drew water from the well' (Francis Jones, *The Holy Wells of Wales*, 1954).

Rowan in Ireland keeps the dead from rising. For the same good purpose it was planted in graveyards in Yorkshire and in Wales; and in the Highlands it was built into coffins and biers (31b). Rowans were also planted around houses – as they are still, the old compelling protective reason transformed to pleasure in the tree. In Wales, if you were foolish enough to step into a fairy circle, only a stick of rowan laid across the circle prevented you from staying there a year and a day (110). In his *Sylva* (1664), John Evelyn wrote that fletchers commended Rowan for bows, next to yew; and that it was used for walking-sticks on account of its power against 'fascinations' – either way a plant preferred for its supernatural force.

Witch-wands, or divining rods, were made of Rowan in Yorkshire and Lancashire, though the magic Hazel is usually preferred. In Scottish fishing-boats a piece of Rowan wood fastened the halyard (35). In Cornwall a piece carried in the pocket averted both ill-wishing and the elf-shot afflictions of rheumatism or lumbago (37, 192).

They said in Northumberland:

Woe to the lad
Without a rowan-tree gad

There, and in Yorkshire and Herefordshire, rowan goads and whipstocks were necessary to prevent a team of oxen or horses from being overlooked or being too restive and coming to an accident. In Herefordshire yoke-pins were carefully made of Yew or Rowan. The goad and the pin worked well in a story told of a Herefordshire farmer. In the ruins of the now vanished Penyard Castle above Ross-on-Wye he knew that there were barrels of money, guarded, in a vault behind iron doors. He harnessed ten yoke of oxen to the doors. They pulled and pulled, and managed to open them a little. The farmer could see the barrels, and a jackdaw on top. He thought the treasure would soon be his, when the doors slammed and clanged, and a voice in the air uttered:

Had it not been
For your quicken-tree goad
And your yew-tree pin
  You and your cattle
  Had all been drawn in (92).

For a last Rowan or Quicken story, turn up the border ballad *Willie o Douglas Dale* (36), which is not unlike the affair of Diarmuid and Grania. Willie has gone off with the King's daughter to the green wood, where she feels her time coming. She says, obviously wanting to ward off the supernatural dangers which attend childbirth,

O had I a bunch o yon red roddins
  That grows in yonder wood
But an a drink o water clear
  I think it would do me good.

Rowan berries and spring water – both were standard in their efficacy. In the Pitt Rivers Museum at Oxford, Rowan loops from Yorkshire are preserved, with Rowan crosses from Aberdeenshire, which in 1893 were put into the openings of houses against witchcraft.

Why is *Sorbus aucuparia* a Quick Beam or a Quicken? Beam is 'tree', Quick apparently the adjective in the sense of 'lively', endowed with life, the quick and the dead. Dictionaries go so far, and are then puzzled for the meaning, which is transparent: the tree endowed with life, and a very active and effective life. In OE the *cwicbeam* is usually taken to be the Aspen, *Populus tremula*, whose leaves are always on the move; but where *cwicbeam* occurs in magical texts – for instance, in a ritual for restoring fertility to bewitched land, on which four quickbeam crosses were laid (88), I feel sure that the plant intended and employed was *Sorbus aucuparia*. The tree is also Whitty and Wicken, Witchen, Witchwood, Witch-hazel, Witch-wicken. Originally these witch names go back to the OE *wice*, *wic*, for a tree with pliant branches or plant timber (cf. Wych-elm, used like *Sorbus aucuparia*, for bows). But

> Roan-tree and red thread
> Haud the witches a' in dread:

in a world of witch-mania, the sense of 'witch' was undoubtedly felt in the *wic* names for the tree. Evelyn, for instance, interpreted Witchen in that way.

28. Whitebeam. *Sorbus aria* (L.) Crantz. 26, H 3; *Sorbus rupicola* (Syme) Hedl. 20, H 6

Local names. CHESS-APPLE (i.e. 'cheese-apple'), Lancs, West; HEN-APPLE, Mor; HOAR WITHY, Hants; IRON PEAR, Wilts; MULBERRY, Aber; QUICKBEAM (cf. *Sorbus aucuparia*), Herts; SERVICE-BERRY, Mor; WHIP CROP, I o W; WHITTEN, WHITTENBEAM (OE *hwītingtrēow*, the tree which goes white), Hants; WHITE-RICE (OE *hris*, 'brushwood'), Hants, I o W; WIDBIN PEAR TREE (i.e. whitebeam?), Bucks; WILD PEAR TREE, Derb.

Sorbus aria of the older floras is divided up into several subspecies. *Sorbus rupicola* is the Whitebeam of northern and Scottish rocks, producing the crimson Hen-apples or Mulberries; *S. aria* (L.) Crantz., producing scarlet fruit, is the finer tree of the southern chalk – of woods or scrub, for instance, on Salisbury Plain or Cranborne Chase, or in Surrey. The name Whitebeam ('white tree') looks old, but has not been recorded before the eighteenth century. A white tree it is, or a silver tree, when a wind tosses up the underside of the leaves, against the sombre green alongside or against the darkness of yew trees. Hoar withy, 'grey willow', occurs as *hāra wīthig* in Anglo-Saxon charters, dealing with land, for instance, in Surrey, Berkshire, and Hampshire – probably for *Sorbus aria* (67).

You can eat the fruits – when they start going rotten like a medlar. John

Evelyn found them 'not unpleasant' (though he preferred the fruit of *Sorbus torminalis*) when mixed with new wine and honey into 'a conditum of admirable effect to corroborate the stomach' (70). The tree he praised for giving the husbandman 'an early presage of the approaching spring, by extending its adorned buds for a peculiar entertainment', and for its hard, durable timber, with which one of his rooms was wainscoted. The timber was used by wood engravers, and for gunstocks, etc., but above all else in machinery. Before the days of reliable cast iron, so it has been said, machine wheels everywhere in Europe were cogged with Whitebeam. England substituted cast iron, and the Continent followed (125).

29. French Hales. *Sorbus devoniensis*. E. F. Warburg. 3, H 3

Several of the subspecies of *Sorbus* are among the small band of plants peculiar to Great Britain, clinging to peculiar places, such as the Avon Gorge or the dark woods which overhang Culbone church and the Severn Sea, in Somerset. *Sorbus devoniensis* is also endemic, and a common tree in Devonshire, where the brown fruit have long been picked, sold, and eaten. Hales is apparently Halse, i.e. Hazel.

For 'French' to mark a somewhat unusual plant, cf. French Grass for *Onobrychis viciifolia*, or French Willow for *Chamaenerion angustifolium*.

30. Wild Service Tree. *Sorbus torminalis* (L.) Crantz. 53

Local names. CHEQUER-TREE, Suss, Kent; CHEQUER-WOOD, Kent; HAGBERRY (see *Prunus padus*), Yks; LEZZORY, LIZZORY, Glos; SHIR, Surr; WHITTY-BUSH (see *Sorbus aria*), Worc; WHITTY-PEAR, Worc. The fruits are WHITTY-PEARS (Worc), and CHEQUERS (Suss, Kent) from their pitted surface.

The griping *Sorbus*, whose small brown fruit settle the *tormina* or colic – fixing matters, like the medlar, 'when the belly is too soluble' (Gerard).

How much are the Chequers still eaten? Bromfield, in his *Flora Vectensis* (1856), says they were tied up in bunches and sold in the shops and markets of Sussex and the Isle of Wight, mainly to children. They were much employed in physic in Gerard's time and later. Evelyn recommended them for the gripe, and recommended the water distilled from the flower stalks and leaves for consumption, green sickness in virgins, and ear-ache (70). Gerard had no great opinion of the fruit except in medicine – 'if they yeelde any nourishment at all, the same is very little, grosse, and cold; and therefore it is expedient not to eate of these, or other like fruits, nor to use them otherwise then in medicines'. But his warning was disregarded.

'Service Tree' is the introduced *Sorbus domestica*, anciently grown in English gardens. Service from 'serves', plural of 'serve' from OE *syrfe*, which is from the Latin *sorbus*.

31. Crab Apple. *Malus sylvestris* (L.) Mill. 99, H 40

Local names. CRAB, CRAB APPLE, Corn, Dev, Dor, Som, Wilts, War, Ches, Lincs, Yks, Lakes; SCRAB (William Turner, *Herbal*, 1562; 'In the Southe countre a Crab tre, in the North countre a Scarb tre'), N Eng, N'thum, Scot; BITTERSGALL, Som; GRIBBLE, Dor, Som; GRINDSTONE APPLE (used in sharpening reap-hooks), Wilts; SCROGG (i.e. a scrubby tree or bush), Dur, N'thum, Berw, Rox; SOUR GRABS (i.e. crabs), Som; WILDING-TREE, Shrop.

Shakespeare's roasted crabs have hissed in the bowl, Crab Apple jelly is delicious, so is Crab Apple wine; and like the Quince, two or three crabs improve an apple tart. But our native apple is at once too different from all the cultivated apples (which belong to the introduced *Malus sylvestris* ssp. *mitis*) and too close to them. The cultivated apple is the Tree of Eden, the tree of magic, Pomona's tree; and the tree which on 17 January is wassailed with song and cider to give a heavy crop.

The crab name of the sour native apple comes perhaps from a 'crabbed' tree, a tree of awkward character which gives the wrong fruit. 'Crabbed', in that sense, from the crustacean, the hard creature of awkward walk and claws. Crabs have been found in an oak coffin of the Early Bronze Age in Denmark. Cultivation of the apple in central and northern Europe may go back to the Neolithic period (107).

There are Irish and Scandinavian stories of magic apples able to revive and rejuvenate, for example the apples curative of fatal illness or wounds which the three sons of Tuírenn stole from the Garden of Hisberna (Joyce, *Old Celtic Romances*, 1920), and the apples which the goddess Ithunn kept for the gods to eat when they grew old. Loki betrayed the apples to the giants, the gods withered and wrinkled with age, and Loki was forced to get the apples back (193).

XXXIV. Crassulaceae

1. Rose-root. *Sedum rosea* (L.) Scop. 45, H 17

When you climb to it on the mountain ledges, this looks rather a dull plant, succulent, creeping out of the rock on a tough, woody stock, and bearing flowers of a mean yellow. Cut the stock, and, as Gerard wrote, it gives out a delicious smell, 'like the damaske Rose'. From mountains or a sea-cliff (such as the black cliffs of Moher in Co. Clare which are speckled with clumps of *Sedum rosea*), this smell of the damask-rose is the last thing to be expected.

The plants in Gerard's garden came from Ingleborough, where *Sedum rosea* still grows out of the crumbling shale. The Irish call it *lus na laoch*, the 'hero's plant'. In Banffshire it had the ironic name of mixed sweetness and opprobrium, Priest's Pintel (for which see *Arum maculatum*). Welsh guides up Snowdon (where it grows plentifully around the wild lake in Cwm Glas – even on the island in the lake) have called it the 'Snowdon Rose'.

2. Orpine, Livelong, Midsummer Men. *Sedum telephium* L. 95, H 24

Local names. ARPENT, ARPENT-WEED, Hants, Herts; HEAL-ALL, N Ire; HEALING LEAF, Scot; JACOB'S LADDER (from the regular, alternate leaves), Kent; LIVELONG, Hants, Norf; MIDSUMMER MEN, Som, Wilts, N'thum; ORPHAN JOHN or HARPING JOHNNY, E Ang; ORPIES, Dur, N'thum, Berw, Rox; ORPY-LEAF, Scot; SOLOMON'S PUZZLE (formerly a London flower-seller's name).

'I remember, the mayds (especially the Cooke mayds and Dayrymayds) would stick-up in some chinkes of the joists, etc., Midsommer-men, which are slips of Orpins. They placed them by paires, sc: one for such a man, the other for such a mayd his sweet-heart, and accordingly as the Orpin did incline to, or recline from the other, that there would be love or aversion; if either did wither, death.' That is John Aubrey's account in the seventeenth century (8); and across Europe this fine plant was strong in magic and medicine. The time for this divinatory use was St John's Eve. John Parkinson was also familiar with the power of Orpine: 'The leaves are much used to make Garlands about Midsommer with the Corne-Marigolde-flowers put upon strings to hang them up in their houses, upon bushes and May-poles, etc. *Tragus* [Hieronymus Bock, the German botanist, who published his *Neu Kreutterbuch* in 1539] sheweth a superstitious course in his country, that some use after Midsommer day is past, to hang it up over their chamber doores, or upon the walles, which will be fresh and greene at Christmas, and like the Aloe spring and shoote forth new leaves, with this persuasion, that they that hanged it up, shall feele no disease so long as that abideth greene' (140). In France, where it was christianized as *herbe de la Sainte-Vierge*, Orpine was one of the plants

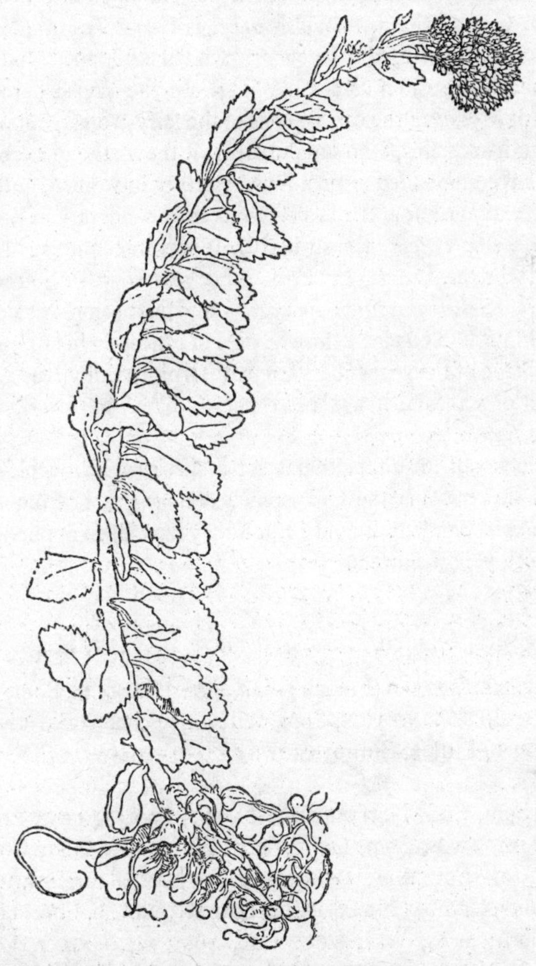

31 Orpine *Sedum telephium*

collected before St John's Eve and purified over the bonfires (see *Hyper-icum perforatum*). In Brittany there are beliefs similar to those mentione by Hieronymus Bock. Orpine is picked, and if it withers before the year i out, someone of the household will die before the New Year; and i Gironde, if a witch, male or female, entered the room, the potted Orpin would at once wither (188).

Aubrey's memory of his childhood in Wiltshire and the wide distribu-tion of Orpine in cottage gardens and on the sites of old cottages (especiall in Cornwall and Wales) suggest that the powers of Orpine were no les regarded in Great Britain. A Welsh name is *Cerwr Taliesin*, 'Taliesin' Cress' – after Taliesin, the legendary wizard and poet, and becaus Orpine leaves were eaten as a salad (124). In the Welsh herbal of th Physicians of Myddvai, the earthly sons of the fairy-woman out of the lak in the Black Mountains, who taught them all the virtue of herbs, *Sedum telephium* is recommended against fever, sterility in women, and too pro-fuse menstruation (61); and it has been a cottage vulnerary in England (i French *herbe à la coupure* ('gash herb'). It was in repute against cance ('for Ulcers in the Lungs, Liver, or other inward parts', according t Culpepper's *English Physitian Enlarged*), and Napoleon is said to hav interested himself in Orpine before he died of cancer on his Atlantic islan (72). The shape of the root tubers 'signed' Orpine with virtue against th King's Evil or scrofula: it was a *scrophularia* like *Scrophularia nodosa* o *Ranunculus ficaria* (q.v.).

Orpine looks an unsuitable name for a species with purple-pink an purple-red flowers. Via the Old French *orpiment*, it is from the Lati *auripigmentum*, 'pigment of gold'. But it may have been applied at first t a *Sedum* with yellow flowers.

3. Wall-pepper, Golden Moss. *Seum acre* L. 111, H 40

Local names. BIRD'S BREAD (French *pain d'oiseau*), CANDLES, Som; CREEPIN CHARLIE, Dev; CREEPING JACK, Som, Wilts, Ches; CREEPING JENNY, Dev Som; CREEPING SAILOR, Shrop; CROWDY KIT-O'-THE-WALL, Dev; FRENCH MOSS, Bucks.

GINGER, Som, Suff, Norf; GOLDEN CARPET, Som; GOLDEN STONECROP, Berks; GOLDEN MOSS, Som, Oxf, War, Yks; GOLDEN DUST, Corn, Suff; HUNDREDS-AND-THOUSANDS, JACK-IN-THE-BUTTERY, Som; LITTLE HOUSE-LEAK, N'hants, Cumb; LONDON PRIDE, LOVE ENTANGLED, Som; LOVE-IN-A-TANGLE, Corn; MOSS, War; MOUSETAIL, Som; PICKPOCKET, Dor; PIG'S EARS, Dev, Som; PLENTY, Dor; POOR MAN'S PEPPER, Suss, Notts.

QUEEN'S CUSHION, Rox; ROCKCROP, Corn; ROCK-PLANT, Dev; STAR, Som; WALL-GINGER, Som, Yks; WALL-GRASS, Dev; WALL-MOSS, Yks; WALL-PEPPER, Som, Glos, Yks, Ork; WALLWORT, Yks; WELCOME-HOME-HUSBAND, Suff; WELCOME-HOME-HUSBAND-THOUGH-NEVER-SO-DRUNK, Dor, Suff.

A Stonecrop (OE *stāncrop*) is the plant of a 'crop', i.e. a flower head or bunch, on the stone; and since this cheerful stonecrop deserts its natural shingle and sand and chalky soil and rock, and trails on a stone roof or hangs from a stone wall, it is an old favourite – for itself, and for its employment in folk-medicine.

As 'Welcome-home-husband-though-never-so-drunk', *Sedum acre* has the longest and most cheerful of English plant-names. In the sixteenth century it had a string of names, including Wall-pepper, from the acrid taste, Pricket and Prick-madam (see *Sedum reflexum*), Mousetail, Stonehore, Little Stonecrop, and Jack of the Buttery. Dr Prior treated this last name with a tall pedantry (149). He invented (?) a *Bot-theriacque*, a theriac or treacle against the botfly larvae of horses and cattle, of which *Sedum acre* was an ingredient. Then he turned 'Jack of the Buttery' round to 'Buttery Jack', which he said was the word *Bot-theriacque* corrupted. He never stopped to think of butteries at the back of the house, of lean-to roofs covered with *Sedum acre*'s golden welcome to the tipsy. But then consider Dr Prior's interpretation of Jack-by-the-hedge, *Alliaria petiolata* (p. 75).

4. Yellow Stonecrop. *Sedum reflexum* L. 67, H 21

Local names. CREEPING JENNY, Heref; GINGER (cf. *Sedum acre*), Kent; INDIAN FOG, Ire; INDIAN MOSS, Donegal; LOVE-IN-A-CHAIN, Cumb; LOVE-LINKS, Scot; PRICK-MADAM, Cumb.

Long ago introduced and naturalized on old walls. The leaves used to be eaten as a tender spring salad. Prick-madam was the name used in the sixteenth and seventeenth centuries (Gerard, Lyte, etc.), adapted from the French *trique-madame*. A *trique* is a cudgel or stick, and the flowering stems stand upright. As *trique-madame*, the allied *Sedum album* was given in France for an aphrodisiac.

John Evelyn called *Sedum reflexum* Trip-madame, and in notes on gardening he made between 1688 and 1706 (Evelyn Library, Christ Church, Oxford) he includes it among *Plants for the Kitchin-Garden*, along with Alexanders, Wood-sorrel, and Samphire.

5. Houseleek, Sengreen. *Sempervivum tectorum* L. (? 112, H 38)

Local names. AYEGREEN, Lancs, West; BULLOCK'S EYE, N Eng; FOOSE, N'thum, Dur, Berw, Rox, Angus, Aber, Mor; FUETS, N'thum, Scot; FULLEN, N'thum, N Eng; HEALING BLADE, HEALING LEAF, Clack; HEN-AND-CHICKENS, Som; HOCKERIE-TOPNER, Dumf; HOLLICK, Corn; HOUSE-GREEN, N'hants, War, Ches; HOUSELEEK, Suff, N'hants, Yks, West; JUPITER'S BEARD, Dev, Som; MALLOW-ROCK, Som; POOR JAN'S LEAF, Dev.

SENGREEN, Dev, Dor, Som, I o W, Kent, War, Shrop, Scot; SINGREEN, Wilts, Hants, I o W, Suss, Kent, Bucks, Worc, Shrop; SILGREEN, Dev, Dor, Som, Wilts, Glos, Oxf, War, Heref, Worc; SUNGREEN, Wilts, Suss; SYPHELT, Cumb; THUNDER-PLANT, Som; WELCOME-HUSBAND-THOUGH-NEVER-SO-LATE (cf. *Sedum acre*), Dor.

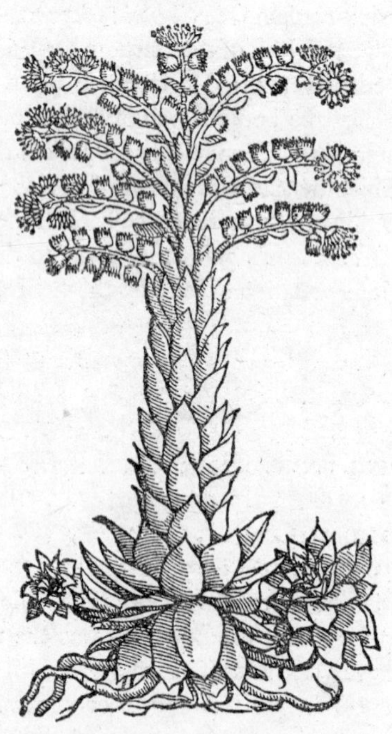

32 Houseleek *Sempervivum tectorum*

The typical Houseleek on the roofs of Europe has partly aborted stamens, and probably has no wild existence anywhere (38). Its home and its origin are lost in antiquity. Turn to Theophrastus, born in *c.* 372 B.C. He says of Aizoon, the Ever-living (*Sempervivum*): 'It grows upon flat shores, on top of earthy walls, and particularly on roof tiles, when there is enough sandy earth.' Dioscorides wrote in the first century A.D. that 'it grows in mountainy places, and some plant it on the tiles above their houses'. Thus the record of Houseleek on the roof is more than 2,000 years old.

The Romans called it *Iovis caulis*, 'Jupiter's plant', or *Diopetes* '(the plant) fallen from Zeus' or Jupiter. It had fallen to protect the house, an ever-living, evergreen badge of fire insurance, a plant given by Zeus or Jupiter to protect the house from the lightning he wielded. For us, in so far as we are Germanic, *Sempervivum tectorum* was presumably the plant of Thunor, the Scandinavian Thor, the god whose name we pronounce in thunder and who may have owned temples at Thunderfield in Surrey, Thundersley in Essex, Thundridge in Hertfordshire (66). A German name for *S. tectorum* is *Donnerbart*, Thunor's Beard or Thunder-beard; and there is wide belief in its insuring, reassuring, or protective power – indigenous belief, one may suppose, reinforced by classical authority.

Houseleek is one of its names in the equivalent Dutch, German, Danish, Swedish, and Norwegian (*tagløg*, 'roof leak'). One OE name was *hāmwyrt*, 'homestead plant'. Sengreen, Singreen, OE *singrene*, means 'evergreen', a name also applied to the Periwinkle, which is *Sinngrün* in German.

In Welsh, *llysiau pentai*, 'Cottage herb'; in Irish (the Irish still understand that it protects the house from fire, according to Seán ó Súilleabháin's *Handbook of Irish Folklore*) it is *luibh an toiteán*, 'fire herb'. The French call it *joubarbe* – which was used in England in the thirteenth century (*jubarbe*) from the mediaeval Latin *Jovis barba*, Jupiter's beard. 'Old wryters', explained William Bullein in his *Booke of Simples*, 1562, 'do call it Iovis barba, Iupiter's Bearde, and holde an Opynion supersticiously that in what house so ever it groweth, no Lyghtning or Tempest can take place to doe any harme there.' His own countrymen held the same opinion superstitiously, as he must have realized. Houseleek is still common on the roof, or on the porch; and was carefully cultivated above the thatch, which burns so easily. When John Ray travelled through Lincolnshire in 1661 he observed Houseleek all along the clay ridges and clay corners of the thatched houses – 'whether for ornament or use we did not enquire' (156).

Houseleek protects against fire from the sky *and* fire in the body. In folk-medicine that secondary use survives, even if the primary use in many or most places is now forgotten. In Wiltshire in 1952 a middle-aged farm labourer's wife told me how I should make Singreen ointment with the leaves, as a cooler: unconsciously she was deferring to Dioscorides who wrote of the Houseleek against ulcers, burnings, scaldings, and inflammations. Either for medicine or to protect the houses, or both, Houseleek was taken to the New World. 'Houseleek prospereth notably,' John Josselyn wrote in *New England's Rarities*, 1672. I have never seen it on an American roof or porch, but a correspondent from New Brunswick tells me it is much grown in Canadian and New England gardens and that always it is called Hen-and-chickens, a name (see above) taken with the plant from the west of England.

Bullock's Eye, in the list of names, was known to Gerard; and is the *bouphthalmon* of Dioscorides. Syphelt is the old name Cyphel from the Greek *kuphella*, the hollows of the ear. In the Middle Ages the plant was often called Erewort and employed against deafness.

6. Pennywort, Navelwort. *Umbilicus rupestris* (Salisb.) Dandy 58, H 40

Local names. BACHELOR'S BUTTONS, Dev; CORN-LEAVES (applied to corns, warts), Worc; COWS, Corn; CUPS-AND-SAUCERS, Dev, Dor, Som, Wilts; CUT-FINGER (a vulnerary), Worc; DIMPLEWORT, Dev; HAPPENNIES-AND-PENNIES, ICE-PLANT, Dev; JACK-IN-THE-BUSH, Scot; KIDNEYWEED, KIDNEY-WORT (used for the kidneys and against the stone); LADY'S NAVEL, Som; LOVER'S LINKS, Scot; LUCKY MOON, Dor; MAID-IN-THE-MIST, Scot; MILK-THE-COWS, Corn; MONEY-PENNY, Dev; NAVELWORT, Som, Worc, Leic, Yks; NIPPLEWORT, Suss.

PANCAKES, Dev, Som; PENNY CAKE, Corn, Dev, Pemb; PENNY CAPS, Dev, Som; PENNY COD, Corn; PENNY FLOWERS, Som; PENNY GRASS, Ire; PENNY HATS, Dev; PENNY LEAVES, Dev, Som, Ire; PENNY PIES, Corn, Dev, Dor, Som, Suss; PENNY PLATES, Dev; PENNYWALL, I o M; PRINCE'S FEATHER, Yks; ROYAL PENNY, Som.

In the south-west every child has picked off the walls the fat, dimpled, shining penny leaves and played with them. The leaves are glabrous and rather pale: it was the glabrous and pale silver penny they were compared to, not the coppers of our time.

Navelwort was identified with the *Kotuledon* of Dioscorides, the *Umbilicus Veneris*, or 'Venus's navel', of the Romans. This Dioscoridean plant the Romans employed in spells to procure love, and more curiously: *Huius foliorumque succus cum vino circumlitus aut collatus in pudendis constrictionem laxat*. In English *Umbilicus Veneris* discreetly became Lady's Navel, and English doctors recommended it (as Dioscorides had done) against the stone, and against cuts, chilblains, and inflammation. The Irish call it *coinneal Mhuire*, 'Mary's candle', a name also given to mullein and foxglove.

So different was sixteenth-century London that Gerard knew the Navelwort or Pennywort growing out of the stones of Westminster Abbey – 'over the door that leadeth *from Chaucer his tombe to the olde palace*'. Readers were oddly told by Culpepper (49) that it grew 'sometimes on the bodies of them that are decayed and rotten'. Compare Dead Man's Finger for *Orchis mascula*, Dead Man for *Orobanche minor*, and the pretty name of Corpse-flower for *Lathraea squamaria*. There *is* some resemblance between these plants and the *Umbilicus*.

XXXV. Saxifragaceae

The Saxifrages write a fragment of the social history of Englishmen. Two of them, *Saxifraga granulata* and *S. tridactylites*, are lowland species. The others climb into rock, mountain, and cloud. Pleasure in the wilderness of mountains – our modern pleasure – would have seemed extraordinary, if not lunatic and dangerous, to mediaeval Englishmen, indeed to most Englishmen before the era of Wordsworth. The mountain flowers, humble in their wet colonies or cushions, lived and sparkled without names or notice. When the great botanist John Ray justified the works of God to man, or the science of man to God, in his book *The Wisdom of God manifested in the Works of the Creation* (1691), which was a religious and scientific best-seller for more than half a century, he had to defend mountains. He demonstrated the 'great use, benefit, and necessity' of them against the feeling that they were warts and superfluous excrescences, 'signs and proofs that the present earth is nothing else but a heap of rubbish and ruins'. Knowing more than most about mountains, and mountain plants, he argued that every mountain was an alembic for distilling water to the plains, that mountains were 'very ornamental to the earth, affording pleasant and delightful prospects'.

A few curious botanizers had taken to the mountains in the north of England or in Wales, but Ray's delight in mountains was novel. The painters and the poets had to follow the botanizers before the rest of the polite world took pleasure and found emotional satisfaction in the high wilderness. The Saxifrages (even in the rock garden) have become an emblem of mountain holidays, but it is instructive to see how late they were discovered. Here are some dates of discovery or first record:

1. *Saxifraga nivalis*, discovered by Thomas Johnson on a journey into Wales, in 1639.
2. *S. stellaris*, discovered on the same journey 'upon the moyst Rockes at Snowden'.
3. *S. hypnoides*, recorded for the mountains of Lancashire in Parkinson's *Theatrum Botanicum*, 1640.
4. *S. aizoides*, found in Westmorland by John Ray in 1661.
5. *S. oppositifolia*, found on Ingleborough by John Ray in 1668.
6. *S. spathularis*, recorded from Co. Kerry, 1697.
7. *S. hirsuta*, found in Co. Kerry by Edward Lhwyd in 1699.
8. *S. hirculus*, recorded for Cheshire in 1724.
9. *S. rivularis*, found on Ben Nevis in 1790.
10. *S. cernua*, recorded from Ben Lawers in 1794.
11. *S. rosacea*, recorded from Wales in 1796.
12. *S. cespitosa*, recorded from Wales in 1800.

To find a dozen species it took the best part of 200 years; and by the time *Saxifraga cespitosa* was made known in 1800, the young J. M. W. Turner had tramped after drawings through the mountain country of North Wales, Scotland, and the Lake District, where the young William Wordsworth had established himself at Grasmere. By this time, as well, though the rock garden was not born, the vivid, purple-flowered *Saxifraga oppositifolia* had been taken from its mountain crags and snow to London, where at Covent Garden plants could be purchased at a shilling or two shillings each (50).

The polite traveller into the 'awful works of nature' now expected to be told of mountain flowers in the books written for his guidance. Tours into Wales informed him of 'the Botanic Garden of Snowden' above the 'baleful waters of Llyn Idwal', and told him where to look for the various species of Saxifrage (see, for instance, Nicholson's *Cambrian Traveller's Guide*, 3rd ed., 1840, or W. R. Dingley's *Excursions in North Wales*, new ed., 1839). Unfortunately the travellers must have increased the rarity of rare species.*

1, 2. St Patrick's Cabbage. *Saxifraga spathularis* Brot. H 12, *Saxifraga hirsuta* L. H 3

These, and the hybrids between them, hit the eye in the south-west of Ireland on the mountain rocks or low by the rivers, among ferns and moss

* In Germany most of the mountain Saxifrages are protected by law. They can be picked, but not uprooted.

and under the lichen-bearded trees. If you are fishing, your cast and your flies catch among the flowering stems. The names for them are either St Patrick's Cabbage or Leaf of St Patrick. They look at first like neat and healthy specimens of the London Pride, the garden hybrid between *S. spathularis* and *S. umbrosa* from the Pyrenees.

3. Dovedale Moss. *Saxifraga hypnoides* L. 55, H 11

This is the commonest of the upland species – in Derbyshire, DOVEDALE MOSS; in Cumberland and Yorkshire, LADY'S CUSHIONS; in Durham, QUEEN'S CUSHION; in Yorkshire, EVE'S CUSHION; and INDIAN MOSS sometimes in the garden.

4. Purple Saxifrage. *Saxifraga oppositifolia* L. 35, H 7

This deserves some livelier name, as one of the most exquisite of natives – perhaps 'Mountain Emperor' or 'Snow Purple', or 'Ingleborough Beauty' after the Yorkshire mountain where it was discovered (see above for *S. oppositifolia* as a garden or pot plant).

5. Meadow Saxifrage, Dry Cuckoo. *Saxifraga granulata* L. 89, H 4

Local names. CUCKOO-FLOWER, Yks; DRY CUCKOOS (in contrast to Water Cuckoos, *Cardamine pratensis*), Wilts; FAIR MAIDS, FAIR MAIDS OF FRANCE, Bucks; FIRST OF MAY, Ches; LADY'S FEATHERBEDS, Suss; PRETTY MAIDS, Berks; SASSIFRAX, Som; SNOW-ON-THE-MOUNTAINS, Wilts.

The better of the two lowland species, which John Parkinson once knew in London, 'in grassie sandy places on the backeside of Grayes Inn where Mr Lambes conduit heade standeth' – not far from the modern and squalid Lambs Conduit Street (140). This is a candid, delicious flower to meet on the Wiltshire downs, in May, every blossom shaking slightly in the breeze; and it is always worth picking a small bunch, since the blossoms will last a week or two in a vase.

Meadow Saxifrage was used against the stone (see Parsley Piert). In French it is *casse-pierre* or *herbe à la gravelle*; and it is Gerard's White Saxifrage or White Stonebreake. The little bulbs or granulations by which it lives through the winter, the 'little knoppes lyke pearles', as Turner called them in his *Herbal*, were sold under the name of *Semen Saxifragae*

Albae. Boiled in wine, they cleansed the kidneys, broke the stone, and were 'singular against the strangurie and all other griefes and imperfections in the raines' (80). So first of all the Saxifrages, *S. granulata* entered the English garden for use; and stayed for charm – 'Pretty Maids' all of a row – when the medicament was forgotten.

6. Golden Saxifrage. *Chrysosplenium oppositifolium* L. 109, H 40

Gerard included this little plant, which crawls along runnels and wet rocks, with *Saxifraga granulata*, contrasting it as the Golden Saxifrage or Golden Stonebreake. English names for it are Creeping Jenny (Suss) and, appropriately, Buttered Eggs (Wilts). Also Lady's Cushion. In the Vosges it is said to be eaten as a salad under the name *cresson de roche* (120).

XXXVI. Parnassiaceae

1. Grass of Parnassus. *Parnassia palustris* L. 90, H 33

Inspired, one may suppose, by its beauty, the Flemish botanist Mathias de l'Obel called this plant *gramen Parnassi*, the grass of the holy mountain of Apollo and the Muses. A northern species, rare and patchy in the south; though it was from the south – not from a mountain, but from Oxfordshire – that de l'Obel gave the first British record in 1570. If you complain that *Parnassia palustris* is not a grass, at least it sends up its straight flowering stems among the grass, cupping each one broadly and beautifully with a white eye.

Yet 'white' is too flat a description. The five petals are delicately veined with green; and across each petal lies a small staminode, faintly like the end of a frog's leg, each of its threads tipped with a glistening yellowish gland. Moreover the flowers smell of honey. On the eastern border the Grass of Parnassus has been called by as drab a name as 'White Buttercup'. Gerard named it 'White Liverwort', since it was administered (*flos hepaticus Parnassi*, 'liver flower of Parnassus') as a liver medicine. A decoction of the leaves settled the stomach, and boiled in wine or water the leaves broke and drove forth the stone.

XXXVII. Grossulariaceae

1. Red Currant. *Ribes sylvestre* (Lam.) Mert. & Koch. 91, H 3

Local names. GAZEL, Kent; RIZZLES, RUSSLES, Kirk, Wigt; WINEBERRY, Yks, Cumb, Lakes, Scot.

Wild or no? Native or introduced? 'Probably native by streams in woods, etc.' (37), but birds guzzle the red currants of the garden and drop them in the hedges, where you sometimes find *Ribes sylvestre* for mile upon mile. *Ribes spicatum* and *Ribes alpinum* of limestone woods and rocks are certainly native.

Turner, in 1548, said that *Ribes* in some parts of England was called a 'Rasin tree', which gives the clue to Red Currant and Black Currant. As they became known in cultivation, the fruit were equated with small raisins or currants; and currants in the Middle Ages were 'raisins of Corauntz', the Greek raisins from Corinth. Lyte, in 1578, wrote of Red Currants as 'Bastard Corinths' or 'Red Gooseberries'. Wineberry was an obvious name when Red Currant Wine (a delicious wine) became popular. Gazel is from the French *groseille*, a red currant (see Gooseberry below).

The taming and diversification of *Ribes sylvestre* began in the north-west of Europe; and began rather late, since the ancients did not cultivate this species, any more than they cultivated the Black Currant or the Gooseberry.

If you have wild or escaped plants growing near you, pick some of the flowering branches in the spring and put them indoors. These green flowers in their arrangement along the stem are never looked at, but they are uncommonly beautiful.

2. Black Currant. *Ribes nigrum* L. 89, H 4

Local names. BLACKBERRY, Lincs, Yks, Cumb, Berw; GAZEL, Suss, Kent; SQUINANCY-BERRY, Ess, Lancs, Cumb; WINEBERRY, Scot (for Gazel and Wineberry, see *Ribes sylvestre*).

A native in Great Britain and across northern Europe. Probably the fruits were used in folk-medicine against a sore throat (a quinsy or squinancy) long before the Black Currant entered the garden. Black Currants were made into wine, syrup, and jelly for the quinsy. Indeed many of the confections we eat for pleasure our ancestors devised for medicine.

The Black Currant's medicinal repute has been vindicated. During the last war the commercially bottled syrup gave Vitamin C to children, and has come to stay, making a good as well as a healthy drink – which is excellent with gin. It has a venerable predecessor in the French *cassis*, the liqueur of the *cassis* or Black Currant.

Dried leaves of Black Currant transform a pot of Indian tea, and give it the taste of a green tea from China.

3. Gooseberry. *Ribes uva-crispa* L. 101, H 6

Local names. CARBERRY, Yks, N Eng; CATBERRY (of wild gooseberries), West, Cumb; DABBERY, Kent; DAYBERRY, Corn, Dev, Kent; DEBERRY, Dev; DEWBERRY, Dev (Dayberries are usually the wild fruit); FEABERRY (also FABES, FAPES, FEABS, FEAPS, THAPES), S Eng, E Ang, War, Shrop, Ches, Derb, Leic, Lancs, Yks.

GOGGLE, Lincs; GOLFOB, Derb; GOOSEGOG, Som, Hants, Mddx, E Ang, N'hants, Ches, Lincs, Yks; GOOSEGOB (i.e. gob, something gobbled by geese), Derb, Lancs; GROSER, N'thum, Scot; GROSET, Scot; GROSSBERRY, Yks; GROZZLE, GRUZEL, Rox, Dumf; GRIZZLE, Dumf; HONEY-BLOB, Scot; WINEBERRY, Yks.

Like the Red Currant and Black Currant, the Gooseberry came late into cultivation, though it was familiar in British gardens by the sixteenth century. Gooseberries were mixed into a sauce with sorrel and sugar to be eaten with young geese, but the name does not derive from 'goose' *plus* 'berry', except by assimilation to both words. Gooseberry, Groser, Groset, Grozzle, Grizzle, descend from the French *groseille* (or its Patinized form *grossula*), which in turn descends from the Frankish *krûzil*, the basic meaning of which is something crisp. In French a red currant (see p. 205) is *groseille*, a gooseberry *groseille à maquereau* (mackerel), from the use of green gooseberries in fish sauce.

XXXVIII. Droseraceae

1. Sundew. *Drosera rotundifolia* L. 108, H 40

Local names. FLYCATCHER, Dev; FLY-TRAP, Som; ILES (i.e. liver rot in sheep, 'caused' by Sundew, cf. French *herbe de l'igler*, Danish *iglegräs*), Corn; MOOR-GLOOM, Yks; MOOR-GRASS, Yks, Cumb; OIL PLANT, Donegal; RED ROT, Som, N Eng, Scot; STICKYBACKS, Som.

The gland at the tip of each leaf tentacle glistens with 'dew', the dew attracts the midge, the midge is caught, and the tentacles bend towards the centre of the leaf for *Drosera*'s meal. The herbalists observed only the dew, the secretion, and none of the process. The leaves took their fancy: and dew was a medicinal and a magic substance. When the cautious William Turner considered the Sundew in his *Herbal* – this 'little small herbe that groweth in mossey groundes and in fennes and watery mores' – he would have none of this without experiment. 'Our Englishmen now-adayes set very much by it, and holde that it is good for consumptions and swouning, and faintnes of the harte, but I have no sure operience of this, nether have I red of anye olde writer what vertues it hath, wherefore I dare promise nothing of it.'

That was published in 1568. But the dew was too powerful. From the earliest times plants for medicine (or magic) had been picked before sun-rise, before the heat of the sun dried off that liquid which so increased their efficacy (57). Here was a plant beaded with a dew which defied the hottest sun. 'The later Phisitians [who were not so delivered from the old superstitious precepts as some of them liked to believe] have thought this herbe to be a rare and singular remedie for all those that be in a consumption of the lungs, and especially the distilled water thereof: for as the herbe doth keepe and hcld fast the moisture and dew, and so fast, that the extreme drying heate of the sun can not consume and waste away the same; so likewise men thought that heerwith the naturall and lively heate in mens bodies is preserved and cherished.' That was Gerard some thirty years after Turner. He disagreed; adding, however, that such vain opinions were greatly strengthened by another quality of Sundew – that 'cattle of the female kinde are stirred up to lust by eating even of a small quantitie'. Sundew was held to be strengthening and nourishing, 'especially if it be distilled with wine, and that liquor made thereof which the common people do call *Rosa* [i.e. Ros, dew] *Solis*'. The plant was not only Sundew or *Ros Solis*, but Youthwort, no doubt from the aphrodisiac and strengthening power of the distillation.

In *The Haven of Health* (1548), Thomas Cogan had given a full recipe for this new *Rosa Solis*, made not only of Sundew, but of sugar, mace, ginger, nutmeg, cinnamon, aniseed, liquorice, and dates, and now and then with rose petals as well; or else with ale, Rosemary, Sage, Thyme, Chamomile, Marjoram, Mint, Avens, Fennel, Dill, Pelletory, Lavender, Hyssop, roses, and spices. He added, since there was some doubt of the nature of the herb, that great store of it grew in the Lancashire peat-diggings, where the people called it 'Youth Grass'.

What, in fact, you can see is the birth of a once celebrated liqueur, the French *rossolis*, the Italian *rosoglio*, the birth of one more of our confections which began in utility, if not in pain, and concluded in pleasure. Long after the sixteenth century *Drosera* was collected in Great Britain for the manufacture of *Ros Solis*; longer still in France, where *Ros Solis* was a home-made liqueur. John Aubrey mentioned that Sundew was collected around Malmesbury in Wiltshire, 'which the strong-water men there do distill, and make great quantitys of it' (7). Culpepper stated of the drink made with *Aqua vitae* and spices that it was taken to good purpose 'in qualms and passions of the Heart' (49).

As though it were dew itself, the sparkling secretion of the *Drosera* glands was used against sunburn and freckles, and the Irish boil the leaves in asses' milk (137) which the children then swallow against whooping-cough (asses' milk, since the famous *Regimen Sanitatis Salernitanum* of the Middle Ages recommended it above all other kinds). Yet, as Gerard knew, the rosy-red leaves were accused of giving sheep the liver rot because they are caustic and biting. The pale *Hydrocotyle vulgaris* was White Rot, the rosy-leaved Sundew, as a comrade of damp places, became the Red Rot. The discovery that the Sundews and other plants are insectivorous gave a shock to romantic beings who personified Nature as a kindly and innocent creature. 'I will not be always paying her compliments,' Ruskin exclaimed after learning of these carnivores; 'the nasty things she turns out!'

XXXIX. Lythraceae

1. Purple Loosestrife. *Lythrum salicaria* L. 98, H 40

Local names. EMMET'S STALK (an emmet is an ant), Som; GRASS POLLY (the name introduced in Johnson's revision of Gerard's *Herbal*, 1633, for the rare *Lythrum hyssopifolia*), Som; LONG PURPLES, Dor, Som, Wilts, N'hants; RAGGED ROBIN, Wilts; RED SALLY (i.e. willow), Lancs; SOLDIERS, Norf; WILLOW-STRIFE, Som. Irish names include BRIAN BRAW, FOX TAIL, STRAY-BY-THE-LOUGH.

Long Purples is one of the most famous of English plant names, from Shakespeare's description of the drowning Ophelia. But Shakespeare

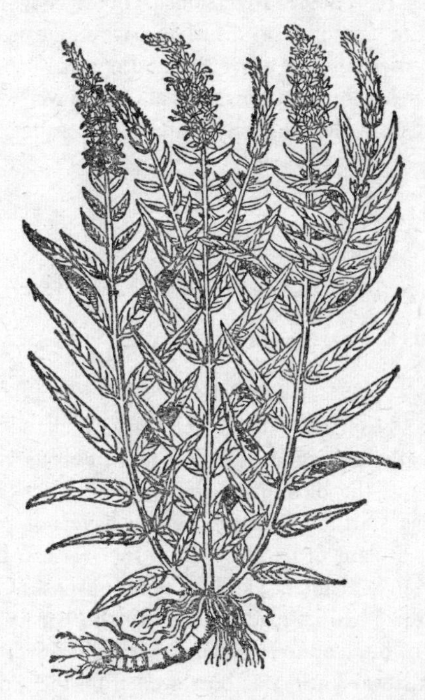

33 Purple Loosestrife *Lythrum salicaria*

meant the bawdier plant, *Orchis mascula* (q.v.). Millais, when he painted Ophelia, included accurate representations of *Lythrum salicaria*, which does belong to places where 'a willow grows aslant a brook'. He and his friend Holman Hunt spent a long time searching for the right situation, which they found on the Ewell River, near Worcester Park Farm, in Surrey. There he completed 'the exact composition of arboreal and floral richness he had dreamed of', the ecological imitation of a streamside, leaf by leaf and flower by flower, having only to insert Ophelia, or Elisabeth Siddal, floating in the celebrated cold bath in Gower Street. Perhaps

Tennyson (see his early poem *A Dirge*) had assured him that Long Purples were *Lythrum salicaria*; or he might have read in John Clare's *The Wild-flower Nosegay* of 'fine long purples shadowed in the lake'.

Purple Loosestrife was coined by William Turner (1548). Dioscorides had described a *lusimachion* with yellow flowers, taken by the herbalists to be *Lysimachia vulgaris*; Pliny described a *lysimachia* with purple flowers, taken to be *Lythrum salicaria*. Pliny added that it was named after Lysimachos, King of Thrace, Macedonia, and Asia Minor; but since *lusimachos* also means 'ending' or 'dissolving strife', Turner made the one species Yellow Loosestrife, the other Purple Loosestrife.

In damp Irish fields Purple Loosestrife and Corn Marigold often grow and flower together, a splendidly sharp and bold combination.

XL. Thymelaeaceae

1. Mezereon. *Daphne mezereum* L. 31

Local names. MEZELL, MAZELL, Hants; PARADISE PLANT, Som, Glos; RED-BERRY LAUREL, Yks.

Paradise Plant – and so it looks in one of its rare woodland stations or outside a cottage, neatly and leaflessly flowering in February or March, the scent as exquisite as the blossom. Later on the scarlet (and poisonous) berries are just as startling. Cottage people have not planted *Daphne mezereum* only as a plant of paradise. Both the *Daphnes* have been used against cancer, in folk-medicine; and for other purposes Cortex Mezerei, Mezereon Bark, once had its place in the British Pharmacopoeia.

'Apothecaries of our countrie name it *Mezereon*,' wrote Gerard, and this name in mediaeval Latin was borrowed from the Arabic *māzarjūn*, used in his medical writings by the great Arabian physician and philosopher Avicenna (980–1037). The plants of Mezereon which Gerard grew in his Holborn garden came from Poland. He did not know of it as wild in Great Britain, where, indeed, it was first recorded (in woods in Hampshire) as late as 1752.

2. Spurge Laurel. *Daphne laureola* L. 61

Local names. COPSE LAUREL, Hants, I o W; FOX-POISON, Lincs; LAUREL-WOOD, Glos; WOOD LAUREL, Som, Hants, I o W, Glos, Bucks; STURDY LOWRIES, Dur.

Daphne mezereum's drab cousin – poisonous, too, and black-fruited. When Cortex Mezerei was recognized in the British Pharmacopoeia, it was allowed either from Spurge Laurel or Mezereon (73). Spurge Laurel bark was a cottage cancer cure. Also the plant was used as a horse medicine, for which it was collected in the Sussex woods and sold in the markets at Chichester and Portsmouth (21). 'Lowry' is an older form of Laurel.

The flowers of Spurge Laurel are also fragrant, though the scent has a slightly sour or green element.

XLI. Onagraceae

1. Codlins-and-Cream, Great Hairy Willow-herb. *Epilobium hirsutum* L. 100, H 40

Local names. APPLE-PIE, Som, Glos, Herts, Ess, Suff, Ches, Yks, N'thum; BLOOMING SALLY, Ire; CHERRY-PIE, Dor, Som, Worc, Notts; CODDLED APPLES, N'hants, Notts, Lincs; CODLINS, Dev, Glos, Oxf, N'hants, Yks, Cumb; CODLINS-AND-CREAM, Dev, Dor, Som, Wilts, Glos, Suss, Surr, Berks, Oxf, N'hants, Worc, Staffs, Ches, Notts, Yks, Cumb; CURRANT-DUMPLING, N'thum; CUSTARD-CUPS, Shrop, Ches.

FIDDLE-GRASS, Yks; GOOSEBERRY-PIE, Dev, Dor, Som, Wilts, Suff; GOOSEBERRY-PUDDING, Suss; LOVE-APPLE, War; PLUM-PUDDING, Ches; RAGGED ROBIN, Som; SOD-APPLE, Som, Wilts; SUGAR-CODLINS, Som, Wilts; WILD PHLOX, Dev; WILD WILLOW, Wilts.

Long ago John Ray explained that *Epilobium hirsutum* was called Codlins-and-cream 'from the smell of the leaves a little bruised'; and this has been repeated a hundred times in book after book by botanists who never crushed the leaves and smelt them. They have no characteristic smell, nor have the flowers, although Sowerby in *English Botany* attributes the name to the flower scent, which he calls very 'transitory'.

A first clue may be in Gerard's *Herbal.* There the species is called 'Codded Willow herbe' – 'the flower groweth at the top of the stalke, comming out of the end of a small long codde'. 'Codded Willow herbe' no doubt suggested 'Codlin Willow-herb', the willow-herb of the codlin or cooking apple. Codlins were often boiled in milk and then eaten with cream, so, as a next stage, the rosy and white combination in the flowers (rosy petals, creamy-white stigma) may have suggested the name Codlins-

and-cream, which in turn suggested other fruit names, Apple-pie, Gooseberry-pie, and Cherry-pie, which is well fitted to the cherry colour of the flowers. The boiling of apples in milk also suggested (see above) the names Coddled Apples and Sod-apples. 'Coddle' and 'sod' are synonymous with 'boil'.

The name 'Willow-herb' was first used by Lyte in 1578, from the likeness of *Epilobium* leaves to leaves of the willow growing in the same habitat. The Lancastrian botanist, Thomas Penny, recorded another name in the sixteenth century: *Epilobium hirsutum* was also Milner Flower, the flower of the miller, growing by his mill or miln – the plant soft and pubescent as the miller himself (151).

2. Fireweed, Rosebay Willow-herb. *Chamaenerion angustifolium* (L.) Scop. 108, H 11

Local names. APPLE-PIE (see *Epilobium hirsutum*), Som; BLOOD VINE, Hants; BLOOMING SALLY, Ire; CAT'S EYES, Shrop; EYEBRIGHT, Dev; FIREWEED, Dor; FLOWERING WITHY, Berks; FRENCH SAUGH, Lanark; FRENCH WILLOW, Som; RANTING WIDOW (i.e. widdy, willow), Ches; TAME WITHY, I o W; WILD SNAPDRAGON, Glos.

Gerard first recorded Rosebay Willow-herb as a British species. He grew it from plants 'in Yorkshire in a place called the Hooke, neere unto a close called the Cow pasture'. In his garden it was 'very goodly to behold, for the decking up of houses and gardens'.

Gerard's description is delightful: 'The branches come out of the ground in great numbers, growing, to the height of sixe foote, garnished with brave flowers of great beautie, consisting of fower leaves a piece, of an orient purple colour. The cod is long . . . and full of downie matter, which flieth away with the winde when the cod is opened.'

Seeds by the thousand fly with the wind; but the Rosebay Willow-herb had to wait a long while before it could spread and become that familiar splash of colour in the landscape which we enjoy. Railways, industry, industrial waste land, and the felling of woodland – and then destruction by bombs – at last turned a local plant (and a plant of gardens) into a common one known to everybody.

It gave indications of spreading a century ago. 'Quite recently the Rosebay Willow-herb has become numerous in several parts of the Vale of Severn, and promises to spread, incited to take possession of new-made roads and embankments. I have observed it by the side of a diverted road near Shatterford, and in the cutting of the Birmingham and Gloucester Railway, near Croome Perry Wood' (E. Lees, *The Botany of Worcestershire*, 1867).

Rosebay Willow-herb likes ground where there has been a fire; and it was once possible, after the bombing of London in the Second World War, to see a splendour of mixed flame around St Paul's – rose-purple and yellow, *Chamaenerion* and *Senecio squalidus*. New habitats cannot be the whole story. By some genetic change Rosebay Willow-herb appears to have become a more vigorous, adaptive, and conquering species.

In the United States, as Fireweed (though the name may have been exported from the west country – see the list above – and then reimported), it has found a thousand acres made bare and habitable for every one acre in Great Britain. For example, it grows in marvellous splendour around Seattle, where you can buy Firewood Honey. According to Linnaeus, the leaves can be eaten as a vegetable, and the young shoots can be served up like asparagus.

'Rosebay' added to 'Willow-herb' comes out of the name *Chamaenerion*, bestowed by Gesner – the *nerion*, or Oleander (for which the English name was Rose Bay), which grows on the ground – *chamai*. The rosy flowers and the leaves are the only points of a faint resemblance between Oleander and Rosebay Willow-herb. Really the plant deserves a name which is less awkward and bookish. Fireweed is too weedy. The French call it *Antonine* or *herbe de Saint-Antoine* (which suggests St Anthony's Fire), the Swedes *Himmelgraes*, Herb of Heaven.

3. Enchanter's Nightshade. *Circaea lutetiana* L. 99, H 40

Local names. DRAGON-ROOT, Ire; MANDRAKE, Dev; PHILTREWORT, Yks; WITCH-FLOWER, Som.

In Germany *Circaea lutetiana* was linked with witchcraft, but in Great Britain there is no evidence of any connection with witches or enchanters before the publication of the herbals. The Flemish botanist de l'Obel (1538–1616) tried to identify the *kirkaia* of Dioscorides, and the *circaea* of Pliny, which was employed as a charm. It was not difficult to conclude that a species with such a name had to do with Circe, Homer's beautiful witch who transformed the crew of Ulysses into pigs. At first de l'Obel favoured the witch-plant *Solanum dulcamara* (q.v.). Then he changed his mind and decided for this plant, which is still the *Circaea lutetiana*, the Parisian or Lutetian Circaea, the Circaea of the botanists of Paris. Once de l'Obel had made his identification, *Circaea* acquired the English name of Inchaunter's Nightshade, which occurs for the first time in Gerard.

A plant known to the Anglo-Saxons as *aelfthone*, used in a medicine against elf-sickness, has been equated with *Circaea* (169, 88), but only by

34 Enchanter's Nightshade *Circaea lutetiana*

a guess. In France its names include *herbe à la magicienne*, *enchanteresse*, *herbe aux sorciers*, or *herbe aux sorcières*, with Christian counterparts, *herbe de Saint-Simon*, *herbe de Saint-Étienne*. In Germany it is *Hexenkraut* and *Heiliges Stephanskraut*. It does, as Gerard remarked, flourish in 'obscure and dark places'. Wrongly, too, and persistently, he went on, there were people who attributed unto *Circaea* 'the vertues of Mandragoras' (see the Devonshire name above), following a Latin version of Dioscorides in which *mandragora* was also named *circeon*.

XLII. Hippuridaceae

1. Mare's Tail. *Hippuris vulgaris* L. 101, H 40

Local names. BOTTLE-BRUSH, Hants, Yks; BRUSHES, Dor; CAT'S TAILS, Dev, Som, Oxf, E Ang; JOINT-GRASS, Dev, Herts; OLD MAN'S BEARD, Dor; PADDOCK'S PIPES (i.e. toad's pipes), Cumb; WITCH'S MILK, Lancs, Scot.

Equisetum species in shops (see Gerard) were *cauda equina*, 'horse-tail'. *Hippuris vulgaris*, thought to be the female kind among the larger male Horse-tails, was distinguished as *cauda equina femina*. This has been transformed (in books) into the tail of a female horse, Mare's Tail. In English neither *Hippuris vulgaris* nor any *Equisetum* appears to have been known as a Horse-tail, outside botanical literature. Cat's Tail is the commonest name. They have been associated far and wide with fairies, goblins, toads, snakes, and the Devil (Old Man's Beard).

Some *Equisetum* names are: FAERIES' SPINDLES, TROWIE-SPINDLES ('trow' is goblin or devil), Shet; SNAKE-PIPES, Som, Ess, Norf, Cam; TOAD'S PIPES, Lincs, Lancs, Yks, West, Cumb, N'thum; PADDOCK'S PIPES, Berw, Rox, Lanark.

XLIII. Loranthaceae

1. Mistletoe. *Viscum album* L. 54

Local names. CHURCHMAN'S GREETING, Som; KISS-AND-GO, Dor; MASSLIN, Suff; MISLE, Scot; MISLIN-BUSH, E Ang.

Species of Mistletoe are revered from Europe to Japan, New Guinea, and West Africa. Here is a plant lifted abnormally from the earth and growing upon another plant, an excrescence, living mysteriously, it seemed, without food and by its own vitality; which is the chief factor. Thus Polypody fern was powerful in medicine because it would grow upon trees, and especially upon the strong, powerful, and revered oak-tree. An elder growing out of the willow was more powerful than an elder growing normally in the ground, and was used to make amulets against epilepsy (14).

One distinction is noticed after another, one reinforces another, building up a complex of attributed powers and virtues. Mistletoes are still green (if they are not deciduous species) when all the leaves of the host have dropped – they are evergreens like the magical Houseleek or Periwinkle. Their shape is peculiar, their winter berries are conspicuous; and

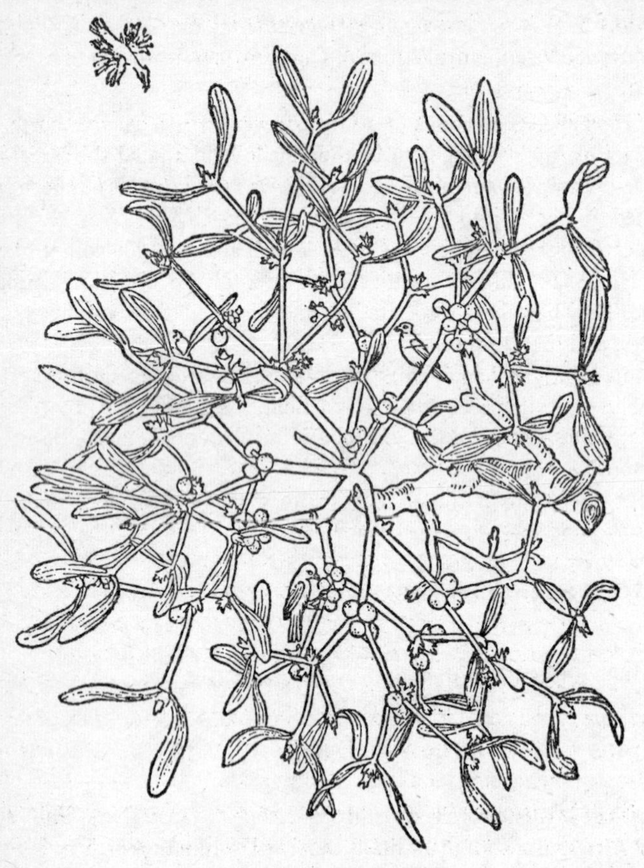

35 Mistletoe *Viscum album*

in the way the berries of *Viscum album* are arranged, they have a likeness to the male parts. So Mistletoe induces fertility ('Some women have worn it about their necks, or on their arms, thinking it will help them to conceive', wrote William Coles in *Adam in Eden*, 1657), women receive a kiss under the Mistletoe, it is an aphrodisiac, it improves the apple crop, it overcomes the death-like trances of the epileptic (not a fancy altogether, since Mistletoe has an active principle which is antispasmodic and reduces blood pressure). Its *mana* is so strong that it keeps evil and witchcraft, the craft and subtility of the devil, at bay; it will open a lock, and used as a divining rod it reveals treasure. In Scandinavian mythology, as described in the Prose Edda, it was Mistletoe, by Loki's cunning, which killed Balder the Good. Frazer in *The Golden Bough* interpreted Mistletoe as the sacred heart of the oak, and Balder as the manifestation of the Mistletoe-bearing oak tree, who could not be killed except by the heart.

England and all Europe inherited Pliny's account of Mistletoe and the Gallic Druids, which makes it difficult to disentangle native beliefs about the plant from beliefs due to the *Historia naturalis*. Pliny numbered three kinds of Mistletoe; including *viscum*, which grew on oaks, wild pear, etc. To the Gauls and their Druids, oaks and oak groves were sacred, and everything as well which grew upon the oak (cf. Houseleek as the plant dropped from Jupiter). Mistletoe upon an oak was a sign that the god had chosen that particular tree; and since oak Mistletoe is rare, Pliny went on, it was cut with great ceremony by a priest in white robes, with a gold sickle, and in such a way that it fell on to a white cloak (never touching that ground which is foreign to Mistletoe). Then two white bulls were sacrificed. Mistletoe was believed to cure barrenness in animals and to be a good antidote to poisons. Elsewhere in the *Historia naturalis* Pliny gave more information about the medicinal virtues attributed to Mistletoe, especially if it was cut from an oak at the new moon, and without iron (the taboo against the use of iron in gathering simples was ancient and wide-spread through Europe): 'If it has not touched the ground, it will cure epilepsy, and carried by women, will make them fertile.'

The suggestion that Pliny's druidic *viscum* was the continental *Loranthus europaeus*, of oak, lime, and sweet chestnut, can be ruled out. It ignores that *viscum* was rare on oak trees and prized upon that account. Also *Loranthus europaeus* is not evergreen.

English customs must be independent of Pliny's record, yet influenced by it in later times. The name we use hints at the picking and employment of *Viscum album*. In OE it was *mistiltān*, the part for the whole: *mistil*, the kind of plant, plus *tān*, a twig – and rather more, a special twig, used in magic. In one Anglo-Saxon incantation against venom, *tān* is the word for the 'nine twigs of glory', with which Woden broke the adder into nine parts (88). In England Mistletoe belonged to the Christmas cycle, which lasted till Candlemas; it did not belong exclusively, or perhaps originally, to Christmas Day itself. The customs and beliefs must have been limited somewhat by the geographical range of the plant, which is abundant in the counties of the Severn, scattered elsewhere, absent from most of the northern counties, from Scotland, and from Ireland. In Herefordshire, the heart of the Mistletoe country, a Dr H. G. Bull busied himself nearly a hundred years ago upon a local investigation, the results of which (in the *Transactions of the Woolhope Naturalists' Field Club*, 1864) have been raided in a score of books without acknowledgment. He found that Mistletoe grew upon some 34 per cent of apple trees in the Herefordshire orchards. In Herefordshire cottages and farmhouses the Mistletoe bough was cut on New Year's Eve, 'and hung up in state as the clock strikes twelve; the old one, which was hung throughout the year, is at the same time taken down and burnt'. A correspondent described the custom in Worcestershire, where the bough was cut 'by the last male domestic that has entered the family'. Dressed with nuts, apples, and ribbons, it was 'suspended in the centre of the room, sometimes with a cord attached to a pulley, to allow of its being lowered for the lady to pick a berry' – after she had been kissed. The berry was then thrown by the woman over her left shoulder. In Worcestershire the bough was hung on Christmas Day and left till the following Christmas. It was held to be efficacious against witches and goblins.

The *Woolhope Club Transactions*, in the volume for 1933–5, print another paper on Mistletoe by H. E. Durham; this mentions instances of Mistletoe growing in Herefordshire on oak (e.g. at Stretton Sugwas, Moccas, and at Pengrove within the bounds of Hereford itself) and one case of holly Mistletoe, another of an infested mulberry tree. The gardens of Holm Lacy Court, five miles from Hereford have, or then had, Mistletoe on birch, hawthorn, pear, lime, apple, locust, red buckeye, and *Zelkova*.

Herefordshire uses in medicine agreed with Pliny, more or less. Mistletoe was given to sheep, especially *after* lambing, and to cows after calving; also to human beings for epilepsy (sometimes when mixed with Spurge Laurel). For doctoring man or animal, Herefordshire people picked Mistletoe from hawthorns and maples. To hang up in the house 'for good luck through the year', the best Mistletoe came from apple trees and poplars. Hawthorn itself is magically effective. So is the Apple tree, while Maple, sometimes regarded in the country as a kind of oak, may also have had an apotropaic power and have been used on May Day. A case was quoted from Worcestershire of Mistletoe being given to a pig which a neighbour had 'overlooked'. As a plant of the Christmas cycle, at any rate, it was a companion to the evergreen Holly and the evergreen Ivy, its berries white to their berries of red and black. In his *Adam in Eden*, William Coles remembered two lines of an 'old carol':

Holly, and Ivy, Misseltoe
Give me a red apple and let me go.

How did a local custom become national and commercialized, as it is today? The answer is the bedevilling mania in the eighteenth and early nineteenth centuries for Druids, and all to do with them – 'druidic' temples, 'druidic' megaliths, and the Druids' plant. The extreme protagonist of the Druids was William Stukeley (1687–1765), who peopled Avebury and Stonehenge with these shadowy priests, whom he made into patriarchal forerunners of Christianity. In his garden at Grantham he contrived a druidic temple out of an old orchard, an imitation Stonehenge of two circles, one of pyramidal evergreens, one of nut trees. In the middle was 'an antient apple tree oregrown with sacred mistletoe'. He concluded that the Druids had liked Mistletoe for its beauty, its unusual mode of growth, its maturity in winter, 'when all nature lyes dormant', and because they found it, on account of that third reason, 'a type of the expected Messiah'. Also he fancied that every prehistoric bronze axe found in Great Britain must have been used by Druids in slicing Mistletoe from the sacred oak (171, 172).

Druids and Mistletoe increased their fame together, and the plant became an article of Christmas export from Herefordshire. The demand increased, until supplies had to be augmented from Normandy.

Stukeley, John Ray, and other botanists also busied themselves with experiments in propagation. A statement of Pliny's had become a dogma: 'Whenever Mistletoe is sown, it fails altogether to sprout, which it will only do when it has passed through birds – particularly through pigeons and thrushes. That is its nature: if it is to grow, it must first of all be ripened in the guts of a bird.' Ray encouraged the endeavours of a London apothecary who grew Mistletoe on a white poplar (155). William Derham, F.R.S., germinated the seeds 'even in the Bark of Oak' (59); and countless attempts must have been made to grow Mistletoe on the Druids' tree. In the nineteenth century nearly twenty mistletoe oaks were known, most of them in Monmouthshire, Gloucestershire, and Herefordshire – some of them, perhaps, the victims of triumphant experiment, and one of them suspiciously alongside a 'druidic' chamber-tomb. In fact, Mistletoe is not at all difficult to propagate: you simply stick the berries on a branch of one of the more usual hosts, tying a bandage or a strip of butter muslin around berry and bough (to protect the berry, but also to show where the berry was placed).

The paganism of Mistletoe, it is often said, kept it out of mediaeval carvings in church or from church decorations. But one must remember again the limited range of Mistletoe as a common plant. A parson of 1780 or 1830, knowing his Pliny or his Stukeley and how Mistletoe had been implicated in pagan worship, might have frowned it from the decorations at Christmas. But in the Middle Ages, if Hawthorn and the May Lord (see p. 182) could be carved on shrine or pillar, Mistletoe as a counter against demons might have been carved as well – where Mistletoe customs were strong enough and where there was point enough in sanctifying its power. Mistletoe is supposed to be carved with oak leaves in Bristol Cathedral around the tomb of Thomas, Lord Berkeley, who died in 1243. In fact the eaves are maple, and the mistletoe leaves and berries are maple keys – a pity since Mistletoe would have fitted the tomb of a nobleman who lived on the banks of the Severn in the mistletoe country. However, at Bilston church in Staffordshire (again in a mistletoe district), the plant during the seventeenth century was blessed and distributed at Christmas time by the incumbent (18); and William Stukeley declared that 'till lately' – he was writing in the seventeen-fifties – Mistletoe had been carried to the high altar of York Minster on Christmas Eve (172). The *mistil* q-mistletoe derives from the Old High German *mist*, 'dung', this 'dung twig' having grown from mistletoe seed left on branches in the droppings of birds (especially the missel-thrush, as mentioned by Aristotle) who had gobbled up the berries.

XLIV. Cornaceae

1. Dogwood. *Cornus sanguinea* L. 68, H 12

Local names. CAT-TREE, CAT-WOOD, Shrop; CORNALEE (cf. Dutch *kornoelie*), Ches; CORNWOOD, Dor; DEVIL-MAY-CARE, Dor; DOG-TIMBER, Dev, Dor, Som; DOGWOOD, Wilts, War; GATTER-BUSH, Kent, E Ang; GATTER-TREE, Lincs; GATTERIDGE (for these goat names see *Euonymus europaeus*), E Ang, S Eng; PRICKWOOD, Som, Bucks, E and S Eng; SKEWER-WOOD, Bucks, Heref; SKIVER-TIMBER, Som; SKIVER-TREE, Corn, Dev, Dor, Som, I o W, Glos, Bucks; SKIVER-WOOD, I o W, Shrop; SNAKES' CHERRIES (the black fruit), Som; WIDBIN, Bucks.

A splendid shrub on the chalk hills, when the leaves turn to burgundy or claret in the autumn. Otherwise the Dogwood has a bad name. It smells

36 Dogwood *Cornus sanguinea*

rather unpleasant. Its black fruit have been no good to anyone ('We for the most part call it the *Dogge berry tree*, because the berries are not fit to be eaten, or to be given to a dogge' – Parkinson in *Theatrum Botanicum*, 1640), and the skewers (skivers, pricks) which gipsies made of the wood have gone out of fashion. Evelyn recommended the timber, not only for skewers, but for millcogs, pestles, spikes, and wedges – it is 'of so very hard a substance as to make wedges to cleave and rive other wood with instead of iron' (70). It made arrows, ramrods, toothpicks, pipe stems, and gunpowder (125). But the names, compounded with cat, dog, goat (several of which the Dogwood shares with *Euonymus europaeus*, *Viburnum opulus*, and *Viburnum lantana*), speak of a contempt we no longer have to share.

XLV. Araliaceae

1. Ivy. *Hedera helix* L. 112, H 40

Local names. BENTWOOD, Berw; BINDWOOD, Scot; HIBBIN, I o M; IVIN, Ches, Derb, Lincs, Lancs, Yks, Cumb, N'thum, N Eng; IVY, Dev, Wilts, Hants, Glos, Kent, N'thum, N Eng, Scot, etc.; IVERY, Ess, Herts, E Ang, Rut, Notts, Lincs, Scot; LOVE-STONE, Leic.

Ivy was mournful and magic. When the other flowers are over, the wasps come around the green flowers of Ivy in the late autumn, in competition with the flies. Ruins, owls, and Ivy go together – or they did until the Ministry of Works decided that ruins were all to be historical exhibits; picked, pickled, pointed, and sterilized.

Tintern Abbey is now clean and dull, and with our eyes shut we find it beautiful by the force of tradition. The tradition began with Wordsworth, the early travellers and the painters, when Tintern was hung with Ivy. 'The shapes of the windows are little altered; but some of them are quite obscured, others partially shaded, by tufts of ivy, and those which are most clear are edged with its slender tendrils, and lighter foliage, wreathing about the sides and the divisions; it winds round the pillars; it clings to the walls; and in one of the aisles, clusters at the top in bunches so thick and so large, as to darken the space below ... nothing is perfect; but memorials of every part still subsist; all certain but all in decay; and suggesting, at once, every idea which can occur in a seat of devotion, solitude, and desolation ... No circumstance so forcibly marks the desolation of a spot once inhabited, as the prevalence of nature over it.' In ruins 'an intermixture of a vigorous vegetation intimates a settled despair of their restoration' (Thomas Whatley, *Observations on Modern Gardening*, 1770).

Old-fashioned but true, these remarks on ivied Tintern. We are forgetting the symbolic Ivy. We have forgotten the Ivy which once had a magical power. So Ivy is no longer the *sine qua non* of Christmas decoration, and the inevitable companion of the Holly. A pity, because you do not have to purchase Ivy at an outrageous cost from the greengrocer, and you can fix it easily around the house, where the umbels, with or without berries, make delicious starry patterns against the wall.

Ivy's function at Christmas has been described under Holly. A few more things may be added. In the Highlands and Islands it kept evil away from the milk, the butter, the animals. An old man in Uist told Alexander Carmichael, author of the delightful *Carmina Gadelica* (vol. ii, 1928), how he used to swim out to an island in a lake to fetch Ivy, Rowan, and Honeysuckle for this protection. Circlets of Ivy alone, or of Ivy plaited with Honeysuckle and Rowan, were hung over the lintel of the byre and put under the milk vessels. Presumably this was done on the night before May Day. In France Ivy is often counted one of the herbs of St John, collected – see *Hypericum perforatum* – before the strengthening *feux de joie* were lit on St John's Eve (188); and mention has already been made of drinking from ivywood cups against whooping-cough (see page 127).

XLVI. Hydrocotylaceae

1. Pennywort, White Rot. *Hydrocotyle vulgaris* L. 112, H 40

Local names. FAIRY TABLES, Ches; FARTHING ROT, Norf; FLOWKWORT (i.e. 'causing' the flukeworms of liver rot in sheep), Norf; PENNY ROT, Shrop; ROT-GRASS, N'thum; SHEEP ROT, War, Cumb, N'thum, Caith; SHILLING ROT, Ayr; WATER ROT, Ches; WHITE ROT, Shrop, Ches.

'It is necessary that a shepeherde shoulde knowe what thynge rotteth sheepe,' wrote Fitzherbert in *The Boke of husbandrie*, 1523. First he accused Lesser Spearwort, *Ranunculus flammula*, then *Hydrocotyle*: 'An other grasse is called peny-grasse, and groweth lowe by the erthe in a marsshe grounde, and hath a leafe as brode as a peny of two pens, and never beareth floure' (the flowers are minute and need searching for). Tudor pennies were still made of silver, which made the pale leaves more penny-like than they appear to us. *Drosera rotundifolia*, *Pinguicula vulgaris*, and *Pedicularis palustris* were also fancied to cause the liver rot.

XLVII. Umbelliferae

Similarity of form has made it difficult at a glance to distinguish one umbelliferous species from another, so there is much overlapping, or overspilling, of the English names. The hollow dry stems were useful for a score of purposes and they are known by various words often applied to the species themselves – 'kecks', in general currency in England, 'kelk' in Wilts, Suss, Kent, Surr, Yks, Dur, N'thum, N Eng; 'kesh' or 'kewse' in Lincs, Lancs, Yks, and the Lakes; 'keglus' in Ches; and 'keggas' in Corn, etc.

1. Sanicle. *Sanicula europaea* L. 110, H 40

Common, and a choice plant, especially when it is dominant in a wood or copse. It is 'Sanicle' as a healing herb – 'Sanicula *a sanandis vulneribus*,

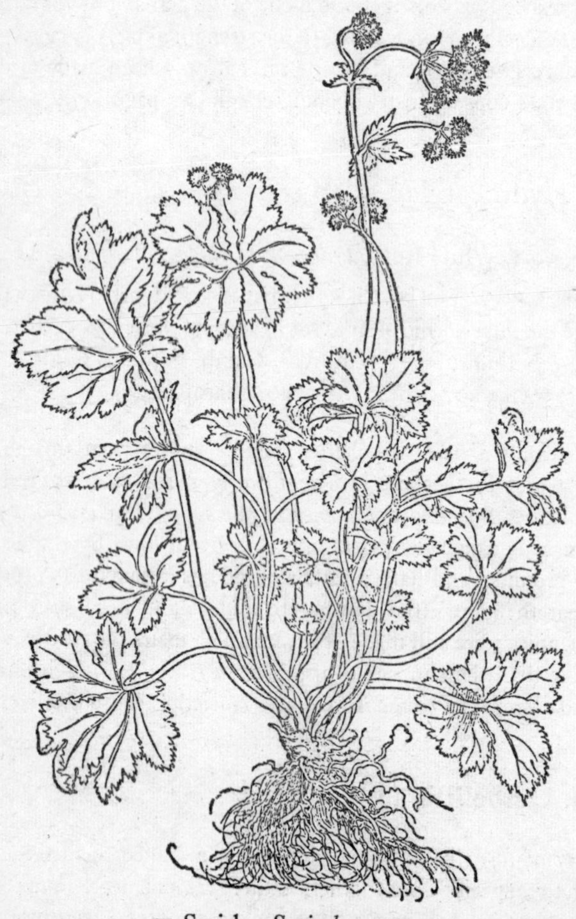

37 Sanicle *Sanicula europaea*

of healing woundes'. It was given in a 'wound drink' for making sound 'all inward wounds, and outer hurts' (81). A fifteenth-century wound drink was made of Sanicle, Yarrow, and Bugle, stamped in a mortar and tempered with wine. 'This is the vertu of this drynke: bugle holdith the wound open, mylfoyle clensith the wound, sanycle helith it.' But it was emphasized that Sanicle must not be given to anyone with a wound in the head, or a broken skull, for fear of killing him (54). According to Bullein, in his *Booke of Simples* (1562), the juice was also valuable for 'wounded or bruised oxen, milche kyene and horses'. In the Orto Botanico at Padua, one of the oldest physic gardens in the world, it is pleasant to find Sanicle carefully tended between the roots of the tulip tree.

Sanicle appears to be the solitary English name, except for Wood Elder (Somerset). In Normandy it was named *herbe de St-Laurent* after the saint and martyr who was grilled to death on a gridiron, and was therefore invoked for burns and scalds (116).

2. Sea Holly. *Eryngium maritimum* L. 54, H 19

Falstaff in *The Merry Wives of Windsor* (V. v) goes to Windsor Forest to meet Mistress Ford:

Mistress Ford. Sir John! art thou there, my dear, my male deer?
Falstaff. My doe with the black scut! Let the sky rain potatoes; let it thunder to the tune of 'Green Sleeves', hail kissing-comfits and snow eringoes; let there come a tempest of provocation . . .

Candied eringoes he meant – candied roots of Sea Holly, which were good for those 'that have no delight or appetite to venery' – 'nourishing and restoring the aged, and amending the defects of nature in the yoonger' (81). Careful wives grew Sea Holly in the physic garden, though it is properly a plant of sand dunes, with roots which grope through the sand for five or six feet, and waxed leaves which reduce the amount of respiration.

Grecian Eringoes now commence their Fame
Which worn by Brides will fix their Husbands' Flame
And check the Conquests of a rival Dame

– according to René Rapin in his poem on gardens (150).

Colchester in Essex was the centre of the eryngo trade. Candied with sugar and orange flower water, Colchester eryngoes were sold from about 1600 until as late as the eighteen-sixties.

3. Rough Chervil. *Chaerophyllum temulum* L. 103, H 20

Rough, blotched stems and white flowers which come out along the roads in July, when the fine show of Cow Parsley is dying away. A few names have been recorded – Cherril in Yorkshire, Cow-mumble and Sheep's Parsley in Suffolk.

4. Cow Parsley. *Anthriscus sylvestris* (L.) Bernh. 112, H 40

Local names. ADDER'S MEAT, Som; BAD MAN'S OATMEAL (i.e. the devil's oatmeal), BUN, Yks; DEIL'S MEAL, Scot; DEVIL'S MEAT, Yks; DEVIL'S OATMEAL, Surr; DEVIL'S PARSLEY, Ches; NAUGHTY MAN'S OATMEAL, War; CICELY (from Latin and Greek *seselis*, Greek, *seseli*. See *Myrrhis odorata*), Derb, Yks; CISS, Lancs; CISWEED, Yks; CONEY PARSLEY, Suss; COW CHERVIL, Som; COW-MUMBLE, Ess, Norf, Cam; COW PARSLEY, Hants, Glos, Bucks, Oxf, Ess, Norf, Cam; COW-WEED, Ess.

DA-HO, N Ire; DOG PARSLEY, Herts; DOG'S CARVI, Shet; ELDROT, Dor; ELTROT, Dor, Wilts; OLDROT, GIPSY CURTAINS, GIPSY FLOWER, Som; GIPSY LACES, Dor, Som; GIPSY'S PARSLEY, GIPSY'S UMBRELLA, Som; HA-HO, HI-HOW, Ire; HARE'S PARSLEY, Som, Wilts; HEMLOCK, Ches, Lincs, Yks; HONITON LACE, Dev; JUNE FLOWER, Som; KADLE DOCK, Ches; KEDLOCK, Derb; KEESHION, N Ire; KELK, Wilts, Suss, Kent, Surr, Yks, Dur, N'thum; KELLOCK, Lincs; KESK, Cumb; KEWSIES, Lincs.

LADY'S LACE, Som; MY LADY'S LACE, Dor; LADY'S NEEDLEWORK, Glos; MAYWEED, Worc; MOONLIGHT, Wilts; QUEEN ANNE'S LACE (? transferred from the Virgin, or St Anna, the mother of the Virgin); QUEEN ANNE'S LACE HANDKERCHIEF, Dor; RABBIT'S FOOD, Bucks; RABBIT'S MEAT, Dev, Som, Suss, War, Lincs, Yks; SCABBY HANDS, Som; SCAB FLOWER, Cumb; SCABS, Som; SHEEP'S PARSLEY, Kent, Norf; SWEET ASH (i.e. ache, parsley), Glos; WHITE MEAT, WHITEWEED, Yks; WILD CARRAWAY, Banff; WILD PARSLEY, Lincs, Rad.

To most of us now an innocent-seeming plant, whitening the road verges in June, filling up pathways, and narrowing green roads to a single track – a plant symbolic altogether of the English countryside and summer travel:

> The Junes were free and full, driving through tiny
> Roads, the mudguards brushing the cow parsley.

But it was connected with the Devil. Many of its names are applied as well to the ultra-poisonous Hemlock; and Reginald Scot, in his *Discoverie of Witchcraft*, wrote of witches who had hog's dung and Charvil about them, perhaps meaning this plant of lace and moonlight. In Norwegian (cf. Dog's Carvi, from Shetland) *Anthriscus sylvestris* is *hundkjeks*, 'dog kecks'.

5. Shepherd's Needle. *Scandix pecten-veneris* L. 103, H 38

Local names. ADAM'S NEEDLE, Som, N'thum, Berw; BEGGAR'S NEEDLE, Som, Shrop, Midlands; CLOCK NEEDLE, Bucks; COMB, Suff; CRAKE NEEDLE (i.e. crow needle), Yks, N Eng; CROW NEEDLE, Som, I o W, Bucks, N'hants, Lincs; CROW-PECKS, Som, Wilts, Hants; DEIL'S DARNING-NEEDLE, Berw, Lanark; DEIL'S ELSHIN (i.e. awl), Berw; DEVIL'S DARNING-NEEDLE, Som.

ELSHINS, Berw; HEDGEHOGS, Dor, Som, Suss; LADY'S COMB (see below), Donegal; LONG BEAKS, Dor; NEEDLE-POINTS, Ess; OLD WIFE'S DARNING-NEEDLES, Yks; OLD WOMAN'S NEEDLE, Hants; PINS AND NEEDLES, Som; PUCK-NEEDLE, POOK-NEEDLE, Hants, Suss; SHEPHERD'S COMB, Yks; SHEPHERD'S NEEDLE, Shrop, Lanark; TAILDERS, Corn; TAILOR'S NEEDLES, Corn, Dev; WILD PARSLEY, Bucks; WITCH'S NEEDLES, Berw.

When the long, erect fruits develop among the legs of the wheat, Shepherd's Needle looks so peculiar that a farm population could not help noticing it and naming it. Turner in 1548 invented the dull 'Corne chervel', and there is a resemblance to Chervil before the needles have shaped themselves – as Gerard wrote, '*Scandix*, or *Pecten Veneris*, doth not much differ in the quantitie of the stalkes, leaves, and flowers from Chervill'. Gerard goes on that, after the flowers, 'come uppe long seedes, very like unto pack needles, orderlie set one by another like the great teeth of a combe'. Pliny's *pecten Veneris*, 'comb of Venus', may have been our species. Gerard's names include Ladies Combe, a Christian counter-name to Venus's comb, and Shepheard's Needle, which is simply a translation of the apothecary's *Acus pastoris*.

An obvious plant to give to goblin (Puck, or Pook), devil, and witch. '*Acus demonis*', says Banckes's *Herball* of 1525, 'This is an herbe that men call Poukenell.' The 'Old Wife' and 'Old Woman' of the Yorkshire and Hampshire names may be equivalent to witch.

6. Sweet Cicely. *Myrrhis odorata* (L.) Scop. 72, H 16

Local names. ANISE, Dur; WILD ANISE, Cumb; ANNASEED, Cumb; COW-WEED, Yks; MYRRH, Cumb, Aber; ROMAN PLANT, Lancs; SWEET BRACKEN,

Lakes; SWEET CICELY, Yks, Scot; SWEET CIS, Yks; SWEET FERN, Ess; SWEET HUMLICK (i.e. hemlock), Berw; SWEET WITHY, I o W; SWEETS, N Eng; SWITCH, Lancs.

Whether a native or no, this ancient pot-herb with sweet-scented leaves has gone wild in Scotland and the north, often in colonies along the mountain roads. There it is a fine plant, though in the garden it looks too plainly umbelliferous. Cicely is a confusion in its form with the girl's name, though it comes by the Latin from the Greek *seselis* or *seseli*. *Myrrhis odorata* was equated both with the *seseli* and the *murrhis* of Dioscorides. According to Culpepper, the candied roots, like the roots of *murrhis*, were thought to prevent infection by the plague. Gerard wrote of the leaves as 'exceeding good, holsome, and pleasant among other sallade herbes, giving the taste of Anise seed' – if you can bear that – 'unto the rest'.

He ate the boiled roots with oil and vinegar, and considered them very good for old people who are 'dull and without courage', rejoicing and comforting the heart, and increasing their 'lust and strength'. The unripe seeds were also eaten with oil, vinegar, and pepper. In Westmorland Sweet Bracken was not only used in puddings but also 'for rubbing upon oaken panels, which, when dry, being rubbed again with a cloth, receive a fine polish and agreeable scent' (*Troutbeck, by a Member of the Scandinavian Society*, 1875).

In South Wales I have often noticed Sweet Cicely growing in graveyards, set, I think, as a plant of memory and sweetness around the headstones (though see *Sambucus ebulus* for graveyard herbs).

7. Upright Hedge-Parsley. *Torilis japonica* (Houtt.) DC. 109, H 40

Local names. DEVIL'S NIGHTCAP, War; HOGWEED, Glos; HONITON LACE, Dev; LACE-FLOWER, Som; LADY'S LACE, Som; LADY'S NEEDLEWORK, MOTHER-DEE (cf. *Melandrium album* and *Conium maculatum*), Ches; PIG'S PARSLEY, Som; RED KEX, Yks; SCABBY-HEAD, Ches.

For the mixture of good and evil names, see *Anthriscus sylvestris*, the Cow Parsley, and the companion plant of white lace along the roads, though *Torilis japonica* blossoms as the true Cow Parsley fades. It is the roadside lace of high summer.

8. Alexanders. *Smyrnium olusatrum* L. 72, H 40

Local names. ALEXANDERS, Corn, Scot; ALICK, Kent; ALLSANDERS, Corn; ALSHINDER, Scot; HELLROOT, Dor; MEGWEED, Suss; SKIT, SKEET, Corn; WILD CELERY, I o W.

Its green stems glistening in the sunlight, Alexanders is a relic of old cultivation as a pot-herb or vegetable – a naturalized plant from the Mediterranean, which is happiest and most frequent by the sea. In mediaeval Latin the name was the *petroselinum Alexandrinum*, the parsley of Alexandria. 'Our Allisanders', Parkinson wrote in *Theatrum Botanicum* (1640), 'are much used to make broth with the upper part of the roote, which is the tenderest part, and the leaves being boiled together, and some eate them either raw with some vinegar, or stew them, and so eate them, and this chiefly in the time of Lent, to helpe to digest the crudities and viscous humours [which] are gathered in the stomacke, by the much use of fish at that time.' It had already naturalized itself outside the garden. By 1562 Turner had found Alexanders 'in Ilandes compassed about by the sea', notably on Steep Holm in the Bristol Channel (see pp. 53-4), where there had been a small religious house in the Middle Ages. In England and in Ireland you find it often by ruins of castles and abbeys. Scully's *Flora of Co. Kerry* (1916) records Alexanders at Old Castle, Ventry, the ruins at Castlelough, the abbey ruins at Ballinskelligs and Abbydorney, etc. Caleb Threlkeld, in *Synopsis stirpium Hibernicarum* (1727), mentioned that a soup was still made of Alexanders, Watercress, and nettles, which the Irish matrons called 'Lenten Potage'. Evelyn, in some gardening notes he compiled between 1688 and 1706 (Evelyn Library, Christ Church, Oxford), still included it among the 'plants for the Kitchin-Garden'.

This old vegetable might be worth cultivating again. When they are boiled, the young shoots have a pleasant, unusual substance, and a pleasant taste to begin with, though the aftertaste is decidedly bitter and forbidding. But then old gardening manuals advise that Alexanders should be earthed and blanched, as we earth the celery by which it was replaced.

9. Hemlock. *Conium maculatum* L. 108, H 40

Local names. BAD MAN'S OATMEAL (i.e. devil's oatmeal), Yks, Dur, N'thum; BREAK-YOUR-MOTHER'S HEART, Dor; BUNK, Norf; CAISE, Yks; CARTWHEEL, Som; DEVIL'S BLOSSOM, Dev; DEVIL'S FLOWER, Som; GIPSY FLOWER, GIPSY CURTAINS, HARE'S PARSLEY, Som; HECH-HOW, Scot; HEVER, Dor; HONITON LACE, Dev.

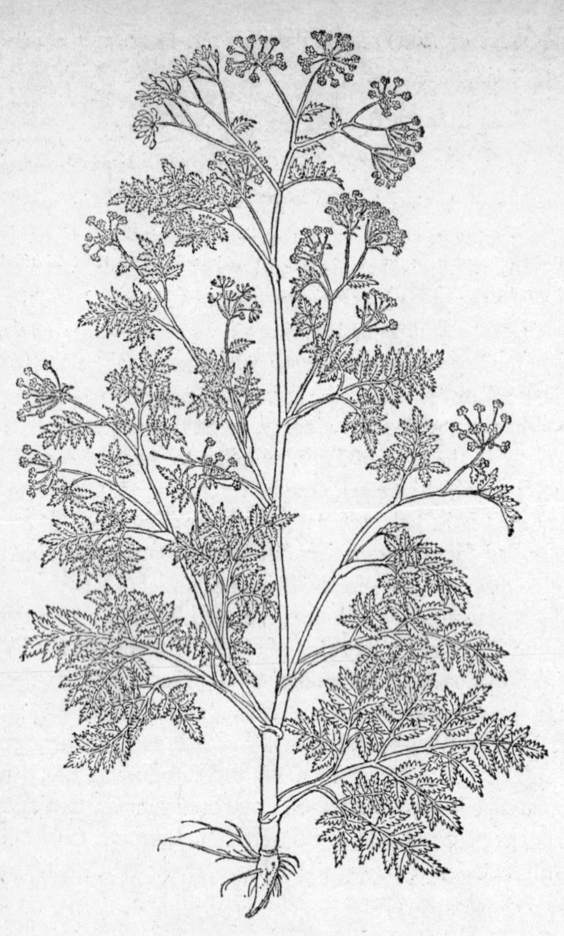

38 Hemlock *Conium maculatum*

HUMLOCK, Norf, Lincs, Yks, Cumb, N'thum, Scot; HUMLY, Rox; KAKA, Ork; KAKEZIE, Dev, Som; KELK, Wilts, Suss, Kent, Surr, Yks, Dur, N'thum, N Eng; KESH, Lancs, Yks, Lakes, N Eng; KEWSE, Lincs, Lancs, Yks; KEXIES, Som; KOUSHE, Lincs, Lancs; KOUSHLE, Lincs.

LACE FLOWER, LADY'S LACE, LADY'S NEEDLEWORK, NOSEBLEED, PICK-POCKET, SCABBY HANDS, STINK FLOWER, Som.

Imagine the last hours of Socrates, if it was *Conium maculatum* that put him to death: 'The general symptoms are salivation, bloating, dilation of the pupils, rolling of the eyes; laboured respiration, diminished frequency of breathing, irregular heart action; loss of sensation, convulsions, uncertain gait, falling, and at the end complete paralysis.' The mind usually remains clear to the end (123, 134). Obviously Hemlock has been recognized as a devil's plant, under names which have spilled over to *Anthriscus sylvestris* and *Torilis japonica*.

Every herbal uttered its warnings against Hemlock – 'one of the deadly poisons,' Gerard observed, 'which killeth by his colde qualitie'. The blotched stems have a deadly look, as though they bore their own signature of destruction and mortification, and children have been poisoned by making them into whistles and blow-pipes. About the effect of Hemlock upon animals William Coles included a tall story in his *Art of Simpling*, 1656: 'If Asses chance to feed much upon Hemlock, they will fall so fast asleep that they will seeme to be dead, in so much that some thinking them to be dead indeed have flayed off their skins, yet after the Hemlock had done operating they have stirred and wakened out of their sleep, to the griefe and amazement of the owners.'

10. Fool's Watercress. *Apium nodiflorum* (L.) Lag. 98, H 40

Local names. BILDERS, Dev; BROOKLIME, War; COW-CRESS, Hants; FOOL'S WATERCRESS, Som; PIE-CRESS, Dev; SWINE'S CRESS, Yks, Ork; WATER-CASE, Corn.

The name Fool's Watercress is not altogether justified. *Apium nodiflorum* often grows with Watercress and may sometimes be picked and confused with it. But in the west of England country people know the difference and they deliberately cook *Apium nodiflorum* with meat in pies or in pasties.

11. Earthnut, Pignut. *Conopodium majus* (Gouan) Lor. & Barr. 112, H 40

The two general names are Earthnut (in various forms, e.g. Arnut in Scot, Ire, and N Eng) and Pignut, which has been recorded from Corn, Dev, Som, Wilts, Ess, Suff, Hunts, N'hants, War, Shrop, Ches, Derb, Notts, Lincs, Lancs, Yks, West, Scot.

Also BAD MAN'S BREAD (i.e. the devil's bread), DEIL'S BREAD, DEIL'S OAT-MEAL, Yks; CAIN AND ABEL, Dur; CATNUT, Yks; CUCKOO POTATO, Donegal; CURLUNS, Kirk, Wigt; FAIRY POTATOES, Ire; FARENUT, VARENUT ('fare', a young pig, OE *fearh*), Corn; FERN-NUT, Corn; GERNUT (cf. *gernotte* in Norman French), Yks, N Eng; GOURLINS, Scot; GOWLINS, Inv; GROUND-NUT, Som, Herts; GROVENUT, Corn.

HARENUT, Dor, Lancs, Yks, Ire; HOGNUT, Dev, Som; HORNECKS, Scot; JACK-DURNALS, Cumb; JACK-JENNETS, Yks; JACKY-JOURNALS, JOCKY-JURNALS, Cumb; KELLAS, KELLY, KILLIMORE (Cornish *keleren*), Corn; KNOTTY MEAL, Inv; LUCY ARNUT, Fife; LOUSY ARNUTS, Perth, Aber; ST ANTHONY'S NUT, Som; SCABBY HANDS, Cumb; SWINE-BREAD, TRUFFLE, Inv; UNDERGROUND NUT, Corn, Dev; YOWE-YORNUT, YOWIE-YORLIN, Cumb.

The 'nuts' or tubers, wrote a Victorian botanist, are 'better fitted to the digestion of the respectable quadrupeds whose name they share, than for Christian bipeds of tender years'. These bipeds still enjoy the hunt after pignuts, tracing down the white stem which breaks so easily, getting their fingers under the tuber, scraping it with a pocket knife and eating it for its crisp substance and clean taste.

Herbalists recommended the nuts for medicine and sometimes for food – for instance, John Pechey in his *Compleat Herbal of Physical Plants*, 1694, who stated that the nuts, peeled and boiled in fresh broth with a little pepper, were pleasant and very nourishing, and a dish which 'stimulates venery'. Richard Jefferies once complained that no one had cultivated the Earthnut and improved it into a good vegetable. It has been said this cannot be done – that the Earthnut will not thrive in tilled land (25).

In Ireland the Earthnut belongs to the fairies, and especially to the leprechaun, the fairy shoemaker (137). In Scotland, if you eat too many of the Lousy Arnuts, lice will crowd into your hair.

12. Burnet Saxifrage. *Pimpinella saxifraga* L. 107, H 37

A herbalists' plant, 'for the propertye that it hath in breaking of the stone in a mannis bodye', as Turner explained in his *Herbal*. Shropshire names for it are Bennet and Old Man's Plaything. For 'saxifrages' and the stone, see *Aphanes arvensis*.

13. Goutweed, Ground Elder. *Aegopodium podagraria* L. 112, H 40

Local names. ASHWEED ('ache' = parsley), Som, Wilts, Shrop; BISHOP'S ELDER, I o W; BISHOP'S WEED, Dor, Som, N'thum, Scot, N Ire; DOG ELLER (i.e. elder), Ches; DUTCH ELDER, Wilts; DWARF ELDER, Hants, I o W; FARMER'S PLAGUE, N Ire.

GARDEN PLAGUE, N Ire; GOAT'S FOOT (from *aegopodium*), Dev; GROUND ASH (see Ashweed above), Corn, Som, War, Ches, Lincs; GROUND ELLER (see Dog Eller above), Hants, I o W, War, Worc, Lincs; JACK-JUMP-ABOUT, Oxf, Herts, N'hants, War, Worc; JUMP-ABOUT, Oxf, War; KESH, Cumb; POT-ASH (see Ashweed above), Dev; WHITE ASH, Som; WILD ELDER, Bucks, Lincs; WILD ESH, Cumb.

Almost certainly introduced in the Middle Ages as a pot-herb and a medicinal plant against gout. But though gardeners grow it no longer, it refuses to desert the garden and stays on resolutely as a weed. It was already a bad weed in the sixteenth century. Gerard described how it 'groweth of it selfe in gardens without setting or sowing, and is so fruitfull in his increase that where it hath once taken roote, it will hardly be gotten out again, spoiling and getting every yeere more ground, to the annoying of better herbes'. William Coles, in his *Adam in Eden*, 1657, already uses the name Jump-about. Lyte, in 1578, called *Aegopodium podagraria* by the name Ashweed (i.e. ache or parsley weed), which must have suggested that it was an ash growing on the ground – Ground Ash. The leaves are not unlike those of Elder, so the plant also became Ground Elder, Dog Elder, Dwarf Elder, Dutch Elder, etc. In Gerard's time the name Bishop's Weed was applied to the umbellifer *Ammi majus*, which was grown in the physic garden. Bishop's Weed was then transferred to the not dissimilar *Aegopodium*.

Nuisance or no, Ground Elder makes spicy and tolerable eating if the leaves are boiled like spinach.

14. Samphire. *Crithmum maritimum* L. 33, H 20

Local names. CAMPHIRE, Cumb; CREEVEREEGH, Donegal; PASSPER (from French *perce-pierre*, 'pierce rock'), Scot; ROCK SEMPER, Yks, N'thum; SAMPER, I o W; SEMPER, Yks; SHAMSHER, Corn.

Samphire and the earlier 'Sampere' or 'Sampier' come from the French *herbe de St-Pierre*, herb of the fisherman saint whose name was Petros, or rock; the herb which grows out of seaside rock, and which was therefore good (in salad or as a pickle) against the stone and troubles of the kidney and the bladder.

Shakespeare may have been prompted to his lines in *King Lear* about the samphire gatherer hanging half-way down the Dover cliff by reading Gerard's *Herbal* – 'Rocke Sampier groweth on the rocky cliffes at Dover', in which Gerard was following William Turner, who in 1548 had written of the Crithmus, Sampere or *creta marina*, which 'groweth much in rockes and cliffes beside Dover'. Yet Shakespeare must have seen the gatherers at work. Robert Turner, in 1664, wrote of Samphire on the cliffs of the Isle of Wight, 'where it is incredibly dangerous to gather; yet many adventure it, though they buy their sauce with the price of their lives' (181); and two centuries later W. A. Bromfield described the samphire harvest which the island cliffsmen often took from the 'meads' or 'greens', the shelves half-way down the vast walls of chalk. For the privilege of collecting Samphire and gulls' eggs from the 600-foot cliffs, the Lord of the Manor of Freshwater exacted a yearly rent. The islanders still made a sauce of minced Samphire and butter, and still ate much of the pickled Samphire. But most of it, collected towards the end of May, was dispatched in casks of sea-water to London wholesalers, who paid four shillings a bushel (22, 21). Thomas Cogan, in *The Haven of Health* (1584), says that Sampere in brine was supplied from the Isle of Man.

On the Yorkshire coast Samphire was cooked and eaten cold with bread. See also *Salicornia stricta* and *Aphanes arvensis*. John Evelyn, in manuscript notes on gardening, compiled between 1688 and 1706 (Evelyn Library, Christ Church, Oxford), mentions 'Sampier' among the 'plants for the Kitchin-Garden'.

15. Hemlock Water Dropwort. *Oenanthe crocata* L. 98, H 37

Local names. BELDRUM, Pemb; BENDOCK, Kent; BILDERS, Corn, Dev, I o W; COWBANE, Yks; DEADMAN'S CREESH, Dumf; DEAD TONGUE, Lancs, West, Cumb; ELTROT, Som, Wilts; FIVE-FINGERED ROOT, Pemb, Glam; WATER HEMLOCK, Suss, Cumb; WATER SAPWORT, Ang; WILD RUE, Donegal.

Exceedingly poisonous. When Ehret, the great flower draughtsman of the eighteenth century, was drawing *Oenanthe crocata*, 'the smell, or effluvia only, rendered him so giddy that he was several times obliged to quit the room, and walk out in the fresh air to recover himself; but recollecting at last what might probably be the cause of his repeated illness, he opened the door and windows of the room, and the free air then enabled him to finish his work without any more return of his giddiness' (119). *Oenanthe crocata* often kills horses and cattle, though it had its virtues as well. The Cumberland people, who called it Dead Tongue, boiled it and used it to poultice the galled backs of their horses (177). *Oenanthe aquatica*, less common, is also poisonous and has been called Horsebane in Somerset, and Deathin (a name as well for the more local *Cicuta virosa*, the Cowbane) in Scotland.

16. Fool's Parsley. *Aethusa cynapium* L. 107, H 39

Local names. COW PARSLEY, Som; DEVIL'S WAND, Dor; DOG POISON, Som; FALSE PARSLEY, Shrop; FOOL'S PARSLEY, W Eng, Suff, Worc, Ches, Rut, Yks, Banff; KELK, Wilts, Suss, Surr, Kent, Yks, Dur, N'thum, N Eng; LACE CURTAINS, PIG DOCK, Som.

A Fool's Parsley for those who mistake it for genuine parsley, which it does not resemble very closely. The fool's folly would hardly be fatal. Its power may vary, but it seldom does much harm either to man or his animals (123). Gerard called it 'Thinne leafed wilde Hemlocks', remarked on the 'naughtie smell', and considered it to have properties like Hemlock.

17. Fennel. *Foeniculum vulgare* Mill. 21, H 20

Local names. FINKLE, Kent, Lincs, Yks, N Eng, Scot; SPIGNEL (properly *Meum athamanticum*), Som.

This may only be a naturalized physic herb. The *Regimen Sanitatis Salernitanum*, the mediaeval guide to health from the famous medical school at Salerno in Italy, ensured that Fennel would be in every garden:

> In Fennel-seed, this vertue you shall finde,
> Foorth of your lower parts to drive the winde.
> Of Fennel vertues foure they do recite,
> First, it hath power some poysons to expell,
> Next, burning Agues it will put to flight,
> The stomack it doth cleanse, and comfort well:
> And fourthly, it doth keepe and cleanse the sight,
> And thus the seede and hearbe doth both excell.

(Sir John Harington's version, *The Englishman's Doctor*, 1608.)

From the gardens this 'powerful' herb has escaped to the warm fringes o the sea. Crush the leaves, which are like a tangle of green hair, and they may remind you of that horrifying stench of liquorice powder, from which modern children with upset insides are fortunately delivered, though *Pulvis Glycyrrhizae Compositus* is still in the *British Pharmacopoeia*. Of the ingredients, senna and sulphur do the trick, liquorice is supposed to take away the taste, and the Fennel prevents griping.

Perhaps it is the link with the vilest looking, vilest smelling, vilest tasting of nursery medicines which has caused Fennel to be neglected in English cooking, though cookery books still recommend fennel sauce with mackerel. At one time, according to the *Compleat History of Drugs*, 1712, translated from the French of Pierre Pomet, confectioners used to candy 'the clusters of the green fennel' and sell them for sweetening the breath.

18. Pepper Saxifrage. *Silaum silaus* (L.) Schinz & Thell. 71

An odd plant, growing, as Turner remarked in his *Herbal* in 1568, in which it was recorded for the first time, 'in ranke medowes' – meadows which are rather damp and undisturbed. The leaves are cut and shiny, the flowers dirty yellow, and the root is dark grey or black, like a miniature of an elephant's trunk. It was named *Saxifraga Anglicana*, 'for that it groweth more plentifully in England then in any other countrey'. 'This kinde of Saxifrage,' Gerard wrote, 'our English women Phisitians have in great use, and is familiarly knowne unto them, vouchsafing that name unto it of his vertues against the stone.' For the rock and stone breakers, see *Aphanes arvensis*.

19. Spignel, Meu, Baldmoney. *Meum athamanticum* Jacq. 31

Local names. BADMINNIE, Scot; BALDMONEY, Yks, West; BAWDRINGIE, Perth; HOUKA, N'thum; MICKEN, Scot; MUILCIONN (Gaelic *muilceann*, *muilcionn*, 'fell wort'), Scot.

Another local umbellifer, this time of the mountain country from Scotland to Wales. It was identified with the *meon athamantikon* of Dioscorides. Meum in Latin gave the name Meu, which Turner said was used by the apothecaries. Both Spignel and Baldmoney, or Bawdmoney, are obscure names. Highlanders chew the aromatic roots (29), which were dried and employed as a carminative, a stimulant, and a spice in cookery.

20. Lovage. *Ligusticum scoticum* L. 29, H 5

Local names. SEA PARSLEY, Scot; SHEMIS, W Coast of Scot, Heb.

Livid green and glittering and obvious to the eye, like Alexanders, though not so tall a plant. At one time Lovage was picked off the seaside rocks in Scotland and eaten against scurvy, but it makes rather a poor vegetable.

21. Wild Angelica. *Angelica sylvestris* L. 112, H 40

Local names. AIT-SKEITER (i.e. oat-shooter, from children shooting oats through the hollow stems), Mor; GHOST-KEX, Yks; GROUND ASH, N Eng, Berw; GROUND ELDER, Ches; JACK-JUMP-ABOUT, N'hants (for these three names see *Aegopodium podagraria*); JEELICO, N Eng; KEDLOCK, Ches, Lancs; KEGLUS, Ches; KESK, Cumb; KEWSIES, Lincs; SKYTES, Scot; SMOOTH KESH, Cumb; SPOOTS, Shet; SWITIKS, Shet; WATER KESH, Cumb; WATER SQUIRT, Som.

Neglected among the plant population of high summer, though it is an umbellifer with distinction and size – not coarse like Hogweed, yet tall and bold, and slightly feminine in its pink and purple stems and sharply notched leaves. The umbels seem to have been dipped in claret.

22. Sulphurweed, Hog's Fennel. *Peucedanum officinale* L. 2

This very rare plant, with broad umbels of yellow flowers and finely divided leaves (not unlike the leaves of Fennel) and a root which smells of sulphur, still grows by the sea in the two counties where Gerard found it before 1597 – Essex and Kent, on either side of the Thames estuary.

The discovery must have excited Gerard, for here, not so far away from London, was the *Peukedanos*, to which Dioscorides ascribed a long list of virtues: 'The roote is thicke and long; I have digged up rootes thereof as big as a man's thigh, blacke without, and white within; of a strong and greevous smell, and full of yellow sap or liquor, which quickly waxeth hard or drie, smelling not much unlike Brimstone, called *Sulphur*, which hath induced some to call it Sulphorwoorte; having also at the top toward the upper face of the earth, a certaine bush of haire, of a browne colour, among which the leaves and stalkes do spring foorth.'

A good description. It was the root itself and the yellow juice which were valuable for medicine.

23. Masterwort. *Peucedanum ostruthium* (L.) Koch. 28, H 4

Local names. FELLON-GRASS, Yks, Lakes, Cumb, Rox; FELLON-WOOD, FELLONWORT, Cumb.

The *Imperatoria*, the master plant, master wort, from southern Europe, introduced into physic gardens, and now naturalized beside streams and in damp meadows, though mainly in the north. A master wort of medicine, this not very exciting plant was alexipharmic, or an antidote to poison, sudorific, and 'a great Attenuater and Opener'.

Also, to explain the north-country names, 'the rootes and leaves stamped, doth dissolve and cure all pestilentiall carbuncles and botches, and such other apostemations and swellings' (Gerard). The demand was great, and Masterwort was regularly supplied by eighteenth-century market-gardeners.

24. Wild Parsnip. *Pastinaca sativa* L. 62, H 27

Local names. COW-CAKES, Rox, Loth; COW-FLOP, Corn; KEGGAS (Cornish *kegys*, umbelliferous plant), Corn.

Parsnip is properly the cultivated *Pastinaca sativa*, the name coming ultimately from the Latin *pastinaca*, a carrot, a something dug from the ground. *Pastinare* is to dig, and *pastinum* is a garden fork. *Pastinaca* gave the Old French word *pasnie*, transformed into the ME *passenep*, as though the garden root were a *nape*, or turnip.

The wild Parsnip of the chalk hills has a mean little root. As Gerard wrote, it is 'small, hard, woodie, and not fit to be eaten'; yet he maintained that wild seeds set in the garden gave sweeter and larger roots than seed from the cultivated plant. Philip Miller tried the experiment for several years and found that he could never raise garden parsnips from wild seed. The wild plants retained their 'usual toughness, dark-green colour, and slender roots' (130).

Contrariwise, the garden parsnips easily escape and revert.

25. Cow Parsnip, Hogweed. *Heracleum sphondylium* L. 112, H 40

Local names. ALDERDROTS, Som; BEAR'S BREECH, Som; BEAR SKEITERS (see *Angelica sylvestris*), Mor; BEE'S NEST (see *Daucus carota*), Som; BEGGARWEED, Beds; BILDERS, Corn, Dev, Som; BILLERS, Dev; BROAD KELK, Yks;

BROADWEED, Dor; BULLERS, Som; BUNDWEED, Suff; BUNEWEED, Yks, Scot; BUNNEN, Yks, Scot; BUNNLE, Cumb, Lanark; BUNWORT, BUNNERT, Yks, Scot.

CADDELL, CADWEED, Dev; CAMLICK, Suff; CATHAW-BLOW, Cumb; CAXLIES, Som; CLOGWEED, Som, Wilts, Glos; COWBELLY, COWBUMBLE, Som; COW-CAKES, Scot; COW CLOGWEED, Glos; COWFLOP, Corn; COW-KEEP, Fife; COW-KEEKS, Berw; COW-MUMBLE, Suff, Norf, Lincs; COW PARSNIP, Staff; CUSHIA, Yks, N Eng.

DEVIL'S OATMEAL, War; DRYLAND SCOUT, N Ire; ELTROT, Corn, Som, Wilts, Hants; GEAGLES, Corn; GIPSY'S LACE, Som; HARDHEADS, Glos; HEMLOCK, Banff; HOGWEED, Kent, Bucks, Herts, Suff, Norf, Staff, Yks, Cumb; HUMPY-SCRUMPLES, Dev; KECK, Som, Suss, Ches, Leic, Yks; KEDLOCK, Ches, Lancs; KEGGA, Corn; KEGLUS, Ches; KEKSI, Shet; KESH, Lancs, Yks, Lakes, N Eng; KESK, Cumb; KEWSE, Lincs, Lancs, Yks; KISKIES, Corn; LIMPERSCRIMP, LUMPERSCRUMP, LIMPERNSCRIMP, Dev, Som; LISAMOO (Cornish *les-an-mōgh*, 'pig weed'), Corn; MADNEP, Dev.

ODHRAN, N Ire; PIG'S BUBBLE, Som; PIG'S COLE, PIG'S FLOP, Dev; PIG'S FOOD, Dor; PIG'S PARSNIP, Som, W Eng, Shrop; PIGWEED, Dev, Oxf; RABBIT'S MEAT, Dev, Som, Suss, War, Lincs, Yks; RABBIT'S VITTLES, Som; ROUGH KEX, Corn; SCABBY HANDS, Som; SKEETS, Corn; SNAKE'S MEAT, Dev; SWEET BILLER, Dev; SWINEWEED, N'hants; WIPPUL-SQUIP (hollow green stalks for drinking cider), Dev, Som.

Used up and down the country as a pig food. To see the village people bringing home bundles of this free harvest for the sty is still no uncommon sight. Sowerby's *English Botany* says that the sprouting leaves and shoots taste like asparagus, so perhaps the pigs have a good palate.

Cow Parsnip is a book name invented by William Turner (1548): 'It may be called in Englishe Cow-persnepe or rough Persnepe.'

26. Wild Carrot. *Daucus carota* L. spp. *carota* 110, H 40

Local names. BEE'S NEST, Som; BIRD'S NEST, Som, Wilts, War, Yks, Scot; CAX, Dor; CROW'S NEST, Beds; CURRAN-PETRIS, Scot; ELTROT, Hants; FIDDLE, Lincs; KEGGAS, Corn; KEX, Som; PIG'S PARSLEY, Corn, Dor; RANTIPOLE, Wilts, Hants.

Gerard says that the name Bird's Nest was taken into English from the German, and he explains how the umbel 'is drawne togither when the seede is ripe, resembling a birdes nest'. The reddish or purple flower in the middle of the umbel attracted the herbalist, and was considered 'a certain remedy for the Falling-Sickness' (141). Imagine the labour of collecting a handful! This purple floret is mentioned by Dioscorides.

Garden carrots descend, not from the English plant, but from the Mediterranean *Daucus carota* ssp. *sativus*.

XLVIII. Cucurbitaceae

1. Bryony. *Bryonia dioica* Jacq. 59

Local names. ACHE, Corn; CANTERBURY JACKS, Kent; COW'S LICK (given as a horse and cow medicine), Norf; DEAD CREEPERS, Lancs; DEATH WARRANT, Dor; DOG'S CHERRIES, Som; ELPHAMY, N Eng; HEDGE GRAPE, Worc; HOP, Glos; WILD HOP, Yks, I o W; VINE, I o W; JACK-IN-THE-HEDGE, Hants.

MANDRAKE (cf. German *Alraunwurzel*), Dev, Dor, Som, Wilts, I o W, Herts, War, Heref, Shrop, Ches, Leic, Lincs, Yks; MURREN, Hants, Norf, Yks; POISONING BERRIES, Yks; ROWBERRY, Som; SNAKEBERRY, Suff; TETTER-BERRIES (i.e. for use against ringworm), Som, Hants; WILD CUCUMBER, War; WILD VINE (cf. French *vigne sauvage*), Som, Wilts, Hants, Worc, Ches; WOMAN DRAKE, Lincs (where *Tamus communis* is Mandrake).

As the home-grown Mandrake, Bryony acquired a chief virtue of the true *Mandragora officinalis*, which was one of the most anciently famous herbs of magical power, the *mandragoras* of the Greeks, the *nam-tar-gir* or 'male Plague-god plant' of the Assyrians (30). The true Mandrake was supposed to help women to conceive. 'Great and strange effects are supposed to be in Mandrakes, to cause women to be fruitfull and beare children if they shall but carie the same neer unto their bodies.' Gerard went on to ridicule the claim, but people knew the Bible story (Genesis xxx. 14–17) of Rachel and Leah and the Mandrakes, so the large roots of the native Bryony did service for the exotic, expensive roots of *Mandragora*. 'The roote is verie great, long, and thicke, growing deepe in the earth, of a white yellowish colour, extreme bitter, and altogither of an unpleasant taste. The Queenes chiefe Chirurgion, Master *William Goodorous*, a very curious and learned gentleman, shewed me a roote heerof, that waied halfe an hundred waight, and of the bignesse of a child, of a yeere old' (Gerard). The root was child-shaped, or could be trimmed to look like a child; and Bryony mandrakes in human form, sometimes sown with grass

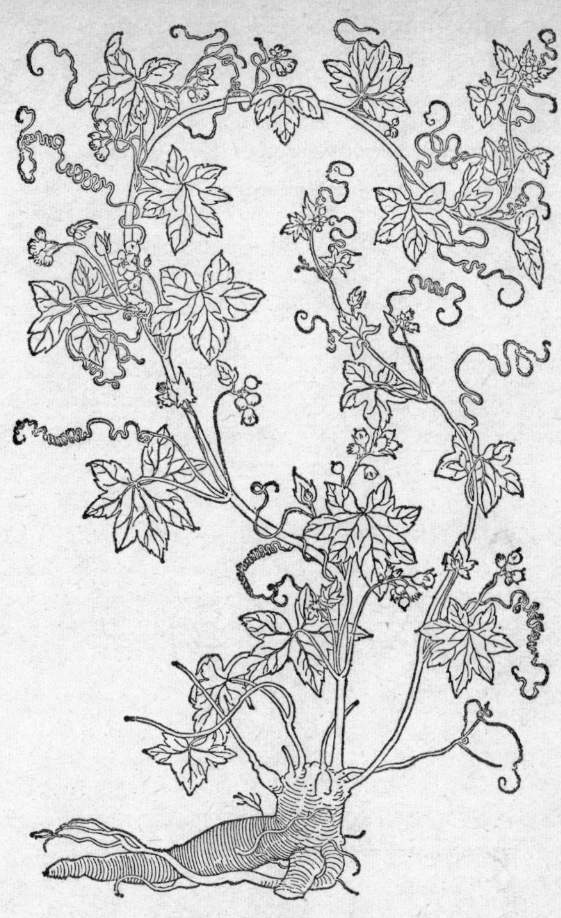

39 Bryony *Bryonia dioica*

seed to give them hair, used to be suspended in the English herb-shops, even as late as the last century (108). Hieronymus Bock, in his *Neu Kreutterbuch* (1539), had written of vagabonds fashioning these monstrous Bryony mandrakes.

The Welsh called this magical plant *eirin Gwion*, the plums of Gwion, the wizard who was reborn as Taliesin. His plums have the filthiest smell when they are overripe.

XLIX. Aristolochiaceae

1. Asarabacca. *Asarum europaeum* L. 13

'It groweth in Englande onely in gardines that I wotte of,' wrote William Turner in his *Names of Herbes*. Asarabacca is one of those engaging rarities which may or may not be native, though it was much cultivated, and used medically, in obedience to Dioscorides. Native or an escape, it creeps about in woods and shady hedges, an inconspicuous plant with leaves of a cyclamen shape and little brown flowers; humble but scarce. Asarabacca

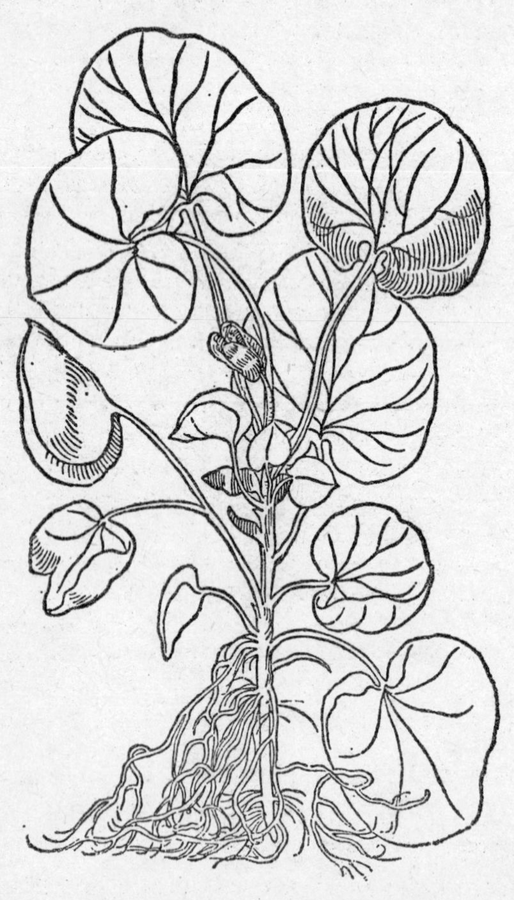

40 Asarabacca *Asarum europaeum*

'purges violently, upwards and downwards, Flegm and Choler. 'Tis diuretick also, and forces the Courses: wherefore Wenches use the Decoction of it too frequently, when they think they are with Child' (141).

In the *De Materia Medica*, Dioscorides described it very precisely under the name *asaron*. He also wrote of a *baccharis*, which some herbalists took to be *Asarum europaeum* as well, though the two descriptions do not tally. And Virgil in the Eclogues wrote of a *baccar* which grew with ivy, in the way of *Asarum europaeum*. As if to compromise and resolve the matter, apothecaries squashed the two names into one, to give the strange word Asarabacca. William Turner said that the English also called it Folefote – a translation of the herbalists' *ungula caballina*, from the shape of the leaves (cf. Coltsfoot for *Tussilago farfara*).

2. Birthwort. *Aristolochia clematitis* L. 12

A strange species of a strange genus, which was introduced from central or southern Europe into the physic garden and then escaped. It survives here and there as a reminder of old practice and belief. The ancient Greek name, *aristolochia*, comes from words meaning 'best birth': Birthwort resisted poison, encouraged conception, helped delivery, purged the womb after the child was born, and repelled demons – since birth is a dangerous time when both the mother and the child are open to supernatural malice.

Birthwort tells its own tale. The greenish-yellow flower, or perianth, constricts into a tube, then opens into a globular swelling at the base. The swelling was interpreted as the womb, the tube as the birth passages. Like helps like; and by the sympathetic magic eventually formalized into a Doctrine of Signatures, its childbed function was assigned to *Aristolochia* in remote antiquity.

Yet one may ask which came first, the discovery of the virtues or the recognition of a signature explaining them and confirming them? As it happens, virtues and signature agree, because *Aristolochia clematitis* does have an abortive effect.

L. Euphorbiaceae

1. Dog's Mercury. *Mercurialis perennis* L. 108, H 13

Local names. ADDER'S MEAT, Herts; BOGGART-FLOWER, Yks; BOGGART-POSY, Yks; DOG FLOWER, DOG'S MEDICINE, GREEN WAVES, Som; SNAKE'S

BIT, Suss; SNAKES' FLOWER, Dor, Som; SNAKE'S FOOD, SNAKE'S MEAT, Som; SNAKE'S VICTUALS, Suss; SNAKEWEED, Shrop.

The Annual Mercury or French Mercury, as it was called by Gerard (*Mercurialis annua*), was a herb useful for making enemas. The Mercury of the Woods, though no doubt it would have served us well, was an inferior plant, a dog's plant, not used in physic, in Gerard's time at least. Both species are poisonous – 'emetic and dangerously purgative, causing irritant and narcotic symptoms' (123); * so country people have not ineptly named this gloomy crop-plant of damp woods and leaf mould and dead twigs after boggarts and snakes. Annual Mercury was also known as Kentish Balsam.

For the name Mercury see *Chenopodium bonus-henricus*. *Böser Heinrich*, Bad Henry, the evil goblin, the evil boggart, is one of the German names for Dog's Mercury.

2. Caper Spurge. *Euphorbia lathyrus* L. 44

Local names. CAPER BUSH, I o W; CAPER PLANT, E Ang, Yks.

Possibly native in woods, more common as an escape from gardens. Books call the Caper Spurge a garden weed. Perhaps, but it should be, and can be, a garden treasure, a plant noble and curious in form, stiff and regular with its opposite leaves, peculiar for the tinge of blue in its green; a plant in fact with aesthetic claims to replace its discarded medicinal virtues.

Euphorbia lathyrus is the 'Spourge gyant' of Turner's *Herbal*, the Garden Spurge of Gerard – 'best knowne of all the rest, and most used'. The latex of all Spurges, wrote Gerard, 'is a strong medicine to open the bellie', and there was the same power in the seed and the roots, but he was against their employment. All were dangerous, and it was well to remember the old worn proverb: 'Deare is the honie that is lickt out of thornes.' Caper Spurge, all the same, was used long after Gerard, as well as long before him. It was the Catapuce which grew in the garden in the *Nonnes Preestes Tale* of Chaucer, one of the 'laxatyves' which Chantecleer's wife, Pertelote, so fussily advised for him after his bad dream and before his capture by the fox; and it was one of the plants which survived, perhaps from monastic cultivation, on Steep Holm, the island of the famous Peony (pp. 53–4). Spurge means a purging plant. Via the French, 'purge' comes from the Latin *purgare*; 'spurge' from the Latin *expurgare*.

* Though *Mercurialis annua* is not poisonous when cooked (134).

How often have the caper-like fruits of *Euphorbia lathyrus* been passed off as capers? That this is frequently done is probably a story of the books, though there is a case on record of women who ate them with boiled mutton and endured a terrible griping and burning of the stomach, though they recovered (123). A plant children should be warned against, if it grows in the garden.

3. Sun Spurge. *Euphorbia helioscopia* L. 112, H 40

Local names. CAT'S MILK, Worc; DEVIL APPLE TREE, Clack; DEVIL'S CHURN-STAFF, Shrop, N Ire; DEVIL'S KIRN-STAFF, Lanark, Ayr; DEVIL'S MILK, Mddx, Worc; FAIRY DELL, Dor; KIRNSTAFF, Ches, Lancs, Yks, Cumb, Kirk, Wigt; LITTLE GUID, LITTLE GOODIE (i.e. the Littlegude, the Devil), N'thum, Scot; MAD WOMAN'S MILK, Bucks; MAMMA'S MILK, Bucks; MILK-WEED, Ess, Herts, E Ang; MILKWORT, Som, Ess; MOUSEMILK, Yks.

POTATOES-IN-THE-DISH, Dor; SATURDAY NIGHT'S PEPPER, Wilts; SATURDAY'S PEPPER, Wilts; SEVEN SISTERS (from the seven branches of the stem), Ire; VIRGIN MARY'S NIPPLE, Dev; WART-GRASS, Derb, Yks, Cumb; WART-WEED, Som, Glos, Ess, Suff, Norf, Cam, Yks, Cumb; WARTWORT, Shrop; WETWEED, Norf; WHITLOW-GRASS, Lincs; WRET-WEED (i.e. wart weed), Norf, Suff, Scot.

Familiar everywhere as a cultivation weed. It exudes milk, in shape it is not unlike the spindle of a wooden churn; but the milk is the Devil's, acrid and poisonous, and the churn-staff also is the Devil's. Other names are explained by Gerard – the juice or milk 'cureth all roughness of the skinne, mangines, leprie, scurffe, and running scabs, and the white scruf of the head. It taketh awaie all maner of wartes, knobs, and the hard callouses of Fistulaes, hot swellings, and Carbuncles'.

Sun Spurge was the name used by Gerard and William Turner. *Euphorbia helioscopia* was taken for the *helioskopion*, the 'sun gazer' of Dioscorides.

4. Cypress Spurge. *Euphorbia cyparissias* L. 46

A pretty Spurge of the garden which escapes, but may also be native here and there. It has some charming names. 'Among women,' wrote Gerard, it was 'Welcome to our house.' This has been explained as a quibble on Cyparissias, the Spurge mentioned by Pliny with which the Cypress Spurge was equated – 'Sip ere ye see us', or help yourself to a drink before you are asked (149). It seems far-fetched. Yet Clip-me-Dick (Ches, Lancs) and Kiss-me-Dick (Ches) certainly look as if they were Cyparissias transformed. Perhaps Cyparissias, without calling on tankards and 'Sip ere ye see us', originally suggested 'Come to our house'. A Yorkshire name is Welcome Home Husband. Cf. the names for *Sedum acre* and *Sempervivum tectorum*; also the modern transformation of Mesembryanthemum (*Carpobrotus edulis*) into Sally-my-handsome. The names may also have to do with the welcome of the honey-like scent of the flowers.

A French name for this species is *rhubarbe du paysan*, 'Countryman's rhubarb' – for obvious reasons.

5. Wood Spurge. *Euphorbia amygdaloides* L. 55, H 3

Local names. BIBLE LEAF (cf. *Hypericum androsaemum*), Dev; DEER'S MILK, Hants (in the New Forest); DEVIL'S CUP AND SAUCERS, GREEN GROWER, Som; MARE'S TAIL, Ire; POTATOES-IN-THE-DISH, Dor; VIRGIN MARY'S NIPPLE, Som.

The latter-day herbalist Dr Fernie, in *Herbal Simples*, pleasantly described the Wood Spurge. It suggested to him 'a clever juggler balancing on his upturned chin a widely branched series of delicate green saucers on fragile stems, which ramify below from a single rod. Each saucer is the bearer again of subdivided pedicels which stretch out to support other brightly verdant little leafy dishes; so that the whole system of well-poised perianths forms a specially handsome candelabrum of emerald (cup-like) bloom.'

LI. Polygonaceae

1. Knotgrass. *Polygonum aviculare* L. 112, H 40

Local names. ARMSTRONG, Suss; BEGGARWEED, Beds; BIRD'S TONGUE, Som, N Eng; BLACKSTRAP, Hants; CLUTCH, Som; CRAB-GRASS, E Ang; CRAB-WEED, Ess; DEVIL'S LINGELS (i.e. thongs), N'thum, Berw; FINZACH, Banff; HOGWEED, Beds, E Ang; IRON-GRASS, Herts; MAN-TIE, Dev, Som; NINETY KNOT, Shrop; PIG-GRASS, Shrop, Notts, Lincs, Yks; PIG-RUSH, Shrop; PIG-WEED, Hants, Worc (cf. French *herbe à cochons*); PINKWEED, Dor, Yks.

RED LEGS, W Eng, Norf; RED ROBIN, Som, Suss; REDWEED, Dev, Som; SNAKEWEED, Som; SPARROW-TONGUE, Som, Norf; STONEWEED, Hants, Suff; SWINE'S GRASS, N'thum, Berw, Rox; TACKER-GRASS (i.e. like a shoemaker's waxed thread), Dev, Som; WAY-GRASS, Kent; WILLOW-GRASS, Yks; WIRE-GRASS, Glos; WIREWEED, I o W, Kent, Suff, Norf.

Called by Turner, in 1538, Swyne gyrs and Knotgyrs. Other names going back to the sixteenth century include Bird's Tongue, Knee Herb, Cumber Field. Knotgrass was used in medicine, especially for ulcers and sores; and according to Gerard it was given to pigs, 'when they are sicke, and will not eate their meate'.

2. Bistort. *Polygonum bistorta* L. 102, H 20

Local names. ADDERWORT, Som, Wilts; EASTER GIANT, Cumb; EASTER LEDGER, West; EASTER LEDGES, Yks, West, Cumb; EASTER MAGIANTS, EASTER MANGIANTS, EASTER MENTGIONS, Yks, West, Cumb, N'thum; GENTLE DOCK, Notts; GOOSE-GRASS, Som; MEEKS, Notts; PASSION DOCK, Derb, Yks, N Eng; PATIENCE DOCK, Ches, Derb, Lancs, Yks, N Eng; PENCUIR KALE, Ayr; POOR MAN'S CABBAGE, Lancs; RED LEGS, War, Shrop; SNAKEWEED, Som, Hants, Ches, Lanark, Banff.

Bistort is an uncommonly beautiful sight in the mountain meadows of the north, flushing the grass with pink, with the spiked knaps or ears, 'set full of small whitish flowers, declining to carnation'. First, the names. It is Bistort (for which William Turner suggested 'twise writhen') from *bistorta*, in reference to the contorted rhizome, which also suggested a coiled adder or snake – Adderwort, Snakeweed, names used by Lyte and Gerard. By the principle of sympathetic magic, it was given against snake bite and poison.

Easter Ledger Pudding, in which *Polygonum bistorta* is an ingredient, is still eaten in the Lake District at Easter time, or properly in Passion-tide, the last two weeks of Lent, which explains 'Passion Dock' and 'Patience Dock' (though the names may be influenced by the medicinal Dock, *Rumex patientia*, which was known as Patience). Easter Giant, Easter Magiant, Easter Mangiant, Easter Mentgions – the Easter food, from the French *manger*, to eat, as in 'blancmange' and the obsolete word 'manger', 'mangerie', for a banquet.

'Easter Ledger' or 'Easter Ledges' reveals still more of the history of *Polygonum bistorta*. Turner, in 1548, gave one of the names used in

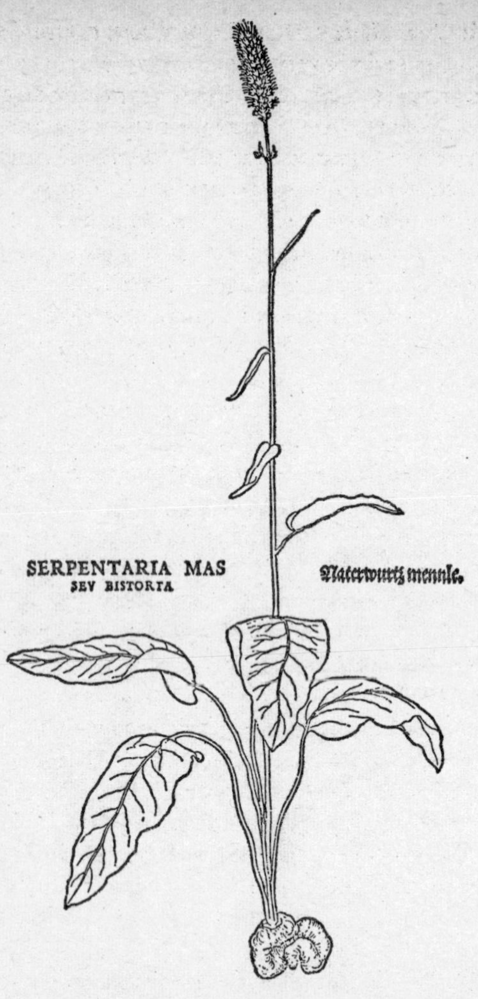

41 Bistort *Polygonum bistorta*

England as Astrologia, which was corrupted by the time of Lyte's *Herbal* (1578) to Oysterloyte. Astrologia, since the plant was an Easter dish, must have suggested that it was Easter-logia, Easter Ledger. But then why was it eaten, and why was it called Astrologia? Aristologia comes from the French *aristoloche*, from Aristolochia, properly the name of *Aristolochia clematitis* (q.v.), that ancient plant of the 'best birth', antidote to poison and demons, encourager of conception and delivery. And if one turns to *bistorta* in *The Grete Herball* of 1526, one discovers that Bistort as well 'hath vertue . . . to cause to retayne and conceyve'.

Here no doubt is the origin of the Easter Ledger pudding, which continues to be eaten in the Lakes: 'To helpe to conceyve, make electuary of powdre of bistorte in quantyte of halfe a pounde, and swete smellynge spyces of the same weyght.' Now change to another piece of evidence, the wonderful tapestry of the captured unicorn, which hangs in the Cloisters at New York, and which was woven across the Channel, probably for the marriage of Francis I of France, in 1514. Here in his final tapestry the delicate unicorn recovers from its wounds inside a fence, in a meadow diapered with flowers (1). Symbol in part of the consummation of the marriage, the unicorn is tethered by a gold chain to the pomegranate tree of fertility. Against its flank and below its hind legs are depicted two of the plants of desire, the Early Purple Orchis and the aptly named Lords-and-Ladies (*Arum maculatum*); and touching a white foreleg is – *Polygonum bistorta*, the plant of virtue in retaining and conceiving.

To complete the tale leading from the unicorn tapestries, which are one of the greatest of all works of art, to the farm kitchen in the Lakes, here is a recipe for Easter Ledger Pudding, this Aristolochia or Best Birth Pudding, as it is prepared now in Cumberland:

'Pick young Easter Ledge leaves, and drop them with leaves of Dandelion, Lady's Mantle, or Nettle into boiling water and cook for 20 minutes. Strain and chop. Add a little boiled barley, a chopped egg (hard-boiled), butter, pepper, salt. Heat in a saucepan and press into a pudding basin. Serve with veal and bacon.'

Of the other ingredients, Nettle and Dandelion were cleansing herbs, Lady's Mantle (q.v.) a herb of powerful *mana*, a wound herb, and one which gave protection against elves.

3. Persicaria. *Polygonum persicaria* L. 112, H 40

Local names. ADAM'S PLASTER, Newfoundland; ALICE OF SAUCY ALICE, Norf; ARSESMART (see *Polygonum hydropiper*), Wilts, Lincs; CRAB'S CLAW, Dor, Som; DEVIL'S PINCH, Dor; VIRGIN MARY'S PINCH, Berks; FLOOERING SOORIK, Shet; SOUROCK, Donegal; LADY'S THUMB, USA; LAKEWEED, Ches; LAMB'S TONGUES, Dev; LOVER'S PRIDE, Suss; PIG GRASS, Lincs; PINCH-WEED, Oxf, N'thum; RED JOINTS, Dor; RED KNEES, Ches, Lancs; RED LEGS, Som, Ches (cf. French *pied rouge*); REDSHANK, Lincs, Yks, Cumb, Lakes, N'thum, N Ire; REDWEED, Ches; SMARTWEED, Norf; STONEWEED, Som, Wilts; USELESS, Scot; YELLOWIN GIRSE, Shet.

Distinguished by Turner in his *Herbal* (1569) as 'Arsmert which hath the blacke spotte in it'. This dark spot on the leaves has given rise to legends. Thus in Gaelic *Polygonum persicaria* is the *lus chrann ceusaidh*, the 'herb of the crucifixion tree': it grew under the Cross and the leaves were spotted by drop after drop of Christ's blood (128). Or by English belief it is 'Pinchweed' and 'Virgin Mary's Pinch' and 'Useless', because the Virgin pulled up a plant, left her mark on the leaves, and threw the plant away, saying 'Useless' (192). Or else the Devil pinched it and made it useless, since it lacks the peppery strength of *Polygonum hydropiper*.

To Gerard this Persicaria or Peechwort (from the leaves) was the Dead Arsmart, because 'it doth not bite those places as the other doth', leaving only 'a little sower smacke' upon the tongue. However, Shetlanders used *Polygonum persicaria* for a yellow dye.

4. Water Pepper. *Polygonum hydropiper* L. 108, H 40

Local names. ARSESMART, Dev, Som, Wilts, I o W, War, Ches, Lincs, Cumb; BITY-TONGUE, Cumb; BOG GINGER, Pemb; LAKEWEED, Ches, I o M; PEPPER-PLANT, Yks; REDSHANKS, Lincs, Yks, Cumb, Lakes, N'thum, N Ire; SMARTARSE, Dev, Som; SMARTWEED, Norf, USA; WATER PEPPER, Som.

Regarded as the *hudropeperi* of Dioscorides, this plant had many uses, for sores, ulcers, swellings, toothache, jaundice; and for repelling fleas, which (if John Minsheu is to be trusted in his *Ductor in Linguas*, 1617) gave rise to the name Arsesmart, which served for the plant until the advent of the politer botanists of the nineteenth century.

It is Arsesmart, according to Minsheu 'because if it touch the taile or other bare skinne, it maketh it smart, as often it doth, being laid unto the bed greene to kill fleas'. Cf. the German names *Flöhkraut*, *Flöhpfeffer*, 'flea herb', 'flea pepper'.

5. Black Bindweed. *Polygonum convolvulus* L. 111, H 40

Local names. BEARBIND, Staffs; BETHWINE (cf. *Clematis vitalba*), Hants; BINDWEED, E Ang, Ches, Cumb; BLACK BUCKWHEAT, War; CORNBIND, Yks, N Eng; DEVIL'S TETHER, Ches, Yks; DODDER, Ches; GOATWEED, Wilts; HAY-GOB, War; HAYRIFF, Dor; LAP-LOVE, Midlands, Rox; LAPWEED, Ches; LILY, Hants, I o W; SPADES (from the shape of the leaves), Berw; TETHER-DEVIL, WILD HOP, Ches.

This species and the Pale Persicaria, *Polygonum lapathifolium* (110, H 38), were both 'utility plants': their seeds have a starch content and were eaten. Both kinds are weeds of agriculture, growing with corn; and archaeologists have pointed out that 'under primitive conditions the distinction between "wild" and "domesticated" plants is often slight', and that 'a multitude of gradations in status may exist between wild, protected, and fully domesticated species', no clear division being recognized 'between weeds and the main crop' (40). A seed and a seed impression of Black Bindweed have been found with wheat grains and with barley grains of the late Bronze Age from Morayshire. Seeds and seed impressions of Pale Persicaria have been unearthed with Bronze Age barley in Scotland and with Iron Age wheat in the excavation of Maiden Castle n Dorset (107). Bronze Age pottery impressions of Black Bindweed seeds have also been recorded from Wiltshire and Sussex (98).

More extraordinary was the analysis of the stomach contents of the Tollund Man, whose excellently preserved corpse was recovered from the Tollund Bog, near Silkeborg, in Jutland, Denmark, in 1950. He had apparently been strangled, and then thrown into the bog, where the peat preserved him to the least wrinkle (though the weight of the peat bent his nose). For his last meal he had eaten, so it seems, a gruel of barley, linseed, and the seeds of Gold of Pleasure, Pale Persicaria, Black Bindweed, Corn Spurrey, Fat Hen, and also Heartsease. The Tollund Man could be given only a rough date: he must have lived not earlier than 400 B.C. or later than A.D. 400 (97). In Donegal *Polygonum lapathifolium* is known as 'Bloodweed'.

6. Buckwheat. *Fagopyrum esculentum* Moench. Casual

Local names. BRANK, Ess, Suff, Norf, Yks, Bucks, Herts, Norf; CRAP, Suss, Worc; FAT HEN, Bucks; FRENCH WHEAT, Corn, Herts, Shrop, Staffs; WILLOW-WIND, Wilts.

The seeds, dark, triangular, and sharp-edged, resemble beechmast, so in German the plant was named *Buchweize*, 'beech wheat', which gave the English 'Buckwheat', and of which *Fagopyrum* is a translation in a mixture of Latin and Greek. Older continental names for it mean 'heathen corn', 'Tartar corn', 'Saracen corn', 'Greek corn' (cf. the English name French Corn); and the crop appears to have come into Europe from central Asia – one of those food crops which do well on a poor acid soil and which come quickly to seed.

Between the sixteenth and nineteenth centuries Buckwheat was widely cultivated in Great Britain, giving a poor man's flour as well as food for cattle and hens. 'Bucke Wheate', according to Gerard, 'may very well be placed among the kindes of graine or corne, for that oftentimes in time of necessitie bread is made thereof, mixed among other graine,' and he added that Buckwheat bread 'is of easie digestion, it speedily passeth through the belly, but yeeldeth little nourishment'.

Smith in his *English Flora*, 1829, says that buckwheat flour was used for making crumpets.

Early colonists took Buckwheat to North America, a good crop for the poor soils of New England. America has not abandoned it as we have, and in the drug-store or the cafeteria no breakfast is more delicious than a golden pile of Buckwheat cakes soaked with maple syrup. Ruddy fields of Buckwheat, or fields ruddy and white when the crop is in flower, were once familiar and pretty in the English countryside, as they still are in some parts of north-east America or in Russia. As a wild plant, it is only a casual with us, sometimes from a rare crop of Buckwheat, more often from grain which was given to pheasants.

7. Sorrel. *Rumex acetosa* L. 112, H 40

Local names. BREAD-AND-CHEESE, Dev; BROWN SUGAR, Som; COCK SORREL, Yks; CUCKOO'S MEAT, Ches; CUCKOO'S SORREL, Ire; CUCKOO'S SORROW, Hants; DOCK SEED, Dev; DONKEY'S OATS, Dev; GIPSY'S BACCY, Som; GREEN SAUCE, Corn, Dev, Glos, War, Ches, Derb, Leic, Notts, Lincs, Lancs, Yks; GREEN SNOB, War; GREEN SORREL, Bucks; LAMMIE SOUROCKS, Rox; LONDON GREEN SAUCE, Lancs; REDSHANK, Rox; RED SOUR LEEK, N Ire.

SALLET, Bucks; SOLDIERS, Som; SOORIK, Shet; SORROW, Glos, I o W, Suss, S Eng; SOUR DOCK, Dev, Dor, Som, Shrop, Ches, Lancs, Yks, Cumb; SOUR DOCKEN, E Ang, Yks, Cumb, N'thum; SOUR GRAB, Corn, Dev, Som; SOUR GRASS, Dev, Norf, Lincs, Yks; SOUR LEAVES, Som; SOUR-LICK, Rox, N Ire; SOUR SAB, Corn, Dev; SOUR SALVES, Dev; SOUR SAPS, Corn, Dev; SOUR SAUCE, Corn, Shrop, Ches, Lincs, Yks; SOUR SODGE, Bucks; SOUR SOG, Corn; SOUR SOPS, Corn, Dev; SOUR SUDS, Dev; SOUROCK, Scot, Ire; SOW-SORREL, Herts; TOM THUMB'S THOUSAND FINGERS, Kent.

A cooling, acid plant, used once much as we use lemons, especially for a green sauce with fish. Leaves were boiled and eaten, were added to ale or decoctions against fever. 'The juice heerof in summer time is a profitable

42 Sorrel *Rumex acetosa*

sauce in many meates, and pleasant to the taste. It cooleth an hot stomacke; mooveth appetite to meate; tempereth the heate of the liver, and openeth the stoppings thereof' (Gerard).

The green sauce is made by pulping the leaves and mixing them with sugar and vinegar; the boiled leaves also – boiled with next to no water – are said to go well with pork or goose in lieu of apple sauce (72). The juice was used for taking rust marks out of linen. Turner's names in 1538 were Sorell and Sourdoc – Sorell is from the Old French *surele*, in turn from *sur*, 'sour'.

8. Broad-leaved Dock. *Rumex obtusifolius* L. 112, H 40

Local names. BATTER DOCK, Shrop; BULMINT, BULWAND, Shet; BUTTER DOCK, Corn, Ches, Shrop, Lanark; BUTTER DOCKEN, Lakes; CELERY-SEED, Suss; CUSHYCOWS, Berw; DOCKEN, N Eng, Scot, Ire; DOCTOR'S MEDICINE, Som; DONKEY'S OATS, Dev; KETTLE DOCK, Lancs; LAND ROBBER, Wilts; RANTYTANTY, Ayr; REDSHANK, Scot; SMARI DOCK, Scot; SOUR DOCK, Donegal.

The Docks are not well distinguished by their vernacular names. Dock in the south, Docken in the north, Scotland, and Ireland, fit any species; and there are other general names: Buneweed in Scotland (which is not only applied to docks), Phorams (Ire), Doodykye (N'thum). *Rumex obtusifolius*, the common Dock with the blunt, wide leaves, is the one children generally search for to rub on nettle stings, to a rhyme many hundreds of years old (Troilus, in Chaucer's *Troilus and Criseyde*, when Pandarus says he should resign himself to the loss of Criseyde and love elsewhere, since the town is full of ladies, replies:

> But canstow pleyen raket to and fro,
> Netle in, dokke out, now this, now that, Pandare?

This species is Butter Dock, Butter Docken, Smair Dock from using the leaves to wrap butter.

Rumex sanguineus L., the Red-veined Dock (104, H 40), has been called Blethard (Derb), Bloodwort (Som, N Eng – Gerard's name), and Bloody Dock (Som). The Great Water Dock, *Rumex hydrolapathum* Huds. (81, H 36), finest plant of all and a bold element of river scenery, especially on the Thames, should have a host of names if plants were christened generally for their effect. In Cheshire this species as well has been known as Bloodwort. The two species, introduced as pot-herbs, *Rumex alpinus* L. (15, in the north) and *Rumex patientia* L. (here and there in the south), have both been called Monk's Rhubarb. More commonly *R. patientia* is Patience; while the very wide leaves of *R. alpinus* were useful for wrapping butter, so it has been named Butter Docken and Butter Leaves in Cumberland. Also Rhubarb (Derb).

Dock or Docken has been loosely applied to other broad-leaved plants from the Burdock to the Yellow Water-Lily. The Burdock and the Butterbur have both been called (and no doubt used as) Butter Docks. One of the more fascinating of a scarcely fascinating group is the rare Shore Dock,

Rumex rupestris Le Gall. (5), which grows on the Isles of Scilly and elsewhere, spreading by corky fruits which float round on the tide.

LII. Urticaceae

1. Pellitory-of-the-Wall. *Parietaria diffusa* Mert. & Koch. 94, H 39

Local names. BILLY BEATIE, Ire; WALL SAGE, War.

George Crabbe wrote in one of his earliest poems, a little Gothically:

> Owls and ravens haunt the buildings,
> Sending gloomy dread to all;
> Yellow moss the summit yielding,
> Pellitory decks the wall.

And old damp walls – for instance, of some of the church towers in the Gower, in South Wales – can be wonderfully rich with Pellitory, among the ferns, tinting the surface with a pale red. But it was less a symbol of ruin and romanticism than a herb for medicine, for troubles of the bladder and the stone (since it grew from stones). Also for coughs and burns and inflammation.

The older name was Paritorie or Parritory, through the Old French, as so often, and ultimately from the Latin adjective *parietarius* (*paries*, a wall). So Pellitory-of-the-Wall is tautological, though it was necessary to distinguish the plant from the Pellitory of Spain, the small composite of the Mediterranean, *Anacyclus pyrethrum*, which gives a valuable root.

2. Nettle. *Urtica dioica* L. 112, H 40

Nettle is the universal name, though sometimes in the form ETTLE (Wilts, Hants, Glos, N'hants, War, Heref, Worc). Also COOL FAUGH, Donegal; DEVIL'S LEAF, DEVIL'S PLAYTHING, Som; HEG-BEG, Scot; HIDGY-PIDGY, Dev; HOKY-POKY, Dev; JENNY NETTLE, I o M; NAUGHTY MAN'S PLAYTHING (i.e. the devil's), Som, Suss; TANGING NETTLE, Yks.

Woodland is the natural home of the Nettle, but it travels round with man, grows out of his rubbish, gets a hold where he has disturbed the ground, clings to the site of his dwellings long after the dwellings themselves have disappeared. So not unnaturally in the Highlands and Islands nettles were believed to grow from the bodies of dead men (31a).* The settlers in New England in the seventeenth century were surprised to

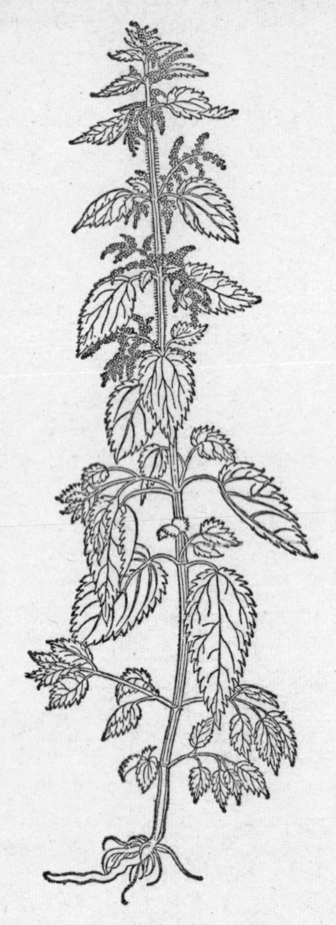

43 Nettle *Urtica dioica*

find that this old friend and enemy had crossed the Atlantic with them. In *New-Englands Rarities*, 1672, Nettle is included by John Josselyn in his list of 'such Plants as have sprung up since the English planted and kept cattle in New England'. It was 'the first Plant taken notice of'.

* In Denmark nettle clumps are supposed to grow from the shedding of innocent blood.

Friend and enemy: The friendship is ancient – nettles for medicine, which leave their trace in nettle soup and the eating of young nettles (they are good) like spinach:

> The Nettles stinke, yet they make recompense,
> If your belly by the Collicke paine endures,
> Against the Collicke Nettle-seed and honey
> Is Physick: better none is had for money.
> It breedeth sleepe, staies vomit, fleams doth soften,
> It helpes him of the Gowte that eates it often.

Which was sound advice from the Middle Ages, in Sir John Harington's version of the *Regimen Sanitatis Salernitanum* (94).

Perhaps older still is the use of nettle fibres for cloth. Nettle cloth was certainly made in Scotland as late as the eighteenth century. The Scottish poet, Thomas Campbell (1777–1844), wrote of sleeping in nettle sheets in Scotland and dining off a nettle table-cloth. Nettle cloth was made in eighteenth-century Denmark, where the nettles were cultivated on useless meadow land, and in Norway as well, where there is evidence of cultivation in place-names. Nettle crops might also explain some of the English place-names, which contain the OE *netele*. The fabric goes back before history. Nettle cloth was found in a Danish grave of the later Bronze Age, wrapped around cremated bones. There are modern records of nettle weaving from the Tyrol as late as 1917, and from Silesia eastwards later still. Nettle fabrics have nothing coarse about them. They combine fineness with strength. Perhaps nettle fibre was driven out by flax (see the Swedish periodical *Folk-liv*, 1942, Tom. vi).

The names Devil's Leaf, Devil's Plaything, suggest a nettle lore similar to that in Denmark, where nettles marked the dwelling place of elves and nettle stings were a protection against sorcery. Nettles also prevented milk from being affected by house trolls or witches.

3. Roman Nettle. *Urtica pilulifera* L. (? Extinct)

A weed of southern Europe. It was found growing at Romney, in Kent, and the Elizabethan antiquary, William Camden, in his *Britannia*, told how it came there: At *Rom*ney, the 'Roman' town, Julius Caesar had landed with his soldiers. Knowing that Britain would be cold, they introduced the nettle for rubbing and chafing their limbs. Alas, it had been called the 'Roman Nettle' before it was discovered at Romney, and before the tall story was invented. Gerard had already grown the Roman Nettle, this *Romana Urtica*, this *Roomsche Netelen* of Low German, in his physic garden. Later the Roman Nettle was found in East Anglia; and there Sir Thomas Brown had a share in spreading it. 'I have founde it', he wrote in 1668, 'to grow wild at Golston by Yarmouth, and transplanted it to other places.'

A variety of the Roman Nettle, *U. pilulifera* var. *Dodartii*, was grown in eighteenth-century gardens for practical joking. Known as Spanish Marjoram, and having leaves without teeth and no appearances of a nettle, the unwary were invited to smell it in the flower bed or in bunches of flowers, and were severely stung (130).

LIII. Cannabinaceae

1. Hop. *Humulus lupulus* L. 100, H 23

The Hop twined and twirled in English hedges and thickets long before it became one of the ingredients of English beer. It is a native. Yet in many places it must also be an escape from cultivation, since hop growing was not always concentrated so much in a few counties – chiefly Kent and Worcestershire – nor was brewing always so exclusively industrial. Farmers and squires, for instance, made their own beer and grew their own hops. Beer is originally a northern drink, the Hop a plant cultivated and diversified in northern Europe. *Humulus* was latinized from a northern word for hop, which in its Anglo-Saxon form was *hymele* or *humele*. The word Hop and the practice of flavouring and preserving beer with the strobiles or female flowers of *Humulus lupulus* we borrowed from the Netherlands:

> *Hops, Reformation, bays and beer*
> *Came to England all in one year –*

which is approximately, but not exactly, true. Hops had been cultivated abroad from the ninth century. English cultivation and brewing with hops began at the end of the Middle Ages, and waxed and became general in the sixteenth century. Making the distinction of his time between ale without hops and beer with them, Gerard in 1597 was near enough to the innovation to write: 'The manifold vertues in Hops [cleansing and opening the body, purging the blood, etc.] do manifestly argue the holsomnesse of Beere above Ale; for the Hops rather make it a Phisicall drinke to keep the body in health, then an ordinarie drinke for the quenching of our thirst.'

Hops met with a good deal of conservative opposition. However, they enabled beer to be kept longer, so that the brewing industry on a large scale and brewing fortunes are really founded on *Humulus lupulus*. Before hops the 'wort' derived from the malt was often flavoured or given bitterness by other plants – Bog Myrtle, Yarrow, Ground Ivy (so its old names of Alehoof and Tunhoof), Sage, Alecost (*Chrysanthemum balsamita*), etc.

In this country and abroad young hop shoots have long been cooked as a vegetable. Gerard thought them 'more toothsome than nourishing'. The toothsomeness is not of the highest quality.

LIV. Ulmaceae

1. Wych Elm. *Ulmus glabra* Huds. 109, H 32

Local names. BOUGH ELM, Yks; CHEWBARK, Berw; DRUNKEN ELM (? from its looser habit), Lincs; ELM-WYCH, N'thum; EMMAL, Cumb; HALSE (i.e. hazel), Som; HERTFORDSHIRE ELM, Herts; HOLME, Yks, Cumb; HORNBEAM, Som; HORN-BIRCH, Surr; QUICKEN, War; SWITCH-ELM, Yks; WITAN ELM, Shrop; WYCH HALSE, Corn, Som; WYCH-TREE, Som; WYCH-HAZEL, Dev, Som, Wilts, Worc, Ches; WYCHWOOD, Yks, Cumb.

A pretty tree, with more character or personality than the English Elm. 'Wych' has no reference to witches: it means switchy, pliant (see *Sorbus aucuparia*). The twelfth-century writer, Giraldus Cambrensis, said that the Welsh bow was made of wild elm, which must be this species, and not yew (104); and it was a common material for the English bowyer, although he preferred yew from overseas. Wych Elm gives excellent, tough, durable timber, still in demand. Furniture has been made of it, and threshing floors in the old thatched barns, since Wych Elm boards did not splinter under the knock of the flail. It is still used for the shafts of carts and wagons.

2. English Elm. *Ulmus procera* Salisb.

Local names. ELEM, Corn, Dev, Dor, Som; ELVEN, Suss, Kent, War, Worc; EMMAL, Cumb; NORFOLK ELM, Norf; HOLM, Yks, Cumb; WARWICKSHIRE WEED, War; WILTSHIRE WEED, Wilts.

Elm, oak, ash – the three great materials of the age before steel, the age of mere iron and wood. The uses were innumerable. Since elm timber endures under water, it was the first material of drain pipes, or water pipes, made of elm trunks hollowed out and fitted one into another. Millers used elm for sluices, shipwrights used it for keels. It was used for troughs and sluices in the coastal saltings. Open-air pumps on the farm were fashioned of elm. The portion in the well did not easily decay, the portion above ground did not split when the water froze. Elm piles were laid under bridges and buildings – for instance, under the old Waterloo Bridge. If you were well-to-do, oak was the timber for beams, rafters, floors in a new house. If you could not afford oak, elm was the substitute. The baker's dough trough was made of elm boards. In the farmyard elm was essential for mangers, cow-stalls, bins in the granary, the weather-boarding of granary or barn or board fences. It was hard and cheap material for many articles of furniture, from milking stools, with legs of ash, to coffers and chairs. A fundamental of civilization, of life and of death, for at last you were – and you are – buried in elm; and preferably English elm, not the imported Japanese elm, boards of which are apt to give and break when they are bent on the sides of the coffin.

Farmers no longer need to be so careful to keep a good supply of elm in their own hedges; and they sell off the trees to the timber merchant, to go eventually into coffins, chair seats, railway buffers, vegetable crates (though elm crates are no longer so pleasing to the market-gardener, who prefers the whiteness of soft wood). And more elms by the thousand are cut into sullen and slow firewood.

However, to quote William Gilpin, 'the picturesque eye scorns the narrow conceptions of a timber merchant'; and the English Elm gives to English scenery much of its rich, loaded, heavy personality. Climb to any escarpment from which the vales or plains can be viewed, and elm is the predominant tree, compact towers of dark green in the summer, yellow in the autumn, and dark brown in the winter – towers receding into blue haze or rising out of the mist. Gilpin, the lawgiver of picturesque beauty, thought the elm lacked a 'strong character', except in full foliage. He was right. There is a certain enlarged weediness about the elm, which no one has ever thought of sanctifying like the oak. But in summer, he added, no tree 'is better adapted to receive grand masses of light', in which it was superior to the oak or the ash or any other tree (86). How well his keen eye was vindicated within thirty years by the elmy landscapes of Constable!

Elm is an OE borrowing from the Latin *ulmus*.

LV. Myricaceae

1. Bog Myrtle, Gale. *Myrica gale* L. 87, H 39

Local names. BOG MYRTLE, I o W, Scot; CANDLE BERRIES, DEVONSHIRE MYRTLE, DUTCH MYRTLE, Som; FLEA-WOOD, N'thum; GALE (in various forms – Gawan, Gold, Goyle, etc.), Corn, Dev, Som, Lincs, Lancs, Yks, Cumb, Scot, N Ire; GOLD-WITHY, Hants, I o W; GOLDEN OSIER, I o W; GOLDEN WITHY, Hants, I o W, S Eng; MOOR MYRTLE, Yks; MOSS WYTHAN, Cumb; MYRTLE, Scot; SCOTCH GALE, Scot; SWEET, Yks, N Eng; SWEET GALE, Lincs, Yks, Renf; SWEET WILLOW, Suss; SWEET WITHY, I o W; WITHYWIND, Hants.

Leland, the King's antiquary, when he came to Axholme in Lincolnshire in the fifteen-thirties, wrote that the fenny part of the district 'berithe much Galle, a low frutex, swete in burning'. Gale, in fact, is the proper English name for *Myrica gale*, from the OE *gagol*, which entered a good many place-names, such as Gailey in Staffordshire, the *gagol leāh*, or clearing, or Galsworthy in Devon, which was probably the *gagol ōra*, the bank where Gale grows (66).

A useful as well as a sweetly resinous shrub. It provided faggots for the cloam oven, it kept fleas away, and Highlanders slept on flea-proof beds of the Bog Myrtle (128), it was put among linen to repel moths, it gave a yellow dye; and more important, it was one of those plants which gave a flavouring to ale or beer before the popularization of hops (see *Humulus lupulus*). 'It is tried by experience that it is good to be put in beare both by me and by diverse other in Summersetshyre,' William Turner reported in his *Herbal.* Gerard said that 'Gaule' gave a headiness to beer or ale, which was then 'fit to make a man quickly drunke'. Gale-beer was long brewed in Yorkshire. On the adverse side are Irish beliefs that Gale had dwindled to a low shrub because the Cross had been made of it (137), that it was an unlucky plant not to be used for cattle switches because Jesus was scourged with it by Pilate before he was delivered to crucifixion.

Bog Myrtle must have been much more common before the reclamation of wet land, the draining of the Fens, etc. Where it is locally dominant, for instance in parts of the wet, sandy basin of the New Forest, it sends out a delicious fragrance, especially in the flowering months of April and May.

LVI. Betulaceae

1. Birch. *Betula verrucosa* Ehrh. 110, H 28; *Betula pubescens* Ehrh. 105, H 40

Local names. PAPER BIRCH, Wilts; RIBBON-TREE, Lincs; both names from the peeling of the bark; BEGH (Irish *beith*), Donegal. The Silver Birch, with silvery bark in contrast to the brownish bark of the second species, is *B. verrucosa*.

44 Birch *Betula pubescens*

The Welsh poet Gruffydd ab Adda ap Dafydd in the fourteenth century wrote a delightful poem 'To a Birch-tree, cut down, and set up in Llanidloes for a Maypole'. The translation by Kenneth Jackson begins: 'No more will the bracken hide your sturdy seedlings, where your sisters stay; no more will there be mysteries and secrets shared, and sleep, under your dear eaves. You will not conceal the April primroses from the proud intruder's eyes' (106). A birchen Maypole: perhaps Birch was chosen for its magic power. The Irish still believe that the Birch is disliked by fairies, in the West of England crosses of Birch were hung over cottage doors to repel enchantment (137, 192), and Gerard wrote of decorating houses with Birch, and 'beautifying the streetes' with it at Rogation-tide. It was an apotropaic plant, like the Hawthorn or the Mountain Ash. As though Birch were a passport back to life, the dead sons who return in the ballad of the *Wife of Ussher's Well*, are wearing birchen hats from a tree which

> . . . neither grew in syke or ditch
> Nor yet in ony sheugh;
> But at the gates o' Paradise
> That birk grew fair eneugh.

The magic may have extended into some of the practical uses. John Evelyn, in *Sylva*, stated that arrows, bolts, and shafts, 'our old English artillery', were made of Birch. The *mana* of Birch may have given them an added power. And was there magic allied to a more common sense in choosing Birch for besoms and birching? 'Its twigs are used for Bessoms and Rods, the one for the cleanly Housewife to sweep down the Cobwebs, and the other for the magisterial Paedagogue to drive the Colt out of the Man' (177). Was evil as well as rubbish swept out of the house with Birch and with that other magical plant, the Broom (q.v.)? Was evil or the Devil thrashed out of children by the magic power, as well as the whippiness, of Birch (the material also of the *fascis* carried by the Roman lictor)?

Birch makes poor timber, excellent as it is for plywood; but where other trees are scarce, the uses have been manifold, as in the Highlands. For one thing, the Highlanders made candles from the bark; and Northumbrian fishermen went out spearing at night with birchbark candles or torches (141). Rolls of birch bark have been found in excavating Mesolithic sites.

2. Alder. *Alnus glutinosa* (L.) Gaertn.   111, H 40

Local names. Widely known from Devon to Scotland as ALLER or ALLER-TREE. The form differs: ELLAR, OLLER, OLERN, OLER, OWLER, OWLORN. OE *alor*, which is in many place-names, e.g. Allerford, Som; Allerwash ('alder swamp'), N'thum; Aldershot (earlier Alreshete), 'alder copse'.

Also, AAR, Scot; ALLS-BUSH, Dev; ARL, Dev, Som, Glos, Heref, Worc, Shrop, Rad; ARN, Scot; AUL, Heref; DOG-TREE, Lancs, Yks, N Eng; HALSE-BUSH, Dev; IRISH MAHOGANY (from the reddish timber), Ire; WALLOW, WULLOW, Shrop; WHISTLEWOOD, N'thum, N Eng.

Not much emotion has gathered around the Alder, perhaps because it was a tree of swamp and marsh and impenetrable valley floors, which needed the exorcism of natural history. Yet once enjoyed, an alder swamp along a Cornish stream, for example, remains perennially and primaevally enchanting – the trees alive and dead, moss-bearded and lichen-bearded, the soil and the water like coal slack and blacksmith's water, in between the tussocks of sedge.

Alder has been used, like the Elm, for water pipes and wooden pumps and piles under bridges and houses (much of Venice rests upon alder). Catkins, twigs, and bark give a black dye, which was a poor man's dye, according to Gerard: 'The barke is much used of poore countrie diers, for the dying of course cloth, caps, hose, and such like into a blacke colour.'

Alder burns slowly and was found to give one of the best charcoals for gunpowder. It remains the proper material for sabots abroad and clogs in Great Britain, sincc the wood is a poor conductor of heat, keeping the feet warm and cosy. Not a lucky tree in Ireland. It is a bad thing to pass by an alder on a journey (137), and the Irish dislike the felling of alders, since the timber cuts white, then changes to a tint of blood (a reddish orange). Hence the name derives from a Germanic root meaning 'reddish yellow'.

LVII. Corylaceae

1. Hornbeam. *Carpinus betulus* L. 78?

Hornbeam – the tree whose wood is tough like horn. Gerard also knew it as the Hardbeam. Other names are: HORSE BEECH, Hants, Suss, Kent; HURST BEECH (hurst, i.e. small wood; but there may be confusion between Hurst Beech and Horse Beech), Hants, Suss, Kent, Norf; WHITE BEECH, Ches; WYCH HAZEL, Ess.

Coppices or woods of pollarded Hornbeam can be places of unusual mystery – the hornbeam portions of Epping Forest, for example, which

William Morris had in mind when he wrote his fine poem, *Shameful Death*, about the murder of the brave Lord Hugh:

> He did not strike one blow,
> For the recreants came behind,
> In a place where the hornbeams grow,
> A path right hard to find,
> For the hornbeam boughs swing so,
> That the twilight makes it blind.

Gnarled trunks and the twisted intricacy of the branches make a hornbeam wood a good shelter for goblins as well as murderers. Hornbeam timber is stronger than oak, and 'in time it waxeth so hard, that the toughness and hardnes of it, may be rather compared unto horne then unto wood'. But it never became a universal timber, since the tree in the wild state is more or less confined to the south-eastern counties. It was used for yokes, wood-screws, mallets, cogging in mills, and threshing-floors in barns, as another of the timbers which would stand up to the flail.

As a 'tensile' tree (read the eulogy in Evelyn's *Sylva*), Hornbeam is superlative for hedges, and was much in demand for espaliers, pleached alleys, and mazes in the formal garden.

2. Hazel. *Corylus avellana* L. 111, H 40

Local names. COBBEDY-CUT, COBBLY-CUT, Corn; FILBEARD, I o W, Glos, Oxf, N'hants, War, Heref, Worc, Shrop, Ches, Leic; HALENUT-TREE, Corn; HALE, Corn, Som; HALSE, Dev, Som, Ire; HASKETTS, Dor; HAZEL (OE *haesel*, German *hasel*), Bucks, Ess, N'hants, Derb, Lincs, Lancs, Yks, Cumb, N'thum, Scot, Ire; NUTTALL, Corn, Dev; VICTOR-NUT, Corn; WITCHHALSE, Corn; WOODNUT, Yks. The catkins are known as LAMB'S TAILS from Corn to N'thum.

A magical tree in Great Britain and through most of Europe, which was supernaturally protected like the Rowan (q.v.), and yet ambivalently powerful against all enchantment. This is well shown in the superb ballad of Hind Etin. He is a demon or elf who guards the hazels. The May Margret comes to the wood for nuts:

> She had na pu'd a nut, a nut,
> A nut but barely ane,
> Till up started the Hynde Etin,
> Says, Lady, let thae alane.

And he takes her by the yellow locks, ties her to a tree, and threatens to kill her. Then he relents, makes her his wife, and has seven children by her (36).

Kinsmen of Hind Etin are the Melsh Dick of the north country and the Churn-milk Peg of Yorkshire, boggarts who protect the unripe nuts in the

45 Hazel *Corylus avellana*

wood. In Ireland the Hazel was the Tree of Knowledge. Hazel sticks there are useful against snakes, Hazel protects you against spirits and evil, and abduction by fairies, it is included in the 'Summer', the apotropaic plants (see Hawthorn and Rowan) brought to the house on May Day (137). It is the universal wood for hurdles, which are protective. A hazel-nut in an Irish pocket wards off rheumatism or lumbago, which is an elf-shot disease. A double nut, two on one stalk, cures toothache in Devonshire, where it is called a 'loady nut'. In Scotland the double nut, or St John's Nut, was thrown at witches (137, 192). Since Hazel was a tree of knowledge, it was the proper material for all rods of power. It was, and still is, one of the proper woods for divining.

The divining rod seems to have been introduced into Great Britain by German miners in the sixteenth and seventeenth centuries (see 'The Folklore of Dowsing', by Theodore Besterman, *Folklore*, XXXVII. ii, 1926). Rowan, Mistletoe, and Birch were also used for the rods, which indicated not only water, but minerals and treasure, and criminals – all by an occult virtue, John Evelyn wrote, which 'is certainly next to a miracle, and requires a strong faith' (70). Reginald Scot in *The discoverie of Witchcraft*, 1584, writes about the divining rod in looking for treasure: three crosses must be made on a hazel wand, and certain words must be said over it, which were 'both blasphemous and impious'.

Two other things may be added, that the Hazel was a mediaeval symbol of fertility, and that the charcoal for primitive gunpowder, in Roger Bacon's recipe of the thirteenth century, was made of young Hazel. If this was an extension of Hazel magic, it was magic only in the sense of the 'bright flash and thundering noise' which was the point of Roger Bacon's mixture, since gunpowder was at first a super-toy and not an agent of death. Bacon also advised charcoal made from willow.

Nut or twig or tree, Hazel was decidedly pagan. It needed christianizing, which it duly received in mediaeval Normandy. So the name Filbert or Filbeard. Filbeard was a fifteenth- and sixteenth-century form of Filbert, used, for example, by Thomas Tusser in his *Five Hundred Pointes of Good Husbandrie*, in 1573; and Filbert is a shortened form of *noix de filbert*, in Norman patois; i.e. *noix de Saint-Philibert*, the nuts of the Benedictine who died in 684 and founded the great abbey of Jumièges on the Seine. St Philibert's feast day is 22 August, when the hazel nuts are ripe. Hazel nuts, in fact, became the property of a saint, instead of the property of a Hind Etin or a Melsh Dick, or their counterpart in Normandy.

LVIII. Fagaceae

1. Beech. *Fagus sylvatica* L. Native in SE. England

Historians sometimes have an odd way of building large superstructures upon some isolated and curt statement in a classical author, which they treat as scientific gospel. Caesar, who paid two brief visits to Britain, stated in *De Bello Gallico* that the country had no beech trees. He saw too little of the country to know. However, Caesar's authority has been sufficient, and it was always concluded that *Fagus sylvatica* must have been introduced after Caesar's time, during the Roman occupation. Beech pollen in datable deposits of peat has now shown that, in fact, *Fagus sylvatica* was flourishing long before Caesar ever crossed into Kent. No matter. Caesar could not have been wrong: when he used the word *fagus*, it has been argued, he did not mean the Beech, but the Sweet Chestnut (104, 68). Yet *fagus* has given the names for Beech in several languages, from Italian to Welsh. And when other Latin authors wrote *fagus*, they certainly meant Beech and nothing else. Ovid, for instance, wrote of cutting names on the Beech and of the names growing – which can be seen on a thousand smooth beech trunks today – as the trunks grew.

A native, then, on the south-eastern chalk and limestone. Yet from early times it must have been planted elsewhere, for instance, at Beckwith in the West Riding of Yorkshire, which was *Becwudu*, or 'beech wood', about A.D. 972. It has also been suggested that the river Fowey in Cornwall comes from a Cornish equivalent of the Breton word *fau*, and the Welsh *ffawydden*, which are beech names derived from *fagus* (66); and, as a fact, there are fine stands of Beech along the banks of the middle Fowey.

Beech would have been worth planting more to provide pigs with mast than to provide men with timber. It makes a timber stronger but less durable than oak, and, like elm and alder, enduring best under water or in waterlogged soil. Millers used it for sluices, etc. Beech piles as well as alder piles underlay old Waterloo Bridge; and there are beech piles under Winchester Cathedral, which was built within thirty or forty years of the Norman Conquest (15). We still seat ourselves on beechen chairs made round High Wycombe in a natural beech country; and wooden soles for shoes (and sabots abroad) are still manufactured of beech wood. Yet Beech is in the hierarchy of ornament rather than the hierarchy of timber.

Gilbert White, loving the Beeches of the Hampshire hangers around Selborne, considered *Fagus sylvatica* 'the most lovely of all forest trees, whether we consider its smooth rind or bark, its glossy foliage, or graceful pendulous boughs'. The aesthetic William Gilpin did not agree (86). He found the trunk picturesque for its irregular fluting, its knobs, its contrast of rough and smooth, its moss; but its skeleton was very deficient: 'The branches are fantastically wreathed, and disproportioned; twining awkwardly among each other; and running often into long unvaried lines, without any of that strength and firmness, which we admire in the oak; or of that easy simplicity which pleases in the ash: in short, we rarely see a beech well ramified. In full leaf it is equally unpleasing; it has the appearance of an overgrown bush.' Not altogether a false criticism.

2. Oak. *Quercus robur* L. 102, H 21

The Oak – *āc* in OE – has been too necessary and familiar a tree to allow any other general names. The name, in fact, is similar in other Teutonic languages. The word 'acorn' is curiously derived. *Æcern* in OE, from *æcer*, a field, it seems to have meant the produce of the field, to have been a general term for mast, which was then confined to the mast of greatest importance for herds of swine – the oak mast. Popular etymology turned it into 'acorn', as if to imply the 'corn' or fruit of the Oak. The Oak has been nicknamed Tom Paine in Yorkshire, and called Macey-tree in Somerset – 'maceys' being acorns; which have also been known in Somerset as Jove's Nuts. The Durmast Oak, *Quercus petraea* L. 99, H 24, has been distinguished as Maiden Oak (Hants) and White Oak (Hants, I o W).

Anciently pre-eminent among European trees, the Oak was sacred, sacredness reinforcing its strength, and strength reinforcing its sacredness. So the word 'oak' still automatically prompts many notions – antiquity, durability, long life, 'hearts of oak', etc. There is no doubt of an oak cult in Indo-Germanic religion, of the connection of the oak with the god of thunder, whether Zeus or Thor, or Thunor. It was the special tree of all Europe. Maximus of Tyre, in the second century A.D., wrote of the Celts worshipping the Oak as a symbol of Zeus. Pliny (see Mistletoe) explained the sacredness of oak and oak groves in Gaul. St Boniface (680–754), the Anglo-Saxon apostle who converted the heathen Germans, had the oak of the thunder god destroyed at Geismar in Hesse. Possibly it was the thunder god's tree, because, in fact, the oak is more often struck by lightning than other trees; and perhaps there is evidence of ancient – still more ancient – respect in oaken ashes and charcoal identified in prehistoric cremation – even from the Isles of Scilly where oaks were not likely to

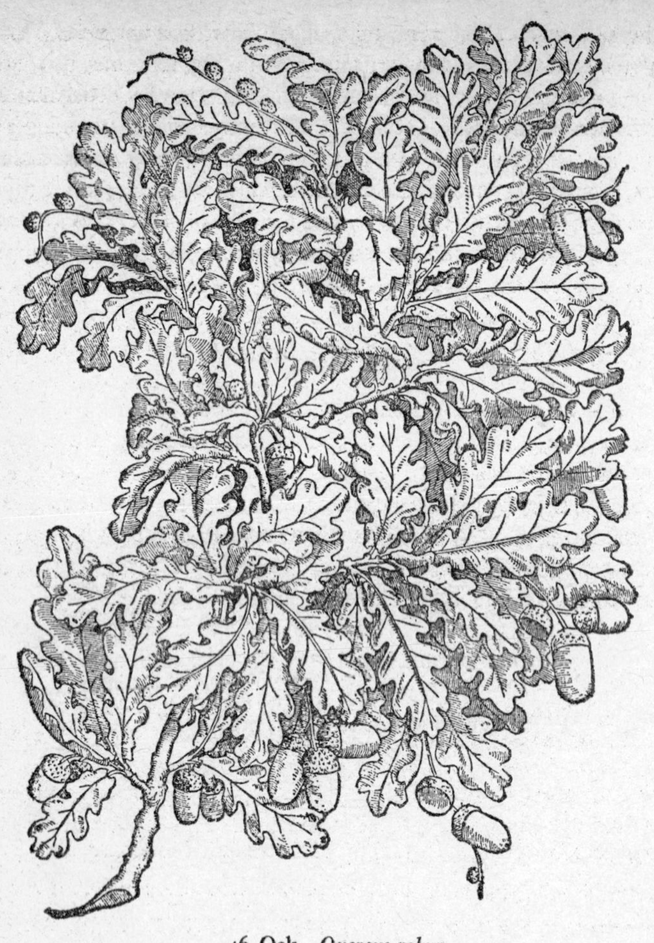

46 Oak *Quercus robur*

have grown (personal information; cf. also *Antiquaries' Journal*, XIV, 1934). Bronze Age coffins in Denmark were made of the trunks of oak trees, and oak is still a coffin timber.

However much modified and transformed, the veneration has come down through the centuries. Oak, ash, and thorn were trees to swear by. When the hero of the ballad of *Glasgerryon* came to his lady's chamber, where he had been preceded and impersonated by his foot page, he swore

> a full great oath
> By oak and ash and thorne

that he had never been there before (36). John Evelyn told an extraordinary tale of an oak in Staffordshire. It was a venerable oak, and 'upon oath of a bastard's being begotten within reach of the shade of its boughs (which I can assure you at the rising and declining of the sun is very ample) the offence was not obnoxious to the censure of either ecclesiastical or civil magistrate' (70). In Ireland acorns are the pipes smoked by the *geanncanach* or *leprechaun* (137). For centuries in European medicine oak gave a special virtue to other plants growing on its trunk or its branches – not only to Mistletoe, but to the Polypody fern. Polypody of the Oak was preferred to Polypody of the Wall. So Gerard says, 'that which groweth on the bodies of olde Okes is preferred before the rest: in steede of this most do use that which is found under the Okes, which for all that is not to be termed *Quercinum*, or Polypodie of the Oke'.

As if the sacred oak had been transferred to Christianity, there are holy oaks in English place-names – Cressage in Shropshire, which was *Cristesache*, i.e. Christ's Oak, in Domesday Book, or Holy Oakes in Leicestershire (66), not to mention a number of Gospel Oaks on parish boundaries, under which the gospel was read at Rogation-tide, when the bounds were beaten. So in Herrick's poem:

> Dearest, bury me
> Under that holy oke, or Gospel Tree;
> Where, though thou see'st not, thou may'st think upon
> Me, when you yearly go'st Procession.

Oak veneration seems to have received a necessary blood transfusion from Charles II, by way of the Boscobel Oak which saved him after the battle of Worcester. May 29 became Royal Oak Day, Oak-apple Day, Oak-ball Day, or Shick-shack Day, a celebration which can hardly have begun with Charles. In Oxfordshire a piece of oak, with an oak-apple, if possible, was worn before midday. This was the Shick-shack. In the afternoon the Shick-shack was replaced by 'monkey-powder' or ash leaves. And if either was worn in the evening, the wearer was beaten with nettles (192). Lady Raglan (*Folklore*, L. i, 1939) has linked the ceremonies of Royal Oak Day, at the end of the May cycle, with the Green Man of May Day, at the beginning of the cycle. The heads so frequently carved in churches, with leaves around them, sometimes issuing from their mouths, represent the Green Man, the Jack-in-the-Green, or Robin Hood or King of May, the sacrificial victim. The leaves depicted are sometimes hawthorn, but usually oak. The 'man in the oak' is one of the supernatural beings mentioned by Reginald Scot in 1584 in *The discoverie of Witchcraft*.

Another story about an oak and a king, but this time Charles I, was told by William Borlase, in his *Natural History of Cornwall*, in 1758. At the time of the battle of Braddock Down, Charles I set up his standard in an oak tree in Boconnoc Park. 'After the King's murder,' the leaves suddenly changed to white, and continued so 'till about thirty years since'. Braddock, the parish of the battle, probably means 'broad oak', so there may have been some earlier tradition there of a venerated tree.* The oak, too, was among the several trees of which the Cross was supposed to have been cut – one of the 'vulgar errors' sceptically discussed by Sir Thomas Browne.

The rhyme of the Cauld Lad of Hilton should be remembered (92). He was a hobthyrst or house goblin of the north of England. The servants were determined to be rid of the Cauld Lad, since his mischief-making outweighed his help. The Lad realized their game, and in the night he was heard to sing:

Wae's me! wae's me!
The acorn is not yet
Fallen from the tree,
That's to grow the wood,
That's to make the cradle,
That's to rock the bairn,
That's to grow to a man,
That's to lay me.

An oak does not grow in five minutes any more than it is felled at one chop. And perhaps only a child rocked in an oak cradle would have power to lay the goblin.

Even as timber the oak now has only a shadow of its old power and a tithe of its old ubiquity. When the Neolithic settlement of Great Britain began, oak was the natural cover across vast areas. The forests dwindled. Through the later centuries oak was overcut for house-building, boat-building, and ironically for charcoal, to smelt the iron which would supplant it so much as a key material. When oak became scarce it was replaced for some purposes by elm, a poor substitute; and then by imported deal; and finally by cast-iron, steel, and concrete. Through the Middle Ages the oak fed the Englishman's swine. It tanned his leather, fought his naval battles, carried his merchant cargoes, spanned his houses, churches, and cathedrals, and did a thousand other jobs. For the first time

* On 29 May an oak bough is still placed on top of the church tower at St Neot, not far from Boconnoc and Braddock.

in history we can manage without oak. The reputation remains, but the worshipful and powerful tree has declined into a patriarch on half-pay.

LIX. Salicaceae

1. White Poplar, Abele. *Populus alba* L. Introduced

2. Grey Poplar. *Populus canescens* Sm. 68 (often planted)

Local names. ABBY, Som; ARBALE, Som; DUTCH ARBEL, Som; LADY POPLAR, Ches; SILVER-LEAVED TREE, Som; SILVER POPLAR, Som; WHITE BACK, E Ang; WHITE WOOD, Herts.

Most of the names probably apply to both species, the introduced *Populus alba* (which is not naturalized) and *Populus canescens*, considered native in some woods, though more familiar as a planted tree.

> The grey lawns cold where gold, where quickgold lies!
> Wind-beat Whitebeam! airy abeles set on a flare!

One may suspect that Hopkins was using the charming word Abele for aspen. Strictly the Abele is *Populus alba*, much imported from Holland in the seventeenth century, together with the Dutch name *abeel*, which by origin is a diminutive of the Latin *albus*, 'white'. A few trees of the White Poplar were known to Gerard in the sixteenth century – 'in some places heere and theere a tree'.

The white soft timber of both species does not take fire easily, so English carpenters used it for flooring boards (125), particularly upstairs. Poplar was *popler*, *poplere*, *popul-tre* in ME, from the Latin *populus*. Poplar species are 'Popple' in Cornwall, Essex, East Anglia, and Rutland, 'Poppilary' in Derbyshire, Cheshire, and Lancashire.

3. Aspen. *Populus tremula* L. 111, H 36

Local names. APS, Corn, Dev, Som, Wilts, Hants, I o W, Glos, Suss, Kent, Surr, Herts, War; ASPEN-TREE, Corn; ASP, Som, Wilts, War, Heref, Worc, Ches, Yks, Cumb, N'thum, Ire; OLD WIVES' TONGUES, Rox; PIPPLE (ME *popul-tre*), Dev, Som, I o W; QUAKIN ASH, Scot; QUAKIN ESP, N Ire;

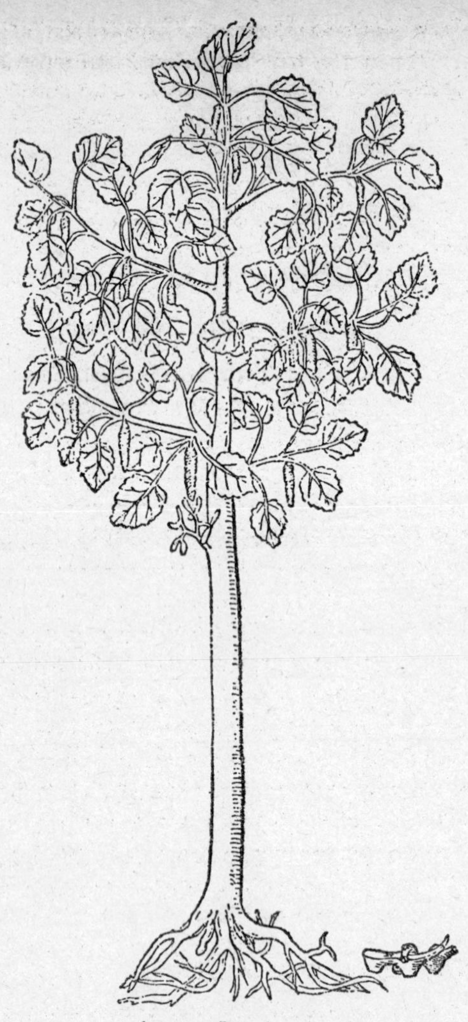

47 Aspen *Populus tremula*

SHAKING ASP, Ches; SNAPSEN, I o W; WOMAN'S TONGUE, Berks, which is also the meaning of the Manx name *chengey ny mraane*.

A sturdy native, known by similar names in the other northern languages. The Anglo-Saxons called it *æspe*, and there are many *æspe* place-names. Gerard Manley Hopkins may be quoted again, on the aspens which were cut down at Binsey, near Oxford:

My aspens dear, whose airy cages quelled,
Quelled or quenched in leaves the leaping sun,
All felled, felled, all are felled;
Of a fresh and following folded rank
Not spared, not one
That dandled a sandalled
Shadow that swam or sank
On meadow and river and wind-wandering weed-winding bank.

As far as words could reproduce it, Hopkins caught the rainy, trembling, gentle noise of aspen leaves, which suggested another thing to Gerard, and to country people, the clack of tongues and gossip: 'In English Aspe, and Aspen tree, and may also be called Tremble, after the French name, seeing it is the matter whereof womens toongs were made, as the Poets and some others report, which seldome cease wagging.'

Legends were invented to explain the trembling of the leaves. In the Scottish islands the Aspen was a cursed tree, since it held up its head when other trees bowed during the procession to Calvary, and since the Cross was made of it. Curses, stones, and clods were flung at Aspens, and crofters and fishermen would avoid using aspen wood for their gear (31a). The Welsh also believed that the leaves could never rest, since Christ was crucified on an Aspen. The legend in Brittany was that the Aspen not only refused to bow, but declared it was free of sin and had no cause to tremble and weep, whereupon it immediately began to tremble and will go on trembling till the last day (160). In Germany they said it was cursed by Jesus on the flight into Egypt, because it refused to acknowledge him. In Russia it was the tree of Judas.

If the timber had been tougher, harder, more durable, and more valuable, perhaps the legends would have been different.

4. Black Poplar. *Populus nigra* L. 48

Local names. CATFOOT POPLAR (in reference to dark knots in the timber), Lancs; COTTON-TREE (from the female catkins), Suff; WATER POPLAR, Som; WILLOW POPLAR, Cam.

Possibly native (though widely planted), possibly introduced, though earlier than *Populus alba*, and perhaps for medicinal reasons. Gerard writes of *Unguentum Populeon*, an ointment made from the buds, used for inflammation and bruises, and 'very well knowne to the Apothecaries'. This popilion, according to a recipe of the fifteenth century, was made of poplar buds to which Henbane, Orpine, Houseleek, Plantain, Endive, Violet, Watercress, and *Solanum nigrum* or *Solanum dulcamara* were added (54).

A tall Black Poplar, hung with scarlet catkins and backed with a cloudless sky, looks like some tree out of a tropical garden. On the ground these male catkins might be fat scarlet grubs. Not surprisingly they have been called Devil's Fingers, and are supposed to bring ill fortune if they are picked up.

5. Bay Willow. *Salix pentandra* L. 52, H 34

Local names. BLACK WILLOW, Ire; SWEET WILLOW, Cumb; WILLOW BAY, Staffs; FRENCH SALLY, Donegal.

A lovely northern species, distinct in its wide dark leaves (Black Willow), and in the smell of the catkins and the leaves (Sweet Willow, Bay Willow), which have the sweet pungency of bay leaves. It flowers late, and the catkins are the May Goslings of Yorkshire.

Most of the willows were not distinguished from one another until after the revival of botany. All were lumped under the common names of Willow (OE *welig*, *wylig*, forms of a word common to the Teutonic languages), which was more generally given to the trees; Sallow, Sally, Saugh (OE *sealh*, Latin *salix*), the common name for the smaller, bushier species; and Withy (OE *wīthig*), particularly, yet not exclusively, applied to *Salix viminalis* and other kinds useful in making baskets. Osier, as a name for basket willows, was borrowed from the French. Willows are bitter, and implied the bitterness of grief, as so often in Elizabethan and Jacobean poetry, in which those who are jilted wear the willow garland. So in *The Seeds of Love*, the famous song supposed to have been written in the seventeenth century by a Mrs Fleetwood Habergham of Lancashire, who was 'ruined by the extravagance, and disgraced by the vices, of her husband':

> In June there's a red rose-bud, and that's the flower for me.
> But often have I plucked at the red rose till I gained the willow-tree.
> The willow-tree will twist, and the willow-tree will twine, —
> O I wish I was in the dear youth's arms that once had the heart of mine

Ophelia very properly drowned herself by a willow, on which she tried to hang her garland of flowers.

Willows were hollow and tasted bitter, because the child Christ had been whipped with a withy wand by his mother, to correct his naughtiness. However, the bitter taste, or the bitter principle, was turned to good account. Willows, since they grow in wet country, were considered good for the common disease of such districts, the ague or malaria. In Gerard's day this was not taken very far: 'The greene boughes with the leaves', he wrote, 'may very well be brought into chambers, and set about the beds of those that be sicke of agues: for they do mightily coole the heate of the aire, which thing is a woonderfull refreshing to the sicke patients.' But country people also drank bitter infusions of willow bark against the ague, until at last in the nineteenth century the remedy was investigated and salicin was isolated; this led to the discovery of salicylic acid, eventually synthesized, and of acetyl-salicylic acid or aspirin.

6. White Willow. *Salix alba* L. 107, H 38

Frequently known as Saugh, or Saugh-tree, sometimes as Popple, but very much the White Willow, from the hairs on the leaf. The timber is valuable for many things, including artificial legs and arms and polo balls – and cricket bats, from the Cricket Bat willow, which is *Salix alba* var. *coerulea.*

7. Crack Willow. *Salix fragilis* L. 105, H 23

Local names. CAT'S TAILS, Som; SNAP WILLOW, Kent.

Look for the roots, which go down into the water like a tangle of brilliant red veins. One of the splendid willows along the Upper Thames. 'Crack Willow' and 'Snap Willow' from the immediacy and readiness with which large or small branches crack away.

8. Almond Willow. *Salix triandra* L. 73, H 14

Local names. FRENCH WILLOW, Suss, E Eng; KIT WILLOW, N'hants; SNAKE SKIN WILLOW, Wilts.

Almond Willow (a book name) for smelling and tasting of almonds, Kit Willow from the brilliant catkins, Snake Skin Willow from the way it sloughs the reddish layers of bark. One botanist in love with willows has written that in April the Almond Willow rivals the 'mimosa' of the flower shops.

9. Common Osier. *Salix viminalis* L. 110, H 40

'Wilger' in Devon. Withy and Osier are the general names, though they are applied to all the basket-making kinds.

10. Great Sallow, Goat Willow. *Salix caprea* L. 111, H 40

Local names. BLACK SALLY, Wilts, Shrop; GOSLING-TREE, Wilts; PALM-TREE, Wilts, Oxf, Yks; PALM WILLOW, Leic; PALMER, Dor; SALLY, W Eng, Wilts, Glos, Suss, Kent, Herts, Suff, N'hants, War, Heref, Worc, Shrop, Yks, I o M, Ire; SALLY WITHY, Wilts, Heref, Shrop; SAUGH or SAUGH-TREE, E Ang, Ches, Lincs, Lancs, Yks, Dur, Cumb, N'thum, Scot, Ire.

Blossoming early in March and April, this is the tree or shrub which gives 'palm' for Palm Sunday, for the catkins are a brilliant yellow in a world still without leafage. Palm Sunday is the Sunday before Easter, celebrated formerly by processions in honour of Christ's triumphal entry into Jerusalem on his ass, when the people 'cut down branches from the trees, and strewed them in the way' (Matt. xxi. 8). The branches in the Holy Land were taken to be palm, for which Sally branches with catkins were a lovely, if dissimilar, substitute. In fact, the catkins of other willow species as well as *Salix caprea* are known as 'palm', from Devonshire to the Scottish border, and the names of *Salix caprea* apply as much, no doubt, to the Common Sallow, *Salix atrocinerea* Brot. 112, H 40.

From Devonshire to the north the catkins are known as Lamb's Tails. The faint duskiness in the yellow gives them the look which goslings have as soon as the dampness of the egg has dried off, so they are also called GOOSE-CHICKS, Dev; GOOSE-AND-GOSLINGS or GEESE-AND-GOSLINGS, N'hants, Shrop, Derb, Lincs; GOOSE-AND-GUBBLIES, Shrop; GEESE-AND-GULLIES, Shrop, Ches; GOSLINGS, Suff.

Sally in Ireland has power against enchantment. It is lucky to take a sally rod with you on a journey, and the butter will come if a girdle of white sally – a peeled rod – is placed around the churn (137). In Donegal the small *Salix repens* is called 'Cran-commer'.

LX. Ericaceae

1. Strawberry Tree. *Arbutus unedo* L. H 4

'Arbutus groweth in Italy, but hath leaves like the Quicken tree, a fruite lyke a strawbery, wherefore it may be called in English strawbery tree, or an arbute tree' (William Turner in 1548). Nearly a hundred years later it was found flourishing in the damp air of the south-west of Ireland (148), a Mediterranean species in its most northern station. At Killarney, and elsewhere in Kerry, Cork, and Sligo, the Arbutus is often a small tree rather than a shrub. Unfortunately the charcoal-burners found this *caithne*, as it is called in Irish, worth cutting, and it was much reduced.

The fruit, of a 'gallant red colour', Gerard described as 'somewhat harsh' in taste, 'and, in a manner, without any relish': they hurt the stomach, he added after Dioscorides, and caused headache. A decoction from the leaves and the flowers, according to Pechey's *Compleat Herbal of Physical Plants*, 1694, was considered 'an excellent Antidote against the Plague, and for Poysons'.

2. Bearberry *Arctostaphylos uva-ursi* (L.) Spreng. 34, H 8

Local names. BEARBERRY, Aber; BRAWLINS, N Scot; BURREN MYRTLE (from growing on the Burren limestone), Ire; CRANBERRY, N Eng; CRANEBERRY-WIRE, Aber; CREASHAK, Ross; DOG-BERRY, Aber; GNASHICKS, Banff, Mor; NASHAG, Caith; RAPPER-DANDY, N Eng, Berw, Scot. Also BLANCH-NOG and MOANAGUS, Donegal.

This trailing undershrub of the Highlands and the Islands contains tannin. Since the eighteenth century the leaves have been used in official medicine as a tonic for troubles of the bladder.

3. Ling, Heather. *Calluna vulgaris* (L.) Hull. 112, H 40

This, and the commoner species of *Erica*, are HEATHER in Ches, Lincs, Yks, West, Cumb, N'thum, Scot, Ire; HADDER or HEDDER in E Ang, Yks, West, Cumb; HEATH in Dev, Som, Hants, Lancs; and LING from E Ang and Shrop to Scot, and in Ire. Other names for *Calluna vulgaris* are BAZZOM, Corn, Dev; BISSOM, Som (for use as besoms or brooms); BLACK LING, Yks; BROOM, Dev, Som, Bucks, Yks; DOG HEATHER, Aber; GRIG, Corn, Norf, Heref, Shrop, Ches, Wales; GRIGLANS (Cornish *grüglon*, 'heather bush'), Corn; GRIGLUM, Corn; HE-HEATHER ('She-heather' is *Erica tetralix* or *Erica cinerea*), N'thum, Berw; MOUNTAIN MIST, Som; RED HEATH, RED LING, Hants.

The more we have taken to the wilderness for romantic relaxation, the more Heather in blossom (often mixed with Highland cattle or the granite peaks of Dartmoor) has become a favourite symbol of bad artists and makers of the picture postcard. In the upland zone of Great Britain Heather has been a necessity of life time out of mind. It makes a springy bed, besoms for the house and the hearth, thatch for the roof, fuel, baskets, rope, an orange dye, and much else. It feeds the sheep and gives a dark honey to the bees. Highland settlers in North America took their heather beds with them, and so introduced Heather to the New World (159). Heather was also one of the plants used (see *Humulus lupulus*) for flavouring ale. 'About Shenston, as I was inform'd by the worthy Mr *Frith* of *Thorns*, they frequently used the *Erica vulgaris*, *heath* or *ling*, instead of *hopps* to preserve their *beer*, which, as he also told me, gave it no ill *tast*' (Robert Plot, *Natural History of Staffordshire*, 1686).

4. Bog Heather. *Erica tetralix* L. 111, H 40

Local names. BELL HEATH, Som, Hants; BROOM, Bucks; HONEY-BOTTLE, Som, Wilts; CROW LING, Yks; FATHER-OF-HEATH, Yks; SHE-HEATHER (see *Calluna vulgaris*), N'thum, Berw; WIRE LING, Yks.

The *Erica* species are very obviously distinguished from *Calluna vulgaris* by the bell-shaped corolla.

5. Bell Heather. *Erica cinerea* L. 112, H 40

Local names. BELL LING, Yks; BLACK HEATH, Hants; CARLIN HEATHER (Carlin, from Old Norse *kerling*, means witch or hag), Yks; CAT HEATHER, Scot; CROW LING, Yks; SHE-HEATHER (see *Calluna vulgaris*), N'thum, Berw.

6. Cowberry. *Vaccinium vitis-idaea* L. 73, H 19

Local names. CLUSTERBERRY, Derb; COWBERRY, Yks, Cumb, N'thum; CRANBERRY-WIRE, Kincard, Aber, Banff, Mor; CROWBERRY, Yks; FLOWERING BOX, Ches; LINGBERRY, Yks, Cumb; KEADYA-ATCHIN, Donegal; MOONOG, Ire.

A pretty evergreen shrublet of the moors and mountains with hanging, racemes of pink-white flowers and red fruit, which are sharp in taste, edible and seldom eaten – in Great Britain, at least – though they make an excellent jelly.

48 Bilberry *Vaccinium myrtillus*

7. Bilberry, Whortleberry. *Vaccinium myrtillus* L. 102, H 40

Local names. ARTS, Som, Wilts; BILBERRY, Heref, Worc, Shrop, Ches, Derb, Leic, Lincs, Yks, Ire; BLACKBERRY, Yks; BLACKHEARTS (i.e.-hurts), Hants; BLAEBERRY (Old Scandinavian *blaa*, dark blue; in Norwegian *blaaber*), Shrop, Lancs, Yks, West, Cumb, N'thum, Scot, Ire; BLUEBERRY, Yks, Cumb, Ire; BRYLOCKS, Scot; COWBERRY, Som; CROWBERRY, Mor; FRAUGHAN, FRUOGS (Irish *fraochóg*, *fraochán*), Ire.

HARTBERRY (i.e. hurt-), Dor, Som; HEATHERBERRY, Co. Leitrim, Co. Donegal; HURDS, Co. Sligo; HURS, BLACK HURS, Co. Cork; HURTLEBERRY, Dev, Som; HURTS, Corn, Dev, Som, Hants, Glos, Surr, Suss, Pemb, Co.

Cork, Co. Sligo; WHORTS, Corn, Dev, Som, Hants, Suss, Surr, Co. Cork, Co. Tipperary, Co. Waterford; WIMBERRY (i.e. wine berry), Glos, Heref, Shrop, Ches, Derb, Lancs, N Eng. Also MOSS-BERRY and CORASEENA and MULBERRY, Co. Donegal.

The berries make good eating in tarts, in jelly and jam, or raw with cream. In Ireland they are intimately associated with Domhnach na bhFraochóg or bhFraochán, Fraughan Sunday (or Bilberry Sunday, Blaeberry Sunday, Whort Sunday, Hurt Sunday, Heatherberry Sunday), the nearest Sunday to 1 August, which is celebrated by open-air jollifications, on hill tops, by holy wells, rivers, or lakes. The fun includes picking bilberries, for which baskets are woven out of rushes. Fraughan Sunday marks the beginning of harvest (whether of corn or potatoes), and is held to be a relic of the festival of Lughnasa, one of the four quarterly feasts of the year (Samhain, 1 November; Imbole, 1 February; Beltaine, 1 May; Lughnasa, 1 August), in honour of Lugh, the brilliant young god of many skills, including agriculture, who was the father of Cú Chulainn.

His feat marked the end of hunger and the beginning of harvest and fruit-gathering. Lughnasa is said to have been adopted by the English in their Lammas (i.e. loaf-mass), 1 August, the festival of the first fruits, on which a loaf was offered to God. All of this is explained by Máire Mac Neill in *The Festival of Lughnasa*, Irish Folklore Commission, 1962.

Bilberries are used in Scotland and Ireland for dyeing. In the Hebrides the leaves used to be dried as a substitute for tea.

8. Cranberry. *Oxycoccus palustris* Pers. 76, H 34

Local names. BOGBERRY, Ire; CRANBERRY WIRE, Cumb; CRANE, N'thum, Scot; CRAWNBERRY, Cumb; CRONE, CRONEBERRY, Lancs, West; FEABERRY (usually means Gooseberry), Shrop; MEABERRY, Yks; MOSSBERRY, Yks; MOSS-MINGIN, Scot.

Crane-berry, because the unexpanded flower and the slender corolla resemble a crane's head and neck? Or because *Oxycoccus palustris* grew in bogs which were haunted by cranes? A much commoner plant in the Midlands and the south before the draining of the Fens and the reclamation of other boggy land. We have come to think of cranberries as American – the fruit of the American *Oxycoccus macrocarpus*. However, the name and the various uses, cranberry sauce, cranberry pie, etc., were transferred from our own *Oxycoccus palustris*.

Lyte, in his translation of the herbal of Dodoens, 1578, called the fruits 'Marrish Whorts' and 'Fenberries'.

LXI. Empetraceae

1. Crowberry. *Empetrum nigrum* L. 77, H 34

Local names. BERRY-GIRSE (i.e. -grass), Shet; BLACKBERRY, Caith; CRAKE-BERRY, N Eng; CRANE, Yks; CRAWBERRY (i.e. crow-), Yks, N'thum; CRAW-CROOKS, N'thum; CRAWCROUPS, Scot; CROW LING, Yks; HEATH, Derb; HEATH-URTS, Som; KNAUPERTS, Banff; LINGBERRY, Yks, Cumb; MOONOGS, N Ire; MONNOCS-HEATHER, N Ire; WIRE-LING, Yks. Donegal names are CORISRAAN, DEER'S GRASS, and SHE-HEATHER.

The black fruit, which cluster round the stem, are eatable, but poor eating, suitable for crows or crakes. In Norwegian the name is *kraekling.*

LXII. Plumbaginaceae

1. Sea Lavender. *Limonium vulgare* Mill. 42

Gerard was the first to record Sea Lavender, from Gravesend, the Isle of Sheppey, and the Essex marshes. He also recorded the Rock Sea Lavender, *Limonium binervosum* (G. E. Sm.) C. E. Salmon – 'a kinde of Limonium like the first in each respect, but lesser, which groweth upon rocks and chalkie cliffes', which he could never find 'in any other place, but upon the chalkie cliffe going from the towne of Margate downe to the sea side'. As for the names – 'the people neere the sea side where it groweth do call it Marsh Lavander, and Sea Lavander'. The seed was employed against the colic, dysentery, strangury, and fluxes of the blood.

2. Thrift, Sea Pink. *Armeria maritima* (Mill.) Willd. 82, H 27

Local names. ARBY, Ork; BRITTONS (Cornish *bryton*), Corn; CLIFF ROSE, Dev; CUSHION PINK, Som, Wilts; CUSHIONS, Dev, Som; FRENCH PINK, Dev; LADY'S CUSHION, Dor, Som, Ches; LADY'S PINCUSHION, Hants; MARSH DAISY, Cumb; MIDSUMMER FAIRMAID, Som; PINCUSHION, Dev, Som;

PINK, Dev; PROFOLIUM, Hants; ROCK ROSE, Dev; SCAWFELL PINK, Cumb; SEA DAISY, Corn, Dev, Suss, Scot; SEA PINK, Corn, Dev, Staffs, Ches, Lancs, Yks, Cumb, Scot, Ire; SEA TURF, Dev; TAB-MAWN (Cornish *tam-ōn*), Corn.

A superlative beauty of sea cliffs and islands. On the Atlantic island of Annet in the Isles of Scilly, the whole surface is cushioned with Sea Pink. A journey across Annet is like a dream of walking on soft rubber which has squirted into flower. Sea Pink was already a garden favourite in the sixteenth century, 'for the bordering up of beds and bankes', Gerard wrote, 'for the which it serveth very fitly'. In physic Sea Pinks had no use, 'neither doth any seeke into the nature thereof, but esteeme them only for their beautie and pleasure in gardens'.

Lyte, in 1578, printed the names Quishion (i.e. cushion) and Ladies Quishion, Gerard used Sea Grass, Our Ladies Cushion, Sea Gilloflower, and Thrift, which is of Scandinavian origin and chiefly a garden name and a book name (in Norwegian *Armeria maritima* is usually *strandnellik*, i.e. 'strand pink'). From garden edgings and seaside holidays, Thrift is familiar enough to have been chosen for a punning emblem on the twelve-sided threepenny-bit of George VI.

A Gaelic name for *Armeria maritima* is *tonn a chladaich*, 'beach wave'.

LXIII. Primulaceae

1. Bird's Eye Primrose. *Primula farinosa* L. 14

Local names. BIRD EEN, Cumb, West; BIRD'S EYE, Yks; BOGBEAN, Yks; BONNY BIRD EEN, Cumb.

If one made a selection of the dozen most exquisite natives of the British Isles, this Bird's Eye would have to be included. Its headquarters are from Yorkshire to the Scottish border. Visit the Craven Highlands in the West Riding. In May and in June the small neat flowers of *Primula farinosa* decorate every bank, every slope, every corner between the grey lumps and outcrops of limestone. Finding them for the first time, a southerner feels like a plant collector on the Chinese mountains. Through all his life, through all his journeys in China and the mountains of the world, *Primula farinosa* was a favourite with the great collector and the great protagonist of the rock garden, Reginald Farrer, who grew up in the shadow of Ingleborough, in the West Riding, a country of this primrose.

The lilac flowers at the end of their pedicel have a yellow throat: 'In the middle of every small flower appeereth a little yellowe spot,' wrote Gerard, who gave the first record of *Primula farinosa*, 'resembling the eie of a bird, which hath mooved the people of the north parts (where it aboundeth) to call it Birds eine.'

This was the primrose Christopher Smart was celebrating from his childhood recollections of Teesdale, in Co. Durham, when he wrote in *A Song to David* (1763), lines which unite the Bird's Eye, waterfall, and water-polished limestone:

> The grass the polyanthus cheques;
> And polish'd porphyry reflects,
>   By the descending rill.

2. Cowslip, Paigle. *Primula veris* L. 101, H 38

Local names. BUNCH OF KEYS, Som; COVE-KEYS, Kent; COWFLOP, Dev, Som; COW PAIGLE, Herts; CARSLOPE, Yks; COOSLOP, Lincs; COWER-SLOP, Shrop; COWSLAP, Herts, N'hants; COWSLOP, Dev, E Ang, N'hants, Shrop, Ches; COWSLUP, War, Worc; COW'S MOUTH, Loth; COW STRIPPLING, COW STROPPLE, COW STRUPPLE, Yks, Cumb, West, N Eng; CREIVEL, Dor; CREWEL, Dev, Dor, Som, Wilts; CUCKOO (*coucou* is also a French name), Corn; CULVERKEYS, Som, Kent, N'hants; FAIRIES' BASINS, Som; FAIRIES' FLOWER, Som; FAIRY BELLS, Som; FAIRY CUPS, Dor, Som, Lincs; FRECKLED FACE, Som.

GOLDEN DROPS, Som; HERB PETER, Som, Ches; HODROD, Dor; HOLROD, Dor; HORSE BUCKLE, Wilts, Kent; KEYS OF HEAVEN (cf. German *Himmelschlüssel*), Dev; LADY'S BUNCH OF KEYS, Som; LADY'S FINGERS, Scot; LADY'S KEYS, Som, Wilts, Kent; LONG LEGS, Som; MILK-MAIDENS, Lincs; ODDROD, Dor; PAIGLE, Dor, Kent, Mddx, Herts, E Ang, Beds, Hunts, N'hants, Ches, Lincs, Yks, N Eng, Pemb; RACCONALS, Ches; ST PETER'S HERB, Yks; ST PETER'S KEYS, Som; TISTY-TOSTY (properly a cowslip ball), Dev, Dor, Som, Wilts, Glos, Heref; TOSTY (properly a cowslip ball), Som, Wilts, Glos, Worc, Pemb.

Cowslip is not the most elegant of names. It is a polite form of Cowslop, in OE *cū-sloppe*, *cū-slyppe* 'cow dung', 'cow pat', obviously from a conception that the cowslip sprang up in the meadow wherever a cow had lifted its tail.

49 Cowslip *Primula veris*

The *Herball* printed by Richard Banckes in 1525 called it 'cousloppe' or *Herba Petri. The Grete Herball* of 1526 called it Cowslyp, Pagle, Saynt Peterworte, and Herbe Paralysy. Gerard knew it also as Palsywort. The mediaeval *Regimen Sanitatis Salernitanum* had commended the Cowslip as a cure for palsy or paralysis, a cure suggested perhaps by the trembling or nodding of the flowers. 'Primrose' (i.e. *Primula veris*) and lavender boiled together in ale were advised 'for trembelynge hand and handis a slepe' in a manuscript of the fifteenth century (54). The nodding flowers also suggested the bunch of keys which were the badge of St Peter. One legend of northern Europe is that Peter let his keys drop when he was told that a duplicate key to Heaven had been made. Where the keys fell the Cowslip broke from the ground (175, 160). Cowslop or no, the deliciously coloured and deliciously scented flowers make the best and most delicate of all country wines.

Welsh names include *dagrau Mair*, 'Mary's tears'.

3. Primrose. *Primula vulgaris* Huds. 112, H 40

Local names. BUCKIE-FAALIE, Caith; BUTTER ROSE, Dev; DARLING OF APRIL, Som; EARLY ROSE, EASTER ROSE, FIRST ROSE, GOLDEN ROSE, GOLDEN STARS, Som; LENT ROSE, Dev; MAY-FLOOER, Shet; MAY SPINK, Scot; PIMROSE, Corn, Dev, Wilts, Hants I o W, Heref, Shrop, Ches, Notts, Lancs, Yks, Scot; SIMMERIN (cf. Norwegian *kusymre*, 'cow anemone'), Yks.

Now, if not so much in the past, the favourite of favourites, the symbol of spring and early summer, wild and in the garden; the flower of banks and hedges and sheltered cliffs, where it blossoms down to the edge of the sea, and of railway embankments, and of Easter decorations in churches filled with the mixed scent of Primrose, damp moss, and cold architecture.

The Primrose is more picked, perhaps, than any other wild flower, except the Bluebell, which makes it unfortunately scarce around London and the great cities. How can you define or diagnose its formal attraction? Gerard Manley Hopkins, who felt flowers more keenly and clearly than any other Englishman, whether poet or botanist or painter, gave an answer in his journal of 1871. 'Take a *few* primroses in a glass and the instress* of – brilliancy, sort of starriness: I have not the right word – so simple a flower gives is remarkable. It is, I think, due to the strong swell given by the deeper yellow middle.'

The name is apt. 'These are very well known to be the Ladies of the Spring, being the first that flower' (Robert Turner, the herbal signaturist, in his *Botonologia*, 1664). Primrose is from the Mediaeval Latin *prima rosa*, the 'first rose' of the year. Yet in early literature it is not easy to tell whether *prima rosa*, Primerose, Prymerolle, meant *Primula vulgaris* or *Primula veris*. In 1548 William Turner needed to distinguish firmly between the kind 'called in barbarus Latin *Arthritica*, and in Englishe a Primerose', and the kind which was 'Paralysis, and in Englishe a cowslip,

* For the meaning of 'instress' see *Endymion nonscriptus* page 440.

or a cowslap, or a Pagle'. Gerard called *Primula vulgaris* the 'common white field Primrose'. Parkinson, in his *Paradisus* (1629), still felt the need of definition. 'I doe therefore . . . call those onely Primroses that carry but one flower upon a stalke, and those Cowslips, that beare many flowers upon a stalk together constantly.'

In Ireland the Primrose is the *samhaircin*, the May flower (the actual name by which it is known in Donegal, as in Shetland), the spring harbinger (137).

4. Common Oxlip. *Primula veris* L. x *vulgaris* Huds.

Local names. BEDLAM COWSLIP, N'hants; BULLSLOP, Ches; COVEY-KEYS, Kent; COW SINKIN, Cumb; COWSLIP, Dor, Kent, E Ang, Midlands, Herts; FAIRIES' KEYS, Dev; FIVE FINGERS, E Ang; LADY'S FINGERS, Suff, Norf, Yks; MILKMAID, Yks; OXSLIP, Suff, Norf, Yks; RACCONALS, Ches; RAG JACK, Lincs; RAG ROSE, Lincs; SUMMERLOCKS, Yks.

The coarse hybrid between Primrose and Cowslip, which lacks the charm of either parent, or of the woodland Oxlip, or Paigle, of the eastern counties, *Primula elatior* (L.) Schreb. Such names as Oxlip, Bedlam Cowslip, or Bullslop show how the difference was felt.

5. Water Violet. *Hottonia palustre* L. 54, H 2

Local names. CAT'S EYES, Som; CUCKOO-FLOWER, Som; FEATHERFOIL, Som, Cumb; WATER YARROW, Yks.

A scarce beauty, recorded first by Gerard, who found it 'in the water ditches adjoining to Saint George his fielde neere London'. He called it Water Violet, after the German, or Water Gilloflower. 'Water Violet hath long and great jagged leaves, verie finely cut or rent like Yarrowe, but smaller; among which come up small stalkes a cubite and a halfe high, bearing at the top small white flowers like unto stocke Gilloflowers, with some yellownesse in the middle. The rootes are long and small like blacke threds, and at the ende where by they are fastned to the ground they are white, and shining like Chrystall.' A neat and lively description, except that the flowers are lilac with a yellow throat.

6. Yellow Pimpernel. *Lysimachia nemorum* L. 110, H 40

Yellow Pimpernel is Gerard's name, to distinguish it from *Anagallis arvensis*. A Wiltshire name is Star-flower. In Irish *seamar Mhuire*, 'Mary's clover'. In Gaelic *lus Cholumcille*, 'St Colum Cille's plant'.

7. Creeping Jenny. *Lysimachia nummularia* L. 84, H 25

Local names. CREEPING JANE, Wilts; CREEPING JENNY, Dev, Som, Wilts, Glos, Mddx, Herts, War; JENNY CREEPER, Som; MEADOW RUNAGATES, N'hants; MONEYWORT, Berks; MOTHERWORT, Dev, Hants; ROVING SAILOR, Hants; STAR, Wilts; STRING OF SOVEREIGNS, N Ire; WANDERING JENNY, Cumb; WANDERING SAILOR, Dev, Dor.

William Turner, in his *Names of Herbes*, 1548, invented 'Herbe 2 Pence' and 'Two penigrasse' on analogy with the German *Pfennigkraut*. The pennies were not the golden flowers, but the leaves running two by two up the trailing stems.

Creeping Jenny was employed against wounds and the bloody flux, 'and all other issues of blood in man or woman' (Gerard). So the name Motherwort, i.e. 'womb plant'.

8. Yellow Loosestrife. *Lysimachia vulgaris* L. 91, H 40

Local names. WILLOW-HERB, WILLOW-WORT, Som; YELLOW SAUGH, Yks.

'Lysimachia is of two sortes. The one is described of Dioscorides, and it hath a yealowe floure. Some cal it Lysimachiam luteam, it groweth by the Temes syde beside Shene, it may be called in Englishe yealow Lousstryfe or herbe Wylowe. The other kynde is described of Plinie, and it is called Lysimachia purpurea, it groweth by water sydes, also and maye be called in Englishe red loosstryfe, or purple losestryfe' (William Turner, in *The Names of Herbes*, 1548). At more than one point along the Thames (above Lechlade, for example) Yellow Loosestrife and Purple Loosestrife grow together.

'Herbe Willow', from the willow-like leaves. Gerard and Lyte reverse the order and give the name invented by Turner as Willowherb. Loosestrife translates the Greek *lusimachion*, which means 'ending strife'. There was a belief that *lusimachion* ended strife between horses and oxen yoked to the same plough, but it was also held by Pliny that the plant was named after Lysimachos, King of Thrace.

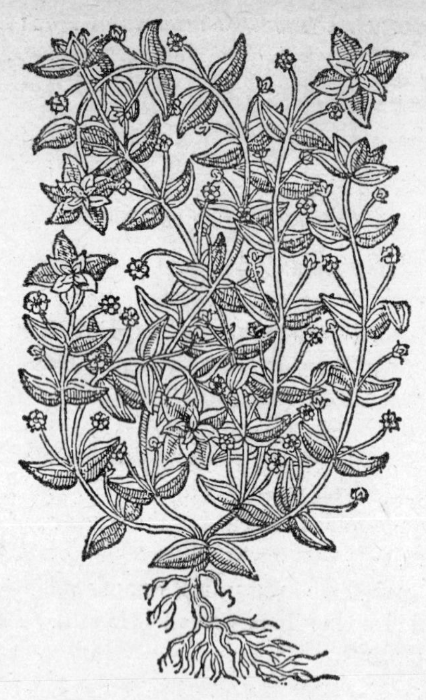

50 Scarlet Pimpernel *Anagallis arvensis*

9. Scarlet Pimpernel. *Anagallis arvensis* L. 112, H 40

Local names. ADDER'S EYES, Herts; BIRD'S EYE, Wilts, Bucks, Oxf; RED BIRD'S-EYE, Som; BIRD'S TONGUE, Norf; CHANGE-OF-THE-WEATHER, CRY BABY, CRY BABY CRAB, Som; DROPS-OF-BLOOD, Wilts; EYEBRIGHT, Corn; GRANDFATHER'S WEATHERGLASS, Som; JOHN-GO-TO-BED-AT-NOON, Dor, Som, N'hants, Ches.

LADYBIRD, LAUGHTER BRINGER, LITTLE JANE, LITTLE PEEPER, NUMPINOLE, Som; OLD MAN, Wilts; OLD MAN'S FRIEND, OLD MAN'S GLASS EYE, OLD MAN'S WEATHERGLASS, OWL'S EYE, PHEASANT'S EYE, Som; PLOUGHMEN'S WEATHERGLASS, Wilts; POOR MAN'S WEATHER-GLASS, Som, Hunts, N'hants, War, Ches, Cumb, Lanark, USA; RED CHICKWEED, USA; REDWEED, Dor.

SHEPHERD'S CALENDAR, Dev; SHEPHERD'S CLOCK, Glos, Bucks; SHEPHERD'S DELIGHT, Som, Lincs; SHEPHERD'S DIAL, Mddx; SHEPHERD'S GLASS, Norf, Rut; SHEPHERD'S JOY, Dor, Som; SHEPHERD'S SUNDIAL, Suff; SHEPHERD'S WARNING, Som, Lincs; SHEPHERD'S WATCH, Ess, Norf, Cam; SHEPHERD'S WEATHERGLASS, Dev, Som, Wilts, N'hants, Notts, Lincs, Yks;

SNAPJACK, Som; SUNFLOWER, Cumb; TOM PIMPERNEL, Yks; TWELVE O'CLOCKS, Som; WEATHER-FLOWER, Dor; WEATHERGLASS, Wilts, Bucks, Lakes; WEATHER-TELLER, Som; WINK-AND-PEEP, WINK-A-PEEP, Shrop, Staffs, Ches, Lancs.

> Al day ageyn undern and non
> He wyl hym spredyn and on-don,
> And ageyne the ewene-tyde
> He lokyth hym-self be every syde;
> He growyth be the erthe lowe,
> Nyh every man wyl hym knowe. (168)

The date of those lines from a ME poem is about 1400. Everyone has indeed known the Scarlet Pimpernel for centuries as a combination of clock and weatherglass. The scarlet flowers open at about 8 am. and close about 3 pm. Sensitive to temperature and the dampness of the air, they also close in wet or humid weather. Clearly a plant with *mana*. In Ireland it is *seamair mhuire*, 'blessed herb'. It has power to move against a stream. If you hold it it gives you second sight and hearing, you can understand the speech of birds and animals, you can 'see a circus cock pulling a wisp when everybody else thinks it is pulling a beam of wood' (137). A Welsh name in the sixteenth century was *gwlydd Mair*, 'Mary's stalk'. Pimpernel is from the Mediaeval Latin *pipinella*, by way of the Old French *pimprenele*, properly the name of the Burnet Saxifrage, *Pimpinella saxifraga*.

Anagallis arvensis was taken to be the *anagallis* of Dioscorides, of which he said there were two kinds, the male with red flowers (cf. 'Tom Pimpernel' above), the female with blue flowers (i.e. *Anagallis arvensis* ssp. *foemina* (Mill.) Schinz & Thell.): 'Pimpernell with the blew flower helpeth up the fundament that is fallen downe; and . . . red Pimpernel applied, contrariwise bringeth it downe' (Gerard). Scarlet Pimpernel was used against toothache and snake bite, inflammation of the kidney, and liver troubles, and also against melancholy (cf. the names Laughter Bringer, Shepherd's Joy, Shepherd's Delight); and in the eighteenth century it was considered a specific against hydrophobia (Bruch, *Dissertatio de Anagallide*, 1768). One charming detail to notice about Scarlet Pimpernel is the way the capsule splits and then hinges backward like a crested skull-cap, to release the seeds.

The leaves of the Scarlet Pimpernel give some skins a dermatitis.

LXIV. Loganiaceae

1. Butterfly-Bush, Summer Lilac. *Buddleja davidii* Franch. Introduced

The Butterfly Bush deserves inclusion as a late-comer to English gardens now frequently naturalized. It was sent back from western China by the French missionary and botanist, Jean André Soulié, who was tortured and shot by Tibetan monks in 1905. In level gardens the Butterfly Bush never looks quite at home. Outside the garden, on a rocky slope, a railway embankment, on a ruined wall or among the debris of bomb devastation, it takes on a new character, to the eye, as though it had found a substitute for its Chinese mountains.

LXV. Oleaceae

1. Ash. *Fraxinus excelsior* L. 111, H 40

A beautiful and a powerful tree. On the practical side the Ash provides the toughest and most elastic of British timbers (15), immensely useful even in our age of steel and alloy. It is a farmer's wood, in a score of ways, in carts and wagons, in durable fencing made of ash poles nailed to uprights of larch, in handles for pick, spade, fork, pitchfork, etc. It makes billiard cues, hockey sticks, oars, cricket stumps; an indispensable timber, close-grained and smooth to the hand. 'In short, so useful and profitable is this tree, next to the Oak,' John Evelyn wrote in his *Sylva*, 'that every prudent Lord of a Manor should employ one acre of ground with Ash to every twenty acres of other land, since in as many years it would be more worth than the land itself.' And in our own time it is one of the principal hard woods grown by the Forestry Commission.

It may be a shame to use Ash as firewood, but, as Evelyn stated, it is 'the sweetest of our forest fuelling, and the fittest for Ladies chambers', burning even while it is green.

> Ash, baneful weapon
> In the hand of a warrior

says a twelfth-century Irish poem (105): it was the wood for spears, for the best practical reasons, and perhaps as an extension of its power. It was a tree against evil. Yggdrasil, the world-tree of Scandinavian mythology, was an Ash (or a Yew? – see page 30). Pliny remarked on the antipathy between the Ash and serpents. 'The leaves of this tree', Gerard could not forbear quoting from the *Natural History*, 'are of so great a vertue against serpents, as that the serpents dare not be so bolde as to touch the morning and evening shadowes of the tree, but shunneth them a farre off.' In

51 Ash *Fraxinus excelsior*

Ireland ash wood is burned to banish the Devil (137). In Devonshire ash faggots were burned at Christmas, no doubt for the same reason, first of all; although the explanation was that his mother gave the new-born Christ his first washing and dressing by a fire of ash wood.

Ash and human birth were linked in other ways. According to the northern myths of the creation (in Snorri Sturlason's *Prose Edda*) the three gods Odin and Vili and Ve, whose brothers, after making the earth and the heavens, found two tree trunks by the sea, and turned them into the first man and first woman, from whom all other humans are descended. The tree trunk of the man was an ash (so he was called *Askr*, Ash-tree) (193).

John Lightfoot wrote, in his *Flora Scotica* (1777): 'In many parts of the Highlands, at the birth of a child, the nurse or midwife, from what motive I know not, puts one end of a green stick of this tree into the fire, and, while it is burning, receives into a spoon the sap or juice which oozes out at the other end, and administers this as the first spoonful of liquor to the new-born babe.' It was a way of giving the child the strength of the Ash and protecting it from witches and goblins. Ash keys were carried against witchcraft, and few things were more lucky than an 'even ash', a leaf with an even number of leaflets. Holding an even ash in your hand, you meet your lover before the day is out (192).

Illnesses and warts can be handed over to the Ash:

> Ashen tree, ashen tree
> Pray buy these warts of me.

Gilbert White, in the *Natural History of Selborne*, gives the classic description of curing a child of rupture by the Ash. In Selborne there were pollard Ashes which had been cleft and held open with wedges, while the ruptured child was passed through, stark naked. The tree was then 'plastered with loam, and carefully swathed up'. As the Ash healed, so the rupture was healed. In *Small Talk at Wreyland* Cecil Torr wrote of Devonshire parents who had passed a ruptured baby, at daybreak, three times from east to west through an Ash split and held open with chunks of oak. The trunk was then bandaged. Torr asked the father why they had done this – 'and he seemed surprised at the question, and said: "Why, all folk do it". I then asked him whether he thought it really did much good, and the reply was: "Well, as much good as sloppin' water over'n in church".' This was in 1902. There are several ash saplings split in this way and then healed, with a long scar down the bark, in the museum at Taunton. Gilbert White also described the 'shrew-ash'. Shrews (or fairy mice) ran over the cattle and stiffened their limbs. The cure was to stroke the cattle with the branch of an Ash which had been the tomb of a living shrew. A hole was augured in the trunk and the shrew plugged in. In Staffordshire bad luck followed if a branch was broken off an Ash tree (75). Lightning, too, runs to the Ash, as to the Oak, if not so frequently:

> Avoid an Ash,
> It courts the flash,

a mark, no doubt of its power.

In brief, in Great Britain and elsewhere in Europe the Ash was at one time a worshipful sacred tree, in company with the Oak. Ash is still the favourite for walking-sticks: as an element of the tree's power and sacredness existed in the ashen spear, so it probably existed (cf. the use of Rowan, Holly, Whitethorn, and Blackthorn for walking-sticks) in the ashen stick or staff – an odd thing to remember next time you buy an ash plant. In Scotland it was believed that a herding stick of Ash would never injure the cattle (75), which may be one of the reasons why Ash is the proper wood for the Shepherd's crook.

A Donegal name for Ash is 'whinshag'.

2. Privet. *Ligustrum vulgare* L. 99, H 4

Local names. BLACK TOPS, BLUE POISON, Som; DOG DRAKE, Dor; PEVIT, PIVOT, Corn; PRIM, Suss, Kent, N'hants, Lincs, Yks; PRIMPRINT, Som; PRIVY, Suff, War, Shrop, Ches, Yks, Scot; PRIVY SAUGH (i.e. privy willow), Yks, Scot; SKEDGE, SKEDGEWITH, SKIDGEY (Cornish *skeswedhen*, *skesjweden*, *skeswyth*, 'shade tree'), Corn; SKERRISH, Corn.

Turner's names in 1548 were Privet (OE pryfet) and Primprint, words of obscure origin. Gerard advised gargling with the juice and decoction of Privet leaves against the 'swellings, apostemations, and ulcers of the mouth and throate', and he mentions that Privet, already, was used to make garden hedges in London. Most privet hedges now are *Ligustrum ovali-folium* Hassk., from China and Japan, a species with longer, wider leaves, which are less tough than the small leaves of *Ligustrum vulgare*.

LXVI. Apocynaceae

1. Lesser Periwinkle. *Vinca minor* L. 85

2. Greater Periwinkle. *Vinca major* L. Introduced

Local names. BACHELOR'S BUTTONS, Som; BLUEBELL, BLUE BETSY, BLUE BUTTONS, Dev, Som; BLUE JACK, BLUE SMOCK, Som; COCKLE, Dev, Som, Glos; COCKLE SHELLS, Som; CUT-FINGER, Dor, Oxf; DICKY DILVER, Som, Suff; GWEEANS, Corn; OLD WOMAN'S EYE, Dor; PINPATCH, Suss; SENGREEN (i.e. evergreen, cf. *Sempervivum tectorum*), Som, I o W; TUTSAN (i.e. all heal, cf. *Hypericum androsaemum*), Suff.

Lesser Periwinkle may have been introduced (though Turner mentioned it in his *Herbal* (1562) as wild in the west country), the Greater Periwinkle was certainly introduced. Culpepper distinguished between the two in his *English Physitian Enlarged*, remarking that those 'Perwincles with the pale blew and those with the white flowers [i.e. *V. minor*] grow in woods, and Orchards, by the Hedg-sides, in divers places of this Land: but those with the purple flowers [i.e. *V. major*] in Gardens onely'. There was good reason for the introduction of the Periwinkles. They were venereal. A medical manuscript of the fourteenth century says that Pervinca powdered with earthworms induces love between husband and wife, if they take it first in their food (99). The German doctor of the early sixteenth century, Hieronymus Braunschweig, wrote that Periwinkle water was good for women with a cold womb (17); and according to Culpepper: 'Venus owns this Herb, and saith, that the Leaves eaten by Man and Wife together, cause Love between them' (49). So *Vinca minor* is among the plants of fertility in the last scene of the Unicorn tapestries (see *Endymion nonscriptus*, *Polygonum bistorta*, *Arum maculatum*, *Orchis mascula*).

In the wonderful French folksong *Sur les Marches du Palais* the little shoemaker tells the girl that there is periwinkle at the corners of the bed of love:

> Aux quatre coins du lit
> Aux quatre coins du lit
> Quatre bouquets de pervenches
> Lon la
> Quatre bouquets de pervenches.

The Periwinkles were also vulnerary herbs, 'used in Fluxes of the Belly, for Dysenteries, the Piles, Bleeding at Nose, and for Wounds with Fluxion', and 'outwardly for overflowing of the Courses, for Loosness and Pains of the Teeth' (141). Hence the names Cut-finger and Tutsan (French *toute-saine*, 'all heal'). In seventeenth-century England bands of Periwinkle were worn around the legs against cramp (181). Hannah Woodly, in *The Queen-like Closet*, 1680, wrote that 'Perywincle hath an excellent virtue to stench bleeding at the nose in Christians if it be made into a garland and hung about the neck'. In Italy a name for Periwinkle is *fiore di morte*, 'flower of death', supposedly because Periwinkle garlands were hung on those who were to be executed (though perhaps as an evergreen symbol of immortality). Such garlands were used in mediaeval England. When Simon Fraser was taken to execution in London in 1306, after the overthrow of the great Sir William Wallace, he is said to have worn a Periwinkle crown (72).

Periwinkle, in OE *pervenche*, goes back to the *Vinca pervinca*, the name of a plant mentioned by Pliny. An early name for it was 'the joy of ground' (168).

LXVII. Gentianaceae

1. Centaury. *Centaurium minus* Moench. 102, H 40

Local names. BLOODWORT, Shrop; EARTHGALL (OE *eorthgealla*, a translation of the Latin *fel terrae*), Som, W Eng; GALL OF THE EARTH, Som; FEVERFEW, Ches, Yks; FEVERFOULLIE, Scot; GENTIAN, Suss, Scot; MOUNTAIN FLAX, Cumb; SANCTUARY, War, Shrop, Ches, Lancs, Yks; SPIKENARD, Som. A Donegal name is DRAMWE-NA-MURROGH.

A bitter herb, 'very wholesome, but not very toothsome', as Culpepper remarked after telling its virtues. It was identified with the *kentaurion* of the Greek medical writers, so named because it was discovered by Chiron the centaur. Centaury opened 'the stoppings of the liver, gall, and spleen' (Gerard), it cleansed and was good against bleeding and fevers – especially malaria (141). 'Gentian', in Sussex and Scotland, because it did the work of *Gentiana lutea*, which was imported by the shops.

Among the Irish, a blessed herb brought into the house for good luck between the Annunciation and the Assumption of the Blessed Virgin Mary – 25 March and 15 August (137). Also a blessed herb in the Isle of Man, the Manx name being *Keym Chreest*, 'Steps of Christ', since it grew where Christ trod on the way to Calvary.

2. Yellow-wort. *Blackstonia perfoliata* (L.) Huds. 65, H 27

Local name. YELLOW CENTAURY, Ches, Yks.

This lovely plant of the chalk downs or the Cotswold limestone, which 'doth grow', Gerard wrote, 'from the chalkie cliffes of Greenhithe in Kent, and such like places', was the *Centaurium luteum*, the yellow Centaury, of the early botanists and apothecaries. 'Another small *Centaury*, which beareth a yellow flower, in all other respects it is like the former, save that the leaves are bigger, and of a darker green, and the stalk passeth through the midst of them. . . In Diseases of Blood, use the red Centaury; if of the Choler, use the yellow' (Culpepper). The flowers shut about two o'clock, Greenwich time.

3. Spring Gentian. *Gentiana verna* L. 4, H 5

Local names. BLUE VIOLET, SPRING VIOLET, Dur.

The Spring Gentian is all too rare, confined to three counties in Ireland and four English counties. Travel to Burren, in Co. Clare: in May the sapphires of the Spring Gentian emerge with fantastic brilliance from the shelves and corners of the grey limestone, more blue than the sky above or the Atlantic down below – a blue without a trace of that purple which makes more than one Gentian (for instance, *Gentianella amara*) rather a disappointment. It was on the west coast of Ireland – though in Galway, not Co. Clare – that the Spring Gentian was first discovered in the seventeenth century. In Germany this is a thunder plant. Storms follow, if you are rash enough to pick it (129).

4. Felwort. *Gentianella amarella* (L.) H. Sm. 89, H 30

'A kind of Gentian,' William Turner wrote in his *Herbal*, 'upon the playne of Salisberrye.' It was important to know that there were English gentians, since the doctors and apothecaries relied for a bitter, tonic, cleansing medicine upon imported supplies of *Gentiana lutea* (the root of which is still in the *British Pharmacopoeia*). 'It is confessed,' Culpepper wrote, 'that Gentian which is most used amongst us, is brought over from beyond Sea: yet have we two sorts of it growing frequently in our Nation, which besides the Reasons so frequently alledged, Why English Herbs should be fittest for English Bodies, hath been proved by the experience of divers Physitians to be not a whit inferior in Vertue.' He meant *Gentianella amarella* and *Gentianella campestris*.

Gentian, because *Gentiana lutea* was named after the *gentiana* of Pliny, a plant so called after Gentius, the pirate king of Illyria, who was defeated and imprisoned by the Romans. Felwort, too, was originally the name of the medicinal and foreign *Gentiana lutea*. In OE manuscripts *gentiana* is more than once translated *feld wyrt*, 'field-wort'. Perhaps this was due to a misunderstanding of *felwyrt*, from *fel*, 'gall', Latin *fel*; or perhaps an original *feldwyrt* was made into 'felwort' in later times by those who knew their apothecary's Latin (cf. *fel terrae* for *Centaurium minus*, and Gerard's synonym 'Bitterwort').

In Shetland *Gentianella amarella* goes by the apt name of Dead Man's Mittens, because 'the half-open buds are like livid finger-nails protruding from the turf' (*Shetland Folk Book*, 1947).

LXVIII. Menyanthaceae

1. Buckbean. *Menyanthes trifoliata* L. 111, H 40

Local names. BOG BEAN (an alteration of Buckbean, from its habitat), N'hants, War, Yks, Cumb, I o M, Ire; BOG HOP (from its use in making beer), N Eng; BOG NUT, Scot; BOG TREFOIL, Yks; BUCKBEAN, Ches, Yks, Cumb, Rox; DOUDLAR (applied to the root), Rox; GULSA-GIRSE ('jaundice grass'), Shet; LUBBERLAB, I o M; THREEFOLD, Yks, Kirk, Wigt; TREFOLD, Shet; WATER TREFOIL, War; WHITE FLUFF, Norf.

'Toward the top of the stalks standeth a bush of feather-like flowers of a white colour, dasht over slightly with a wash of light carnation.' A good description; but Gerard fell short in his words of the combined delicacy of form and tint of Buckbean flowers, which might have defeated Gerard Manley Hopkins. Another botanist is worth quoting: the Buckbean 'is one of the most beautiful plants this country can boast, nor does it suffer when compared with the *Kalmia*'s, the *Rhododendron*'s, and the *Erica*'s of foreign climes, which are purchased at an extravagant price, and kept up with much pains and expence, while this delicate native, which might be procured without any expence, and cultivated without any trouble, blossoms unseen, and wastes its beauty in the desart air. To such as wish to have this plant flower with them in perfection, I would recommend the following mode of cultivation: collect the roots of the plant either in spring or autumn, put them in a large pot (having a hole at the bottom) filled with bog earth, immerse the pot about two-thirds of its depth in water, in which it should continue; the advantage of this method is, that when the plant is coming into flower it may be brought into any room and placed in a pan of water, where it will continue to blossom for two or three weeks. A single root which I treated in this manner, planted in the Spring, produced the ensuing May eight flowering stems, many of which had fifteen or sixteen blossoms on them' (William Curtis, in *Flora Londinensis*).

This beauty of the black moorland waters has been prosaically named and prosaically employed. Unless it was the OE *ramgealla*, 'ram gall', it seems to have had no English name (far more common, as it must have been, in the early centuries) until Lyte, in his translation of Dodoens in 1578, turned the Dutch *bocks boonen*, 'goat's beans', into Buckbean. Leaves and root are bitter. They were given, no doubt as an ancient folk-medicine, for jaundice and ague and rheumatism. Like hops, the leaves were a flavouring for beer in the north country and northern Europe (191; see also *Humulus lupulus*). Known as *pónaire chapaill*, 'mare's beans', the Irish use Buckbean against boils and for purifying the blood (137). In German it is *Fieberklee*, 'fever clover'; also *Biberklee*, 'beaver clover', which suggests the European beaver slipping into the still water by the flowers, or nibbling the soft stems.

'Goat's beans' is a fair description, since the leaves resemble the young Broad Bean, and the all-devouring goat is one of the few animals ready to eat them (191).

LXIX. Polemoniaceae

1. Jacob's Ladder. *Polemonium caeruleum* L. 11

Local names. BLUE JACKET, N Ire; CHARITY, Cumb; JACOB'S LADDER, Dev, Suss, E Ang; JACOB'S WALKING-STICK, Hants; LADDER TO HEAVEN, Lanark; POVERTY, Cumb.

Christopher Merrett gave the first record of Jacob's Ladder as a native plant in his *Pinax*, in 1666. Five years later John Ray, during one of his northern journeys, saw it at Malham Cove in the West Riding, under the vast wall of limestone; and there, though not quite in the same spot, it still makes a blue haze among the grey rubble. Possibly Jacob's Ladder was isolated in its scattered colonies on the limestone during the Late Glacial period.

Jacob's Ladder had long been grown in English gardens as *Valeriana graeca*, or Greek Valerian. 'Valeriana greca . . . thys is oure commune Valerian that we use agaynste cuttes wyth a blewe floure' (William Turner, *The Names of Herbes*, 1548).

As for 'Jacob's Ladder'. Jacob slept with a stone for a pillow, 'and he dreamed, and behold a ladder set up on the earth, and the top of it reached to heaven: and behold the angels of God ascending and descending on it' (Genesis xxviii. 12). *Polemonium caeruleum* has pinnate leaves, going up ladderwise, rung by rung.

LXX. Boraginaceae

1. Hound's Tongue. *Cynoglossum officinale* L. 77, H 14

Local names. GIPSY-FLOWER, Glos, USA; LITTLE BURDOCK, Suff, Norf; NAVELWORT, RATS AND MICE, Wilts; ROSE-NOBLE, N Ire; SCALD-HEAD (i.e. scabies, ringworm, etc.), Suff; STICKY BUDS, Dor.

52 Hound's Tongue *Cynoglossum officinale*

Queer in its soft grey leaves, its gipsy-coloured flowers (described by John Pechey as a 'sordid red', by Culpepper as a 'purplish red', a 'dead colour'), and in its strong smell of mice. Dioscorides recorded a *kunoglosson*, 'hound's tongue', which was good against the bite of dogs, baldness, and burns, for which *Cynoglossum officinale* was prescribed up to the eighteenth century. It was used also against sores and skin diseases (Scald-head, above) and piles.

Taking his cue from a German name, Gerard thought it smelt, not of mice, but of dog's urine, the smell perhaps confirming the use against dog bite. The OE *hundes micgean* may have been the same plant. In English it is also Rib, in Dutch *hondstribbe* or *ribbe*, again perhaps the plant 'ribbe', which was used in Anglo-Saxon magic and medicine. Culpepper quoted from Mizaldus (Antoine Mizauld, *Alexikepus, seu auxiliaris hortus*, 1565), 'that the leaves laid under the Feet, will keep the Dogs from barking at you; Hound's-tongue, because it ties the tongues of Hounds' (49). So Culpepper may have started the English (and Welsh) use of Hound's Tongue as a charm against dogs: a leaf must be placed inside the shoe and underneath the great toe.

The name Rose-noble was perhaps a pun. Cynoglossum may have smelt; but it was a noble rose for medicine, worth all of a rose noble, the gold coin stamped with a rose and first minted by Edward IV. See *Scrophularia nodosa*.

2. Comfrey. *Symphytum officinale* L. 105, H 14

Local names. ABRAHAM, ISAAC, AND JACOB (from the variation in the colour of the flowers), Lincs; CHURCH BELLS, COFFEE FLOWERS, Som; GOOSEBERRY-PIE, Dev, Dor, Wilts, Suff; PIGWEED, Wilts; SNAKE, Dor; SUCKERS and SWEET SUCKERS, Som.

Consolida major, one of the favourite plants for making firm and solid, for consolidating, since it was taken to be the *sumphuton* of Dioscorides, the 'grow-together plant'. Comfrey goes back, by way of the Old French *confirie* and the Mediaeval Latin *cumfiria*, to the Latin name *conferva*, in Pliny, from the verb *confervere*, to grow together. ''Tis an excellent Wound-Herb, is Musilaginous and Thickning, and qualifies the Acrimony of the Humours. 'Tis used in all Fluxes, especially of the Belly; and for a Consumption. The Flowers boyl'd in Red Wine, are very proper for those that make a Bloody Urine. Outwardly applied, it stops the Blood of Wounds, and helps to unite broken Bones; wherefore 'tis called Bone-set' (John Pechey's *Compleat Herbal of Physical Plants*, 1694).

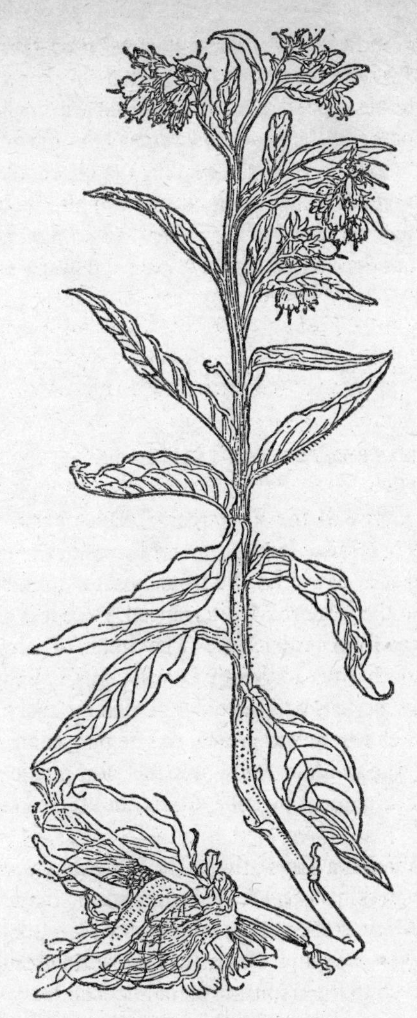

53 Comfrey *Symphytum officinale*

Gerard gives a long list of virtues, of which some are a little unexpected: 'The slimie substance of the roote made in a posset of ale, and given to drinke against the paine in the backe, gotten by any violent motion, as wrestling, or over much use of women, doth in fower or five daies perfectly cure the same, although the involuntarie flowing of the seed in men be gotten thereby.'

Though common enough wild, Comfrey was grown as a necessity in gardens, and 'Comferie with white flowers' was taken over to New England in the seventeenth century (111). There it escaped and established itself. Comfrey leaves dipped in batter and fried are much eaten in Bavaria (see German cookery books under *Schwarzwurz*) – one of the most delicious of all 'wild' foods, both to the eye, for the viridian leaf in the golden batter, and the taste. The leaves and stems can also be boiled.

It is no good picking Comfrey for its striking flowers. Like others of the Boraginaceae, it at once shrivels, stem, leaves, flowers, and all, and looks miserable.

3. Evergreen Alkanet. *Pentaglottis sempervirens* (L.) Tausch 79, H 11

Local names. BIRD'S EYE, PHEASANT'S EYE, Som.

In Devon and Cornwall the Evergreen Alkanet opens its blue, white-throated flowers and raises its dark coarse leaves in the damp hedges and corners. Abroad it is native on the west coast of France, in the Iberian peninsula, and in the western Mediterranean. Here it is almost certainly a denizen, an escape from gardens (see 'The Status of *Anchusa sempervirens* L. in the Plymouth Area of South Devon', J. G. Vaughan, *Watsonia*, Pt. V, 1950). The herbals and the older gardening books have nothing to say of its virtues or why it was grown, so one may conclude it was an old introduction of the Middle Ages, possibly for medicine, possibly for dyeing. It is not infrequent by abbey ruins, and the roots certainly give a red dye (12).

The name Alkanet is a diminutive of the Spanish *alcanna*, which in turn comes from the Arabic *al-henna*, the henna plant, the name of the small tree *Lawsonia inermis*, source of the henna which Egyptian women used to redden their finger-nails and toe-nails. The 'little alcanna', or Alkanet, was the name given to the colouring plant *Alkanna tinctoria*, the red roots of which have been imported from France and Germany for a score of uses, among them (more recently) the tinting of imitation or inferior port, the tinting of ointments, and colouring the liquid in thermometers or in the large decorative bottles of the chemist's shop. Evergreen Alkanet may have been a mediaeval substitute.

J. G. Vaughan believes that the Evergreen Alkanet might have spread more widely, if the seeds were dispersed more efficiently. The four nutlets of each flower simply divide and drop to the ground.

4. Lungwort. *Pulmonaria longifolia* Bor. 3

Local names. ADAM AND EVE (so called by the gipsies in the New Forest), Hants; BLUE COWSLIP, I o W; JOSEPHS AND MARIES, LUNGWORT, SNAKE'S FLOWER, Hants.

First discovered in the seventeenth century by the Hampshire botanist, John Goodyer, this is the wild, native Lungwort of the New Forest (where it makes a fine show in April and May), the Isle of Wight, and Dorset. Compared with *Pulmonaria officinalis*, the nutlets are different, the flowers smaller, and the root leaves are spear-shaped instead of egg-shaped.

5. Lungwort, Jerusalem Cowslip. *Pulmonaria officinalis* L. 26

Local names. ABRAHAM, ISAAC, AND JACOB, Lincs; ADAM AND EVE, BEDLAM COWSLIP (i.e. Bethlehem), Som; BEGGAR'S BASKET, Som, Ches; BOTTLE-OF-ALL SORTS, Cumb; CHILDREN OF ISRAEL, Dor; GOOD FRIDAY PLANT, Som; JERUSALEM COWSLIP, Glos, Bucks, Oxf, Norf, Ches, Cumb; JERUSALEM SEEDS, Dev, Som; JOSEPHS AND MARIES, Dor, Wilts, Hants; LADY MARY'S TEARS, Dor; LADY'S MILK-SILE (i.e. strainer), Ches; LADY'S PINCUSHION, Ches, Yks.

LUNGWORT, Hants, Glos; MARY'S TEARS, Dor; MOUNTAIN SAGE, SAGE OF JERUSALEM, Cumb; SOLDIER AND HIS WIFE, I o W; SOLDIERS AND SAILORS, Som, Suff; SPOTTED MARY, Rad; SPOTTED VIRGIN, Heref; THUNDER AND LIGHTNING, Banff; VIRGIN MARY, Corn, Hants, Lakes; VIRGIN MARY'S COWSLIP, Glos, Worc, Shrop; VIRGIN MARY'S HONEYSUCKLE, Shrop, Ches; VIRGIN MARY'S MILKDROPS, Wilts, Mon; VIRGIN MARY'S TEARS, Dor.

Often naturalized, making a pond of azure in the woods. Since the leaves have white spots, sympathetic magic made it into a medicine 'against the infirmities and ulcers of the lungs'. Gerard also wrote that the leaves (cf. *Symphytum officinale*) were 'used among pot herbs'. It was then planted 'almost every where in gardens', and the names he gives for it are Spotted Comfrey, Sage of Jerusalem, Cowslip of Jerusalem, Sage of Bethlehem, and 'of some Lungwoort'. It was a common fancy through Europe that drops of the Virgin's milk or the Virgin's tears had fallen on the leaves and spotted them (cf. the Milk Thistle, *Silybum marianum*, or Tintoretto's painting in the National Gallery of the 'Origin of the Milky Way', which springs from the milk of Juno's breast).

The double names – e.g. Josephs and Maries, Soldiers and Sailors,

Adam and Eve – originate from the flowers changing from pink to blue, or rather from flowers together on the plant in the pink and the blue stages.

6. Forget-me-not. *Myosotis palustris* L. ssp. *palustris* 108, H 40

Local names. BIRD'S EYE, Som, Hants, N'hants; BUGLOSS, Dev; FORGET-ME-NOT, Dev, Som, Ches, Yks, Cumb; LOVE-ME, Yks; MAMMY-FLOOER, Shet; ROBIN'S EYE, Hants; SNAKE-GRASS, Yks.

Here is a tangled story of names. Gerard knew no English names for the three species of *Myosotis* of which he gave the first English record, *M. palustris*, and *M. arvensis* and *M. hispida*. Because the cymes are more or less bent over or coiled, these were placed with the Scorpion Grass, the leguminous plant *Scorpiurus sulcata*, from southern Europe, which was identified with the *skorpioeides* in the *De Materia Medica* of Dioscorides, who likened the 'fruit', the coiled pod, to a scorpion's tail. So Scorpion Grass was fixed to the various kinds of *Myosotis* as an English book name.

In Old French the lovely *Myosotis palustris* was called *ne m'oubliez mye*, in Middle High German *vergiz mîn niht*, and the plant was a symbol of love: if you wore it you were not forgotten by your lover – all of which was explained by the German tale of the knight who picked *Myosotis palustris* for his lady as the two of them walked by a river. The knight fell in, but before he was carried away and drowned, he threw the flowers to the lady, crying *vergisz mein nicht*.

Coleridge, who knew his Germany, wrote a poem, *The Keepsake*, which was published in a newspaper in 1802:

> . . . Nor can I find, amid my lonely walk
> By rivulet, or spring, or wet roadside
> That blue and bright-eyed flowerlet of the brook,
> Hope's gentle gem, the sweet Forget-me-not!

But this unfamiliar name required a note: 'One of the names (and meriting to be the only one) of the *Myosotis Scorpioides Palustris*, a flower from six to twelve inches high, with blue blossom and bright yellow eye. It has the same name over the whole Empire of Germany (*Vergissmein nicht*) and, I believe, in Denmark and Sweden.'

The name stuck and spread, driving out Scorpion Grass, especially for all the garden forms of the variable *Myosotis sylvatica*. The French still call Forget-me-not *ne m'oubliez pas*, as well as *aimez-moi* and *plus je vous vois, plus je vous aime.* Forget-me-not (and Remember-me) were also applied to that other Bird's Eye, *Veronica chamaedrys.* Lyte in 1578, and Gerard after him, had used Forget-me-not for the rare *Ajuga chamaepitys.*

7. Blue Gromwell. *Lithospermum purpurocaeruleum* L. 8

Too local to be well known, restricted as it is to limestone and chalk, in Somerset, south Devon, Kent, Suffolk, Monmouth, Glamorgan, and Denbigh. Everyone should try to see the Blue Gromwell. The flowers change from a reddish-purple to a blue as intense and thrilling as that of *Gentiana verna.*

It is very much a plant of the Mendips, glittering among the brambles and the scrub at the fringe of copses. But if you pick Blue Gromwell and put it in water, the blue light goes dull and the leaves shrivel and droop.

8. Gromwell. *Lithospermum officinale* L. 85, H 33

Local names. GREY MILLET, Som; STONY-HARD, N Eng.

Rather a scarce plant, always entertaining to find for the hard, glistening, and grey-white little nuts, which suggested that Gromwell was the *lithospermon*, or 'stone seed', of Dioscorides. He recommended them for expelling the equally hard nutlets of the stone, so they were much used in sixteenth- and seventeenth-century medicine.

Gromwell is a mysterious name, from the Old French *gromil*, in modern French *grémil.* Whatever the origin, *gré-* suggested 'grey'; the last syllable – and the nutlets themselves – suggested grains of *mil*, or millet. Turner, in 1548, gave the English names as Grummel or Grey Myle. Gromwell also became known as *milium solis*, 'millet of the sun'.

The nuts have been investigated for the contraceptive substance they contain.

9. Viper's Bugloss. *Echium vulgare* L. 98, H 26

Local names. BLUEBOTTLE, Norf; BLUE CAT'S TAIL, Herts; BLUE DEVIL (cf. the German devil or goblin names for the species, *Blauer Heinrich*, *Stolzer Heinrich*, blue and proud Henry), Som, USA; BLUE THISTLE, Som; BLUEWEED, Som, Herts, USA; BUGLES (i.e. bugloss), Hants; CAT'S TAIL, Som, Ess, Herts, Norf, Cam; IRONWEED, Beds; OUR LORD'S FLANNEL, OUR SAVIOUR'S FLANNEL, Kent; SNAKE'S FLOWER, I o W, USA; WILD BORAGE, Surr.

A viperish beauty, on bare hillsides, on shingle by the sea, on mine dumps. It was taken to be the plant Dioscorides knew as *echion*, 'viper plant', from *echis*, a viper, because the nutlets resembled a viper's head. Dioscorides recommended it both to cure and prevent snake-bite. According to the *Theriaca*, a poem on venomous animals by the Greek poet and physician Nicander, who lived in the second century B.C., this virtue was discovered by Alcibiades, after a snake had bitten him in his sleep. William Coles, the signaturist, found the anti-poisonous character of Viper's Bugloss confirmed by the stalks, which are 'speckled like a Serpent's Skin' (45).

European doctors also followed Dioscorides in giving this 'most gallant Herb of the Sun' to increase the flow of mother's milk and to ease lumbago. Moreover a syrup of Viper's Bugloss was considered effectual against sadness and melancholy, a virtue transferred from the *bouglosson* (*bouglossos*, 'ox tongue') of Dioscorides. *Bouglosson* the physicians equated with the Common Bugloss of Tudor gardens – the modern *Anchusa azurea* (*A. italica*), which now comforts the heart in a different way, in a score of horticultural forms.

Turner in 1538, in his *Libellus de re herbaria nova*, mentioned Cattestayle (see above) as a countryman's name for *Echium vulgare*. In his *Herbal* in 1551 he said it was also known in some places as 'Wylde Buglosse'. Lyte wrote of it in 1578 as Viper's Bugloss. Introduced into the U.S.A., and a farmer's plague, Americans call it by the Somerset names Blue Devil and Blueweed; also Snake Flower and Viper's Grass, as in the Isle of Wight (131).

LXXI. Convolvulaceae

1. Bindweed. *Convolvulus arvensis* L. 109, H 40

Local names. BEARBIND (plant which binds or winds about the barley, OE *bere*), Som, Kent, Worc, Shrop, Yks; BEDWIND, Hants, Glos, War; BELLBIND, Som, Ess, Norf, Cam; BELLWIND (cf. German *Windglocke*), Bucks; BILLY-CLIPPE, Kent; BILLY-CLIPPER, Shrop; BINE LILY, Dor; CORNBINE (or Cornbind), Bucks, Oxf, N'hants, War, Notts, Lincs, Yks, N Eng; CORN LILY, Yks; DEVIL'S GUTS (cf. German *Teufels Nähgarn*, 'devil's thread'), Som, I o W, Kent, Beds, Shrop, Leic, Notts, Lincs, N'thum; DRALYER (Cornish *draylyer*, 'trailer'), Corn; EARWIG, FAIRIES' UMBRELLA, FAIRIES' WINECUPS, GIPSY'S HAT, Som; GRANNY'S NIGHTCAP, Wilts.

HEDGE BELLS, I o W; HELLWEED, N'hants; JACK-RUN-IN-THE-COUNTRY, Yks; KETTLE SMOCK, Wilts; LADY'S SMOCK, LADY'S SUNSHADE, Som; LAP-LOVE, Midlands, N Eng, Scot; LILY, Hants, Suss; MORNING GLORY, OLD MAN'S NIGHTCAP (i.e. devil's), PARASOLS, ROBIN-RUN-IN-THE-FIELD, ROPE-WIND, Som; SHEEPBINE, Herts; WHITE SMOCK, Dev; WILLOW-WIND, Wilts; WITHWIND (see Withywind), Dor, Som, Wilts, Glos, Hants; WITHYWIND (equivalent to 'string-twist', 'thread-twist', cf. German *Wedewinde*), Dev, Som, Dor, Wilts, Glos, Hants, Berks, Oxf.

'The smallest fragment of the root of this species of *Devil's-guts* (to speak technically to human beings) will very quickly give rise to a perfect plant,' says one of the older floras.

Every gardener knows it, and perhaps more blasphemy is expended on Devil's Guts, Cornbine, Withwind, Withywind, Gravelbind (the name for it in Miller's *Gardeners Dictionary*, 1741), than upon all the other weeds of Great Britain. Neither blasphemy, hoeing, nor selective weed-killers have yet destroyed it. One should speak kindly of its white and pink flowers, all the same.

2. Larger Bindweed. *Calystegia sepium* (L.) Roem & Schult.

Local names. BEARBIND, Kent, Surr, Mddx, Bucks, Herts, Heref; BELL-BIND, Ess, Suff; BELLWIND (cf. German *Windglocke*), Surr, Bucks; BETH-WIND, Hants, Glos, Mddx, Bucks; BIND, E Ang, Lincs; BINE LILY, Dor; CAMPANELLE (cf. French *campanelle* – from *campanella*, diminutive of *campana*, a bell), CREEPING JENNY, CUPS, DADDY'S WHITE SHIRT, Som; DEVIL'S GARTER, Pemb, Ire; DEVIL'S GUTS (cf. German *Teufels Nähgarn*, 'devil's thread'), Norf; DEVIL'S NIGHTCAP, FAIRY TRUMPET, Som; GRAND-MOTHER'S NIGHTCAP, Dev, Suss; GRANNY'S NIGHT-BONNET, GRANNY'S NIGHTCAP, Som; GROUND IVY, Dev.

HARVEST LILY, Surr; HEDGE BELLS, Som, I o W, S Eng, Staffs, Cumb; HEDGE LILY, Som, Hants, I o W; HELLWEED, N'hants; HONEYSUCKLE, Dev; HOLLAND SMOCKS, JACK-RUN-IN-THE-HEDGE, LADY'S SHIMMY (i.e. chemise), Som; LADY'S NIGHTCAP, Som, Wilts, Hants; LADY'S SMOCK, Corn, Dev, Som, Suss, N'hants, War, Notts, Yks; LADY'S UMBRELLA, Som; LILY, Wilts; LILYBIND, Yks; LILY-FLOWER, Hants; LONDON BELLS, Dev; MILK-MAID, Suss, Surr; MORNING GLORY, Som; NIGHTCAPS, Som, Wilts, Lincs; NIGHTSHIRTS, Som.

OLD LADY'S SMOCK, Wilts; OLD MAN'S NIGHTCAP (i.e. devil's), Som, Suss, Surr; OLD MAN'S SHIRT, OLD WOMAN'S NIGHTCAP, OUR LADY'S NIGHT-

CAP, OUR LADY'S SMOCK, Som; PISSPOT (cf. German *Pisspott*, *Pisspöttchen*, Dutch *piespotje*), Suss; ROBIN-RUN-THE-HEDGE, Hants, Ches, Lancs, N Ire; RUTLAND BEAUTY, USA; SHIMMIES (i.e. chemises), Dev, Dor, Som, Wilts; SHIMMIES AND SHIRTS, Dev, Dor, Wilts; SHIMMY AND BUTTONS, Dor; SHIMMY SHIRTS, Dev, Dor, Wilts; SMOCKS, Dor, Som; STRANGLEWEED, TRUMPETS, Som; WAYWIND, Notts; WHITE LILY, Cumb; WHITE SMOCK, Dev; WIREWEED, Surr; WITHWIND (see *Convolvulus arvensis*), Dor, Wilts, Hants, Glos; WITHYWIND, Dev, Dor, Som, Wilts, Hants, Glos, Berks, Bucks, Oxf; WOODBINE, Suss.

For several of the names of this and *Convolvulus arvensis*, compare *Clematis vitalba* and *Lonicera periclymenum*; and see also *Geranium robertianum*.

Less of a weed, and more easy to control as a shallow rooting species, *Calystegia sepium* belongs chiefly to the south of England, though the distribution is imperfectly known. Names recorded from the northern counties no doubt belong strictly to the introduced and naturalized *Calystegia sylvestris* (Willd.) Roem & Schult. Gerard dismissed the Bindweeds with the gardener's impatience: they were not fit for medicine, 'but unprofitable weedes and hurtfull unto eche thing that groweth next unto them'.

Yet was it these two plants which first prompted the Somerset name of Morning Glory? And was Morning Glory transferred by west country settlers to the lovelier and more glorious Ipomoeas of the New World, which are so vivid against the white boards of the frame-house in America?

3. Common Dodder. *Cuscuta epithymum* (L.) Murr. 60, H 3

Local names. ADDER'S COTTON (cf. German *Schlangenseid*), Corn; BEGGARWEED, Dor, Wilts; CLOVER DEVIL, Som; CLOVER DODDER, Ess; DEVIL'S GUTS, Som, Hants, I o W, Glos, Suss, Beds, Worc, Shrop, Cumb, Lanark; DEVIL'S NET, Kent; DEVIL'S THREAD (cf. German *Teufelszwirn*, 'devil's thread'), Kent; EPIPHANY (corruption of *epithymum*), Corn; FAIRY HAIR, Jersey; HAIRWEED, Herts, Beds, Norf; HAIRY BIND, Herts; HELLBIND, Herts; HELLWEED, Suss, Kent, Berks, Herts, Camb, Beds, N'hants, Lanark; LADY'S LACES, Som; MAIDEN'S HAIR, I o W; MULBERRY, Som; RED TANGLE, Norf; SCALD, SCALD-WEED (i.e. scab or scabies weed, cf. German *Grind* (scabies) for this plant), Cam; STRANGLEWEED, Som.

Some of the names doubtless apply also to *Cuscuta europaea*, parasitic mainly on Hop and Nettle.

'This fawning Parasite, and ungrateful Guest, hugs the Herb it hangs upon, with its long Threads, and reddish Twigs; and so closely embraces it, that at length it defrauds the hospitable Herb of its Nourishment, and destroys it by its treacherous Embraces' (John Pechey, in *The Compleat Herbal of Physical Plants*, 1694).

Dodder was considered to be the *epithumon* of Dioscorides, which was purgative; and the winding stems were obviously a signature of the guts. However, 'the nature of this herbe changeth and altereth, according to the nature and qualitie of the herbes wherupon it groweth'. This was an article of faith (cf. the special virtue of Mistletoe of the Oak); and Culpepper wrote: 'He is a Physitian indeed that hath wit enough to chuse his Dodder according to the Nature of the disease and Humor peccant.' Thus Dodder growing on nettles (*Cuscuta europaea*) was long held to be 'a most singular and effectual medicine to provoke urine', *Cuscuta epithymum*, taken from Wild Thyme, was specially good for obstructions, spleenful headaches, and the 'scabbie evil' – the itch or scabies.

Devil's Thread – in Haute-Bretagne it was believed that the Devil spun the Dodder at night to destroy the clover; clover was created by God, Dodder was the Devil's counter-plant (165).

LXXII. Solanaceae

1. Duke of Argyll's Tea-plant. *Lycium halimifolium* Mill. Introduced

The story is that the Duke of Argyll, who delighted in exotic trees and shrubs in his gardens at Whitton in Middlesex, was sent a *Thea* labelled as this species of *Lycium*, and a *Lycium* labelled as a *Thea*. So *Lycium halimifolium* kept the ironical name of the Duke of Argyll's Tea-plant, or Tea-tree (125).

The shrub had been introduced earlier in the eighteenth century, in 1730 or thereabouts. Since it grows quickly and easily, and trails prettily down a hedge or over a wall, and has purple flowers (though not many of them in most seasons) and then scarlet fruit, it went rapidly from garden to garden, from the level of the garden of the 'Treemonger', as Horace Walpole called the Duke of Argyll, down to the cottager's garden. Deserted by fashion, *Lycium halimifolium* has refused to die out. It is naturalized, and it now belongs as much to the village lane or bank as the Greater Celandine.

2. Deadly Nightshade. *Atropa belladonna* L. 44

Local names. BANEWORT, DAFT BERRIES, DEVIL'S BERRIES (cf. German *Teufelsbeeren*), DEVIL'S CHERRIES (cf. German *Teufelskirsche*), DEVIL'S RHUBARB, Som; DOLEFUL BELLS, Dor; DOG BERRIES (cf. German *Lundsbeeren*), Dur; DWALE, Dev, Som, Yks, Scot; JACOB'S LADDER, Ayr; JACOB'S STEE (i.e. ladder), Lincs; NAUGHTY MAN'S CHERRIES, Som, Bucks; SATAN'S CHERRIES, Yks.

54 Deadly Nightshade *Atropa belladonna*

The crumbling slope of a chalk quarry or a limestone escarpment: nettles, elders, burdock, ropes of Old Man's Beard over the nut bushes – and then the brown-violet bells, the black cherries, and the coarseness of Deadly Nightshade, that *solanum lethale* whose narcotic and killing powers were well known. 'This naughtie and deadly plant', Lyte called it in his translation of Dodoens. 'If you will follow my counsell,' wrote Gerard, 'deale not with the same in any case, and banish it from your gardens and the use of it also, being a plant so furious and deadly.' Children have been killed by three berries (123). With rough methods of preparation and control, Deadly Nightshade was too dangerous to use, and this odd narcotic had to wait for nineteenth-century investigators to turn it into a valuable drug – an indispensable drug for the eye specialist, through its action of dilating the pupil of the eye.

The early history is thin for so famous a plant. Pierandrea Mattioli (1501–77), the physician and botanist who grew up in Venice, declared that Venetians knew it as *Herba bella donna*, 'beautiful lady herb', since they used a distilled water of it as a cosmetic, and to excite and enlarge the pupil of the eye. The berries also enticed the ignorant by looking attractive, like the glance of a *bella donna*. Dwale is a name which has lived rather from book to book (and poem to poem) than from mouth to mouth. Applied to *Atropa belladonna* by Gerard, by Turner in 1538, and earlier still, it seems to be descended from a Germanic base meaning 'to send astray', i.e. to stupefy.

3. Henbane. *Hyoscyamus niger* L. 83, H 31

Local names. DEVIL'S EYE, Som; HENPEN, West; HENPENNY, Lakes; HOG-BEAN (a translation of the Greek *huoskuamos*), Cumb; STINKING ROGER, Cumb.

A plant with a long history which provides the British Pharmacopoeia with one of its powerful, if possibly dangerous, drugs. Virtuous doctors use the alkaloid hyoscine (which is derived from the green tops and leaves of Henbane) as a hypnotic and a brain sedative, which works upon the mad, the excitable, or the seasick. The unvirtuous Dr Crippen used hyoscine to murder Mrs Crippen in 1910. Back through the centuries the herbalist, the apothecary, the Tudor physician, Greek and Roman doctors, and before them Assyrian doctors, used Henbane against toothache. A narcotic herb is clearly possessed of *mana*: according to one Assyrian tablet Henbane was placed on the outer door and the hinges 'to prevent sorcery from approaching a man's house' (30). It had a magic reputation in the Middle Ages. The Stockholm Medical Manuscript (*c.* 1400) says:

> Amongis wommen if thou schuldist gon
> And hennebanne hawe the up-on
> This ilk cas schall be-falle,
> It schall hem make to lowe the all.

55 Henbane *Hyoscyamus niger*

If you wanted a rout of hares:

> Harys blod if thou take
> And with jows of hennebanne medele and make
> And in an harys skyn do it bynde
> And lete it so lyn in feld ore lynde
>
> Or sow it on an harys skyn,
> And in what place thou lete it lyn:

Alle the harys ther abowte
Schull gader thedir on a rowte.

In fact,

Of all erbys that growyn on grownde,
To wickyd spiritis it is a wounde;
Ageyns hem fel it is and fers,
That well knowyn nygromanseris (168).

To the signaturists a branch of Henbane in seed resembled a jaw with molars. 'The Husk wherein the seed of Henbane is contained, is in figure like to a Jaw Tooth; and therefore the Oyl of it, or the Juyce by itself, or the decoction of the root with Arsmart in vinegar, being gargled warm in the mouth, is very effectuall in easing the pains of the Teeth' (William Coles, in *Adam in Eden*, 1657, partly echoing Dioscorides). In Normandy it was *herbe Ste Appoline*, after the saint invoked for toothache (116).

Though narcotic and poisonous, it was more easily handled and more widely employed by the Tudor and Jacobean doctors than Deadly Nightshade. The earlier name in English was Henbell, replaced in the fifteenth century by Henbane (cf. the French *mort-aux-poules*). Tobacco was referred to by Gerard and others as Henbane of Peru.

4. Bittersweet, Woody Nightshade. *Solanum dulcamara* L. 100, H 39

Local names. BITTERSWEET, Som, N'hants, Lincs, Cumb, N'thum; DOGWOOD, Lancs; FELONWOOD (the berries were used against felons, or whitlows), Corn, Lancs, Yks, Cumb; FOOL'S CAP, Som; GRANNY'S NIGHTCAP, Dev; HALFWOOD, War, Worc; MAD DOG'S BERRIES, Mor; POISONBERRY, Oxf, N'thum, N Eng, Scot; POISON-FLOWER, Herts; POISONING-BERRIES, Yks; POISONOUS TEA PLANT (? in distinction to *Lycium halimifolium*), Oxf.

ROBIN-RUN-THE-HEDGE, Lancs; SCAW-COO (Cornish *scaw cough*, 'scarlet elder'), Corn; SHADY NIGHT, Lancs; SNAKEBERRY, Suff; SNAKE-FLOWER, SNAKE'S FOOD, SNAKE'S MEAT, Som; SNAKE'S POISON-FOOD, Bucks; TERRYDIDDLE, TERRYDIVIL, TETHER-DEVIL, Ches; WITCHFLOWER, Som.

Gerard's description is delightful: 'The flowers be small and somewhat clustered together, consisting of five little leaves [i.e. petals] a peece, of a perfect blewe colour with a certaine pricke or yellow pointell in the middle:

which being past there do come in place faire berries, more long then rounde, at the first greene, but very red when they be ripe, of a swete taste at the first, but after very unpleasant, of a strong savour; growing togither in clusters like burnished corall.'

German physicians in the sixteenth century introduced the stalks of Bittersweet into medicine. They were used against rheumatism and skin diseases and as a purgative. Bittersweet was also a plant of magic efficacy, like other red-berried species. It was good against witches: in Lincolnshire collars of Bittersweet were hung around overlooked pigs, and elsewhere garlands of Bittersweet and Holly were put around the necks of horses if they were hag-ridden (192, 8). Culpepper observed that Bittersweet was 'excellent good to remove Witchcrafts both in men and beasts; as also all sudden Diseases whatsoever. Being tied round about the neck, it is one of the admirablist remedies for the Vertigo or Diziness in the head that is, and that's the reason (as *Tragus* saith) the people in *Germany* commonly hang it about their Cattels neck when they fear any such evil hath betided them' (49). It was Tragus (Hieronymus Bock), the German physician and botanist (1498–1554), and before him Hieronymus Braunschweig, who called the plant *Amara dulcis*, *Dulcis amara*, *Dulcamarum*, etc., which William Turner (*Herbal*, 1568) made into Bittersweet.

The berries have been found in Egyptian tombs, threaded on thin strips of the leaf of the Date Palm. So threaded, they were part of the collarette on Tutankhamun's Third Coffin, in conjunction with sequins of blue glass (Howard Carter, *The Tomb of Tutankhamun*, 1922).

5. Thorn-apple. *Datura stramonium* L. Casual

Narcotic, poisonous, a weed – and always worth finding as a beauty in flower or fruit. In Somerset it has been called Angel's Trumpets, the name usually given now to the tall, shrubby Thorn-apples from Brazil and the Andes, which are hung about with trumpet blossoms of whiteness and purity up to a foot long. In America *Datura stramonium* is pitilessly hated as Jimsonweed, since it was first a plague at Jamestown, Va.

By no means, though, a worthless species. The leaves and flowering tops give the stramonium of the British Pharmacopoeia. John Pechey, in 1694, maintained that the seed, powdered and taken in beer, causes madness for twenty-four hours, and was given by thieves to those they intended to rob. 'And Wenches give half a dram of it to their Lovers, in beer or wine. Some are so well skill'd in dosing of it, that they can make men mad for as many hours as they please' (141). A milder use for the seed may be recommended – or rather for the prickly green 'thorn-apples' which succeed the ivory trumpets. If you like the effects of light, infuse two or three in surgical spirit: pass a light through the glass and you see the most delectable green fluorescence.

LXXIII. Scrophulariaceae

1. Mullein, Aaron's Rod. *Verbascum thapsus* L. 102, H 37

Local names. AARON'S FLANNEL, Dor; AARON'S ROD, Som, Glos, Lincs, Midlands, Scot; ADAM'S FLANNEL, Som, N'hants, War, Ches, Lincs, Yks; BEGGAR'S BLANKET, Som, Cumb; BEGGAR'S STALK, Cumb; BLANKET LEAF, Dev, Som, War; BLANKET MULLEIN, Ches; BUNNY'S EARS, Dor; CANDLE-WICK, Som; CLOTE, Wilts; CUDDY-LUGS (i.e. 'donkey-ears'), N'thum, Rox; DEVIL'S BLANKET, Wilts; DONKEY'S EAR, Dor; DUFFLE (i.e. a woollen cloth), Suff.

FAIRIES' WAND, Dor; FLANNEL, FLANNEL-FLOWER, Som, Suff; FLANNEL-JACKET, Norf; FLANNEL-LEAF, FLANNEL PETTICOATS, Som; FLANNEL-PLANT, Corn, Hants, I o W; FLUFF-WEED, Som; FRENCH POPPY (i.e. a 'French' kind of Foxglove), Dev.

GOLDEN GRAIN, Dev; GOLDEN ROD, Dev, Som; GOLDILOCKS, N'hants; HAG-LEAF, Som, Bucks; HAG-TAPER, Bucks, Herts, N Ire; HARE'S BEARD, Dor, Som; HIGH TAPER, Som, N Ire; KING'S TAPER (from the Latin *Candela regia*), LADY'S CANDLE, LADY'S FLANNEL, LADY'S TAPER, MOSES' BLANKET, OLD MAN'S FLANNEL (i.e. devil's), Som; OUR LORD'S FLANNEL, OUR SAVIOUR'S FLANNEL, Kent; POOR MAN'S BLANKET, Donegal; POOR MAN'S FLANNEL, Som, Glos, Bucks.

RAG PAPER, Bucks; SEA CABBAGE, Glam; SHEPHERD'S CLUB, I o W, Scot; SHEPHERD'S STAFF, Cumb; SNAKE'S FLOWER, Dor, Wilts; SNAKE'S HEAD, SOLDIER'S TEARS, Dor; SWEETHEARTS, Som; VELVET DOCK, Dev, Som; VELVET POPPY (i.e. velvet foxglove), Corn; VIRGIN MARY'S CANDLE, Ire.

A plant that cannot be overlooked. 'The whole toppe with its pleasant yellow floures sheweth like to a wax candle or taper cunningly wrought' (Henry Lyte, 1578). 'Called of the Latines', wrote Parkinson, 'candela regia, and Candelaria, because the elder age used the stalks dipped in suet to burne, whether at funeralls or otherwise.' The whole top suggested candles and tapers, and Aaron's rod, which 'was budded, and brought forth buds, and bloomed blossoms' (Numbers xvii. 8). The leaves suggested everything that was soft: rabbit's ears, donkey's ears, rag paper, velvet, duffle, and flannel – especially in the cloth-making counties. Welsh names include *tapr Mair*, 'Mary's taper'; *canwyll yr adar*, 'bird's candle'; *clûst y tarw* and *clûst y fuwch*, 'bull's ear' and 'cow's ear'; and, evocative at once of softness and whiteness, *sirkyn y melinydd*, 'miller's jerkin'. Mullein itself, Anglo-French *moleine*, goes back to the Latin *mollis*, soft.

The tall stem also suggested the rod of power, the wand. The Anglo-Saxons learned from their version of the *Herbarium* of Apuleius Platonicus that Mercury had given Mullein to Ulysses, when he came to Circe – 'and he after that dreaded none of her evil works' (42). 'If a man beareth with him one twig of this wort,' the pseudo-Apuleius continued, 'he will not be terrified with any awe, nor will a wild beast hurt him; or any evil coming near.'

Several names hint at the power of Mullein, its connection with the Devil, with the hag:

> The Hag is astride
> This night for to ride,
> The Devil and she together

(though Hag-taper, Hag-leaf, etc., may mean only 'hedge taper', 'hedge leaf', OE *hecg*) – and with fairies, hares, and snakes. In Dorset and Wiltshire children were told that a snake might be hiding under the leaves of this Snake's Flower or Snake's Head. As so often, Mullein belongs in its names both to the Devil and his opposites. Medicinally it was given for coughs, gripes, piles, etc., and for consumption in cattle.

2. Toadflax. *Linaria vulgaris* Mill. 99, H 27

Local names. BACON AND EGGS, Wilts; BRANDY-SNAP, Suss; BREAD AND BUTTER, Som; BREAD AND CHEESE, Wilts; BRIDEWEED (for 'bride', a disease of pigs), Shrop, USA; BRIDEWORT, Shrop; BUNNY RABBITS, Som; BUTTER AND EGGS, Dev, Dor, Som, Wilts, Glos, Suss, Kent, Bucks, Ess, Worc, Yks, Cumb, USA; BUTTERED HAYCOCKS, Yks; CHOPPED EGGS, Som, Cumb; CHURN-STAFF, Ches; DOGGIES, Aber; DOG'S MOUTH, Som, Wilts; DRAGON-BUSHES, Bucks.

EGGS AND BACON, Dev, Som, Wilts, Glos, Norf, Rut, Notts, Yks; EGG AND BUTTER, Dev, Som, Wilts; EGGS AND COLLOP (i.e. bacon), Som, Lancs, Yks, N Eng; FAIRIES' LANTERNS, FINGERS AND THUMBS, FLAXWEED, Som;

FOX AND HOUNDS, Lincs; GALLWORT, Suss; GAPING JACK, GAP MOUTH, LADY'S SLIPPER, Som; LARKSPUR, Bucks; LION'S MOUTH, Dev, Dor, Suss; MONKEY FACES, Dor; MONKEY-FLOWER, Yks; MONKEY-PLANT, Yks; MOUSE'S MOUTH, Wilts.

PATTENS AND CLOGS, PIG'S CHOPS, PIG'S MOUTH, Som; PUPPY DOG'S MOUTHS, Wilts; RABBIT-FLOWER, Dev, USA; RABBITS, Dev, Som; RABBIT'S CHOPS, Dor; RABBIT'S MOUTHS, SEARCHLIGHT, Som; SHOES AND STOCKINGS, Corn; SNAPDRAGON, Yks; SNAPJACKS, SNAPS, SQUEEZE-JAWS, Som; STRIKE, Dor; WILD GAPMOUTH, Som; WILD SNAPDRAGON, Som, Oxf, Lincs; WEASEL-SNOUT, Kent; YELLOW ROD, Ches.

Linaria vulgaris was a weed in flax, which looked not unlike flax itself until the flowers developed. So it was named *Linaria* (*linum*, 'flax'), and *Krötenflacks* in German, which William Turner in 1548 translated into '*todes flax*'. An apothecaries' name was *Urinalis*, since, yellow having suggested yellow, *Linaria vulgaris* had some reputation as a diuretic.

Never be seduced by Toadflax, never allow it in the garden. It spreads incessantly. Every quarter-inch of root breeds a new plant. Every big root breaks and leaves innumerable quarter-inches in the soil. An unhappy introduction into the U.S.A., it is known there as Brideweed, Butter and Eggs, Rabbit-flower, and (very justly) Impudent Lawyer. The Norwegian name is *torskemund*, 'cod's mouth'. The double names Eggs and Bacon, Butter and Eggs, etc., point to the two colours, the yellow and orange, in the flower.

3. Ivy-leaved Toadflax. *Cymbalaria muralis* Baumg. 93, H 40

Local names. AARON'S BEARD, Scot; CLIMBING SAILOR, Dumf; CREEPING JENNY, Dev, Dor, Som, Wilts; CREEPING SAILOR, Som, Suss; CREEPING SEEFER (i.e. sailor), Kirk, Wigt; FLEAS AND LICE, Som; HEN AND CHICKENS, Kent; HUNDREDS AND THOUSANDS, Som; KENILWORTH IVY, USA; LAVENDER SNIPS, Hants; MONKEY-JAWS, MONKEY MOUTHS, Som.

MOTHER OF THOUSANDS, Corn, Dev, Dor, Som, Wilts, Glos, Kent, Worc, Derb, Ches, Yks, Cumb, N Eng, Scot; MOTHER OF MILLIONS, Corn, Dev, Som, Suss, Herts; NANNY GOAT'S MOUTHS, Dev; OXFORD WEED (from abundance on Oxford college walls), Oxf, Berks; PEDLAR'S BASKET, Som, Ches, Derb, Lancs.

RABBIT-FLOWER, RABBITS, Dev; RABBIT'S MOUTHS, Som; RAMBLING SAILOR, Som, Lancs; ROVING JENNY, Dev, I o W; ROVING SAILOR, Corn, Dev, Som, I o W; THOUSAND-FLOWER, Ches; TRAVELLING SAILOR, Hants;

WANDERING JACK, Som; WANDERING JEW, Dev, Suss, West, Ang; WANDERING SAILOR, Dev, Som, Cumb, Scot.

The Ivy-leaved Toadflax arrived, as a garden plant, early in the seventeenth century. William Coys, one of the best amateurs of the time, appears to have grown it before anyone else in his garden at North Ockendon, in Essex. The record is in the papers of John Goodyer: 'Cymbalaria with us in England, where it is sowen runneth and spreadeth on the ground and clymeth and hangeth on walls even as Ivie or Chickweed doth, the branches are verie small round and smooth, limmer and pliant neere like the hampering threeds of Cuscuta . . .

'I never saw this growinge but in the garden of my faithfull good frend Mr William Coys in North Okington in Essex, and in my garden at Droxford [Hants] of seeds receaved from him in Anno 1618' (91).

Garden walls, park walls, boundary walls, went up apace in the seventeenth, eighteenth, and nineteenth centuries, so it was a good time for a rock species in search of new territory. By 1640 *Cymbalaria muralis* was known 'about Hatfield in Hertfordshire' (140). Much planted up and down the country, it must have made separate escapes over and over again. By 1724 it was growing on walls round about the Chelsea Physic Garden (Ray's *Synopsis*, 1724 ed.), and the Oxford plants may have escaped on to the walls from the Botanic Garden, as *Senecio squalidus* was to escape in years to come. Gilbert White planted it on a wall at Selborne in 1779, and in his *Naturalist's Journal* recorded its progress through the following seasons. It spread very quickly. Obviously he regarded it as something exotic and unfamiliar; and he marked the way it was cut by the frosts. (*Journals of Gilbert White*, ed. Walter Johnson, 1931) There are also records of planting this charming exotic as far afield as St Michael's Mount. In just over three centuries it has conquered most of the walls of Great Britain. It must now come to a halt in the age of fences and barbed wire.

4. Figwort. *Scrophularia nodosa* L. 111, H 40

Local names. BRENNET, Som; BROWN-NET, Dev, Som; BROWNWORT, Corn, Som; BRUNNET, CROWDY-KIT, Dev; CUT-FINGER, Surr; FAIRIES' BEDS, Dor; FIDDLE, Yks; HASTY ROGER, West; POOR MAN'S SALVE, Dev; ROSE-NOBLE (cf. *Cynoglossum officinale*), Dur, Kirk, Wigt, N Ire; SQUARRIB, SCARIBEUS, Wilts; STINKING CHRISTOPHER, Cumb; STINKING ROGER, Ches, Cumb, Lakes, Ayr, N Ire.

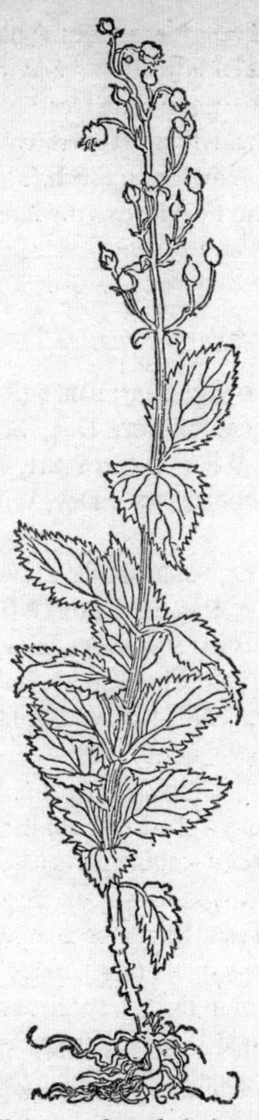

56 Figwort *Scrophularia nodosa*

Another wort for the 'fig' or piles, another signature plant, like the Lesser Celandine or Pilewort and the Orpine (q.v.). The knobs or protuberances on the root signified both piles and the kernels or tubercular glands of the King's Evil (on the neck). So the names 'great Figgewort' (which distinguished it from *Ranunculus ficaria*, the small Figwort), 'Kernellwoort', 'Cervicaria', or 'Throatwort'. Piles and the King's Evil, ulcers, cancer, the itch and worms and red faces were all amended with this rather ugly, brown-flowered and smelly herb, which long continued in folk-medicine. John Pechey describes a plaster for the King's Evil (scrofula), made chiefly of Figwort leaves and Hound's Tongue leaves, and the flowers of Foxglove and White Dead Nettle (141). Note that both Hound's Tongue (q.v.) and Figwort were called Rose-noble, the royal golden coin, in lieu of the King's own touch. The Irish thought of Figwort as the Queen of Herbs, and Foxglove as the King (133). For 'Crowdy-kit' and 'Fiddle' see *Scrophularia aquatica.*

5. Water Betony. *Scrophularia aquatica* L. 79, H 38

Local names. ANGLER'S FLOWER, Som; BABES-IN-THE-CRADLE, Som, Wilts; BISHOP'S LEAVES, Som; BROWN-NET, Dev, Som; BROWN-WORT, Corn; BRUNNET, Dev; CRESSET, Wilts; CROWDY-KIT, Dev, Som; FIDDLES, Dev, Som, War, Lincs, Yks; FIDDLESTICKS, Dev, Wilts; FIDDLE-STRINGS, Som; FIDDLEWOOD, Dev, Yks.

HUNTSMAN'S CAP, Corn; POOR MAN'S SALVE, Dev; ROSE-NOBLE (see *Scrophularia nodosa*), Dur, Ire; SCAW-DOWER (Cornish *scaw dowr*, 'water elder', from the smell), Corn; SQUEAKERS, Dev, Dor; STINKING CHRISTOPHER, Cumb; STINKING ROGER, Ches, Cumb, Lakes, Ayr, N Ire; VENUS-IN-HER-CAR, Som; WASPWEED, N'hants; WATER BETONY, Som, Suss, Kent, Leic, Yks, Ork.

Similarity of names was often due to the similarity of the parts rather than the whole of different species – of the parts used in pharmacy, which were always being handled by the herb-woman, apothecary, or patient, always crossing the counter. Thus leaves of *Scrophularia aquatica* resemble leaves of Betony, *Stachys officinalis* (q.v.), so it was sensible – more sensible than it seems to us – to group the two species and also to distinguish them as Water Betony and Wood Betony. Water Betony was another stinking herb sympathetically clapped on to stinking sores and ulcers, and used also for wounds and bruises.

'Crowdy-kit' – a crowdy was a fiddle. Children make a squeaking, fiddler's noise, by scraping one dry stem of Water Betony across another. No fly-fisherman will need to have the name Angler's Flower explained to him, after hooking the dry stems.

6. Monkey-flower. *Mimulus guttatus* DC. 100, H 27

Local names. GAP-MOUTH, Som; MONKEY-BLOSSOM, Corn; MONKEY-CUPS, Dor; MONKEY-FLOWER, Corn, Som, Wilts, N Ire; MONKEY JACKS, Dor;

MONKEY-PLANT, Dev, Som, Wilts ('Monkey-flower', etc., from the monkey-like features of the blossom. The names have been transferred from the yellow and orange flowered *Linaria vulgaris*).

Monkey-flower came to us from the grimness and sogginess of the Aleutian Islands, off Alaska. The first plants to reach Europe originated on Unalaska Island in the Fox group of the Aleutians, where it is fog more often than sunshine, and where the rain may not stop for 250 days in the year. But the Monkey-flower is tolerant. At the other extreme it will flourish along watercourses in the sunny desert uplands of New Mexico.

English gardeners had *Mimulus guttatus* by 1812. Escaped plants were observed in Wales by 1824, near Abergavenny. Since then it has ramped, and still ramps, along lowland rivers, mill-streams, and the brooks of foothill, moorland, and mountain, often illuminating black countryside with the clearest and most liquid yellow. The great mileage of new canals early in the nineteenth century helped several invaders, not only *Mimulus*, but *Impatiens glandulifera* and *Elodea canadensis* (q.v.).

Mimulus luteus L., another escape, chiefly in Scotland (where the names include Frog's Mouth and Blood-drop Emlet), was introduced from Chile in 1826.

7. Cornish Moneywort. *Sibthorpia europaea* L. 9, H 2

Local names. PENNY-PIES, SMALL PENNY-PIES, Corn.

One of the smallest, most delicate, most lovable of English plants – hairy leaves, hairy stems, flowers rather pink than yellow, but no bigger than pinheads. Look for it above Cornish brooks and rock trickles, trailing and tangled in the dampness with Golden Saxifrage. John Ray the Great first discovered it near St Ives, on 1 July 1662, as he was riding to Land's End (152). For a name, all that his imagination could devise was 'Bastard Chickweed'!

8. Foxglove. *Digitalis purpurea* L. 110, H 40

Local names. BEE-CATCHERS, BEEHIVES, Som; BLOBS, Som, Suff, Ess; BLOODY BELLS, Lanark; BLOODY FINGERS, Som, Heref, Yks, Cumb, Scot; BLOODY MAN'S FINGERS, Heref, Rad; BUNCH OF GRAPES, BUNNY RABBITS, BUNNY RABBIT'S MOUTHS, Som.

CLOTHES PEGS, Som; COTTAGERS, Ire; COVENTRY BELLS, Dor; COWFLOP (see *Primula veris*), Corn, Dev, Som; COWSLIP, Dev; DEADMEN'S BELLOWS (i.e. pillies, male members), N Eng; DEADMEN'S BELLS, Dor, N'thum, Scot; DEADMEN'S FINGERS, Inver; DEADMEN'S THIMBLES, Som; DOG'S FINGERS, Som, Wales; DOG'S LUGS (i.e. ears), Fife; DRAGON'S MOUTH, Suss; DUCK'S MOUTH, Som.

FAIRY BELLS, Som, Ire; FAIRY CAP, Som, Glos, Ess; FAIRY FINGERS, Som,

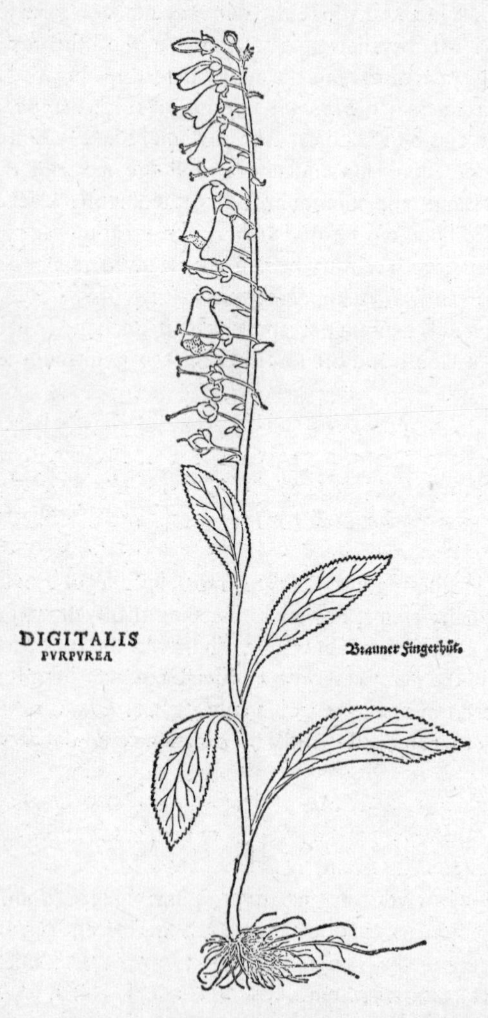

57 Foxglove *Digitalis purpurea*

Ire; FAIRY GLOVES, Dor, Som, Heref, Ire; FAIRY HAT, Dor; FAIRY WEED, Ire; FAIRY'S PETTICOATS, Som, Ches; FAIRY'S THIMBLES, Som, Ess, Norf, Cam, Lanark, Ire; FINGER CAP, Som; FINGER HUT, Dev; FINGER ROOT, Suss, War; FINGERS, FINGERS AND THUMBS, Som; FINGER TIPS, Wilts; FLOP-A-DOCK (cf. Cowflop), Corn, Som, Dev, Wilts; FLOPDOCK, Corn, Dev; FLOPDOCKEN, Yks; FLOP-POPPY (i.e. flowers which 'pop'), Dev; FLOPS, FLOPTOP, Dev, Som; FLOSS-DOCKEN, Yks; FLOWSTER-DOCKEN (i.e. 'showy dock'), Yks; FOX-AND-LEAVES, Ire; FOX-DOCKEN, FOX-FINGERS, Yks; FOX-FLOPS (cf. Cowflop: the plant which has grown from fox droppings), Som; FOXES' GLOVE, Oxf; FOXTER, Scot; FOXY, Ire.

GAP-MOUTH, Som; GOOSEFLOPS (cf. Cowflop, Foxflop), Dev, Dor; GRANNY BONNETS, GRANNY'S GLOVES, Som; GREEN POPS, GREEN POPPIES, Corn; HAREBELL, Ire; HEDGE POPPY, HILL POPPY, Som; KING'S ELWAND, N'thum; LADY'S FINGERS, Dev, Som, Worc, Shrop, N'thum, N Ire; LADY'S GLOVES, Som, Shrop; LADY'S SLIPPER, Som; LADY'S THIMBLE, Som, Norf, N'thum; LION'S MOUTH, Suss; LONG PURPLES, Som; LUSMORE (i.e. 'large plant'), Ire.

POP-BELL, Som; POP-LADDERS, Dor; POP-DOCK, Som, Corn; POP-GLOVE, Corn; POP-GUNS, Som; POPPERS, Hants; POPPY (since the unopened flowers can be popped), Corn, Dev, Dor, Som, Wilts, Hants, Bucks, N'hants, War, Yks; POPPY-DOCK, Corn, Dev, Dor, Som, Wilts, Hants, Bucks, N'hants, War, Yks.

RABBIT'S FLOWER, Dev; SCABBIT-DOCK, Corn; SCOTCH MERCURY, Berw; SNAPDRAGON, Dev; SNAPJACK, SNAPS, Som; SNAUPER, SNOXUM, Glos; THIMBLE-FLOWER, Dor, Som; THIMBLES, Cumb, Ire; THROATWORT, Som; TIGER'S MOUTH, Suss; VIRGIN'S FINGERS, Som; WILD MERCURY, Berw; WITCH'S THIMBLE, N'thum, Scot.

Irish names include: *lus na mban sídhe*, 'plant of the fairy women'; *lus mor*, 'large plant'; *méaracán na mban sídhe*, 'thimbles of the fairy women'; *méaracán sídhe*, 'fairy thimbles'; *méiríní púca*, 'puck fingers'; *coinneal Mhuire*, 'Mary's candle'. In Donegal English *méaracán sídhe* is SHEEGIE-THIMBLES, SHILLY-THIMBLES.

Gaelic names: *meuran nan daoine marbh*, 'dead men's thimbles'; *meuran nan caillich mharbha*, 'dead women's thimbles'; *ciochan nan cailleachan marblia*, 'dead old women's paps'.

Welsh names: *bysedd cochion*, 'red fingers'; *bysedd ellyllon* 'elves' fingers'; *bysedd y cŵn*, 'dog's fingers'; *menyg ellyllon*, 'elves' gloves'.

Foxglove means the glove of the fox, and not the glove of anything else or anybody else (OE *foxes glōfa*). Nevertheless the Foxglove was a fairy's plant, or a goblin's plant, in England (judging by its names) as well as in Wales and Ireland. The belief must have arisen from the tallness of the Foxglove, the glove-shape of the corolla, and the poison of the leaves. An Irishwoman told Lady Gregory that she knew someone who was cutting the *lus mor*, when a fairy voice called out: 'Don't cut that if you're not paid [i.e. if you are cutting it for yourself], or you'll be sorry.' An Irishman remarked to her: 'As for the *lus mor*, it's best not to have anything to do with that,' and another woman maintained that the *lus mor* was the only plant 'good to bring back children which are gone away' – i.e. which have been taken by the fairies (89). It must have been the Foxglove they had in mind, though Lady Gregory believed it was *Verbascum thapsus*, which is also called *lus mor* and is frequently confused with the Foxglove. The fairy women's *lus mor* was powerful in the matter of the children *left* by the fairies as well. Foxglove juice gets rid of a changeling in Scotland, in Ireland (137), in Wales (110). In Scotland Isobel Haldane in 1623 confessed to luring, charming, and traffic with 'the ffarye-folk'. A woman consulted her about a child who was a 'sharg' or changeling. She sent her son to gather 'focksterrie leaves', made tea with them, and gave it to the sharg, who died (145). Perhaps Isobel Haldane was not sure about the right way to test the changeling. In Ireland and Wales the foxglove juice was rubbed on to the child.

The goblin who appears to the lovers in the thrilling Northumbrian poem of *The Gloamin' Buchte* (printed in James Telfer's *Border Ballads*, 1824) looks around with snail-cap eyes, washes its hands in the dew, and then sings a strange mournful song in which it mentions the Foxglove:

O where is tiny Hewe?
  O where is little Lenne?
An' where is bonnie Lu?
  An' Menie o' the Glen?
An' where's the place o' rest?
  The ever changin' hame –
Is't in the gowan's breast
  Or' 'neath the bells o' faem?
    Ay lu lan, lan dil y'u.

The fairest rose you'll find
  May have a taint within –
The flower o' womankind
  May ope her breast to sin.

The fox-glove cupp you'll bring,
The tayle o' shootin' sterne,
An' at the grassy ringe
We'll pledge the bluid o' ferne.
Ay lu lan, lan dil y'u.

And when the blushing moon
Glides down the western skye,
By streamers wing, we soone
Upon her top will lye:
Her hiest horne we'll ride
An' quaffe her yellow dewe,
An' frae her skaddowy side
The burnin' daie we'll view.
Ay lu lan, lan dil y'u.

– after which the goblin vanishes and the shepherd's girl hears only a plitch-platch in the stream.

As usual, this plant of the fairies was powerful in their hands against you, or in your hands against them, provided you took the risk of gathering it. Fairies are supposed to have given the corollas of their powerful plant to the fox. Wearing these foxgloves, the fox could then sneak in magic silence up to the poultry or away from men. Welsh people used bunches of Foxglove to make black crosses on the floor stones with the juice – obviously apotropaic crosses to keep the fairies away (167).

Such a plant of magic and of supernatural names needed its Christian names, and so became *gant de Notre-Dame* and *doigt de la Vierge*, Lady's Glove, Lady's Fingers, Virgin's Fingers, etc. It needed also a name for the botanist or apothecary. Fuchs, in his *De Historia Stirpium*, 1542, therefore coined *Digitalis*, from the German finger and thimble names. But there was still no classical warrant for employing the Foxglove. In the countryside physicians found it was used a good deal as a purge and a vomit in fevers, and against colds, the King's Evil, and dropsy. This led to clinical investigations by William Withering, summed up in his *Account of the Foxglove* (1785). Withering proved that Foxglove acted upon the heart and was a good diuretic. He did not know the exact way in which Foxglove affects the heart-muscle, slowing it down so that the heart fills and empties properly and makes the kidney more efficient by an increased supply of blood. But he knew enough. He had established the power of Foxglove. Further investigation turned the old herb of the fairy women into a major instrument against heart disease.

The Foxglove story is not one of unrelieved magic and medicine. Where English poets compared the young girl's cheeks to the conventional rose, Welsh mediaeval poets and prose-writers likened them to the foxgloves of their upland country. Olwen in the *Mabinogion*: 'Her two breasts were more white than the breast of a white swan; her cheeks more red than the foxglove.'

9. Brooklime. *Veronica beccabunga* L. 112, H 40

Local names. BECCY LEAVES, Dev; BEKKABUNG (Old Norse *bekkr*, 'brook', + *bung*, the name of a plant), Shet; BIRD'S EYE, Dor, Som; BROOKLIME (from OE *brōc*, 'brook', + *hleomoc*, plant-name equivalent to Old Norse *bung*), Dev, Hants, Suff, Derb, Ches, Yks (and general).

COW-CRESS, Ess; HORSE-CRESS (French *cresson de cheval*), Yks; HORSE WELLCRESS (i.e. 'stream', 'water-cress'), Scot; LIMPWORT (? OE *hleomoc*-), Heref; PIG'S GREASE, Dor; WATER BIRD'S EYE, Dev; WATER PURPIE, WATER PURPLE, Cumb, Scot; WELLINK, Cumb, Scot, Ire.

Turner, in 1548, called it Brooklem, which is nearer the OE *brōc hleomoc*. Together with Watercress, Brooklime was a salad plant of northern Europe. The young tops and leaves are not unpleasant, and they were long recommended (by Gerard, for example) against scurvy. Diet drinks were made of Brooklime in the seventeenth century. 'To cure the scurvy: Take of the Juice of Brook-lime, Water-cresses, and Scurvy-grass, each half a Pint; of the Juice of Oranges, four Ounces; fine Sugar, two pounds; make a Syrup over a gentle Fire: Take one Spoonful in your Beer every time you drink' (141).

Lemmington in Northumberland (*Lemocton* in 1201) was the homestead or village where the *hleomoc*, or Brooklime, grew (66), and a plant with this name comes into Anglo-Saxon remedies in the *Lacnunga* (88).

10. Germander Speedwell, Bird's Eye. *Veronica chamaedrys* L. 112, H 40

Local names. ANGEL'S EYES, Dev, Som; BILLY BRIGHT EYE, Ire; BIRD'S EYE (general); BLIND-FLOWER, Dur; BLUE BIRD'S EYE, Suss, Bucks, Oxf; BLUE EYES, Dev, Dor, Som, Wilts; BLUE STAR, Stir; BOBBY'S EYES, Som, Hants; BONNY BIRD-EE, Cumb; BOTHERUM, Ches; BREAK-BASIN (quick falling of the petals), BRIGHT EYE, Som.

CAT'S EYES, Corn, Dev, Som, Hants, Glos, Kent, Ess, Cumb;

DEVIL'S EYES, War; DEIL'S FLOWER, Dumf; DOTHERUM (i.e. 'trembling'), Ches; EYEBRIGHT, Som, Shrop, Midlands, Yks, N'thum, N Ire; EYE OF CHRIST, Wales; FORGET-ME-NOT (see *Myosotis palustris*), Dev, Som, Yks, Scot; GOD'S EYE, Dev, Lincs; HAWK-YOUR-MOTHER'S-EYES-OUT, Dor.

JERRYMANDER, Ches; LADY'S THIMBLE, Lancs; LARK'S EYE, Som; LOVE-ME-NOT, Bucks; MILKMAID'S EYE, N'thum; MOTHER-BREAKS-HER-HEART, Corn; PICK-YOUR-MOTHER'S-EYES-OUT, Dor; POOR MAN'S TEA, Cumb; REMEMBER ME, Yks; STRIKE-FIRE (see below), N'hants; TEAR-YOUR-MOTHER'S-EYES-OUT, Dev; WISH-ME-WELL, Ches.

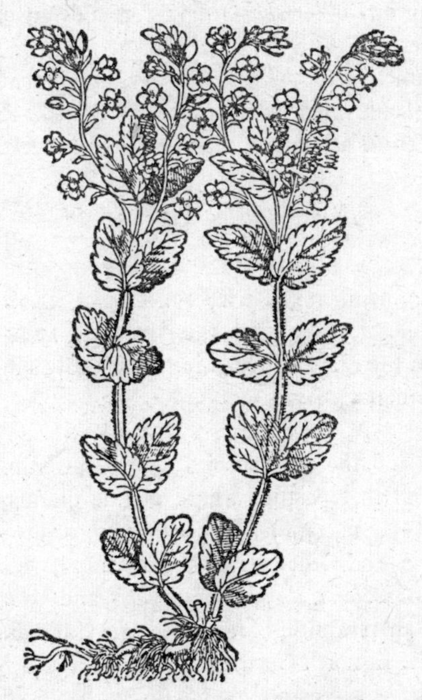

58 Germander Speedwell *Veronica chamaedrys*

Speedwell is a plant of the roadside which speeds you well; so in Ireland Speedwell was sewn on to clothes to keep the wearer from accident (137) (cf. Wish-me-well, above). The blue petals of Germander Speedwell or Bird's Eye tremble and fall too easily – as the 'purses' of the Shepherd's Purse (q.v.) break so easily, and break your mother's heart.

Pick a Bird's Eye, children learn in Somerset, and birds will peck out your own eyes. Or in Dorset, your mother's eyes will be picked out, or hawked out; in Devonshire, torn out. These eyes or flowers of *Veronica chamaedrys* belong not only to Bird, but to Cat, Lark, Bobby (? Robin Goodfellow), Devil, then, *per contra*, Angel, God, and Christ. Destroy the flower, which is so easily done, and somebody will be taking his revenge. In German belief picking this speedwell causes a storm (129), which may be implicit in the Northamptonshire name (Strike-fire). A sinister little plant. Cf. Herb Robert, the Red Bird's Eye.

Veronica chamaedrys was called Wilde Germander in the sixteenth century, Germander coming through mediaeval Latin from the Greek *chamandrua*, a corruption of *chamai*, 'on the ground', and *drus*, 'oak' – as if the Speedwell or any other *chamaedrys* were a dwarf oak. Cf. Ground Ash, Ground Elder.

Some of the names of *Veronica chamaedrys* are applied to the various Veronica species of cultivated ground, such as *V. persica* Poir. (112, H 40), the plant of western Asiatic origin recorded first in Berkshire in 1825.

11. Red Rattle. *Pedicularis palustris* L. 112, H 40

Local names. COCK'S COMB, Scot; COW'S WORT, Notts; DEADMEN'S BELLOWS (i.e. pillies, male members), Berw; HONEY-CAP, HONEYSUCKLE, WILD HONEYSUCKLE, Donegal; MOSS-CROP (i.e. 'moor-'), MOSS-FLOWER, Ches; RATTLEWEED (from the rattle of seeds in the dry capsule), Norf; SUCKIES (from sucking the honey), Ayr.

In Germany this and the next species were *Läusekraut*, 'lousewort' – 'bycause the cattell that pasture where plentie of this grasse groweth become full of lice' (Lyte, 1578). They were supposed to give animals the fluke-worms of liver-rot. Other plants supposed to have imparted the rot include *Drosera rotundifolia*, *Pinguicula vulgaris*, and *Hydrocotyle vulgaris*.

Red Rattle is 'an infirmitie of the medowes', but often a charming element of scenery. In north Cornwall I have seen it in May turning the sea coombe below Morwenstow church to acres of suffused pink. For Deadmen's Bellows, cf. *Ajuga reptans*, *Gentianella amarella*.

12. Lousewort. *Pedicularis sylvatica* L. 112, H 40

Local names. BEE-SOOKIES, Shet; COCK'S COMB, Som, Scot; HONEYSOOKIES, Shet; HONEYSUCKLE, Hants; RATTLE BASKETS, RATTLE PODS, Som; RED RATTLE, Dev, Som; SHACKLE-BOXES, Dev; SOOKIES, Shet (see above).

13. Yellow Rattle. *Rhinanthus minor* Ehrh. 112, H 40

Local names. BABY'S RATTLE, Som; BULL'S PEASE, Donegal; CLOCK, Scot; COCK'S COMB (from *Crista galli*), Som, Glos, Shrop; COW-WHEAT, Cumb; DOG'S PENNIES, Shet; DOG'S SILLER, Scot; FIDDLE-CASES (the dry capsules), I o W; GOWK'S SHILLINGS (i.e. 'cuckoo's), Lanark; GOWK'S SILLER, GOWK'S SIXPENCES, N'thum, Berw, Rox; HAY-SHACKLE, Som; HENPEN, Som, Yks, Cumb, N'thum; HEN-PENNY, Yks, West, N Eng, N'thum; HORSE-PEN, Cumb; HORSE-PENNIES, Derb, Lancs, Yks; LOCUSTS, Bucks (where Yellow Rattle was supposed to be the locusts eaten by St John the Baptist).

MONEY, Som, Bucks, N'hants; MONEY-GRASS, Leic; MONKEY-PLANT, Dor; PENCE, N'hants; PENNIES AND HAPPENNIES, Som; PENNY-GRASS, Som, Lancs, Yks, Ire; PENNY-GIRSE, Shet; PENNY-RATTLE, Som, Suss; PENNY-WEED, Midlands; PEPPER-BOX, POVERTY, POVERTY WEED, PURSES, Som.

RATTLEBAGS, Dev, Dor; RATTLE-BASKET, Som, Wilts; RATTLE-BOX, Shrop, Ire; RATTLE-CAPS, Som; RATTLE-GRASS, Som, Herts, Ire; RATTLE JACK, Lincs; RATTLE-PENNY, Dor, N Eng; RATTLE-TRAPS, Dor; ROCHLIS (? Flemish *rochel*, 'death-rattle'. 'Rochlis' is 'death-rattle' in dialect of Heref and Pemb), Heref; SHACKLEBAGS, Dor, Som; SHACKLE-BASKET, Dor, Som; SHACKLE-CAPS, SHACKLERS, Som; SHACKLES, Dor; SNAFFLES, Kent; SHEPHERD'S PURSE, Cumb; YELLOW RATTLE, Glos, Ches, Scot.

Money in the purse, the box, the basket, and the rattle. Gerard's names included Coxcombe, yellowe Rattle, and Penie grasse (*pengegraes* in Norwegian, *penningagras* in Icelandic). As a semi-parasite, this plant is another of the pretty infirmities of the land. 'What temperature or vertue this herbe is of, men have not as yet been carefull to knowe, seeing it is accounted unprofitable' (Gerard).

14. Common Cow-wheat. *Melampyrum pratense* L. 110, H 39

Local names. HEN PENNY, Lancs; BABY'S RATTLE, Som (cf. *Rhinanthus minor*).

A woodland annual which does not thrust itself forward by any exuberance of size or colour. Yet it deserves more notice. The flowers, by their combined tones of yellow against the green bracts and leaves, are delightfully cool – often yolk gold in the mouth and then pale yellow outside, almost to being white. The neighbours of Cow-wheat in mossy and broken ground are often Wood Rush, Wood Sorrel, and Creeping Jenny. It might be worth collecting seed and growing garden colonies – particularly since *Melampyrum pratense* has a long flowering period from May onwards.

Gerard could find little to pronounce upon the nature or virtues of Cow-wheat, beyond a note that Cow-wheat flour 'mightily provoketh venerie', and according to some writers, caused women to give birth to male children.

15. Field Cow-wheat. *Melampyrum arvense* L. 8

Local name. POVERTY-WEED, I o W, Surr.

The humble clarities of *Melampyrum pratense* are no preparation for this beauty, which stands up in the wheat like a purple, rose, and yellow pagoda. A rare native. John Ray knew of it in the wheatfields of Norfolk. In 1823 it was found in quantity in the Isle of Wight. Above the Undercliff from Ventnor to St Lawrence, there were pagodas by the thousand, among the wheat, among the barley, on the dry banks and the borders of the ploughland. The seeds were too heavy to winnow from the grains of wheat, and they gave a blue tinge and a bitter taste to the flour. Farmers acted furiously against it. The 'Poverty-weed' was rooted up and burnt, yet fifty years ago it was still abundant. In the island now it is confined to a single bank near Ventnor. There are other stations for it here and there, for instance, in north Wiltshire, where it comes up, though in small quantity, season after season.

No wild flower of the British Isles is more improbably flamboyant. Out of light green leaves, the turrets or pagodas of a light pinkish purple, made up of the close-set bracts, rise up on each side of the stem. Then the flowers begin to open. The long tube of the corolla is first pink or purple, then egg yellow, then, on the lip, pinkish-purple once more. Lengthwise, the stems are half green, half black; and in the corn, among the wheat or the oats, it is as if plants of the Red Bartsia (*Odontites verna*) had changed to a new dressmaker (21, 126, 189).

In Normandy in the early summer this *Queue de Renard* is plentiful and splendid, though principally on the fringe of the corn and on the banks near by.

16. Eyebright. *Euphrasia officinalis* L., agg. 112, H 40

Local names. BIRD'S EYE, Som; EYEBRIGHT, Dev, Ches, Yks; FAIRY FLAX, Donegal; JOY, Som; PEEWEETS, Dev; ROCK RUE, Donegal.

In the Middle Ages, and later, *Euphrasia officinalis* was famous for brightening the eyes. 'It preserveth the sight, increaseth it, and being feeble and lost it restoreth the same' (Gerard). Everyone knows the quotation from *Paradise Lost* (Book XI, 411, etc.) in which the Archangel anointed Adam's eyes with Euphrasie and Rue and three drops from the Well of Life, until he could see death and the miserable future of mankind:

> Death thou hast seen
> In his first shape on man; but many shapes
> Of Death, and many are the wayes that lead
> To his grim Cave, all dismal . . .

An eye-plant by signature and sympathetic magic: 'The purple and yellow spots and stripes', Coles wrote in *Adam in Eden*, 1657, '. . . doth very much resemble the diseases of the eyes, as blood-shot, etc.'

'The Oculists in *England* and Beyond-Sea, use the Herb in Sallets, in Broths, in Bread, and in Table-Beer; and apply it outwardly in Fomentations, and other External Medicines for the Eyes. Take of the Water of Eyebright and Fennel, each one ounce and a half: of White Rose-water, one ounce, Prepar'd Tutty [oxide of zinc] two drams, Camphire two grains; mix them, drop two or three drops into the eye, warm, twice a day. This is good when the eye is much bruis'd' (John Pechey, in *The Compleat Herbal of Physical Plants*, 1694).

Welsh names are *golwg Crist*, *llygad Crist*, 'Christ's sight', 'Christ's eye'.

17. Yellow Bartsia. *Parentucellia viscosa* (L.) Caruel 30, H 9

Local names. TWINY-LEGS, TWEENY-LEGS, Dev.

An odd, stiff, sticky, annual species of damp meadows and hillsides, especially in Devon and Cornwall, where the yellow flowers shake in the warm winds off the Atlantic. It is more at home in the warmer countries of the Mediterranean.

John Ray found it in St Columb parish in Cornwall and recorded it in his *Catalogus Plantarum Angliae* in 1670, though the first record had come from the Isle of Wight a few years before.

18. Red Bartsia. *Odontites verna* (Bell.) Dum. 112, H 40

Local names. COCK'S COMB, Yks; HEN-GORSE, Ches; POOR ROBIN, N'thum; SANCTUARY, Ches; TWINY-LEGS Dev.

Here is one of those plants, common, neither minute nor inconspicuous, which are without a strong character. No peculiarities, no beauties, no virtues. A red, dullish, disregarded annual of the cornfields, a weed which has not even incurred the hatred of farmers.

LXXIV. Orobanchaceae

1. Toothwort. *Lathraea squamaria* L. 72, H 30

Local name. CORPSE-FLOWER, Yks.

The flowering shoot is white and scaly, then becomes a livid purple, changing, as it climbs past the flowers, to a pale pink, felted with sticky white hairs. In front of livid, pinkish, or dirty white bracts hang the flowers, each of a pink to purple corolla inside a white, or nearly white, calyx. The corolla rapidly dries and shrivels to a dark brown, so that on a spike at one time fresh flowers on top are succeeded by shrivelled flowers below, adding to the plant's deathliness. In a naked wood, in April, before the bluebells, Toothwort might have grown out of a buried corpse, instead of being a parasite on the roots of hazel, elder, or elm. The clusters are bizarre, but tone well with dead sticks and pale oak leaves.

It is 'Toothwort' as the *dentaria* of the sixteenth-century botanists (Latin *dens*, a tooth) from the thick scales on the rhizome, and more exactly from the capsules, which are ivory-white and shiny and channelled like small fangs. Gerard found that country women called it Lungwort and used it against the cough and 'all other imperfections of the lungs'. But it is not a common plant.

2. Greater Broomrape. *Orobanche rapum-genistae* Thuill. 69, H 5

A root parasite, but not a plant which 'rapes' the Broom – instead, *Rapum genistae* of the old botanists, the Rape or turnip, or tuberous plant, of the Broom. This was the only kind known to Turner and Gerard, who called it 'a certaine bulbed plant growing unto the rootes of broome, bigge belowe, and smaller above'. The whole plant, he added, 'is of the colour of the Oken leaf' – the dead leaf.

The rape of the Broomrape was boiled in wine for kidney and bladder trouble, and the juice of it was used as a vulnerary. It was also candied.

3. Lesser Broomrape. *Orobanche minor* Sm. 57, H 21

Local names. DEAD MAN (cf. *Lastraea squamaria* and *Umbilicus rupestris*), Wilts; DEVIL'S ROOT, HELL-ROOT, Kent; SHEPHERD'S POUCH, I o W; STRANGLEWEED, Som.

Most of the Broomrapes are not of consequence to the farmer. He saw *Orobanche rapum-genistae* when he cut Broom for the oven or cattle food, but *Orobanche minor* he knew and named as the Devil's Root and Hell-root rising from his clover. The rarer species do not touch anyone's interests or interfere with them.

LXXV. Lentibulariaceae

1. Butterwort. *Pinguicula vulgaris* L. 110, H 38

Local names. BEANWEED, Herts; BOG VIOLET, N Eng; BUTTER-PLANT, Selk; BUTTER-ROOT, Yks; CLOWNS, Rox; EARNING-GRASS (i.e. 'curdling-'), N'thum, N Eng, Lanark; YIRNIN-GIRSE, Shet; EKKEL-GIRSE, Ork, Shet; FLYCATCHER, Dor; MARSH VIOLET, Yks; ROT-GRASS, Cumb, N'thum, Berw; SHEEP-ROOT, Rox; SHEEP-ROT, N Scot; STEEPWEED, STEEPGRASS, STEEPWORT, N Ire; THICKENING-GRASS, Ayr; WHITE SINCLES (i.e. Sanicle, healing herb), Herts.

Bog Violet or Marsh Violet – in a way neither of these names is unreasonable. Butterwort flowers glow and sparkle in the wet herbage with a depth of violet or sapphire.

A magical plant in the Scottish islands. On Colonsay in the Hebrides, if you picked Butterwort, it protected you from witches; and if your cows had eaten Butterwort, they were safe from elf-arrows. In a house on Colonsay the women were keeping watch on a new-born child to make sure that it would not be exchanged by the fairies. Two fairies came up to the cradle. 'We will take it,' said one. 'We will not,' said the other; 'we cannot. Its mother ate of the butter of the cow that ate the Butterwort' (128).

The *mana* was no doubt behind other practices – Butterwort against ruptures and as a vulnerary, Butterwort as a cattle cure. 'The husband-mens wives of Yorkshire', wrote Gerard (the plant was also known as 'Yorkshire Sanicle'), 'do use to anoint the dugs of their kine with the fat and oilous juice of the herbe Butterwoort, when they are bitten with any venomous worm, or chapped, rifted, and hurt by any other meanes.' An apotropaic plant, once more, a protector of milk and butter, which are the special objectives of evil, rubbed on cow's udders like the Buttercup. In the north the leaves were also steeped in milk to curdle it or thicken it, a practice known as well to the Lapps. But it is the magic, not the curdling, which explains the name Butterwort. On the debit side *Pinguicula vulgaris*, like *Drosera rotundifolia*, *Hydrocotyle vulgaris*, and other plants, was accused of giving the rot to sheep.

LXXVI. Verbenaceae

1. Vervain. *Verbena officinalis* L. 72, H 26

Local names. BURVINE, N'hants; HOLY HERB, Som; SIMPLER'S JOY, Corn, Som, War; WELSH names are *cas gangythraul*, 'devil's hate' (also given to *Sorbus aucuparia*), and *llysiau'r hudol*, 'wizard's herb', 'enchantment herb'.

Verbenae in Latin were boughs of various use in religious ceremony, but Pliny's *verbena* or *verbenaca* was also a particular plant, otherwise *hierabotane* (mentioned by Dioscorides) or *sacra herba*, 'holy herb'. The holy herb was employed by the Romans in sacrifice and purgation, it was a plant of the Magi, and in Gaul it was used prophetically. Wrongly or rightly, in the Middle Ages this wonder plant was taken to be our *Verbena officinalis*. In the Anglo-Saxon *Lacnunga* (88), with Rue, Dill, Periwinkle, Mugwort, Betony, and other powerful herbs, Vervain was an ingredient in a Holy Salve against the demons of disease.* Its power against the Devil was celebrated, in a clear description, by the poet of the Stockholm Medical Manuscript (about 1400):

> If it be on hym day and nyth
> And he kepe fro dedly synne aryth,
> The devel of helle schal hawe no myth
> To don hym neyther fray ne fryth . . .
>
> Comely be weye and gate
> Thou may it fynde hey in state,
> With heye stalkys, many smale brawnchys,
> Smale bloysh flouris owt of hym lawnchys (168).

* The OE version of the *Herbarium* of Apuleius Platonicus told the Anglo-Saxons that Vervain drove away all poisons and was said to be used by sorcerers.

59 Vervain *Verbena officinalis*

More than a century goes by: in 1528 Margaret Hunt, a wise woman or white witch, whose craft came to her from Wales, was persecuted for administering herbs with incantations, Pater Nosters, Aves, creeds, etc. (112). 'Vervey' she included with Dill and Rue, once again, when she treated sores (for which Dioscorides had recommended the *hierabotane*). Vervain was christianized, as in a charm which made it a plant of Calvary:

> Hallowed be thou, Vervein, as thou growest on the ground,
> For in the mount of Calvary there thou was first found.

Thou healedst our Saviour Jesus Christ, and stanchedst his bleeding wound;
In the name of the Father, the Son, and the holy Ghost, I take thee from the ground.

John White, who recorded the charm in *The Way to the True Church*, 1608, a book disputing 'the principall motives perswading to Romanisme', wrote that the Vervain was picked, crossed with the hand, and blessed with the charm; and then worn against 'blasts'. It comes, too, in stories of witch and of demon or elf lover, linked with Dill, as we have seen, or *Hypericum perforatum* (q.v.):

Vervain and Dill
Hinders witches from their will.

Or the demon lover says (35):

If thou hope to be Lemman mine
Lay aside the St John's grass and the Vervine.

In England Vervain garlands were worn on St John's Eve. In Italy it is *Erba san Giovanni.* A St John's herb in France as well, it is one of the medico-magical plants collected before St John's Eve and purified in the smoke of the *feux de joie*. Gun-flints, adding old magic to new weapons, were boiled with Rue and Vervain, to make them more efficacious (75, 188).

Gerard was uneasy about Vervain's magic reputation. 'Many odde olde wives tales are written of Vervaine tending to witchcraft and sorcerie, which you may reade elsewhere, for I am not willing to trouble your eares with reporting such trifles, as honest eares abhorre to heare.' He goes on that later physicians give Vervain for the plague – in which they are deceived, 'not onely in that they looke for some truth from the father of falshood and leasings, but also because in steede of a good and sure remedie they minister no remedie at all; for it is reported, that the divell did reveale it as a secret and divine medecine'.

Still, the Devil did not prevent doctors continuing to prescribe Vervain for wounds, the stone, the gripes, the bloody flux, malaria, headache, sore throat, cancer, eye troubles, piles, childbirth, and especially for the tuberculous glands of the King's Evil. John Morley's *Essay on the nature and cure of scrofulous disorders commonly called the King's Evil*, a tractate on the virtues of Vervain, was published in 1767, and went into sixteen editions within ten years.

When the doctors gave up Vervain at last, it still had a career in literature and a corner in the antiquarian's mind. Pliny had connected it with the Gauls, the Gauls were ministered to by Druids, Druids and Vervain must have been as close as Druids and Mistletoe (q.v.). The gardening poet, William Mason (1724–97), lived into the Druid mania, and pictured plant and priest in rather pleasant lines:

> Lift your boughs of *Vervain* blue
> Dipt in cold September dew;
> And dash the moisture, chaste and clear,
> O'er the ground, and through the air.
> Now the place is purg'd and pure.

One early Victorian antiquary and cartographer of the Ordnance Survey noticed Vervain in a carpenter's yard not far from the small stone circle at Duloe in Cornwall. The Vervain at once convinced him of the circle's 'druidical' origin.

Notice that Vervain is a plant of waste ground, road verges, etc., and is not so remarkably abundant. The *Flora of the British Isles* considers it a native. History and habitat combine to suggest an old denizen of the early Middle Ages, a plant already 'comely by way and gate' at the end of the fourteenth century, and growing in Gerard's day 'in untilled places neere unto hedges, high waies, and commonly by ditches almost everywhere' (see also S. T. Dunn's *Alien Flora of Britain*, 1905, and the *Geographical Handbook of the Dorset Flora*, 1948, by Professor Ronald Good, who takes it to have been introduced). Miller, in his *Gardeners Dictionary* (1741 ed.), writes of Vervain as a medicinal plant brought into market from the countryside, though 'rarely cultivated in gardens'.

LXXVII. Labiatae

1. Pennyroyal. *Mentha pulegium* L. 52, H 13

Local names. LILY-ROYAL, S Eng; LURKEY-DISH, Ches; ORGAL, Corn; ORGAN, Corn, Dev, Som, Wilts, Hants, Heref, Worc; ORGANY, Dev, Wilts, Hants, Heref, Worc; PUDDING-GRASS, Cumb; PUDDING-HERB, Yks; WHIRL MINT, Hants.

As for hollyoaks at the cottage doors, and honeysuckles and jasmines, you may go and whistle;
But the Tailor's front garden grows two cabbages, a dock, a ha'porth of pennyroyal, two dandelions, and a thistle –

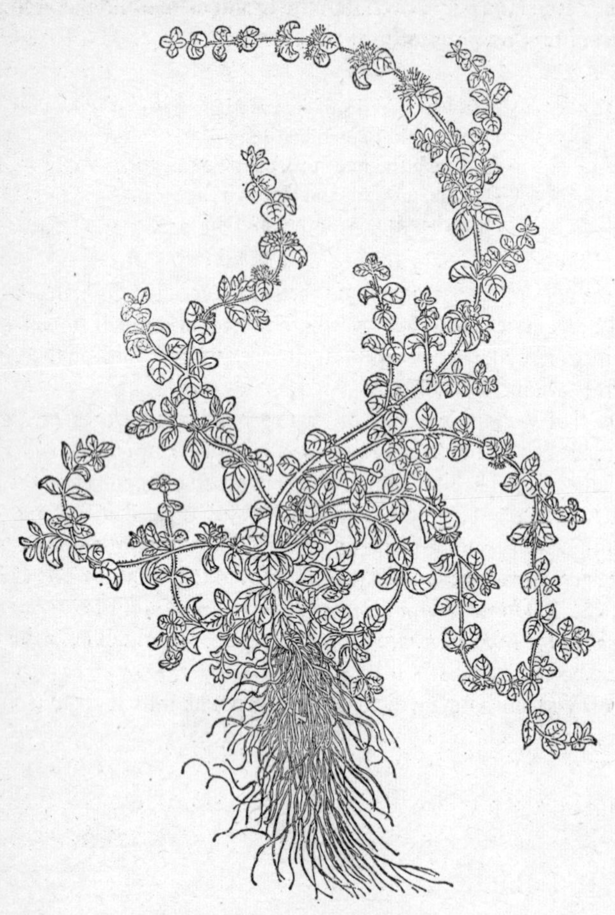

60 Pennyroyal *Mentha pulegium*

which is Tom Hood giving the truth about village gardens in answer to the sentimental Miss Mitford; and a tribute also to the uses of Pennyroyal – 'so exceedingly well knowne to all our English nation', wrote Gerard, 'that it needeth no description.' Poor women brought plenty to sell in the London markets. It was taken to be the fragrant *pulegium* ('flea plant') of the Romans, and the *glechon* of Dioscorides, who gives a very full list of its medicinal qualities. The sensible John Pechey contracted this cure-all into a paragraph:

''Tis used to provoke the Courses, and to help Delivery. 'Tis good for Coughs, for the Gripes, the Stone, Jaundice and Dropsie. A spoonful of the juice given to Children, is an excellent remedy for the Chin-cough [whooping-cough]. For an Hoarseness, take six ounces of the Decoction of it, sweeten'd, at Bed-time. The fresh Herb wrap'd in a Cloth, and laid in a Bed, drives away Fleas; but it must be renewed once a week' (141). As a carminative and antispasmodic, Pennyroyal stayed its ancient course in the *British Pharmacopoeia* until 1867; and long after that was picked or grown for 'Organy-tea'. The leaves, especially in the north of England, were chopped into hog's puddings, black puddings, haggis, etc. (Turner in 1538 called it 'Penyryall' or 'puddynge gyrse').

The names: Gerard gives 'Pennie royall, Pudding grasse, Puliall royall, and of some Organie'. Organ and Organy go back through the Latin to *origanon*, an aromatic herb mentioned by Theophrastus. Pennyroyal from *pulegium*, ultimately; through a diminutive *pulegiolum*, which gave the Old French *puliol*. *Puliol* was used for thyme, *puliol ryal*, 'royal *puliol*', distinguished this special 'thyme'. The Middle English *pulyol ryal* fitted the tongue less easily than Pennyroyal, into which it was corrupted, as though it were a plant worth a royal silver penny (cf. Rose-noble for Hound's Tongue and *Scrophularia nodosa*). The south country name, Lily-royal, is nearer to the old *pulyol ryal*.

The New Englanders took Pennyroyal with them to America (111).

2. Corn Mint. *Mentha arvensis* L. 111, H 40

Local names. APPLE MINT, Ire; LAMB'S TONGUE, Scot.

In this Mint of the cornfields, minty fragrance is replaced by a smell of wet, mouldy gorgonzola. In Ireland sprigs of it are placed in corn stacks to keep away the mice.

3. Water Mint. *Mentha aquatica* L. 112, H 40

Local names. BISHOPWEED, Hants; BISHOPWORT, Wilts, Hants; HORSE MINT, Som, I o W, Glos, War, Lincs, Yks, N'thum; LILAC-FLOWER, Som.

'The red Mynt that groweth by water sydes, and is called of some horse Mynt' (Turner, in *The Names of Herbes*, 1548). Step on this Water Mint as you walk through fens or over wet meadows, and up comes the delicious damp fragrance, different from the smell of Garden Mint and not quite that of Peppermint. 'The savour or smell of the water Mint', wrote Gerard, 'rejoiceth the hart of man' – quite true – 'for which cause they strowe it in chambers and places of recreation, pleasure, and repose, and where feasts and banquets are made.' It was sacred to Aphrodite.

Minstead in Hampshire, and Minety in Wiltshire, along a slow tributary of the Upper Thames, are named after the Water Mint (OE *minte* + *sted*, 'mint place', and *minte* + *ea*, 'mint stream') (66).

4. Peppermint. *Mentha piperita* L. 101, H 34

Mentha piperita is a lucky hybrid between the Spear-mint, *M. spicata* L. emend. Huds., introduced and grown in the gardens, and the native Water Mint, *M. aquatica*. It was not recognized or discovered until late in the seventeenth century, first in Hertfordshire by a Dr Eales, then in Essex by the physician and botanist, Samuel Dale. The new plant was recorded by John Ray in his *Synopsis stirpium Britannicarum* in 1696, as a mint, Peppermint, *sapore fervido Piperis* – 'with the fiery taste of pepper'. An extraordinary description. Did Ray's tongue work like the eyes of a colour-blind man? Was he subconsciously influenced by the name *Sisymbrium*, applied (*a*) to the peppery Water Cress and Lady's Smock, (*b*) to *Mentha aquatica* and *Mentha longifolia* (L.) Huds. (Dioscorides had described a *sisumbrion*, 'like garden mint', and a *sisumbrion heteron*, a water herb called *kardamine* because it tasted like *kardamon* or cress)?

Peppermint it continued to be. Gardeners of the early eighteenth century (e.g. Miller in the *Gardeners Dictionary* or Stephen Switzer in the *Practical Kitchen Gardener*, 1727) speak of Peppermint as a newly introduced and not very common pot-herb – 'by some greatly esteemed for its Heat, to make a simple Water' (Miller). By that time the hybrid of the ditches had entered the London Pharmacopoeia. The new savour began its conquest of the world. Wild supplies were not enough, so cultivation developed in the home counties and the eastern counties, and abroad. The French call Peppermint *menthe d'Angleterre* or *menthe anglaise*; in German it is *Pfefferminze*, but also *Englische Minze*.

5. Gipsywort. *Lycopus europaeus* L. 98, H 40

The story that gipsies stain themselves with *Lycopus europaeus* runs from one book to another, beginning with Lyte's translation of Dodoens, 1578:

'The rogues and runagates, which name themselves Egyptians, do colour themselves black with this herbe.' Then, for example, Gerard in 1597: 'Some also thinke good to call it *Herba Aegyptia*, bicause they that counter fet themselves Egyptians (such as many times wander like vagabonds from citie to citie in Germany) do use this herbe to give themselves a swart colour, such as the Egyptians and the people of Afrike are of.' The gipsies or 'gipsons' were still few in Gerard's England. By the time of Caleb Threlkeld, there were enough of them to be hated and persecuted. He wrote in 1727: 'Some call this the *Gipsy-herb*, because those stroling cheats called Gipsies do dye themselves of a blackish Hue with the Juice of this Plant, the better to pass for *Africans* by their tanned Looks, and swarthy Hides, to babble the credulous and ignorant by the Practice of Magick and Fortune-telling; they being indeed a nasty Sink of all Nations, living by Rapine, Filching, Pilfering and Imposture' (177).

Whether any English gipsy has ever been known to stain himself with *Lycopus europaeus* is another matter. It does give a good black dye, 'which also holdeth so fast, as that it cannot be washed or wiped awaie: in so much as linnen cloth being died heerewith, doth alwaies keep that colour' (Gerard). The people of Colonsay in the Hebrides used this dye (128).

6. Marjoram. *Origanum vulgare* L. 96, H 38

Local names. JOY OF THE MOUNTAIN, Som; ORGAN, ORGANY (cf. *Mentha pulegium*), Dev, Wilts, Worc.

They distinguished in the sixteenth century between the rather tender Sweet Marjoram of the gardens and the Common Organ, or Organy, Wild Marjoram, Bastard Marjoram, which is our native *Origanum vulgare*, 'exceedingly well knowne to all' (Gerard). The Wild Marjoram was something of a cure-all; made into marjoram tea, it helped, for example, indigestion, earache, cough, dropsy, bladder trouble, etc. The New Englanders cultivated *Origanum vulgare* for this indispensable tea, and it is now a denizen of the eastern states (*Watsonia*, I. iii. 1949).

Origanum vulgare was identified with the *agrioriganos*, the wild *origanos* of Dioscorides. The name Marjoram properly belongs to the Sweet Marjoram (*Majorana hortensis*), taken to be the *sampsuchon* or *amaracon* of Dioscorides. From *amaracon* and its Latin equivalent, came the Mediaeval Latin *majorana*, Old French *majoraine*, and Middle English *majoram*.

7. Wild Thyme. *Thymus serpyllum* L. 112, H 38

Local names. BANK THYME, Berks; HORSE THYME, N'hants; MOTHER OF THYME, Worc, Cumb, Ire; MOTHER THYME, Som; SHEPHERD'S THYME, Glos, Bucks, Oxf, War, Worc; TAE-GIRSE (i.e. 'tea grass'), Shet.

''Tis hot and dry. It forces the Courses, and Urine. 'Tis Cephalick, Uterine, and Stomachick. 'Tis good for Spitting of Blood, and Convulsions, and for Gripes. Outwardly applied, it cures Headaches, and Giddiness; and disposes to Sleep' (141). Gerard called it 'Puliall mountaine' (see Pennyroyal), 'Running Time', 'Creeping Time', 'Mother of Time', and 'Our Lady's Bedstraw', as in German (see *Galium verum*). A good choice for the Virgin's bedstraw, since it was a uterine herb (hence probably Mother of Thyme, Mother Thyme, from 'mother' in the sense of womb). Highlanders and Islanders drank tea made of Wild Thyme to prevent bad dreams; and they believed that the smell of Wild Thyme gave strength and courage (128). In France this was one of the herbs of St John (188). (See *Hypericum perforatum.*)

8. Balm, Bee Balm. *Melissa officinalis* L. 50, H 2

The rain may long ago have washed out a chalk cottage in Wiltshire, a cob cottage in Devon or Cornwall, but as likely as not a thicket of Bee Balm will survive. An introduced species, from southern and central Europe. If it is no beauty, it is always worth having in a garden for the lemon fragrance of the leaves. This indeed makes it a cordial herb, which 'removes melancholy and cheers the heart'. Balm tea was taken as tonic and uterine and sedative, though it is by no means as nice as the scent of the leaves would lead you to expect. The Oxford don, Thomas Cogan, thought that the distilled water of Bugloss, Borage, and Balm was an especially good drink for students, since they were liable to melancholy, and since Balm drove away 'heavinesse of minde', sharpened the understanding and the wit, and increased the memory (43).

The early botanists identified Balm with the *melissophyllon*, 'bee leaf', of Dioscorides ('so called, because bees delight in this herb'), and the *apiastrum* of Pliny. Here was another reason for its cottage popularity. 'It is profitablie planted in gardens as *Plinie* writeth ... about places where Bees are kept, bicause they are delighted with this herbe above all others, whereupon it hath been called Apiastrum: for saith he, when they are straied away, they do finde their way home againe by it' (Gerard; who added that hives should be rubbed with Balm to cause the bees 'to keepe togither', and to make others join them).

Balm comes through the French from the Latin *balsamum*, 'balsam'.

9. Wild Clary. *Salvia horminoides* Pourr. 69, H 10

Local names. BLUE BEARD, Som; EYESEED, Lincs.

Several garden sages, especially *Salvia sclarea* of southern Europe, were known as Clarie, from the Mediaeval Latin *sclarea*. *Salvia horminoides* was Wild Clarie. Place the seeds in water: they are mucilaginous and they swell up. These soft items of 'frog-spawn' were then put like drops into the eye to cleanse it, or to remove some irritating speck or mote, or to cure inflammation: 'The seede put whole into the eies, clenseth and purgeth them exceedingly from waterish humours, rednesse, inflammation, and divers other maladies, or all that happen unto the eies, and taketh away the paine and smarting thereof, especially being put into the eies one seede at a time, and no more, which is a general medicine in Cheshire and other countries thereabout, knowne of all, and used with good successe' (Gerard).

Popular etymology made Clarie into 'clear-eye', translated even into *Oculus christi*, 'Christ's eye'. Though the flowers are small and violet, the general habit, the crinkled leaves, the blue-dyed stem, make the Wild Clary rather fascinating. But it is not the best of wild perennials to transfer to a garden. It spreads quickly and roots deeply.

10. Bastard Balm. *Melittis melissophyllum* L. 15

Local in the south-west, but it can be abundant in its localities, especially along the roads and lanes in south Devon and Cornwall. The flowers are large and white and hectically blotched with purple, and sometimes you see them from top to bottom of a tall earthen hedge for two or three miles. It was first discovered in a wood near Totnes in south Devon, and recorded in How's *Phytologia Britannica* in 1650.

'Bastard Balm' sadly and badly distinguishes this much more handsome plant from the Balm, Bee Balm, or Balm Gentle of the gardens, *Melissa officinalis*. In France, from the rather foetid smell, it is *melisse de punaisse*, 'bug balm'.

11. Self-heal. *Prunella vulgaris* L. 112, H 40

Local names. ALL-HEAL, Som, Ches, Yks; BLACKMAN'S FLOWER, Yks; BLUE CURLS, Dev, Dor, USA; BUMBLE-BEES, Yks; CARPENTER'S GRASS, Som, Ches; CARPENTER'S HERB, Som, Glos; FLY-FLOWERS, Glos; HEART O' THE EARTH, E Ang, N'thum, Berw, Rox; HEART'S EASE, Donegal.

HERB BENNET (since it was a useful plant. See *Geum urbanum*), HOOKWEED, Som; HISKHEAD, N'hants; KEANADHA-HASSOG, Donegal; LONDON BOTTLES, Ayr; PICKPOCKET, Ess; PRINCE'S FEATHER, E Ang, N'thum, Berw, Rox; PROUD CARPENTER, Ches; TOUCH-AND-HEAL, WOOD SAGE, N Ire.

The rounded middle tooth of the upper lip of the calyx resembles a hook, and was taken as the signature of a vulnerary herb. Early botanists, including Turner, believed this species to be the *sumphuton petraion* of Dioscorides, which remedied a sore throat and cured wounds. So it was named Brunella, or Prunella, from *Braune*, the German for quinsy. Gerard knew it as Prunell, Carpenter's Herb, Self-heal, Hook-heal, and Sicklewort: 'It serveth for the same that Bugle doth, and in the world there are not two better wound herbes as hath been often prooved.'

12. Betony. *Stachys officinalis* (L.) Trev. 83, H 13

Local names. BISHOP'S WORT (OE *bisceopwyrt*), Som, Glos; BIDNY, Kent; BITNY, Dev; DEVIL'S PLAYTHING (children refuse to pick it), Shrop; WILD HOP, Worc; WOOD BETONY, Ches.

A Welsh name is *cribau S. Ffraid*, St Bridget's Comb.

This famous herb was taken to be the *kestron* of Dioscorides, the Latin *betonica*, and the medico-magical herb on which there has come down to us a Latin treatise, supposedly by Antonius Musa (ed. by Howald and Sigerist, in *Corpus Medicorum Latinorum*). The treatise includes a prayer to Betony:

'Betony, you who were discovered first by Aesculapius or by Chiron the centaur, hear my prayer. I implore you, herb of strength, by him who ordered your creation and ordered that you should be useful for a multitude of remedies. Kindly help in making these seven and forty remedies.'

In the *Herbarium* of Apuleius Platonicus, Betony is recommended once more against forty-seven diseases and situations (though the list shrinks to twenty-nine in the OE version). The *Herbarium* declares it to be good for a man's soul as well as his body. Poison, evil, wicked spirits, witchcraft – Betony has power against them. It came frequently into Anglo-Saxon

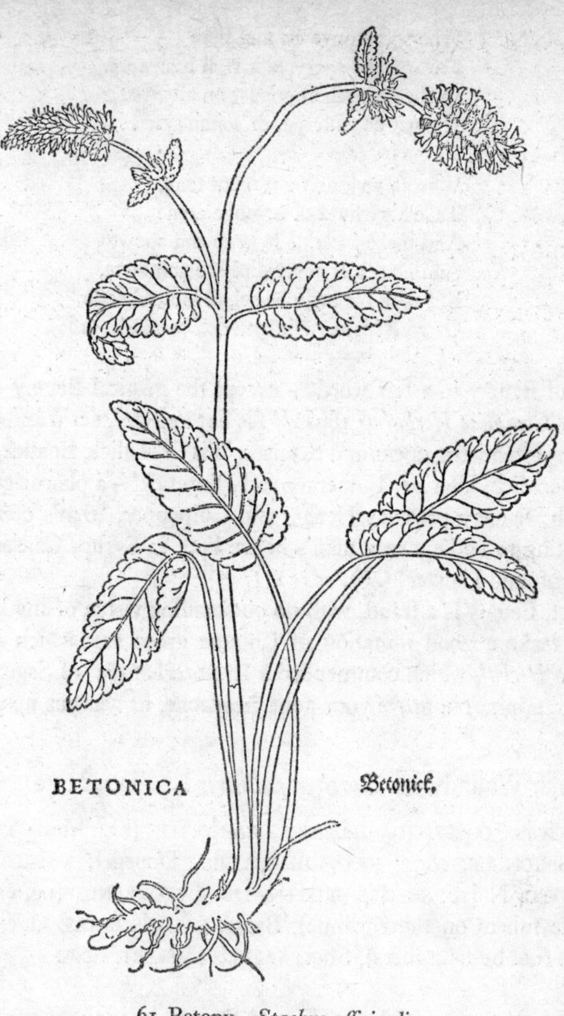

61 Betony *Stachys officinalis*

medicine and magic – against the elf-sickness, for example, in the tenth-century Leachbook, or against the devilish afflictions of the body in the long Holy Salve in the *Lacnunga* (169, 88). In the medical poem in the Royal Library of Stockholm, written about 1400, Betony takes the lead:

> At betonye I wyll be-gynne,
> That many vertewys hath hym with-inne . . .

Who so betonye on him bere
Fro wykked sperytes it wyll him were.
In the monyth of august on all wyse
It must be gaderyd, or sonne ryse . . .

Who so wyll don a serpent tene,
Take a garlund of betonye grene
And make a cerkle hym rownd abowte,
And he schall never on lywe gon owte,
But with his tayle he schall him schende
Or with his mowth hym-self to-rende. (168).

For all of Betony in a few words – except the magical Betony – turn to Pechey's *Compleat Herbal* of 1694: ''Tis hot and dry, acrid and bitter. It discusses, attenuates, opens and cleanses. 'Tis Cephalick, Epatick, Splenetick, Thoracick, Uterine, Vulnerary and Diuretick' – a pharmacopoeia in one herb; 'a very precious Herb', said Culpepper, 'that's certain, and most fitting to be kept in a man's house both in Syrup, Conserve, Oyl, Oyntment, and Plaister' (49).

In fact, Betony is a fraud, with no outstanding virtue of any kind. But it does make a good imitation of Chinese green tea, which *may* (see Pechey's *Herbal*, which commended a 'counterfeit tea' of Sage, Betony, and Groundpine) improve your gout, headache, or nervous upsets.

13. Marsh Woundwort. *Stachys palustris* L. 112, H 40

Local names. COCKHEAD, Lanark; DEA-NETTLE (i.e. 'deaf-'), Cumb; HOUND'S TONGUE, Mor; KEANADHA-HASSOG, Donegal; MASKERT, Scot; ROUGHWEED, N Ire; SHEEP'S BRISKEN, Ire; SWINE ARNOT (i.e. earth nut, from the tubers on the rhizome), Banff; SWINE'S BEADS, Ork; SWINE'S MURRIKS (i.e. bulbs, tubers), Shet; SWINE'S MASKERT, Scot.

Gerard found that Kentish people used *Stachys palustris* as a wound herb. He was on a professional visit to Kent when 'a very poore man in mowing of Peason did cut his leg with the Sieth, wherin he made a wound to the bones, & withal very large and wide, and also with great effusion of bloud, the poor man crept unto this herbe which he brused in his hands, and tied a great quantitie of it unto the wound with a peece of his shirt'. He was soon well, poulticing the wound from day to day with the herb stamped in lard. Gerard saw the wound 'and offered to heale the same for charitie, which he refused, saying, that I coulde not heale it so well as himselfe; a clownish answer I confesse without thankes for my good will, whereupon I have named it Clounes Woundwoort'.

Gerard felt that he had discovered an all-heal. He applied *Stachys palustris* to gentlemen grievously wounded in Elizabethan brawls. One of them was a young man of Gray's Inn, stabbed to the lungs. He cured him; and also a shoemaker's servant in Holborn, who stabbed himself in the stomach and in the throat – 'a most mortall wound . . . in such sort, that when I gave him drinke it came forth at the wound, which likewise did blowe out a candle'.

14. Black Horehound. *Ballota nigra* L. 76, H 25

Local names. DOUBLE DUMB-NETTLE, Wilts; DUNNY-NETTLE, Bucks; HAIR-HOUND, Berw; HENBIT, Som; HOREHOUND, Herts, Berw; STINKING ROGER, Shrop; WIRRAL, Som, Wilts.

The *ballote* of Dioscorides, a stinking plant, prescribed, in obedience to him, against stinking ulcers and the bite of mad dogs. Turner (1548) called it 'blacke horehound' or 'stynkyng horehound'. For Horehound see *Marrubium vulgare*.

15. Yellow Archangel. *Galeobdalon luteum* Huds. 69, H 4

Local names. ARCHANGEL, Som; BEE-NETTLE, Ches, Notts; DEAD-NETTLE, DEAF AND DUMB, Som; DUMB-NETTLE, Oxf; DUNNY-NETTLE, Bucks; LION'S SNAP, Som; SNUFF-CANDLE, Wilts; STINGY-WINGIES, WEASEL'S NOSE, Dor; WEASEL-SNOUT, Som, Glos, Yks; YELLOW DEAD-NETTLE, Som; YELLOW ARCHANGEL, Dev, Yks.

Why were dead nettles, deaf nettles, dumb nettles, named *archangelica* in the Middle Ages, unless from the angelic quality (which is hardly arch-angelic) of not stinging? Weasel's nose, weasel-snout – cf. Galeobdalon, which means 'weasel stench'; not a fair name for one of the prettiest woodland flowers, colony by colony.

16. Red Dead-nettle. *Lamium purpureum* L. 112, H 40

Local names. ARCHANGEL (see *Galeobdalon luteum*), Som, Glos, Leic; BAD MAN'S POSIES (i.e. 'devil's posies'), Yks, N'thum; BEE-NETTLE, Notts, Lincs; BUMBLE-BEE FLOWER, Som; DEA-NETTLE (i.e. 'deaf-'), Worc, Yks,

Cumb; DAA-NETTLE, DEE-NETTLE (i.e. 'deaf-'), Shet; DEAF-NETTLE, Dev, Som, Yks; DUMB-NETTLE, Som; DUNCH-NETTLE (i.e. 'deaf-'), Dor; FRENCH NETTLE, Shrop; OKKERDU, Shet; RABBIT'S MEAT, Shrop; RED ARCHANGEL (so in Turner's *Names*, 1548), Norf; RED BEE-NETTLE, War; STINKING WEED, Cumb.

17. White Dead-nettle. *Lamium album* L. 100, H 25

Local names. ADAM-AND-EVE-IN-THE-BOWER, Som; ARCHANGEL (see *Galeobdalon luteum*), Dev, Som, Glos, Bucks, Leic; BEE-NETTLE, Som, Leic, Notts, Lincs; BLIND-NETTLE, Dev, Som; DEA-NETTLE (i.e. 'deaf-'), Shrop, Yks, Cumb, N'thum; DEAF-NETTLE, Dev, Wilts, Lincs, Yks; DUMB-NETTLE, Som, Wilts, Glos, Wor, Herts, Ess; DUNCH (i.e. deaf), Wilts; DUNCH-NETTLE, Dor, Som, Wilts, Hants; DUN-NETTLE, Shrop.

HELMET-FLOWER, Som; HONEY-BEE, Dev; HONEY-FLOWER, Som; HONEY-SUCKLE, Wilts; SNAKE'S FLOWER, Ess, Norf, Cam; SUCK-BOTTLE, Som, N'hants; SUCKY SUE, WHITE ARCHANGEL, Som; WHITE BEE-NETTLE, War; WHITE NETTLE, Glos; WHITE STING-NETTLE, Dev.

Adam-and-Eve-in-the-Bower: turn the plant upside-down, and beneath the white upper lip of the corolla, Adam and Eve, the black and gold stamens, lie side by side, like two human figures. This is a plant so common, so associated with the Nettle, that its charms are overlooked. Adam-and-Eve give it one distinction. But the flowers have also a great charm of shape, colour, and texture, from the time they lie like soft knobs within the long green teeth of the calyx. For one thing, they are not pure white but white faintly suffused with green. The knob is formed by the upper lip, curled over before its expansion. When it does expand into the hood, look at it with the bare eye or beneath a lens, see how it is felted and fringed with soft white hairs, like a moth. A bunch of White Archangel (as Gerard called it) can look well in a vase, though some of the leaves want cutting away to reveal the flowers

Various uses in medicine, especially against the King's Evil.

18. Motherwort. *Leonurus cardiaca* L. Introduced. Rare

An uncommon physic plant, 'of a verie ranke smell and bitter taste', introduced in the Middle Ages for troubles of the heart, 'for them that are in hard travell with childe' (Motherwort, i.e. 'womb plant'), and as a cattle drug.

Motherwort was naturalized by the early years of the sixteenth century, at least. Turner (1548) wrote of it 'in hedges and about walles'. Gerard said that it did well in gardens, but also that 'it joieth among rubbish, in stonie and other barren and rough places, especially about Oxford'.

19. Catmint. *Nepeta cataria* L. 66, H 18

Local names. CATMINT, N Eng; DOGMINT, Pemb; NEP, NEP-IN-A-HEDGE, Suff, E Ang, Yks, N Eng.

A native, and a sweet herb of the garden. *Herba Cattaria*, *Herba Catti*, since cats like the plant; so Catmint. ''Tis hot and dry. 'Tis chiefly used for Obstructions of the Womb, for Barrenness, and to hasten Delivery, and to help Expectoration. 'Tis used outwardly in Baths for the Womb, and the Itch' (Pechey's *Compleat Herbal*, 1694).

20. Ground Ivy. *Glechoma hederacea* L. 107, H 40

Local names. ALEHOOF, Corn, Dev, Som, Suss, Shrop, Yks; BIRD'S EYE, Bucks, Oxf, N'hants; BLUE-RUNNER, Bucks; CAT'S FOOT, Som, N'thum, N Eng; CREEPING CHARLIE, USA; CREEPING JENNY, Som; DECEIVER, Som, Ess; DEVIL'S CANDLESTICK, Som, War, Yks; FAT HEN, Bucks; GILL, Dev, Som, Glos, War, Worc, Leic; GILL-CREEP-BY-THE-GROUND, Som; GILL-GO-ON-THE-GROUND, Som; GILL-OVER-THE-GROUND, USA; GILL-RUN-BY-THE-GROUND, Som, Bucks, Lincs; GILL-GO-BY-THE-GROUND, Cumb; GRUNDAVY, N'thum, Scot. GROUND IVVINS, Ches, Lancs.

HAY-HOA, Hants; HAY-MAIDENS (i.e. 'hedge-'), Corn, Dev, Dor, Som, Glos, Cumb; HEDGE-MAIDS, E Ang; HEN AND CHICKENS, Bucks; JENNY-RUN-BY-THE-GROUND, Lincs; LION'S MOUTH, Suss; MONKEY-CHOPS, Som; MONKEY-FLOWER, Som; MOULDS, Rut; NEP (see *Nepeta cataria*), Suff; RABBIT'S MOUTHS, RAT'S FOOT, RAT'S MOUTHS, Dev.

ROBIN-RUN-IN-THE-HEDGE, Som, Suss, Worc, Ches, Derb, Leic, Notts, Lancs, Ire; ROB-RUN-UP-DYKE, Cumb; RUNAWAY JACK, Som, Glos; RUN-NIDYKE, Cumb; TUNHOOF, Som, Ess, Suff, Norf, Cam; TUNFOOT, Som; WANDERING JEW, Donegal.

Everybody once knew this bitter, aromatic, and charming 'blue runner', since it was not only a doctor's medicine and a home medicine, strengthening and cleansing, but the chief bitter before the general use and cultivation of hops (see *Humulus lupulus*) – it was the Alehoof, the Tunhoof, the plant called *hofe* in OE, used in tunning the ale. 'The women of our northern parts,' Gerard wrote in 1597, 'especially about Wales and Cheshire, do tun the herbe Ale hoove into their ale.' By that time the practice was old-fashioned enough to deserve comment, at any rate from a Londoner, though hops can never have conquered all of Great Britain at a blow. Either for medicine or still for brewing, or both, Alehoof was taken over to New England (111), where it is now a familiar denizen on the east coast. In Ludlow there was a tradition of eating pork stuffed with the leaves, and there are still people in the cottages who drink Gill-tea.

Ground Ivy – *lierre terrestre* in French, *hedera terrestris* to the older botanists, *eorthifig* to the Anglo-Saxons – is a poor name, but an old one translating the *chamaikissos* of the ancients. A French name for the plant is *courroie de Saint-Jean*, 'St John's Girdle'.

21. White Horehound. *Marrubium vulgare* L. 75, H 12

More frequently an escape from gardens than properly wild. White and woolly and smelly, the OE name for this plant was *hune* (which has nothing to do with dogs) or *har hune*, white or hoary *hune*. It was considered to be Pliny's *marrubium*, and the *prasion* which, with honey, Dioscorides commended against coughs; and also against poison. The Anglo-Saxons used it frequently for the lungs (88), the Tudor herbalists praised it for coughs and consumption, and we can still buy Horehound mixtures and lozenges from the chemist – Syrupus Marrubii in the *British Pharmaceutical Codex* (1949) is laxative and expectorant, and made from the leaves and tops; 'Sirupe made of the greene fresh leaves and sugar, is a most singular remedie against the cough, and wheesing of the lungs' (Gerard).

Certainly both native and introduced, Gerard knew it plentifully 'neere unto olde wals, high waies, beaten pathes, in untoiled places'. It is still grown in cottage gardens, and it must have been cultivated in this country for more than a thousand years.

The old name 'Marvel' has been recorded in Sussex.

22. Skullcap. *Scutellaria galericulata* L. 108, H 38

Another of the bitter labiates, and one which was unnoticed along the rivers, the streams, and ditches of England until the sixteenth century. Then it was grouped with the pretty flowered *Gratiola officinalis*, a south European plant brought over as an almighty purge, and it was therefore called the little *gratia*, or little thanks-be-to-God. But the two plants are not related. Since it was an apparent *gratiola*, Gerard called the Skullcap *gratia dei* and Hedge Hyssop; and recommended it as a febrifuge. As it happens, de l'Obel, who first recorded it for England in 1576, made it not into a *gratiola*, but into a *lysimachia*, a loosestrife – *Lysimachia galericulata caeruleopurpurea*, the 'blue-purple loosestrife [with a calyx] like a small *galerum*', a leather skull-helmet worn by the Romans. In the eighteenth century *galericulata* suggested the happy English name of Skullcap. In Germany it is the *Keppen-helmkraut*, 'cap-helmet plant', in France *toque bleue*.

Every plant has its peculiar localities. For me *Scutellaria galericulata* summons into mind not a flowery bank of the upper Thames, but a muddy trickle and a dirty black pool at one end of the island of Samson, in the Isles of Scilly. There, miles out over the Atlantic, sprawl England's last blue-helmeted plants of *Scutellaria*.

23. Lesser Skullcap. *Scutellaria minor* L. 77, H 17

Lesser Skullcap, like *S. galericulata*, has been used as a nervine. Gerard's was the first record: 'I found it growing upon the bog or marrish ground at the further end of Hampstead heath, and upon the same heath towards London, neere unto the head of the springs that were digged for water to be conveied to London 1590, attempted by that carefull citizen Sir *John Hart* Knight, Lord Maior of the Citie of London: at which time my selfe was in his Lordships company, and viewing for my pleasure the same goodly springs, I found the said plant, not heretofore remembred.' That is to say, near the Highgate ponds, which still have their good flora, which includes *Scutellaria galericulata*, if not *S. minor*.

24. Wood Sage. *Teucrium scorodonia* L. 110, H 40

Local names. GIPSY'S BACCY, Som; GIPSY'S SAGE, Dor; MOUNTAIN SAGE, Ches, Cumb; ROCK MINT, Som; WILD SAGE, Ches, Ire.

Because the Wood Sage waves no brilliant colour, it is rather neglected. Given a rocky position and the right neighbours, Gertrude Jekyll thought of it as a plant for the flower garden. She could see the charm of its form and tones. It is sturdy and tough and does not sag. Its crinkly dark leaves set off the flowers, which are the colour of straw touched with green.

Salvia bosci, or Wood Sage (*sauge des bois*), *Salvia agrestis*, or Wild Sage, to the early botanists, it was not greatly valued, although it was known as a diuretic, a wound herb, a herb against broken veins, ulcers, French pox, and scurvy. People in the Dursley neighbourhood of Gloucestershire still pick the leaves in the spring and dry them for a tea against rheumatism (158). As a bitter, it was used in brewing before ales and beers were hopped (see *Humulus lupulus*).

25. Bugle. *Ajuga reptans* L. 112, H 40

Local names. BABIES' SHOES, Som, Wilts; BABY'S RATTLE, Som; BLINDMAN'S HAND, Hants; CARPENTER'S HERB (cf. German *Schreinerkraut*), Som; CUCKOOS (cf. German *Kuckuck*, *Blauer Kuckuck*, *Kuckucksblume*), Dor; DEAD-MEN'S BELLOWS (i.e. pillies, male members, cf. *Pedicularis palustris*), Berw; HERB-FLOWER, Dor; HONEYSUCKLE, Dor; HORSE AND HOUNDS, Dor; HORSE PEPPERMINT, Wilts; NELSON'S BUGLE, Som; OKERDU, Shet; SELF-HEAL, Som; THUNDER-AND-LIGHTING (cf. thirteenth-century name for Bugle, *thundre-clovere* and the Germans *Gewitterblume*, 'storm flower'), Glos; WILD MINT, Berks; WOOD BETONY (Betony was also a vulnerary), Ire.

Bugle stands bravely on parade in the most cheerful and healthy colonies, in wood or in damp grass. 'It is a blacke herbe and it groweth in shaddowy places and moyst groundes.' William Turner meant not so much black as dark – not the flowers, but the upper stems and leaves, shinily darkened with dark blue, or a green-violet. It was rather better described in *Agnus Castus* in the Middle Ages as having leaves 'sumdel turnyngge to blak' (20). With this black goes the light contrast of the blue corollas – the flowers described by John Pechey as being 'of a Sky and changeable colour'. The tints vary: instead of green dyed with violet or dark blue, the plants can look a mixture of green and brown – Gerard's 'Brown Bugle', or the name 'Wodebroun' which is on record for the fourteenth century. A most lovable and inexhaustible little plant.

Like *Prunella vulgaris*, *Sanicula europaea*, and *Symphytum officinale* (q.v.), Bugle was a popular consound or wound-herb, *consolida media*, Middle Comfrey, in German still *Günsel*, from *consolida*, the consolidating, joining, curing herb. Gerard wrote that it was much planted in gardens. Pechey, in *The Compleat Herbal of Physical Plants* (1694), summarizes its virtue as an interior or external vulnerary, an ingredient of the 'Traumatick Decoction' or wound-drink of the London Dispensatory – also, with Sanicle and Scabious, of an ointment for ulcers, wounds, and bruises.

Bugle is an obscure name with counterparts in German, French, Italian, and Spanish, from the apothecaries' *bugula*. It has nothing to do with the bugle you blow, though it may be the source of 'bugle' meaning the long, tube-shaped, glass bead, black or blue, which is sewn on to clothes – from the glistering dark blue or violet of its own leaves. 'Thunder-and-lightning' and *thundre-clovere* above may also refer to this mingling of colour and shine, and to a belief which has survived in at least one district of Germany that to pick Bugle flowers and bring them indoors will cause fire (129). See, for thunder plants, *Melandrium rubrum* and *Papaver rhoeas*.

A pleasant Austrian name for Bugle is *Blauer Kirchturm*, 'blue steeple'.

LXXVII. Plantaginaceae

1. Great Plantain. *Plantago major* L. 112, H 40

Local names. BIRD'S MEAT, Aber; BIRDSEED, Som; BROAD LEAF, Ches; CANARY FLOWER, Som; CANARY FOOD, Som, Wilts; CANARY SEED, Som; COW-GRASS, N Ire; GREAT WAYBREDE, N Eng; HARDHEADS, Dev, Dor, Wilts, Worc, Lancs, Yks; HEALING BLADE, Scot; JOHNSMAS PAIRS, JOHNSMAS FLOWERS, Shet; LAMB'S FOOT, Cumb; LARK SEED, Wilts; PONY'S TAILS, Dev; POVERTY GRASS, Som; RAT'S TAILS, Norf, Cumb, Scot; RATTEN TAILS, N'thum; RIPPLE-GIRS, Scot; SLANLIS (Irish *slánlus*, 'healing herb'), N Ire; TRAVELLER'S FOOT, War.

WAAVERIN-LEAF, Shet; WABRAN-LEAF, Scot; WAYBERRY, Ches; WAYBREAD (OE *weg-brǣde*, 'way breadth'), Worc, Ches, Lancs, N'thum, N Eng, Ber, Rox, Kirk, Wigt; WAYBROAD, Worc, Lancs; WAYBROAD-LEAF, Worc; WAYBURN-LEAF, Lanark; WAYFRON, N'thum; WIBROW, Ches; WIBROW-WORROU, Ches.

Plantains grow on the path, on the track, on the road, on the doorstep. You tread on them, you crush them, and they go on living. Sympathetic magic therefore suggests that a herb so powerful, tough, and elastic, known by everyone and seen every day, must heal crushing, tearing, and bruising. Dioscorides wrote that Plantain was a vulnerary and was good against ulcers and sores and much else; also that some people wore the root around their necks as an amulet against the King's Evil. So it continued through the centuries and through Europe, a healing herb, still used, in Shetland, for example, against wounds, burns, and sores (166).

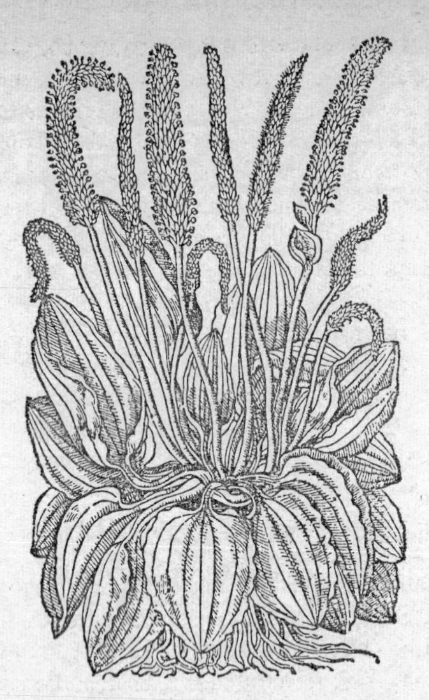

62 Great Plantain *Plantago major*

As a magical herb, the aid of Plantain was invoked in the famous Lay or Charm of the Nine Herbs, which is pre-Christian (169 88,). Here the Plantain is powerful precisely because it stands up to traffic on the green roads. First the Anglo-Saxon medicine-man addresses Mugwort, then Plantain, or Waybread:

> And you, Waybread, mother of worts,
> Open from eastward, powerful within,
> Over you chariots rolled, over you queens rode,
> Over you brides cried, over you bulls belled;
> All these you withstood, and these you confounded,
> So withstand now the venom that flies through the air,
> And the loathed thing which through the land roves.

A herb of medico-magical power, a herb also of divination, whether *Plantago major* or *Plantago lanceolata*. In France *P. major* is one of the important herbs of St John's Eve (see *Hypericum perforatum*); on Johnsmas, St John's Day, in Shetland, two scapes or flowering stems are picked, one for the boy, one for the girl, to foretell if they will love and marry (166). In Berwickshire this was done by removing all the visible anthers, wrapping the two scapes in a dock leaf, and placing them under a stone till the next day. Then if more anthers have appeared, love is certain.

On St John's Day in 1694, John Aubrey was walking in a London pasture behind Montagu House, at noon. He saw twenty-two or twenty-three young women 'most of them well habited, on their knees very busy, as if they had been weeding'. Puzzled, he asked a young man what they were up to: 'They were looking for a coal under the root of a plantain, to put under their head that night, and they should dream who would be their husbands: it was to be sought for that day and hour' (5). But possibly his well-habited young ladies had been told the story of the coal under the plantain out of Mizaldus's *Centuriae memorabilium arcanorum*, 1613. Mizaldus more properly says that the coal is to be found on St John's Eve. Plantain, as one might expect, was christianized – as *sawdl Crist*, Christ's heel, in Wales; as *copóg Phádraig*, (St) Patrick's leaf, in Ireland; *cabbag Parick* and *dvillag Parick*, Patrick's dock, in the Isle of Man.

There is no escaping from *Plantago major*. It goes round the world. It came up spontaneously in the New England settlements, where the Indians, according to John Josselyn, named it English Man's Foot – 'as though produced by their treading' (111).

2. Hoary Plantain. *Plantago media* L. 90, H 18

Local names. ASHY POKES, Wilts; BOOTS AND STOCKINGS, COTTON FLOWER, Som; FIRE LEAVES (see *Plantago lanceolata*), Glos; LAMB'S EAR, Cumb; LAMB'S TONGUE (from *arnoglosson*, the Greek name for Plantain), Som, Suss; LORDS AND LADIES (from love divination by the erect scapes, see *P. major*), Norf; SCENT BOTTLES, Dor, Som.

The beauty among the Plantains. Both *Plantago major* and *Plantago media* have an altogether tense and acute form too little appreciated, because they are too familiar, and now too much detested as lawn weeds. In *P. major*, the green-sepalled flowers, before they open, are rough and close like tarpaulin canvas. The anthers are maroon, drying to an untidy rust. In *P. media* the filaments and the anthers they carry are always alive in the wind, they frizzle the scapes with purple and lilac, and the leaves and stems are silver haired. Tones and form are both delightful.

3. Ribwort, Ribgrass. *Plantago lanceolata* L. 112, H 40

Local names. BASKETS, Wilts; BLACK BENT, Bucks; BLACK BOYS, Wilts; BLACK GIPSIES, Som; BLACK JACKS, Shrop; BLACK MEN, Dor, Som; BLACKSMITHS, BLACKIE TOPS, Som; BOBBIES, BOBBINS, Dor; CARL DODDY, CURL DODDY, Donegal, Scot; CHIMNEY-SWEEPER, Wilts, War, N'hants; CHIMNEY-SWEEP, Som; CLOCK, Bucks; COCKS AND HENS, Dev, Som, N'thum, Ire; COCKS, Ire; COCK-GRASS, Dev, Som; COCK'S HEAD, Som, Suss, E Ang; CONKERS (as for horse chestnuts, see below), CONQUEROR-FLOWERS, DEVIL AND ANGELS, Som.

FECHTERS (i.e. 'fighters'), Scot; FIGHTEE-COCKS, Suss, E Ang, N'hants, Ches, N'thum; FIGHTING COCKS, Dev, Som, Wilts, E Ang, N'hants, Shrop, Ches, Yks; FIRE-GRASS, Som; FIRE-LEAF, Som, Glos; GIPSY, Som; HARDHEADS, Corn, Dev, Dor, Som, Wilts, Worc, Lancs, Yks; HEADMAN, Perth; JACK STRAWS, Yks; KEMPS ('kemp' is to fight in northern dialect, cf. Danish *kjaempe*; Swedish *kämpa* for *Plantago lanceolata*; OE *cempa*, a warrior), N Eng, Scot; KEMPSEED, Selk.

LAMB'S TAIL, Som; LAMB'S TONGUE (*arnoglosson*, Greek for Plantain), Hants, Suss, Shrop, N'thum; LORDS AND LADIES, Norf; MEN OF WAR, NIGGERS, Som; NIGGER'S HEADS, Dev; PASH-LEAF, Pemb; RAT'S TAILS, Cumb; RIBGRASS, Hants, Suss, E Ang, Staffs, Ches, Yks, Cumb, Lakes, N'thum, Kirk, Wigt; RIPPLING-GRASS, Lanark; SOLDIERS, Som, Notts, Ches, Lakes, Scot; SOLDIER'S TAPPIE, Ang; SWEEP'S BRUSHES, Som; SWORDS AND SPEARS, Dor; TINKER-TAILOR GRASS, Som.

The black man and the fighter. The flowers at the end of the long scapes appear black, with a dusting of white from the stamens. Children, as they have done time out of mind, still match one 'Black Man', one 'Soldier', one 'Hardhead', one 'Fighting Cock', against another, until the weaker of them loses its head – snip. The game is widely played across Europe. But how exactly did farmers come to divine whether a rick would burn by twisting the 'fire-leaves' of *Plantago lanceolata*? They estimated the amount of moisture in the hay by the amount in the leaf – practical and reasonable on the face of it, but perhaps another example of the ancient power possessed by the Plantains.

Pollen analysis in Great Britain and Denmark proves an increase of this plantain and other kinds with the increase of neolithic farming and the decrease of the old forest cover (40). It spread with the grass and the herds – a plague, incidentally, for the sensitive, since pollen from *Plantago lanceolata* is one of the causes of hay-fever.

Plantago coronopus, the Buck's-horn Plantain, has the Manx name of *bollan Vreeshey*, (St) Bridget's wort.

LXXIX. Campanulaceae

1. Ivy Campanula. *Wahlenbergia hederacea* (L.) Rchb. 49, H 8

This 'elegant and pretty little flower', delicate ivy leaves and delicate bells of a light blue, is one which the nineteenth-century tourists to Cornwall were always bidden to discover – trailing around the tussocks of sedge or rush, often in company with the Bog Pimpernel, and with Cornish Money-wort festooning a wet rock near by. It belongs to the south, the south-west, and the long western side of Great Britain.

For the first record of it, turn to Johnson's new edition of Gerard, 1633: 'This pretty plant was first discovered to grow in England [not a statement for the Welsh!] by Master George Bowles, anno 1632, who found it in Montgomeryshire, on the dry banks in the high-way as one rideth from Dolgeogg, a worshipful gentleman's house called Mr Francis Herbert, unto a market towne called Mahuntleth and in all the way from thence to the sea-side.' Mahuntleth is Machynlleth. Master George Bowles was a young student of medicine then in his late twenties, who became a distinguished doctor in London. It was the doctor's job to know his plants, though Dr Bowles botanized also for the sake of botany.

In Somerset the Ivy Campanula has been called Witches' Thimbles.

2. Giant Bellflower, Giant Throatwort. *Campanula latifolia* L. 82

Local names. FOXGLOVE, Yks; GOWK'S HOSE (i.e. cuckoo's hose), Scot; WHITE FOXGLOVE (flowers blue or white), Lancs; WILD SPINACH, Yks.

The tallest Bellflower, a fine plant in the mountainy parts of the north of England – for instance, around the limestone uplands of the West Riding, where there are no calcifuge Foxgloves to illuminate the grey rock. White Foxglove, either because the flowers are sometimes white or because the plant exudes a milky latex. The shoots can be peeled, cooked, and eaten like spinach.

3. Throatwort, Nettle-leaved Bellflower. *Campanula trachelium* L. 61, H 5

Local names. BLUE FOXGLOVE, Shrop; COVENTRY BELLS, War.

This was known to Gerard 'in the lowe woods and hedgerows of Kent about Canterburie'. He thought that *Campanula trachelium*, and not the exotic *C. medium* of the gardens, should be called the Canterbury Bell (a name which arose from the likeness of the flower-bells to the St Thomas's Bells, badges made of latten, bought by pilgrims to the Canterbury Shrine of St Thomas à Becket.

A tall plant secreting a yellow latex – the double signature of its value against a severe sore throat and tonsillitis, and in Germany, therefore, *Halskraut*. As counterparts to *Halskraut*, Lyte invented Haskewort and Throatwort. Gerard added Uvula wort, and he used Coventry Bell for *C. medium*.

4. Clustered Bellflower. *Campanula glomerata* L. 54

Local names. DANES' BLOOD, Cam; STARS, Wilts.

The notion of plants growing up from the dead or growing from their blood is not at all uncommon or unexpected. For Danes' Blood, compare *Sambucus ebulus*, which has berries with a bloody juice and goes red and ragged in the autumn. In *Campanula glomerata* the stems are bloody or crimson. Also the flowers are livid or purple-blue. Cf. Deadman's Mittens, the Shetland name for the livid *Gentianella amarella*, or Deadmen's Bellows for *Ajuga reptans*.

5. Harebell, Bluebell. *Campanula rotundifolia* L. 111, H 28

Local names. BELLFLOWER, Dev, Dor, Som; BLAVER, BLAWORT, Scot; BLUEBELL (cf. Icelandic *bláklukka*), Dev, Dor, Som, Herts, Derb, Notts, Yks, Cumb, Scot; BLUEBELLS OF SCOTLAND, Som; BLUE BLAVERS, Rox; BLUE-BOTTLE, Bucks; CUCKOOS, Dev; DING-DONGS ('ding-dong bell'), Dor; FAIRY CAP, Wilts; FAIRY CUP, Dor; FAIRY BELLS, Som; FAIRY RINGERS, Dor; FAIRY THIMBLE, Som; GOWK'S THIMLES OR THUMLES (i.e. 'cuckoo's thimbles'), N Scot; GRANNY'S TEARS, Som.

HAREBELL (i.e. hare, the animal), Ches, Lancs, Yks, Scot, Donegal; HARVEST BELL, N'hants; HEATHBELL, Dor, N'hants; LADY'S THIMBLE, Som,

E Ang, N'thum, Scot, N Ire; MILK-ORT, N Scot; OLD MAN'S BELL (i.e. devil's), Scot; SCHOOL BELL, Wilts; SHEEP BELLS, Dor; THIMBLES, Som, Wilts, Glos, Scot; WITCH BELLS, Scot; WITCHES' THIMBLES, Som, Lanark.

Bluebell of Scotland or no, it was also the Old Man's Bell, the devil's bell, which was not to be picked, the Witch-bell, the Witch-thimble, the Cuckoo's Thimble, and in Gaelic the Cuckoo's Shoe, *bròg na cubhaig*. In Ireland this dangerous and fine etched plant is sometimes *méaracán púca*, thimble of the puca or goblin; and it was a fairy plant in the south-west of England, however much it has now been airyfairy'd. The hare, too, of Harebell is a witch animal. In the Isle of Man it was also *mairanyn ferish*, 'fairies' thimble'.

After this, the exorcizing variant of (Our) Lady's Thimble is to be expected.

6. Venus's Looking-glass, Corn Bellflower. *Specularia hybrida* (L.) A. DC. 52

Corn weeds are becoming rarer, and in some parts one may search mile upon mile of wheat for the open eyes of this little queer one – eyes 'of a bright purple colour tending to blewnesse very beautifull'. The flower perks upon the end of a stiff prismatic capsule. Here, on top, the corolla lies above the protruding teeth of the calyx, like a brilliant floret within a larger flower, or as if two cog-wheels with pointed cogs were laid together. When the long capsule opens, the looking-glasses of the Virgin or Venus are revealed, the seeds which are 'oval or elliptical, pale brown, exquisitely polished, and pellucid like a speculum', and which give the plant its name.

Gerard preferred Venus Looking Glasse, printing also the botanists' *speculum veneris*, and Ladies glasse, from the German name (which is still current) of *Frauenspiegel*.

7. Round-headed Rampion. *Phyteuma tenerum* R. Schulz. 10

Climb Silbury in Wiltshire on a hot August afternoon, climb to the top of this ziggurat of prehistory, and at your feet you may see an unusual insect of sharp blue or violet. Look nearer, and it is more like a violet sea-anemone – air-anemone – closing upon an incautious bee or fly. But it is vegetable, after all, a globe of curving, or incurving, tentacle-like corollas, which is the flower head of Rampion.

The Round-headed Rampion was found here on Silbury in 1634 by

Thomas Johnson and a party of botanizing apothecaries exploring their way from Marlborough to Bath. Not far off more of these violet air-anemones float over the actual ramparts of the Avebury 'temple'.

Rampion is a plant of the chalk downs, from Dorset and Wiltshire, through Hampshire (where the Hampshire botanist, John Goodyer, had found the first British plants only a little while before the botanical troop dismounted at Silbury), and through Sussex and Kent to Surrey, where the Rampion grows near the racecourse on Epsom Downs. 'Rampion', distinguished by 'Round-headed', since nobody could think of a better name than the one which already belonged to the vegetable bellflower, *Campanula rapunculus*. It has also been known as 'Pride of Sussex'.

8. Sheep's Bit. *Jasione montana* L. 85, H 31

Local names. BLUE BONNETS, Dumf; BLUE BUTTONS, Dor, Som, Cumb; BLUE CAP, Yks; BLUE DAISY, Ches; IRON-FLOWER, Ches.

Devil's Bit (*Succisa pratensis*) and Sheep's Bit have an outer resemblance, and both are frequently called Blue Buttons, and in Wales or Cornwall, etc., both may grow within a few yards of each other, Devil's Bit where it is more damp, Sheep's Bit on dryer ground. Besides all the differences of structure, note the pale sandy blue of Sheep's Bit flowers, and the deep mauve to blue of the Devil's Bit.

LXXX. Lobeliaceae

1. Acrid Lobelia. *Lobelia urens* L. 6

Along a road under a high wet hill, west of Axminster in Devon, you come to Lobelia Cottage. Between the road and the scrubby hill-top, this rare plant, in the first and most famous of its stations, grows with a pleasant incongruity among hen-coops and the rusty ends of old bedsteads and old tyres sliced in half for chicken troughs. An Atlantic species of France and Spain and Portugal, it was discovered in Devonshire, and for a long while it was regarded as one of the shire's unique ornaments. 'In describing the Flower of the Axe,' wrote a Devonshire parson, the Rev. Z. I. Edwards, in his *Ferns of the Axe* (1862), 'we speak not of the blushing rose, nor the pale scented violet, nor the yellow cowslip . . . we speak of a flower of a blue colour and an acrid taste, so rare, so peculiar to a certain portion of this locality, that not all England can produce its like again.'

The first man to stumble on it was a local herbalist, William Newberry, in October 1768, according to the *Victoria History of Devonshire*. As for the bitterness of this frail, spare Lobelia, in 1796 Lord Webb Seymour sent a box of plants by coach from Axminster to London addressed to William Curtis, who wished to have it figured in his *Flora Londinensis*. Curtis wrote: 'Mr Sydenham Edwards, my draughtsman . . . having handled a branch of this plant broken off from the main stem, and afterwards rubbed his eyes slightly, had a violent pain and temporary inflammation excited in them thereby; which however soon went off, on washing them with cold water.'

He had no doubt rubbed on his eyes some of the white latex which exudes when you break the plant.

2. Water Lobelia. *Lobelia dortmanna* L. 45, H 21

Violet flowers, or rather mauve flowers, over the blackness of a tarn; yet this Lobelia is never difficult to pick or examine, since it flourishes in shallow water, rooted among stones, not mud. I made its acquaintance first wading in the warm shallows of the lake below the One Man's Pass and the mountain of Slieve League in Donegal, blaspheming as I hooked the flowering stems with trout flies. A plant in this way more cursed, perhaps, than appreciated, until one comes to welcome it as a symbol of wild mountains. In Cumberland it has been called Water Gladiole, in North Wales *bidawglys*, 'dagger plant'.

LXXXI. Rubiaceae

1. Woodruff. *Asperula odorata* L. 109, H 40

Local names. BLOOD CUP, Dor; HAY-PLANT, N Ire; KISS-ME-QUICK, Dor, Som; LADIES-IN-THE-HAY, Wilts; LADY'S NEEDLEWORK, Som; MADDER (cf. French *petite garance*, 'little madder', for several species of *Asperula* used in dyeing), Wilts; NEW-MOWN HAY, Som, Notts; RICE FLOWER (? not in reference to the flowers, but 'rice' in sense of brushwood, ME *ris*, OE *hrīs*, cf. *Galium cruciata*), Som; ROCKWOOD, Dev; SCENTED HAIRHOOF, Yks; STAR-GRASS (from the whorled leaves. Old botanists' name *stellaria*), Cumb, N'thum; SWEET-GRASS, Berw; SWEET HAIRHOOF, Yks; SWEETHEARTS, Som; WITHERIPS, Banff; WOODREP, Scot; WOOD-ROWELL, Yks; WOODY-RUFFEE, War, Yks.

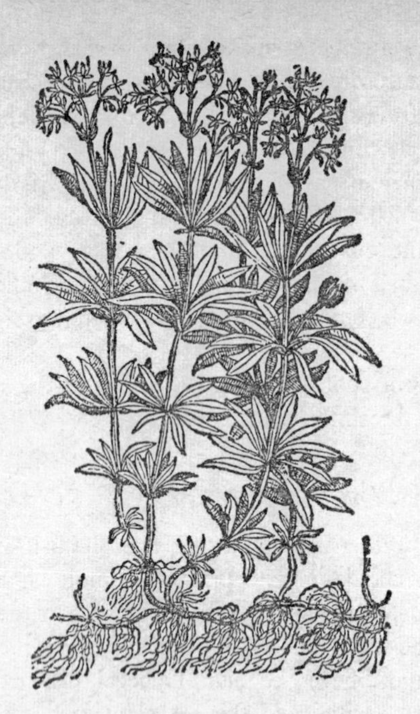

63 Woodruff *Asperula odorata*

For Woodruff there has always been a welcome – for its china flowers and new green leaves in the copse, in the new summer, Kiss-me-quick, Sweethearts, Ladies-in-the-Hay. A song of the late thirteenth or early fourteenth century begins:

> Lenten is come with love to toune
> With all blosmen and with briddes roune
> That all this blisse bryngeth.
> Dayes-eyes in this dales,
> Notes suete of nyhtegales;
> Uch fowl song singeth.
> The threstelcoc him threteth oo;
> Away is huere wynter woo
> When woderove springeth.

Woodruff develops a sweet smell of hay as it dries, and in the fifteenth century 'wodrove' garlands were hung up in the churches – for instance, on 11 June, the feast of St Barnabas (2); and Gerard wrote of the sweet-smelling garlands and bundles of Woodruff brought into the house. French and German names include *fleur de la Ste-Vierge* and *Mariengras*. Floors were strewn with it, beds were stuffed with it, it was placed among the linen, it was put into wine, 'to make a man merrie', according to Gerard. So in the full sense it was a herb of cordiality.

Woodruff entered the complex remedies of the English medicine-man, against headache, for example, in the *Lacnunga* (42, 88), it opened the liver and the gall-bladder (80, 141), it dispelled melancholy; and in France and Great Britain, especially in the Highlands, Woodruff tea has been popular for sweating, against colds, consumption, etc. (29, 85). Tea of the dried flowers is, in fact, delicious.

In OE Woodruff is *wudurofe*, 'wood', with a plant name which perhaps implies roving, creeping, spreading. One of its mediaeval names was Herb Walter, possibly after the French physician of the thirteenth century, Walter Agilor (129).

2. Squinancywort. *Asperula cynanchica* L. 40, H 8

Local names. QUINSYWORT, SQUINANCYWORT, Som; SHEPHERD'S BEDSTRAW, Glos; WAXFLOWER, Dor.

Every time a botanist journeyed from London to Bath, he was tempted to get down from his horse and climb Silbury, as Thomas Johnson and his friends climbed it in 1634 (see *Phyteuma tenerum*). The Flemish botanist de l'Obel must also have been up this '*acclivem cretaceam et arridam montem arte militari aggestum*', this 'steep chalky dry hill raised by military art', as he called it in his *Stirpium adversaria nova* in 1570. On Silbury he found a plant blossoming in July and August which seems to have been *Asperula cynanchica* and which he called *Anglica Saxifraga* (41) – the first record for Great Britain.

Squinancy is the quinsy, sore throat, and this waxy-flowered little perennial of the downs made an astringent gargle.

3. Crosswort. *Galium cruciata* (L.) Scop. 95, H 2

Local names. LADY'S BEDSTRAW (see *Galium verum*), Yks; MAIDEN'S HAIR, N'thum; MAYWORT, Som; RICE (rice = brushwood, OE *hrīs*, see *Asperula odorata*), Dor.

Gerard was the discoverer of Crosswort in Great Britain, during one of his expeditions to the heath and the gravel pits and the country village of

Hampstead: 'I found the same growing in the churchyarde of Hampsteede neere London, and in a pasture adjoining thereto by the mill.'

Crosswort, as he described it so well, is 'a lowe and base herbe, of a pale greene colour, having many square, feeble rough stalks full of joints or knees, covered over with a soft downe: the leaves are little, short, and smal, alwaies fower growing togither, and standing crossewise one right against another, making a direct Burgunion crosse: toward the top of the stalke, and from the bosome of those leaves come foorth verie many small yellow flowers, of a reasonable good savour.'

It was long in repute as an inward or outward vulnerary, taken also in wine against ruptures (141). Another name for *G. cruciata* is Mugweed.

4. Lady's Bedstraw. *Galium verum* L. 112, H 40

Local names. A-HUNDREDFALD, N Eng; BROOM, Shrop; CHEESE RENNET, Som, War, Cumb, Ire; CHEESE-RUNNING, S Eng, Ches; CREEPING JENNY, Som; FLEAWEED, Suff; GOLDEN DUST, Som; HALFSMART (i.e. arsesmart, ? from use as a flea-weed – see *Polygonum hydropiper*), Bucks; HUNDRED-FALD, N'thum.

JOINT-GRASS, Midlands, N Eng; KEESLIP (i.e. cheeselip or rennet), Scot; LADY'S BED, Dev, Scot; LADY'S BEDSTRAW, Dev, Wilts, Notts, Yks, Ork; LADY'S GOLDEN BEDSTRAW, Yks; LADY'S TRESSES, Som; MAIDEN'S HAIR, Yks, N Eng; RENNET, Kent, Herts, Cumb; ROBIN-RUN-THE-HEDGE, Lancs, Yks; STRAWBED, Dev.

A Manx name is *lus-y-volley*, herb of the sweet smell.

'O perilous fyr, that in the bedstraw bredeth', Chaucer wrote about the young wife, her lover, and the elderly husband. Bedstraw was straw, or some other dry plant, covered by a sheet, in lieu of our permanent mattress; a bed easily burnt and renewed when it became too verminous. The phrase 'in the straw' meant 'in childbed' (*Oxford English Dictionary*), and on a bedstraw of bracken and *Galium verum*, according to a mediaeval legend of northern Europe, the Virgin lay during the Nativity. Bracken refused to acknowledge the child and lost its flower, *Galium verum* welcomed the child and, blossoming at that moment, found its flowers had changed from white to gold. A legend rather of Germany and the Netherlands than of Great Britain, which borrowed the name from Germany. Another version says that *Galium verum* was the only plant in the stable which the donkey did not eat. In a valley of the Sudetenland women put Lady's Bedstraw in their beds to make childbirth easier and safer, and

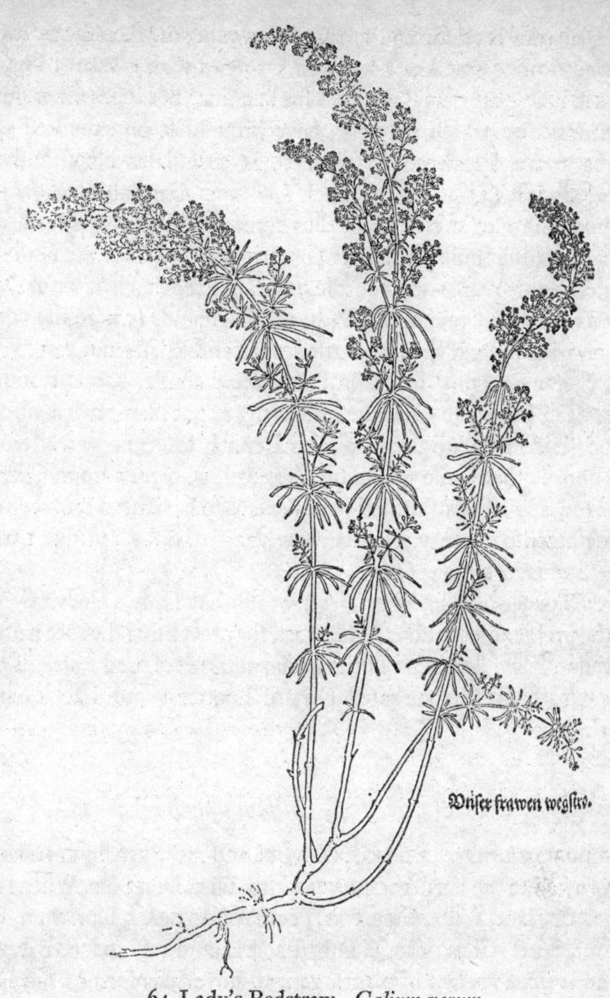

64 Lady's Bedstraw *Galium verum*

since women who have just had children are open to attack from demons, in that state they will not go into a house unless they have some Lady's Bedstraw in their shoes (Marzell, *Wörterbuch der Deutschen Pflanzennamen*, 13, 1954). There is no record of 'Our Lady's Bedstraw' for *Galium verum* until 1527, when it occurs in Laurence Andrew's translation of the *Liber de arte distillandi*, written by the German physician, Hieronymus Braunschweig. The German names include *Unser Frauen Bettstroh*, *Marienbettstroh*, and *Herrgottsbettstroh*.

Lady's Bedstraw is elastic and soft; in the evening or when the air is damp, its myriad flowers smell of honey, and it dries with a scent of hay, like Woodruff. These are genial qualities for bedding; but it has other qualities – an astringency, which may also have brought it into the bed against fleas (see above, Fleaweed, Halfsmart). It coagulates blood and was a common styptic (141) – the French *bon sang*, German *Blutstiel* – and throughout Europe for centuries it has curdled milk for making cheese. It was the *galion*, the 'milk plant', of Dioscorides, who said it was a substitute for rennet. So also *caille-lait*, 'curdle milk', in French, *melklöbe* in Danish, and *lus an leasaich*, 'rennet plant', in the Highlands (where the curdling was done with the addition of nettle leaves and salt) – not to repeat the various cheese or rennet names in English and similar names throughout Europe. 'The people in Cheshire, especially about Namptwich where the best Cheese is made, do use it in their Rennet, esteeming greatly of that Cheese above other made without it' (Gerard, who grew up in Cheshire). The use of Lady's Bedstraw in this way seems to have been rare by the end of the eighteenth century. Highlanders also derived a red dye from the creeping stems or stolons (119).

Robert Turner (*Botonologia*, 1664) wrote that Lady's Bedstraw 'challenges the preheminence above Mugwort, for preventing the sore weariness of Travellers. The decoction of the herb and flowers used warm, is excellent good to bathe the surbated Feet of Footmen and Lackies in hot weather'.

5. Goosegrass, Cleavers, Hayriff. *Galium aparine* L. 112, H 40

The common names are CLEAVERS, CLAVERS, and more usually CLIVERS, used from Cornwall to the border; CLIDERS, Corn, Dev, Dor, Som, Wilts, Hants, I o W; CLITES, Dor, Wilts, Glos, Oxf, Pemb; GOOSEGRASS, Dor, Som, Wilts, Glos, Suff, Norf, Ches, Yks, Cumb, Lanark; and HAYRIFF (OE *hegerife*, hedge, and *rife*, a species of plant), general in various forms from Devon to Yorkshire.

Local names. BEGGAR LICE, Wilts, Hants, Glos, Bucks, N'hants; BEGGAR-WEED, Notts; BLOOD-TONGUE ('being drawne along the toong it fetcheth bloud', Gerard), Ches, N'thum, Scot; BOBBY BUTTONS, Som; BURWEED, Bucks, Herts, Notts; BURHEAD, Notts; CATCH-GRASS, Ches; CATCHWEED, Som; CLADEN, Dor; CLAPPED-POUCH, Som; CLEGGERS, Yks; CLETHEREN, CLIDDEN, CLIDE, CLIMB, Som; CLINGING SWEETHEARTS, Wilts; CLING RASCAL, Dev; CLITCHE BUTTONS, Dev; CLY, Som; CLYDEN, Dor.

DOCTOR'S LOVE, DONKEYS, Som; GENTLEMEN'S TORMENTORS, Suff; GOOSE-

BILL, Som; GOOSE CLEAVERS, Lanark; GOOSE-SHEAR (i.e. grass), Som; GOOSETONGUE, Som, Ches; GOOSEWEED, Som; GOSLING-GRASS, Oxf, N'hants; GOSLING-SCROTCH, Ess, Norf, Cam; GRIP-GRASS, N'thum; GULL-GRASS (i.e. unfeathered gosling grass), Glos, Heref; GYE, Suff, Norf; HEDGEHOGS, Som; HUGGY-ME-CLOSE, Dor; JACK-AT-THE-HEDGE, Ire; JACK-RUN-THE-DYKE, N'thum; KISSES, Som; KISS-ME-QUICK, Dor, Som, Wilts; LIZZIE-RUN-THE-HEDGE, N'thum; LOVE, Som; LOVE-MAN (Greek *philanthropon*, in Dioscorides), Som; LOVER'S KISSES, Som; LOVER'S KNOTS, Wilts.

MUTTON-CHOPS, Dor; PIGTAIL, Notts; PIN-BURR, Beds; RABBIE-RINNIE-HEDGE, Ayr; ROB-RUN-UP-DYKE, Cumb; ROBIN-ROUND-THE-HEDGE, Ayr; ROBIN-RUN-THE-HEDGE, Som, E Ang, War, Lancs, Cumb, N'thum, Scot, N Ire; ROBIN-RUN-THE-DYKE, Cumb, Lakes, N'thum; ROBIN-RUN-UP-DYKE, Cumb.

SCRATCH-GRASS, Herts; SCRATCHWEED, Cam, Hunts, N'hants, Lincs, Yks; SCURVY-GRASS (so used), Ches, Yks; SNARES, Yks; SOLDIER'S BUTTONS, Cumb; STICK-A-BACK, Ches, Lancs, Cumb; STICK-BUTTONS, STICK-DONKEY, Som; STICKLEBACK, Ches; STICKYBACK, Cumb; SWEETHEARTS, Dor, Som, Wilts, Oxf, Yks; TETHER-GRASS, N'thum; TONGUE-BLUIDERS (see BLOOD-TONGUE), N'thum, Berw; TRAVELLER'S COMFORT, Wilts; TRAVELLER'S EASE, Wilts; TURKEY'S FOOD, Som; WILLY-RUN-THE-HEDGE, Ire.

Also DEVIL'S GARTER, STICKY BILLY, N Ire; GARBHLUS, Donegal.

'Therefore shall a man leave his father and his mother, and shall cleave unto his wife': the weed, the hedge plant, the plant of thickets, which *cleaves* stem, leaves, fruit, to the passer-by. Certainly a goose plant, eaten by geese, and chopped and given to goslings, in their early stages out of the egg, a common practice in the Isle of Wight, according to Bromfield's *Flora Vectensis*, 1856. A fifteenth-century leech-book mentions 'an erb that is cald clyvers that yonge gese eten' (54). Here the plant was recommended for piles; and in country medicine it has long been given for skin diseases, scurvy, ulcers, etc., and its roasted seeds have been used as a substitute for coffee.

Books often repeat that the leaves and the stems were used by shepherds to strain hair out of the milk. This is a statement made by Dioscorides.

6. Wild Madder. *Rubia peregrina* L. 25, H 17

Local names. CATCHWEED, Som; EVERGREEN CLIVER (see *Galium aparine*), I o W.

Like Cleavers, only bigger, tougher, and more rasping, darker green, and with all the extra personality of a perennial. A south-western plant on many cliffs and slopes, in Devon, Cornwall, Somerset, South Wales, etc., crawling among the brambles and the wind-bent blackthorn by many paths which drop to the beach. The true Madder, *Rubia tinctorum*, was grown in England back at least to the Anglo-Saxon period, as place-names indicate. The roots of Wild Madder do not give the same brilliant red, but they do give a rosy-pink which is much liked by the arts-and-craftsmen (179).

LXXXII. Caprifoliaceae

1. Danewort, Dwarf Elder. *Sambucus ebulus* L. 88, H 35

Local names. BLOOD HILDER, Norf; DANE BALL, Som; DANE WEED, Som, Suff; DANE'S BLOOD, N Ire; DANEWORT, Berks; DWARF ELDEN, Cumb; DWARFT ELDER, Dev; GROUND ELDER (Greek *chamaiacte*), I o W, S Eng; WALLWORT, Shrop; WATER ELDER (from Latin *Sambucus palustris*), N'hants.

By English legend Dwarf Elder grew spontaneously out of the blood of slaughtered Danes. It is not unreasonably a plant of blood. Luxuriant, growing up to five feet, the furrowed pithy stems turn red in September. So do the leaves and the pedicels below the fruit, which stain the fingers with black blood. Moreover a colony of *Sambucus ebulus* looks ragged as well as red: it has the curious dignity of a ragged regiment, bleeding and forlorn. The leaves go, the blood fades out (or soaks back into the ground), the dry stems break and rot, and there is no sign of the Elder until next year's stems and leaves emerge, and the pink buds form again and blow into white flowers with purple anthers. The leaves are more delicate and fern-like than those of the large elder. Flowers and leaves are foetid, but not to an extreme.

Remains of Dwarf Elder have frequently been discovered in neolithic and later prehistoric settlements in Switzerland (40). Perhaps it was collected to give a blue dye to cloth, perhaps some of its purging medicinal virtues were already known. Dioscorides described it accurately: it was a plant familiar also to the Romans, the Dacians, and the Gauls, it was valuable for purging, dropsy, gout, snake-bite, dog-bite, dyeing the hair black, and so on. There is no prehistoric record for this country, but the

Anglo-Saxons knew it, and made use of it. It was prominent in the herbals as a more energetic *Sambucus nigra*; it kept a place for a while in medicine (for a summary account, turn to Pechey's *Compleat Herbal*, 1694), and in folk-medicine. A way of treating piles among the Anglo-Saxons was to heat a quern stone, lay on top of it and underneath it Dwarf Elder, Mugwort, and Brooklime, and then to apply cold water, having your patient poised so that the steam 'reeks upon the man, as hot as he can endure it'

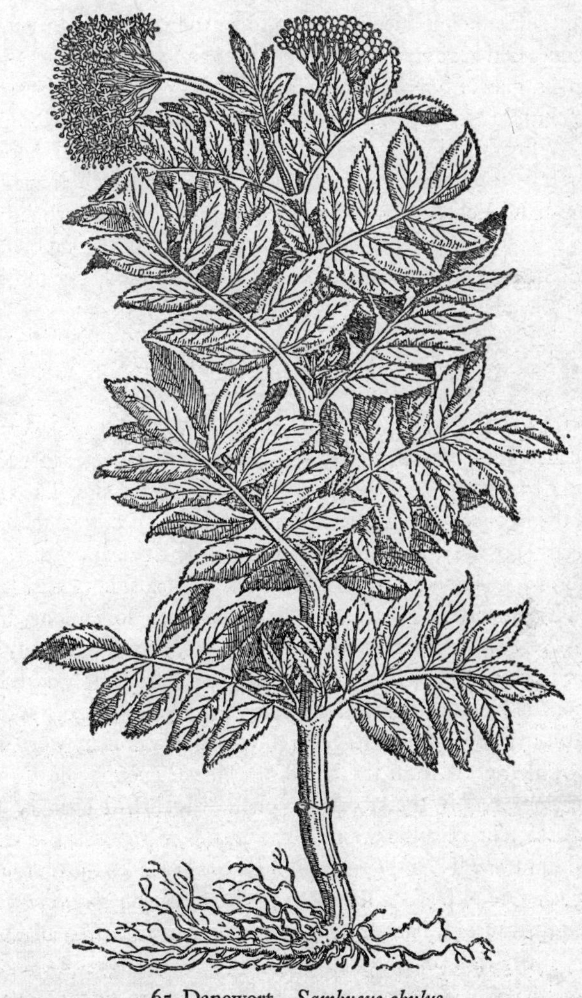

65 Danewort *Sambucus ebulus*

(42). According to the herbal of Dodoens, in Lyte's translation, it was burnt and the smoke drove away 'serpentes and other venemous beastes'. In France it is a St John's Herb to be gathered and purified – see *Hypericum perforatum* – on St John's Eve (188). Welsh names which show appreciation of its virtues are *ysgawen bendiged*, 'blessed Elder', and *ysgaw Mair*, 'Mary's Elder' – from Salesbury's sixteenth-century Welsh herbal (161).

Culpepper remarked that the Dwarf Elder grew wild in many parts of England, 'where being once gotten in to a ground, it is not easily gotten forth again'. Old scattered colonies endure; and very often, as though to proclaim its generation from the dead or from blood, Dwarf Elder is found in churchyards (see new or old county floras, e.g. those for Gloucestershire, Wiltshire, and Hampshire). However, medicinal herbs were deliberately planted in graveyards, in holy ground, which made them more powerful. Belief in the efficacy of the churchyard herb survives in Ireland (137). Compare, too, lines in one version of the ballad of *Tam Lin*. When the demon lover gets Lady Margaret with child, her mother says angrily:

> There grows ane herb in yon kirkyard
> That will scathe the babe away.*

Dwarf Elder sticks to such places – churchyards, waste corners, road verges, old banks and hedges, often upon a mediaeval site. The distribution and the habitats do not suggest a native, but a species which has been introduced – possibly, and probably, during the English centuries before 1066. Abroad it is one of the most frequent of roadside plants.

The OE name was *wealhwyrt*, which survived for a long while as Walwort. Here is a problem: *wealhwyrt* (*wealh*, a foreigner, usually applied to Celt or Roman) means 'foreigner plant', so did this imply a herb which grew from the blood of British foreigners, the Welsh, Cornish, etc.? And long afterwards were the British forgotten, and was Danewort used for the herb which grew from the blood of a newer enemy, the Danes? Or was *wealhwyrt* simply the 'foreigner plant' which had been brought in from overseas, just as the walnut, the *wealh hnutu*, was the exotic or imported nut tree? The *Oxford English Dictionary* takes the first view, not realizing what good reasons there are for believing *Sambucus ebulus* to be a denizen instead of a native. And the dictionary points to an OE form,

* Observe, though, that species found in churchyards are often the species found by roads – plants which colonize ground frequently disturbed.

wael wyrt, which could have been suggested by *wealh wyrt*, and could be translated 'slaughter plant'. However, centuries had to go by before this Walwort was indubitably connected with the dead Danes. There is no record of the name Danewort until 1538. Even then it may not have meant the herb of the Danes.

That a plant should have grown from blood, life giving rise to life, was not extravagant by older ideas. The notion of plants growing from the dead is world wide – European, Babylonian, for instance, South American, North American (175). Tobacco grows from the grave of a bad woman (Finland), lovers are buried side by side, rose-trees spring from each body and twine over the graves; or nettles grow from innocent blood, and so cannot be killed (Denmark). *Hypericum quadrangulum* has red juice, is called *mansblod*, and grows from blood (Finland and Sweden). Mandrake waxes from the blood of a criminal hanged and quartered, poppies have grown from the blood on the battlefield of Landen (1693), and from the blood which soaked Flanders again in the First World War. Yet why was it Danish blood in particular that nourished the Walwort? Supposing that *wealhwyrt* ever carried with it the implication of Welsh or British blood, why was there a switch to the blood of the Danes, from Walwort to Danewort? Or has confusion been at work, not least with the *Oxford English Dictionary*?

Let the botanists speak first of all. William Turner, in 1538, refers to *Sambucus ebulus* as 'Danwort' – here the word makes its bow – or Walwort, names he repeated in 1548, when he added: 'It groweth abroade in Cambryge fieldes in great plentie.' Bullein, in *The Booke of Simples* (1562), writes of 'Ebulus, called Walwort whiche in Suffolke they call Danes weede'. Turner, again, in his *Herbal* in 1568, has 'Daynworte'. Lyte, in 1578, calls it Walwort, Danewort, Bloodwort. Gerard, in 1597, gives Danewort, Wallwort, Dwarf Elder.

Not a whisper – so far, and in the botanical line – of the blood of Danes; and no word of scorn for a tradition that one or other of these writers would have condemned, if he had heard of it. Daynewort, Danewort – but could there have been another meaning and origin for the first syllable? Parkinson, as late as 1640 in his *Theatrum Botanicum*, wrote that the species was Danewort, because it purged, and so produced the 'danes', a strong flux or diarrhoea. By that time, however, the antiquaries had been at work, and they, in their own line, had started off the legend of Danewort and dead Danes. William Camden appears to have been the culprit, basing his crime, or his invention, upon the innocent statement of an earlier antiquary. He must have read the *Historia Regum Anglorum*, written in

Latin by John Rous of Warwick, who died in 1491. Under Canute the Danes had overrun Mercia. Describing this, John Rous had written that 'in villages near Warwick, where the people on the manors were slain, we can actually see the herb *ebulus*, or walwort, growing abundantly from their blood'. This *ebulus*, he goes on, in a piece of word fancy, 'originates in the course of nature from the *ebullition* of human blood'.

In a flash Camden must now have seen an 'explanation' for the alternative name of Danewort; though observe that Rous does not say the plant grew from Danish blood, but from English blood spilt by the Danes. In his new *Britannia* of 1607, Camden writes of *ebulus* growing from the blood of Danes at Bartlow in Cambridgeshire, for which reason it is called 'Danes-blood'. Three more antiquaries, Anthony à Wood, John Aubrey, and Tom Hearne, all swallowed their Camden. 'It is not unknowne but that great store of Daneworth and Walwort groweth at Seckworth [i.e. Seacourt, near Oxford], being testimonies embraced by most of the effusion of men's blood (and particularly of Danes' blood) wheresoever great quantities of it growes' (Wood, *Antiquities of the City of Oxford*, 1661–6). Aubrey wrote that the Danes were slaughtered by the women at Gatton in Surrey, and 'as a confirmation of this tradition the vulgar show the herb called Dane-wort in great plenty, which they fancy to have sprung from Danish blood' (*Natural History and Antiquities of the County of Surrey*, written 1673–92). Even more delighted Aubrey must have been to discover Danes-blood, as he called it this time, after Camden, at Slaughterford in Wiltshire, where (1974) it still grows. At Slaughterford 'there was heretofore a great fight with the Danes, which made the inhabitants give it that name' (*Natural History of Wiltshire*, written 1656–91). Tom Hearne (1678–1735) improved on the story of Seacourt and *Sambucus ebulus* by declaring that the Danes had certainly fought against Wytham Castle, not more than half a mile away (63).

Last of all, try Sweden, where again *Sambucus ebulus* is an old denizen, the isolated colonies of which persist vegetatively.* Linnaeus now takes his part; and one may add that Linnaeus must have read the *Synopsis stirpium britannicarum* of John Ray, in which *Sambucus ebulus* had been entered as Danewort '*quia e Danorum occisorum sanguine ortum fabulantur*' – 'because it sprang, so men say, from the blood of the slaughtered Danes'. Ray had lined up behind the antiquaries. In 1741 (*Kungl. Svenska Vetenskaps-Academiens Handlingar*, vol. ii) Linnaeus reported that the Dwarf

* For much of what follows I have to thank Dr Nils Hylander of the Botanical Garden at Uppsala, and Dr Herbert Gustavson of the Landsmåls-och Folksminnesarkivet, Uppsala.

Elder grew between Kalmar and Kalmar Castle: 'The simple people believe this plant to have grown up out of the blood of the dead, who have been slain in war here, and so they have called it *manna blod*' – 'man's blood' (cf. the Norfolk name Blood Hilder, or the Welsh *llysau gwaed gwyr*, 'men's blood plant'). Linnaeus tells the story again four years later (*Öländska och Gothländska Resa Åhr 1741*, published in 1745), by which time it has grown, and the Danes have entered, and the Dwarf Elder is called *mannaort*, 'man's plant', as well as *mannablod*. 'According to the legend, *mannablod* exists at no place in the world except the castle at Kalmar, where it is said to have sprung from the blood of Swedes and Danes, who fought on this field in ancient battles.' In this new version, is there not a grain of romantic dishonesty? A grain of influence from Rous and Camden, via John Ray?

A later botanist, Elias Fries (1794–1878), seems to have imported Danes' Blood as a name into Swedish. He told Linnaeus's story over again, added *danskablod* to the Swedish names, and, as if he had read the *Natural History of Wiltshire*, continued: 'Quite the same story is told in England, where the plant is called Danes-blood, at a village near Chippenham where there was once a battle with the Danes' (*Kritisk Ordbok öfver Svenska Växtnamen*). Some of this evidence from Sweden was collected and misapplied by the *Oxford English Dictionary* to bolster its own case: the dictionary even declared that *Sambucus ebulus* was also known in Sweden as *valört*, i.e. *waelwyrt*, i.e. 'slaughter wort', though the Swedish word is *vallört*; it is the ordinary name for Comfrey, and it has nothing to do with slaughter.

A long festival of popular and romantic etymology – *ebulus* (Pliny's name for Dwarf Elder) and *ebullire*, to bubble up or bubble out, as of blood, John Aubrey's slaughter and Slaughterford (which really means the 'sloe-tree ford'), *valört* and slaughter wort, and above all, Danes and diarrhoea. The solution seems plain. Folklore and medicine were muddled up: an old tradition maintained that *Sambucus ebulus*, blood-red in the autumn, sprang from human blood, as a plant which purged it became known also as Danewort, the plant which produced the 'danes' or diarrhoea; antiquaries catch hold of the name and the legend, misinterpret the one and enrich the other; and the enriched or altered legend becomes a new article of popular faith. Possibly the 'danes' of Parkinson's *Theatrum Botanicum*, so fiercely induced by taking Dwarf Elder, and the *Dane* of Danewort are both the sixteenth-century word 'dain' for an unpleasant smell (whether, in this case, of the plant or the consequence). Other plants as well have been allowed to share the name or the notion. *Anemone pulsatilla* and *Campanula glomerata* also became Danes' Blood. In Northamptonshire *Eryngium campestre*, rare but flourishing at one time beside Watling Street, became Daneweed. In Berkshire it was believed not many years ago that Bloody Warriors – plants of *Fritillaria meleagris* – had each grown from Danes' blood, drop by drop. In Sweden the tale of *mannablod* and the Danes and the Swedes at Kalmar produced another tale of *Trollius europaeus* growing in a valley where Danish and Dalecarlian blood had been shed in battle.

And for an outcome in respectable verse, read again those lines from Fitzgerald's *Omar Khayyám*:

> I sometimes think there never blows so red
> The Rose as where some buried Caesar bled –

suggested to him in part by the Danes' Blood of *Anemone pulsatilla* seeping up on Fleam Dyke.

2. Elder. *Sambucus niger* L. 112, H 40

Local names. BOUN-TREE, BOON-TREE, N'thum, Scot; BOUR-TREE, Ches, Lincs, Lancs, Yks, Lakes, Dur, N'thum, Scot, N Ire; BORRAL, N'thum, Scot; BULL-TREE, Cumb; DEVIL'S WOOD, Derb; DOG-TREE, Yks.

ELDERN (OE *ellaern*), Wilts, and in general use; ELLER, Suss, Kent, Norf, Ches, Derb, Lincs, Yks, West, N Eng; ELLET, Suss; GOD'S STINKING TREE (since it was used for the Cross), Dor; JUDAS-TREE (since Judas hanged himself on it), Kent; SCAW, SCAW-TREE, SCAWEN, Corn; TEA-TREE, Som; TRAMMON, I o M; WHIT-ALLER, Som.

The Elder grows like a weed, it does not live to a great age, its young stems are not strong, it stinks, yet produces sweet-smelling flowers, and sweet, if cloying elderberries. It makes effective medicine and poor timber, it is neither bush nor tree, neither bad entirely nor entirely good. In country after country there are records of an elder-spirit or an elder-mother who *is* her own tree and protects it at the same time; and traces of the Elder's old power are not yet erased altogether from the English or Irish mind. Elder must not be burnt. If you put it on the fire, you will see the Devil (the Elder spirit originally?) sitting on the chimney-pot, or else the Devil will come down the chimney (e.g. in Warwickshire and Hampshire, *Folklore*, XXIII, 1911, and LV. i, 1944. And notice the Derbyshire name of 'Devil's Wood'). Woodmen still dislike cutting the Elder. In

Ireland the timber may be neither burnt nor used in the making of boats (137). 'It is a bad thing to give a man a scelp [i.e. blow] of that,' an Irishman remarked of Elder. 'If you do his hand will grow out of his grave' (*Irish Naturalist*, March 1914). It was dangerous or foolhardy to make Elder into a cradle. The child would sicken and the fairies would steal it.

Power against you could also be power for you: a question of who held the gun. In the seventeenth century the grazier driving cattle across country, or the carrier with his string of pack animals, would cut an elder stick, keeping a joint on either end, and would take it with him as an amulet against galling, and no doubt other evils (44). Hearse drivers recently are said to have favoured an elderwood handle for their whips, in their dangerous association with the dead (72); and warts and sorrows in the parish where I live are still transferred to elder sticks, which are then buried. Clay from beneath an Elder allays toothache (137), and so on.

Strictly nothing divided Elder's power in magic from its power in medicine. An extract of the berries, John Evelyn wrote, was 'a catholicon against all infirmities whatever'; and in his day an English translation appeared of a treatise on the Elder by the German physician Martin Blochwich, which is part reasonable practice, part magic – e.g. advice on amulets of Elder and the use of elder leaves on May Day against witches. From this *Anatomia Sambuci: or, the Anatomie of the Elder*, 1670, here is the recipe for 'a mucilaginous anodyne liquor': 'Of quick snails, newly taken out of their shelly cottages; of Elderberries dried in the oven, and pulverized; and of common salt, of each as much as you will; put it in the straining bagg, called *Hippocrates* sleeve, making one row upon another, so oft as you please; so that the first be of the snails, the next of the salt, and the last of the berries, continuing so till the bagg be full; hang it up in a Cellar, and gather diligently the glutinous liquor that distils out of it little by little.'

There appears to be evidence that elder-berries were used in Egyptian medicine (26). Elder has fallen from its high estate, yet at the far end of the skein from Egypt, Elder-flower water, Aqua Sambuci Triplex, is still a proper vehicle for eye and skin lotions (*British Pharmaceutical Codex*, 1949 ed.).

In charms, amulets, medicines, the power of the Elder was tapped and controlled; but it was still a fearful and rather unsatisfactory tree in its collapsing branches, its stink, and its weediness. Betwixt and between, the Elder 'tree' asked for explanatory legend. So Judas hanged himself upon an Elder, in token of which, beyond stench and poor timber, the ear of Judas, 'Jew's Ear', the fungus *Hirneola auricula-Judae*, grows from the bark. You can eat these gelatinous ears of Judas (which are quite good), and they were long made into a gargle for sore throats – in which there seems to be a slightly extravagant application of signatures, the stretched throat, the ears of the hanged, and the sore throat, added to the likeness between Jew's Ear and tonsil. The author of the fourteenth-century *Travels of Sir John Mandeville* claimed to have seen the Elder on which Judas swung, alongside the Pool of Siloam.

However, not only did Judas kill himself on an Elder: it was the tree upon which Christ was nailed. In the *Golden Legend*, the most popular of all romantic religious handbooks, or legendary handbooks, in the Middle Ages, Jacobus de Voragine says that before the Passion the Elder was a tree of shame, used for the execution of criminals; a tree of darkness, being dark and without beauty; a tree of death, since men were put to death upon it; and a tree of evil smell, because it was planted in the midst of rotting corpses. This fits the Elder, 'God's Stinking Tree' of Dorset. It would explain all that was good and all that was evil about the Elder, as in the verse given by Robert Chambers in his *Popular Rhymes of Scotland* (1847):

> Bour-tree, bour-tree, crookit rung,
> Never straight, and never strong,
> Ever bush, and never tree
> Since our Lord was nailed t'ye.

If Christ sanctified the actual tree which carried him, in popular legend its kind often suffered the consequences of an evil act.

For the whole complex of reasons, legendary and natural, the Elder is not often regarded without prejudice. 'Rugged and full of chinks,' said Gerard of its bark, 'and of an ill favoured wan colour like ashes.' Yet in bark and leaf and flower, in form and growth and the mixture of tones, the Elder develops an uncommon and unforgettable charm in the right surroundings. Old Elders against grey limestone are lovely as old olive trees on the Karst.

If you dislike most elderberry or elder-flower recipes, try elder-flower pancakes, from Austria – circlets of blossom held by the stem, dipped into batter, fried, and eaten with sugar. They are fragrant as pods of vanilla.

3. Wayfaring Tree. *Viburnum lantana* L. 51

Local names. COBIN-TREE, Dor; COTTONER, Kent; COTTON-TREE (from the tomentose underside of the leaves), Ess, Lancs; COVEN-TREE, Wilts, Bucks;

DOG-BERRY TREE, Lincs; DOG-TIMBER, Dev; DOG-WOOD, Som; RED ROYAL OAK, Dur; SHEPHERD'S DELIGHT, Dor.

TWISTWOOD (from twisting into whip handles), Hants; WHIPCROP, I o W; WHIPTOP, Dor; WHITEWEED, Wilts; WHITEWOOD, Dor, Wilts, Hants; WHITTEN-BEAM (from white underside of leaves, cf. *Sorbus aria*), Midlands, S and W Eng; WHITTEN-TREE, Berks, Midlands, S and W Eng; WHITNEY, Dev.

Gerard knew *Viburnum lantana* on the journey over the chalk from London to Canterbury. It was thought to be the Latin *viburnum*, from which the French *viorne* was derived. Gerard took *viorne*, *viorna* as something which ornamented the way, the road, *via*: so Wayfaring Tree. You can eat the fruit when they go black, with no ill consequence, but also, I think, no great pleasure. For us it is a shrub to admire for autumn effects. Leaves and berries have been used to make a gargle, to fasten the teeth, and settle the stomach, and the leaves to make a black hair dye.

4. Guelder Rose. *Viburnum opulus* L. 105, H 40

Local names. DOG-BERRIES, Ches, Cumb, Donegal; DOG ELLER (i.e. elder), Ches, Lancs; DOG-TREE, Yks; DOGWOOD, Lancs; GATTERBUSH, Kent; GATTERRIDGE (see *Euonymus europaeus*), E Ang; KING'S CROWN (from using it to crown the May King), Glos; MUGGETS, MUGGET ROSE, Dev, Som; SNOW TOSS, Som; STINK-TREE (from the over-ripe berries), I o W; TRAVELLER'S JOY, N Ire; WATER ELDER (cf. the sixteenth-century *Sambucus montana aquatica*), Som, Berks, N'hants, Banff; WAYFARING TREE, N Ire; WHITE ASH, Dor; WHITE ELLER (i.e. elder), Ches; WILD PINCUSHION-TREE, War.

Names of the garden form, *Viburnum opulus* var. *roseum* L.: LADY'S PINCUSHION, War; LOVE ROSE, Herts; MAY BALL, Dor, Suff; MAY ROSE, MAY TOSSELS, Som; MAY TOSTY (for tosty, tisty-tosty, see *Primula veris*), Corn, Dev, Som; PINCUSHION-TREE, Bucks, Oxf, Heref, War; QUEEN'S PINCUSHION, N'hants; SNOWBALL, Dev, Som, Wilts, Hants, Oxf, N'hants, War, Shrop, Ches, Notts, Lincs, Lancs, Yks; TISTY-TOSTY, Dev, Som, Wilts; WHITSUN BALL, Som; WHITSUN BOSS, Glos; WHITSUNTIDE BALL, Som; WHITSUN TASSELS, Som.

Properly the name Guelder Rose belongs to the variety *roseum*, the familiar garden bush, planted so often near the front door of farmhouses and cottages, and splendid with its snowballs of blossom in May. These snowballs consist altogether of sterile flowers – a drawback of the virtue, since the Guelder Rose sets no fruit. Probably the variety was cultivated first at Guelders on the frontier between Germany and Holland:

> *Sambucus* too from *Gueldria*'s Plains will come,
> Drest in white Robes she shows a Roselike bloom,
> Be kind, and give the lovely Stranger Room. (150).

Gerard knew it as Rose Elder or Gelders Rose, admiring its 'goodly flowers of a white colour, sprinckled or dashed heere and there with a light and thinne carnation colour'. The flowers, he went on, had been doubled, 'by Art as it is thought'.

For *Viburnum opulus* Gerard's names were Marrish Elder, Water Elder, Whitten Tree, Ople Tree, Dwarffe Plane Tree. Flowering at the same time as Elder, it can fairly be considered one of the most beautiful of all native shrubs. The larger, outer flowers, which are sterile, are brilliant as new whitewash in the sun. The fertile flowers, pinkish-white in bud, then rather yellow or creamy, produce the red and glossy drupes. Along wet ditches, the flowers toss and sparkle in a May wind. Get them close and they smell like crisply fried, well-peppered trout, if you can imagine that trouty, peppery smell with a touch of sweetness.

5. Honeysuckle. *Lonicera periclymenum* L. 112, H 40

Local names. BEARBIND, Ches; BIND, BINDWEED, Yks; BINDWOOD, Yks, Cumb; BUGLE BLOOMS, Dor; CAPRIFOY (cf. old apothecaries' name *caprifolium*), Dev; EGLANTINE, Yks; EVENING PRIDE, Dev; FAIRY TRUMPETS, Som; GOAT'S LEAF (German *Geissblatt*, whence *caprifolium*), Dev, Som; GRAMOPHONE HORNS, Som; GRAMOPHONES, Som.

HONEYBIND, Oxf; HONEYSUCK, Dor, Som, Hants, Yks; IRISH VINE, Ire; LADY'S FINGERS, Yks, Dor, N'thum, Rox; LAMPS OF SCENT, Som; PRIDE OF THE EVENING, Dev, Dor; SUCKLES, Som; SUCKLE-BUSK, Norf; SUCKLINGS, E Ang; SWEET SUCKLE, Som; TRUMPET FLOWER, Yks; WITHYWIND, Dev; WOODBIND, WOODBINE, Wilts, Glos, Oxf, Lincs, Yks, Cumb, Renf, etc. (a general name); WOODWIND, Glos, Shrop.

A sweet smell by the edge of Buttermere or another of the Lakes, and no sign of its origin till, after another fifty, or a hundred, or two hundred yards, you come to a small tree roped and festooned by Honeysuckle. So William Bullein, in his *Booke of Simples*, in 1562: 'Ah how swete and pleas-unt is Woodbinde, in woodes or arbours, after a tender soft rayne: and now frendly doth this herbe, if I may so name it, imbrace the bodies, armes, and braunches of trees, wyth his long winding stalkes, and tender leaves, opening or spreding forth his sweete Lillies, like ladies fingers, emong the thorns or bushes.' Woodbine, Honeysuckle, hugs more like a killing snake

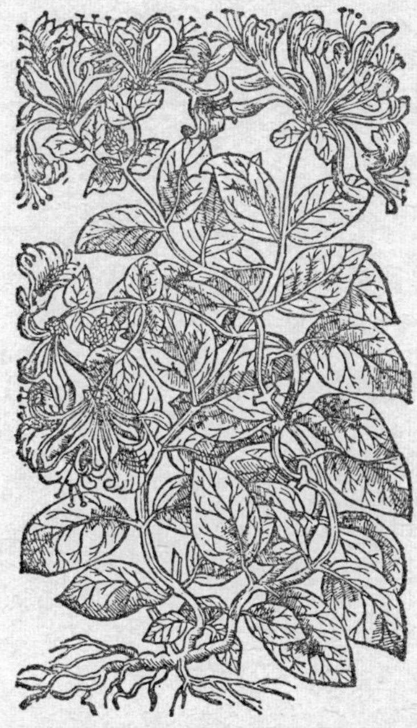

66 Honeysuckle *Lonicera periclymenum*

than a friend, often squeezing saplings into a spiral. But the fragrance on the damp air made the twist into a twist of love, whether for Bullein or Shakespeare:

> Sleep thou, and I will wind thee in my arms.
> So doth the woodbine – the sweet honeysuckle
> Gently entwist.

The corolla with its long tube suggests the lady's finger. The Honeysuckle was also identified with the third flower of the *Song of Solomon*: 'I am the rose of Sharon, and the lily of the valleys. As the *lily among thorns*, so is my love among the daughters.'

Honeysuckle was among the plants which averted the evil powers abroad on May Day, and took care of milk, the butter, and the cows. It was so i the Highlands and elsewhere (31a). A constricting plant could constrict i more ways than one. In the ballad of *Willie's Lady*, the vile witch trie various means of preventing the birth of the Lady's child, including 'bush o' woodbine' planted between her bower and the girl's. The Bell Blind, or house goblin, tells Willie how the spells are to be broken. Onc this restricting, constricting plant and other objects have been removed the child forces a normal entry into the world (36).

LXXXIII. Adoxaceae

1. Moschatel. *Adoxa moschatellina* L. 105, H 1

Local names. FAIRY'S CLOCK, Dev; GLORILESS (translation of *Adoxa*), Som GOOD FRIDAY, Wilts; GOOD FRIDAY FLOWER, Dor, Som; LADY'S MANTLE Som; MUSKWEED, Yks; WHISKERS, Som; WOOD ALONE, Dor; TOWN CLOCK Glos; TOWNHALL CLOCK, Cumb.

The little terminal head of green flowers is well expressed by 'Tow Clock' or 'Townhall Clock'. Four green clock-faces with yellow anther stare out in four different directions. A fifth stares up into the sky. Whe they are wet, or when the air is damp, the whole plant and the flower smell faintly of musk.

LXXXIV. Valerianaceae

1. Valerian. *Valeriana officinalis* L. 111, H 40

Local names. ALL-HEAL, Som; BLACK ELDER, Corn; CAT'S LOVE (since cat like the scent of the dry root), Wilts, Yks; CAT-TRAIL (of the root; used i trapping cats), Yks; CUT-FINGER, CUT-FINGER LEAF, Wilts; DRUNKEN SLOTS Som; FILAERA (i.e. valerian), N Eng, N Ire; HEAL-ALL, Corn, Oxf; HER BENNETT, Som; VALARA, Cumb, Donegal.

Valerian tea is a commoner drink in Germany, and elsewhere abroad than among ourselves. It is taken to calm the head, to prevent hysteria etc. All the same, Valerian keeps a strict medicinal place in the *Britis Pharmacopoeia* (1948), in the shape of a tincture from the roots and rhizo mes. Since it is a common wild plant, herb-women used to collect it in quantity, especially in Cranborne Chase, though it has also been grown for the market in several Derbyshire parishes (191, 73).

If you wish to try the effect of Valerian root on your cat, let the root dry first until the smell develops. Not a very pleasant smell: like new leather and yet foetid, but they liked it well enough to put Valerian roots among the linen in the sixteenth century (186). The smaller Marsh Valerian, *Valeriana dioica* L., 83, has been called Cherry-pie in Wiltshire, Gooseberry-pie in Devon, Dorset, Wiltshire, and Suffolk.

2. Red Valerian. *Kentranthus ruber* (L.) DC. 54, H 12

Local names. AMERICAN LILAC, Dev; BLOODY BUTCHER, Som; BOUNCING BESS, BOUNCING BETSY, Dev, Dor; BOVISAND SAILOR, Dev; CAT BED, Lincs; CONVICT-GRASS, Isle of Portland; DEVON PRIDE, Dev; DRUNKARDS, DRUNKARD'S NOSE, Som; DRUNKEN WILLY, Dev, Som; DRUNKEN SAILOR, Dev.

FOWEY PRIDE, Corn; FOX'S BRUSH, Som, N Ire; GERMAN LAYLOCK (i.e. lilac), Lincs; GIPSY MAIDS, Som; GOOD NEIGHBOURS, Som, Wilts, Glos, Oxf; GOOD NEIGHBOURHOOD, Wilts, Glos, Oxf; GROUND LAYLOCK (i.e. lilac), Lincs; KISS ME, KISS-ME-LOVE, Dev; KISS-ME-QUICK, Dev, Dor, Som, Wilts; KISSING KIND, Isle of Portland; LADY BETTY, Dor; LADY'S NEEDLEWORK, Corn, Som, Worc.

MIDSUMMER MEN, OLD WOMAN'S NEEDLEWORK, Som; PRETTY BETSY, Dor, Suss, Kent, Ess, Oxf, Ire; PRETTY BABY, Donegal; PRIDE OF PADSTOW, Corn; PRINCE OF WALES'S FEATHER, Dor; QUEEN ANNE'S NEEDLEWORK, Som; QUIET NEIGHBOURS, Wilts; RED MONEY, ROGUERY, Som; SAUCY BET, Corn; SCARLET LIGHTNING, Dev, Hunts; SOLDIER BOYS, Som; SWEET BETSY, Som, Kent; SWEET MARY, Bucks.

This cheerful and blowsy plant from central and southern Europe seems to have been introduced to English gardens in the sixteenth century. It was a great ornament to Gerard's garden, and he described it as 'not common in England'. That was 1597. By the early eighteenth century it had become well known. Miller, in his *Gardeners Dictionary*, observed that it seeded itself on the garden walls. Towards the end of the eighteenth century, in 1778, in William Hudson's *Flora Anglica*, it was recorded *in muris antiquis et ruderatis*, 'on old tumbledown walls', in Devon and Cornwall. But it still had not conquered. When Druce published his *Flora of Oxfordshire* in 1886, the Valerian was very rare on the walls of Oxford, and in Berkshire he did not find it wild till 1890.

To look well, it needs a more massive habitat than it commands in the small garden. Blowing a little drunkenly in the wind, how fine it is on the limestone walls of Plymouth. How this 'convict grass' enlivens the grey blocky, desolate scenery of the Isle of Portland, and how it reddens the grey sides of the Cheddar Gorge! Gerard called it Red Valerian (a translation of the *Valeriana rubra* first used by Conrad Gesner) and Red Cow Basil. He ascribed to it no virtues, mentioning only the 'excellent sweete savour of the roots'. In fact the young leaves (eaten in France and Italy), or rather the very young leaves, are not at all unpleasant (they are slightly bitter) if you mix them into a lettuce salad, or if you boil them and then shake them up with butter – though boiling does not destroy the bitterness altogether.

LXXXV. Dipsacaceae

1. Teasel. *Dipsacus fullonum* ssp. *sylvestris* (Huds.) Clapham 77
H 16

Local names. BARBER'S BRUSHES, Dor, Som, Wilts, Ess; BRISTLY HEAD, Berks; BRUSHES, Som, Wilts, Lincs; BRUSHES AND COMBS, Dor, Som; BUTTONS, CHURCH BROOM, CLEAVERS, Som; CLOTHES BRUSH, Som, Wilts; CLOTHIER'S BRUSH, Cumb; COCK'S COMB, Dor; COMB AND BRUSH, Wilts; DONKEY'S THISTLE, FAIRIES' BROOM, FAIRIES' FIRE, Som; GIPSY'S COMB, Yks; HAIRBRUSH, Som; JOHNNY-PRICK-FINGER, Dor; LADY'S BRUSH-AND-COMB, LADY'S BRUSHES, LITTLE BRUSHES, OUR LADY'S BASIN, PINCUSHION, POOR MAN'S BRUSH, Som; PRICKLY BEEHIVES, Dev; SHEPHERD'S ROD, SHEPHERD'S STAFF, Som; SWEEP'S BRUSHES, Dev; PRICKYBACK, Lincs; VENUS' BASIN, Som.

The Teasel is the teasing plant of the clothier (OE *taesan*, to tease) – the name properly belonging to *Dipsacus fullonum* ssp. *fullonum*, the Fullers' Teasel. In the Wild Teasel the bracts are straight, in the Fullers' Teasel they curve back, and so can be used to raise the nap on the newly woven cloth. The Wild Teasel, wrote William Turner in 1548, is called *virga pastoris* – 'shepherd's rod' – by 'the common Herbaries'. The Romans called the Teasel *labrum Veneris*, lip of Venus, or *lavacrum Veneris*, bason of Venus, from the leaves which join around the stem and hold rain water. William Coles sententiously explained matters. It was *labrum veneris* –

'because whores are as ready to be kissed as those hollowe leaves to receive the Raine, and afterwards to card and teare the estates, if not the bodies of their followers' (*Adam in Eden*, 1657).

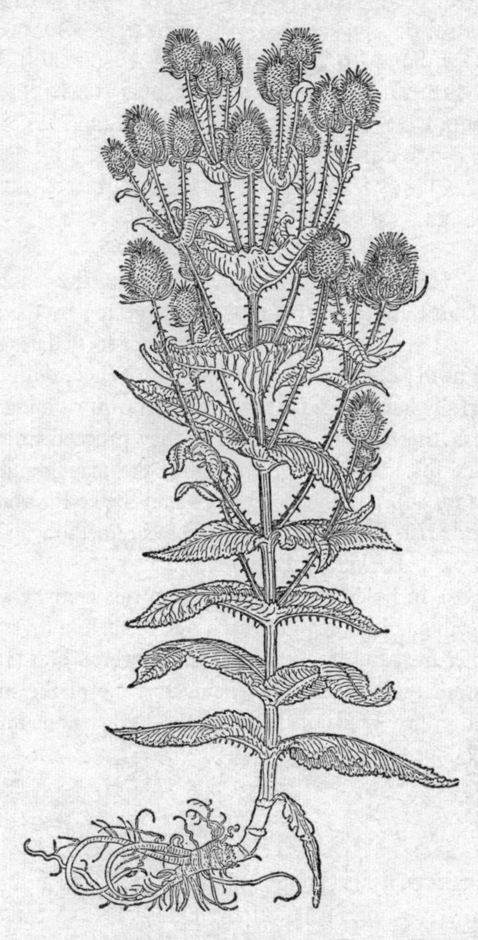

67 Fullers Teazel *Dipsacus fullonum*

2. Field Scabious, Gipsy Rose. *Knautia arvensis* (L.) Coult. 109, H 40

Local names. BACHELOR'S BUTTONS, Dev, Som, Wilts, Glos, Berks; BILLY BUTTONS, Som, Yks; BLACKAMOOR'S BEAUTY, Som; BLACK SOAP, Dev; BLUE

BONNET, Scot; BLUE BUTTONS, Wilts, Cumb; BLUE CAPS, Som; BLUE MEN, Berks; BROADWEED, Dev; CALSCALARY, Dev; CARDIES, Ire; CLODWEED, CLOGWEED, Bucks; COACHMAN'S BUTTONS, Som; CORNFLOWER, Som, Glos; CURL-DODDY (i.e. curly head), N'thum; CUSHIONS, Som.

EGYPTIAN ROSE (i.e. gipsy), I o W; GENTLEMAN'S PINCUSHION, Som; GIPSY ROSE, Dev, Som, I o W, Norf, Yks; LADY'S CUSHION, Kent; LADY'S HATPINS, Dev; LADY'S PINCUSHION, Som; MOURNFUL WIDOW, Dev, Som; MOURNING WIDOW, Som; PINCUSHION, Dev, Som, Wilts, Ess, Norf, Camb, Notts; PINS-AND-NEEDLES, PURPLE BUTTONS, ROBIN'S PINCUSHION, Som; SCABRIL, Ches; SNAKE FLOWER, SOLDIERS' BUTTONS, TEDDY BUTTONS, WILD ASTER, Som.

The *scabiosa herba*, the herb for scabies, the scab, the mange, the itch – Scabious for short. The use was perhaps suggested by the roughness of the stems. 'The Decoction of the Roots taken for forty daies together, or a dram of the Powder of them taken at a time in Whey, doth (as *Mathiolus* saith) wonderfully help those that are troubled with running or spreading scabs, Tetters, Ring worms, yea, though they proceed from the French Pox' (49). The juice was also given against the plague and to consume plague sores (27, 80). For the Somerset name Snake Flower see Gerard: 'The later Herbarists do also affirme that it is a remedie against the bitings of serpents.' As for Bachelor's Buttons (see also *Melandrium rubrum* and *Centaurea nigra*), in Belgium, and perhaps in this country as well, a girl would pick Scabious buttons, give each one of them a lover's name, and then choose her husband by the one which flowered best (160).

Scabious from the scab is a sad name for one of the most obvious, abundant, and pretty weeds of the cornfield. Perhaps one might abandon Scabious for 'Gipsy Rose'.

3. Devil's Bit. *Succisa pratensis* Moench. 112, H 40

Local names. ANGEL'S PINCUSHION, Dor; BACHELOR'S BUTTONS, Hants, Glos; BEE-FLOWER, Hants; BITIN BILLY, Donegal; BLUE BALL, Suss; BLUE BOBS, Hants; BLUE BONNETS, Som, Scot; BLUE BUTTONS, Som, War, Ches, Yks, Cumb; BLUE CAP, Som, Lancs, Yks, N Eng; BLUE HEADS, Shrop; BLUE KISS, Suss; BLUE TOPS, Worc; BUNDWEED, E Ang; CURL-DODDY (i.e. curly head), S Scot, N Ire.

DEVIL'S BIT, Corn, War, Worc, Shrop, Lincs, Yks, Scot; DEVIL'S BUTTON, Corn; FIRE-LEAVES (used to test moisture in hay. See *Plantago lanceolata*), Heref; GENTLEMAN'S BUTTONS, Shrop; GIPSY ROSE, Som; HARDHEADS,

Lancs; HOG-A-BACK, Cumb; LAMB'S EARS, War; LOUGH-SHULE, Ire; PIN-CUSHION, Som; STINKING NANCY, Ches; WOOLLY HARDHEAD, Corn.

Morsus diaboli was the name in mediaeval Latin, of which Devil's Bit(e) is the translation. The fifteenth-century *Ortus Sanitatis* tells a Byzantine story of the Devil acting with such power by means of the root that the Virgin Mary intervened and made this impossible. In his irritation the Devil bit the root off. Other explanations of the abruptly ending root stock (which is very easy to dig up and examine for oneself) were a little different. *The Grete Herball* of 1526 wrote of 'devylles bytte' that the Devil had envy of its virtue and bit the root to destroy it. So the plant cured less than it used to. Still in Culpepper's *English Physitian Enlarged* the ailments for which Devil's Bit was prescribed include the plague, fever, poison, the bite of venomous beasts, bruises, falls, clotted blood, swellings of the throat, wind, worms, wounds, scurf, dandruff, pimples, and freckles.

In Cornwall, if you picked this Devil's Bit or Devil's Button, the Devil would come to your bedside at night (192), and children in Fife spoke a rhyme to the Devil's Bit or Curl-doddy,

> Curl-doddy, do my biddin
> Soop my house and hool my midden,

as if the plant gave them power to summon a brownie to sweep the house and shovel the dung and drudge for them in the brownie's manner. For Bachelor's Buttons see *Knautia arvensis* and *Melandrium rubrum*.

LXXXVI. Compositae

1. Kew Weed. *Galinsoga parviflora* Cav. Introduced

Local names. GALLANT SOLDIERS, SOLDIERS OF THE QUEEN, London.

Far in towards Piccadilly Circus you can find the rather mean plants of *Galinsoga*, thin, long-legged, little-flowered daisies, ray flowers white, disc flowers yellow – annual, weedy, naturalized little cockneys in a waste corner or an uncultivated garden. *Galinsoga* (from Don Mariano Martínez de Galinsoga, Spanish botanist of the eighteenth century) suggested Gallant Soldier, perhaps ironically for the kind of private soldier Kipling described in his Indian stories. It is the least military of plants, though it scrounges and is content with little comfort.

This native of South America was introduced into Kew Gardens in 1796. By 1863 it was established between Kew and East Sheen and Richmond. It liked the asparagus grounds which supplied the London markets, and then began its march towards the West End.

2. Ragwort, Ragweed. *Senecio jacobaea* L. 112, H 40

Local names. AGREEN, Cumb; BALCAIREAN, Ire; BEAWEED, Scot; BENNEL, Ire; BENWEED, Scot, Ire; BINDWEED, Scot; BOHOLAWN (Irish *buadhghallan*), Ire; BOUIN, BOWEN, Cumb; BOWLOCKS, Scot; BUNDWEED, Suff, Scot; CAMMOCK, Hants; CANKER-WEED (applied in cancer), E Ang; CHEEDLE-DOCK, Ches; CROW-FOOT, Shrop; CRADLE-DOCK, Ches; CUSHAG, I o M; DEVILDUMS, Dor; DOG STALK, Yks; DOG STANDARD, Yks, I o M; DOG STANDERS, Worc, Yks, N Eng; ELL-SHINDERS, Berw.

FAIRY HORSE, Ire; FIZZ-GIGG, Berw; FLEA-NIT, FLEA-NUT, Ches; FLY-DOD,

68 Ragwort *Senecio jacobaea*

Ches; GIPSY, Som; GRUNDSWATHE, Cumb; GRUNSEL (see *Senecio vulgaris*), Cumb; JAMES'S WEED, Shrop; JAMES'S WORT, Shrop; KEDLOCK, Lancs; KEEDLE-DOCK, Ches, Lancs; MARE-FART, Ches; MUGGERT, Cumb.

RAGGED JACK, RAGGED ROBIN, Yks; RAGWEED, Herts, Suff, N'hants, Berw, Loth, Ayr, Aber; SCATTLE-DOCK, Lancs; SCRAPE-CLEAN, Lincs; SEAGRUM, I o M; SEGGRUMS, SEGGY, Yks; SLEEPY DOSE, Banff; STAGGER-WORT, STAMMERWORT, I o M; STINKING ALISANDER, N'thum, Stir; STINKING BILLY, Lincs; STINKING DAVIES, Fife; STINKING NANNY, Notts; STINKING WEED, Scot; STINKING WILLIE, Mor, USA; SUMMER FAREWELL, Dev, Glos.

WEEBO, Scot; YACKYAR, Lincs; YALLERS, Som; YARKROD, Lincs; YELLOW BOY, Donegal; YELLOW DAISY, Dor; YELLOW WEED, Berw.

Ragwort, Ragweed, was a plant used by fairies for locomotion – Irish fairies, and the fairies of the Highlands and Islands, who rode sticks of Ragweed across the straits from one island to the next; and also sheltered by Ragweed against the rain (137, 31b). Then as fairies decayed and dwindled, and as the appalling witch mania grew in the sixteenth and seventeenth centuries, so witches continued the old fairy acts, including this one of travelling first class by Ragweed. In Burns's poem the Devil and his hags and his male witches ride together:

> Let *Warlocks* grim, an' wither'd *Hags*
> Tell how wi' you on ragweed nags,
> They skim the muirs an' dizzy crags
> Wi' wicked speed.

A convenient plant to have chosen, common everywhere, stiff, tall, and mildly poisonous, and ragged in its leaves (so Ragwort, Ragweed). Cinnabar moths flutter around, the zebra caterpillars eat the leaves away to the stem; and the leaves stink when they are crushed. Calling Ragwort the *herba sancti Jacobi*, the herb of St James, gave it a cloak of virtue. St James's feast day is 25 July, when the Ragwort is in full blossom. It is St James's Herb still in several languages. The Scots are said to have called this plant Stinking Willie after William Duke of Cumberland. They maintained the ragwort was spread around by his forage in the Culloden campaign.

3. Oxford Ragwort. *Senecio squalidus* L. 54, H 6

'Squalidus' is an insult, since the Oxford Ragwort (like the Rosebay Willow-herb, with which it often grows) adds colour and cheerfulness to squalid surroundings, bomb ruins, railway tracks, railway embankments, and marshalling yards, a healthy and anything but squalid plant.

Historically there is one place for seeing this *Senecio squalidus*: on the lofty dark walls around the colleges of Oxford, around the gardens of Trinity, for example, or between the Bodleian Library and the Turl. It was on these walls that *Senecio squalidus* established itself first of all in the eighteenth century, an Italian, or Sicilian, from the volcanic cinders of Vesuvius and Etna (the Italians call it *Erva de S. Petro*, St Peter's Herb), which had been planted in the Botanic Gardens some time in the seventeenth century. It was first noticed on the walls outside the gardens in 1794.

In the last century and a half two things occurred to make the Oxford Ragwort's fortune – first the railways were built, then the bombs were dropped. Reaching the railway from the walls, the plant began to extend itself from Oxford as a passenger of the old Great Western Railway. It travelled towards London. The clinkers and the stones of the permanent way and the sidings proved as good a home as ever it had possessed. Claridge Druce, the Oxford botanist, has described a journey with some of the seeds, which floated into his carriage at Oxford and floated out again at Tilehurst (64). Indeed you can see this happening in railway carriages any day in high summer.

The spread was slow at first. The third edition of Sowerby's *English Botany* in 1899 only gave 'old walls and waste ground at Bideford, Devon; Oxford; Allersley Church, Warwickshire' as habitats – a little out of date, since by 1890 it had been reported at Swindon. By 1916 the Ragwort had reached Denbighshire. When the bombs laid waste so much of London and elsewhere, it had new ground to colonize and illuminate with its clear yellow blossom and healthy, glabrous leaves.

The Ragwort was noticed in Cork more than a hundred years ago, and it is spreading, though not so fast or widely, through Ireland.

Variable and adaptable, an eager and successful conqueror, this Oxford Ragwort must have its waste ground. It does not spread from town and railway on to village walls, for example. It has not become so universal as *Cymbalaria muralis*, the Ivy-leaved Toadflax, which has been spreading (with willing hands to plant it as well) since the early seventeenth century; nor is it quite as successful as the Red Valerian, *Kentranthus ruber*, which was an Elizabethan gardener's plant before it began to take possession of walls and cliffs. Yet it is seldom possible in any month to make a journey by train without seeing a plant or two cheerfully in blossom.

4. Groundsel. *Senecio vulgaris* L. 112, H 40

Local names. BIRDSEED, CANARY FOOD, CANARY SEED, Som; CHICKENWEED, Yks; CHICKWEED, Som; GRINSEL, Wilts; GROUNDSWELL, GROUNDWILL, Dev; GRUNDY-SWALLOW, Som; SENCION, SINSION, E Ang; SWALLOW-GRUNDY, N Eng; WATTERY DRUMS, Shet; YELLOW HEADS, Som.

Leave a piece of ground after the early potatoes have been removed, do not touch it for two or three months, and it will be swallowed by the Groundsel, close, thick, and untidy. Groundsel is from the OE *grunde-swilige*, 'ground swallower'. So, too, *Rumex obtusifolius* has been called the Land Robber. Turner in 1538 wrote of 'Grunswell' and 'Grundeswell', Bullein in his *Booke of Simples*, 1562, used the names Groundesyll and Senetion: 'The flower of this herbe, hath whyte hayre, and when the wynde bloweth it awaye, then it appeareth like a Bald headded Man, therefor it is called Senecio.' The Latin *senecio*, *senecion* (which gave the Old French *senechion*, Bullein's Senetion, and the Sencion of East Anglia) does probably come from *senex*, an old man.

'This Herb is *Venus* her Mistriss piece, and is as gallant an universal Medecine for all Diseases coming of heat whatsoever they be, or in what part of the body so ever they lie, as the Sun shines upon' (Culpepper).

5. Saracen's Consound, Broad-leaved Ragwort. *Senecio fluviatilis* Wallr. 45, H 12

A colony of *Senecio fluviatilis* by a river or a stream gives something of a shock. Yellow flowers; and obviously a Ragwort, but extra tall, leaves smooth, and toothed and like willow leaves – a plant of considerable charm and dignity. It was a medicinal herb introduced from the Netherlands or Germany in or before the sixteenth century – chiefly a wound-herb, known as Saracen's Comfrey or Saracen's Consound (for 'consound' see *Ajuga reptans*), two names used by Lyte in 1578. 'Among the *Germans*,' wrote Culpepper, 'this wound-herb is preferred before all others of the same quality.' He considered it as valuable a plant as Bugle or Sanicle.

6. Coltsfoot. *Tussilago farfara* L. 112, H 40

Local names. ASS'S FOOT (cf. French *pas-d'âne*), BACCY PLANT, Som; BULL'S FOOT, Dev, Bucks; CALVES' FOOT, Som; CLATTERCLOGS, Cumb; CLEAT, Lincs, Yks, Cumb; CLOTE, Norf, E Ang; COLTSFOOT, Oxf, War, Ches, Cumb, West, Berw, N Ire, etc.; COW-HEAVE, Selk; DISHILAGO (i.e. *tussil-ago*), Scot; DOVE-DOCK, Caith; DUMMY-WEED, DUMMY-LEAF, Herts.

FOAL'S FOOT, FOAL-FOOT (cf. French *pas-de-poulain*), Kent, Suff,

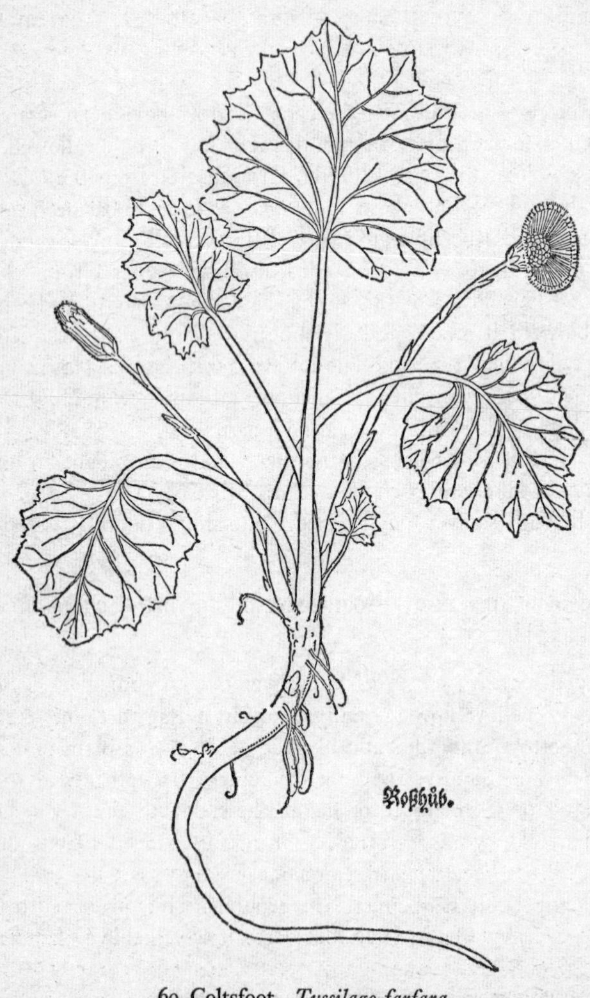

69 Coltsfoot *Tussilago farfara*

N'hants, Derb, Leic, Notts, Lincs, Yks, Dur, West, Cumb, N'thum, Berw, Rox; FOHANAN, Donegal; HOGWEED, Yks; HOOFS, Glos; HORSE-HOOF, Som, N'hants; POOR MAN'S BACCY, Som; SON AFORE THE FATHER (flower appearing before leaf), Cumb, Scot; SOW-FOOT, Yks; SWEEP'S BRUSHES, Som; TUSHALAGIES, TUSHALAN, N'thum; TUSHY-LUCKY GOWAN, Dumf; WILD RHUBARB, YELLOW STARS, YELLOW TRUMPETS, Som.

Bechion, the plant in Dioscorides taken to be Coltsfoot, was smoked against a dry cough. Coltsfoot is still smoked in herbal tobacco, Tussilaginis Folium – coltsfoot leaf – still has its place as a demulcent against coughs in the *British Pharmaceutical Codex* (1949). Pectoral beers and jelly and wine have been made of Coltsfoot leaves, which come up large and downy after the early appearance of the naked flowers. In tinder-box days the down was transformed into tinder – 'wrapped in a Rag, and boyl'd a little in Lee, adding a little Salt-Petre, and after dried in the Sun' (John Pechey, *The Compleat Herbal*, 1694). This tinder was long made and used in the Highlands (29).

Coltsfoot and similar names are from the shape of the leaves.

7. Butterbur. *Petasites hybridus* (L.) Gaertn., Mey. & Scherb. 110, H 40

Local names. BOG'S HORNS, Lincs; BOG RHUBARB, Som, Lincs; BURBLEK, West; BURN-BLADE (i.e. 'stream leaf'), Kirk, Wigt; BUTTERBURN, Bucks, Cam, E Ang; BUTTER-DOCK, Ches; CAP-DOCKIN, Yks; CLEAT, Lincs, Yks; CLOTS, Ches, Yks, N Eng; CLOUTS, Ches; CLUTS, N Eng.

DUNNIES, Hants; EARLY MUSHROOM, Dor; ELDEN, N'thum; ELDIN-DOCKEN, N'thum, N Eng, Berw, Rox; ELL-DOCKEN, N'thum; GALLON (Irish *gallán-mor*), Ire; GIPSY'S RHUBARB, Som; KETTLE DOCK, Ches; POISON RHUBARB, Yks; SNAKE'S FOOD, Som; SNAKE'S RHUBARB, Dor; SON BEFORE THE FATHER (flowers preceding the leaves), Clack.

TURKEY RHUBARB, UMBRELLA PLANT, Som; UMBRELLA LEAVES, Yks; UMBRELLAS, Som; WATER-DOCKEN, Cumb; WILD RHUBARB, Som.

Up in the north no rough little mountain stream is complete without Butterbur – so called from wrapping the big leaves around butter. 'Butter Burre doth bring foorth flowers before the leaves,' wrote Gerard, 'as Coltesfoot doth'; and when the flowering stems push through the soil in early spring, they look like small button mushrooms of a livid and unusual colour (so the Dorset name of Early Mushroom). The leaf, to quote Gerard again, is 'of such a widenesse, as that of itselfe it is bigge and large inough to keepe a mans head from raine, and from the heate of the sunne'. So *petasites* in Greek, the Dioscoridean name, from the kind of wide-brimmed hat called a *petasos.*

The dried roots, powdered and mixed into wine, were taken against fevers, especially the plague, to provoke perspiration and to drive 'from the hart all venom'. In Germany it is still called *Pestwurz* or *Pestilenzwurz.*

8. Elecampane. *Inula helenium* L. 73, H 25

Local names. ALLECAMPANE, Ches, Lancs, Yks; ELICOMPANE, Corn, Ches, Lancs, Yks; VELVET DOCK, I o W; SUNFLOWER, Stir; WILD SUNFLOWER, I o W.

Out on Rathlin Island, between Scotland and Ulster, you find the broad green leaves and the ragged sunflowers of Elecampane. It looks wild, and so on Rathlin do other herbs out of the physic garden – Evergreen Alkanet, Tansy, Alexanders, Sweet Cicely, Soapwort, etc. Out in the Atlantic on Inishboffin, off Connemara, you find Elecampane once again; and it grows, no less of an escape, across the Atlantic – along the roadsides of New England, to which it was taken by the settlers (148 111,). Yet the home of *Inula helenium* was probably in central Asia; from Asia it has worked its way in cultivation through the Near East and across western Europe and then over the Atlantic. Deep-rooted and enduring, a colony of Elecampane will outlast house or garden. De l'Obel and Gerard in the sixteenth century knew it not only as one of the nobler plants in the physic garden, but also much as we know it – in meadows and old orchards, long escaped, long established.

There seems no doubt that our Elecampane is the *helenion* or *inula campana* of Dioscorides, and it was probably the plant which the Anglo-Saxons called *elenan*, *elene*, *ellen*, *eolonan*, or *hors-eolene. The Grete Herball* of 1526 calls *Inula helenium* by several names – *enula campana* (the name in the shops), Elf docke, Scabwoort, and Horshale.

Names, in fact, had led to a rare old muddle. Read Bullein's *Booke of Simples*, 1562, on 'Helenium, called Enula campana, which we common plain people call Alacampane'. It was named Elenium, he says, 'of the lamentable and pitifull teares of Helena, Wife to Menelaus when she was violently taken away by Paris into Phrigia, having this herbe in her hande. Or as other doe say, this noble Helena made a goodly medecine of this herbe, agaynst the deadly Venome, or poysone of Serpentes.' This is two herbs in one, (i) the *helenion* or *inula campana* of Dioscorides, or the *inula* of Pliny, good against the cough and the bite of poisonous creatures, and (ii) Pliny's *helenium*, born of the tears of Helen – which must be a different plant, since it spread over the ground and had leaves like those of the Wild Thyme.

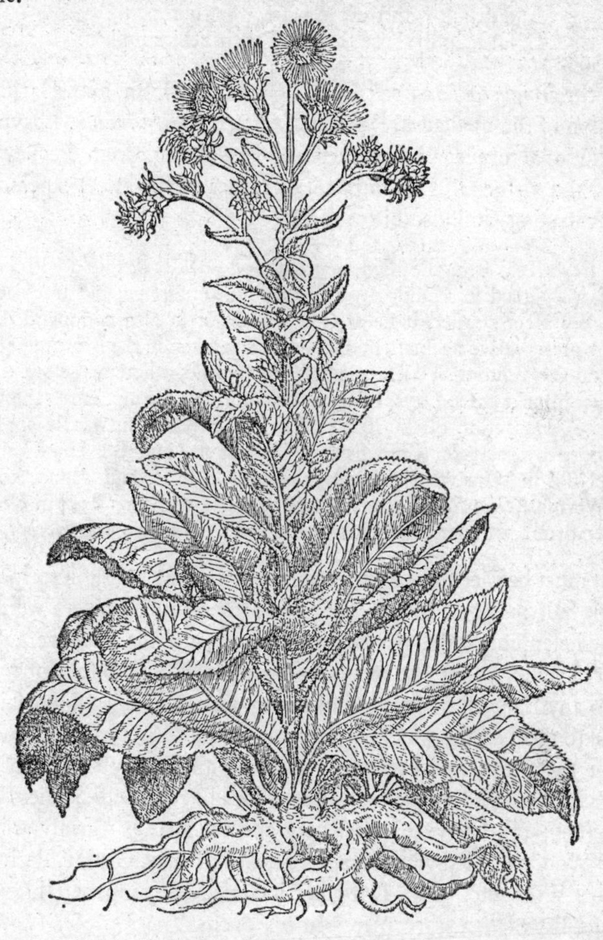

70 Elecampane *Inula helenium*

Inula also appeared to come from *hinnulus*, a young mule, and so was good for horse or mule as well as for man – *horselene*, *hors-eolene*; which again prompted a change into Horshele, Horseheal – *Inula helenium* becoming and continuing through the centuries, and down to the last century, a notable horse-medicine

> Ellecompane strengthens each inward part,
> A little loosenesse is thereby provoken,
> It swayeth griefe of minde, it cheeres the heart,
> Allaieth wrath, and makes a man faire spoken:
> And drunk with Rew in wine, it doth impart
> Great help to those that have their bellies broken.

That is the advice of *The Englishman's Doctor*, 1608, Sir John Harington's translation of the mediaeval *Regimen Sanitatis Salernitanum*. Leaving out the traditional use against snakes and snake-bite, John Pechey summarized the virtues of Elecampane, or rather of the sweet-smelling and aromatic-tasting, and also bitter roots:

'The fresh Root being candied, or dried, and powder'd, mix'd with Hony or Sugar, is very good in a Difficulty of Breath'ng, an *Asthma*, and an old Cough. Being taken after supper, it helps Concoction. It is also commended as an excellent preservative against the Plague. Being taken in the Morning, it forces Urine and the Courses. Half a pint of White-wine, wherein the slic'd Roots have been infus'd three Days, taken in the morning fasting, cures the Green-sickness. A Decoction of the Root, taken inwardly, or outwardly applied, is commended by some for Convulsions, Contusions, and the Hip-Gout. The Roots boyl'd in Wine, or the fresh Juice infus'd in it, and drunk, kills and expels Worms. Wine that is every where prepar'd with this Root in *Germany*, and often drunk, wonderfully quickens the sight' (*The Compleat Herbal*, 1694).

What must be Elecampane entered into many Anglo-Saxon recipes, half medical, half magical. In one complex prescription, prayers were sung over the helenium; it was marked with a knife, and the roots were dug up the next evening, after the medicine-man had been careful not to say a word to anything of an awful kind or any man (i.e. elf, goblin, fairy) he chanced to meet on the way. The elecampane root was then laid under the altar for the night, and eventually mixed with Betony and with lichen from a crucifix. The medicine was swallowed against elf-sickness or elf-disease (169, 42). Chew the fresh root and the taste is warmly aromatic.

With Woodruff, Walwort (*Sambucus ebulus*), Tansy, Orpine, Periwinkle, St John's Wort, etc., John Gardener included Elecampane ('horsel') in the short list of necessary plants in his poem, *The Feate of Gardening*, which he wrote about 1440 (*Archaeologia*, LIV).

9. Fleabane. *Pulicaria dysenterica* (L.) Bernh. 84, H 40

Local names. CAMELS, Dev; CAMMOCK, Hants; HARVEST FLOWER (it blossoms in August and September), Corn; JOB'S TEARS, Som; MARE'S FAT (cf. Mare-fart for *Senecio jacobaea*), E Ang; PIG-DAISY, Dor.

The yellow 'daisy' of the wide, damp, rushy road verges. To Gerard it was a Conyza, the *Konuza* of Dioscorides, which drove away midges and fleas.

10. Cudweed. *Filago germanica* (L.) L. 97, H 34

Local names. CHAFEWEED, N'thum, Berw; CLODWEED, Bucks; HEN AND CHICKENS, Yks; OLD OWL, Suff; OWL'S CROWN, Norf; SON-AFORE-THE-FATHER, Scot.

Turner, in 1548, wrote of this little plant that 'it groweth gladly where as turves have ben digged'. It was a Chafeweed because the soft leaves prevented chafing in horse or rider, a cudweed or quidweed, according to Leonard Mascal's *Government of Cattle*, 1662, because the plant, bruised with a little fat, was put into the mouths of cattle which had lost their cud or quid. Gerard called it 'Herbe impious' or 'wicked Cudweede', 'bicause the yonger, or those flowers that spring up later, are higher, and over top those that came first, as many wicked children do unto their parents.' So the name 'Son-afore-the-father'. Owl's Crown in Norfolk is also applied to *Gnaphalium sylvaticum* L. (108, H 33).

11. Cat's Foot. *Antennaria dioica* (L.) Gaertn. 94, H 39

Local names. CAT'S FOOT, Som, Rut, Yks; CAT'S PAW, Cumb; MOOR EVERLASTING, N Eng.

'The Cotton weede of the hils and stonie mountaines ... so exceeding white and hoarie, that one would thinke it to be a plant made of wool.' Though Gerard was not certainly referring to the Cat's Foot, his description serves, only one must add to the whiteness and hoariness the lovely pink of the bracts, on the female flower heads.

Cat's Foot makes an *immortelle*, an Everlasting (so it has been called the Moor Everlasting) – 'for such is the imperishable nature of our present species, that it retains a perennial bloom through successive years, and constitutes a principal ornament of the dried winter bouquet, for the vase of the saloon, or the head-dresses of our belles'.

12. Pearly Everlasting. *Anaphalis margaritacea* (L.) Benth. 21, H 6

Local names. EVER WHITE, N'hants; LADY-NEVER-FADE, Glos; OLD SOW, Norf.

This garden Everlasting was introduced very early from the New World, and was one of the first of the New World plants to escape. Gerard describes it neatly: 'There is a kind of Cotton weed being of greater beauty than the rest, that hath straight and upright stalks, three foote high or more, covered with a most soft and fine wool, and in such plentiful maner, that a man may with his handes take it from the stalke in great quantitie: which stalke is beset with many small, long, and narrow leaves, greene upon the inner side, and hoarie on the other side . . . The flowers do growe at the top of the stalkes in bundles or tufts, consisting of many small flowers of a white colour, and very double, compact or as it were consisting of little silver scales, thrust close togither, which do make the same very double. When the flower hath long flourished, and is waxen old, then commeth there in the middest of the flower, a certaine browne yellow thrum, such as is in the middest of the Daisie; which flower being gathered when it is yong, may be kept in such manner as it was gathered, I meane in such freshnesse and well liking, by the space of a whole yeere after in your chest or elsewhere; wherefore our English women have called it Live Long, or Live for ever, which name doth aptly answer his effects' (1597).

Before so very long the Pearly Everlasting settled into damp and dismal ground like its old habitat across the Atlantic. Edward Lhwyd, in 1696, found it growing in Monmouthshire, for at least twelve miles along the rocks of the Rhymney river, which runs from the Brecon uplands into the Bristol Channel. By 1696, too, it had been found in Essex. The special counties for Pearly Everlasting are Monmouthshire still, and Glamorgan in the south-west, Selkirk in the Lowlands, and Aberdeenshire in the north of Scotland.

13. Golden Rod. *Solidago virgaurea* L. 111, H 40

Local names. AARON'S ROD, Corn, Som, War, Shrop; CAST-THE-SPEAR, Dor; FAREWELL SUMMER, Som; GOLDEN ROD, Berks, Lancs, Yks, Cumb; WOUNDWORT, Som.

Aaron's Rod – the rod which budded, Numbers xvii. 8 – a name given also to Mullein (*Verbascum thapsus*) and Agrimony, two other plants with yellow flowers in long racemes.

Golden Rod had its fame as an excellent wound-herb, applied externally or taken as a drink for internal wounds; and since there was stabbing and thrusting enough in Tudor London, such herbs were expensive and in demand. Gerard remarked that Golden Rod, brought in from overseas, had once sold at half a crown an ounce. Then the plant was found 'even as it were at our townes end' in Hampstead Wood, 'neere unto the gate that leadeth out of the wood, unto a village called Kentish Towne' – and now 'no man will give halfe a crowne for an hundred weight of it: which plainly setteth foorth our inconstancie and sudden mutabilitie, esteeming no longer of any thing (how pretious soever it be) than whilest it is strange and rare'.

14. Sea Aster. *Aster tripolium* L. 84, H 28

Local names. BLUE CAMOMILE, BLUE DAISY, Kent; MICHAELMAS DAISY, Dev, Som, Ches; MUSK-BUTTON, Glos; SUMMER'S FAREWELL, Dor, Som.

This blue daisy, or blue-purple daisy, of the saltings was already taken into Elizabethan gardens, where it lived a long while, according to Gerard, but waxed 'huge, great, and ranke'. Then the Michaelmas Daisies that we know began to arrive from the New World, with *Aster paniculatus* (*A. lanceolatus* Willd.), which was here by 1640. However, the name Michaelmas Daisy in the garden no doubt applied as much to *Aster tripolium* (which flowers from July into November) as to the introduced kinds. In the wild it can make a handsome plant, especially when it grows on cliffs instead of in marshes. In marshland the tide too often washes its leaves with dirt and scum.

Gerard recommended the roots of Sea Aster as a wound-herb, and against dropsy and poisons.

15. Canadian Fleabane. *Erigeron canadensis* L. 46

A weed, and a North American weed, which has been with us for more than two and a half centuries, and which has spread around most of the globe. In John Ray's *Synopsis stirpium Britannicarum*, 1690, it was recorded as a weed near London, observed first of all by the doctor and naturalist, Sir Tancred Robinson. How had it arrived? The story has been told that the seeds dropped from the stuffing of a bird. Perhaps so. But in France the plant had been cultivated in the botanic garden at Blois by 1653, and it was naturalized in the Midi by 1674. Accident or botanical curiosity must have been the cause, since no one would introduce so mean a plant for the sheer ornament (96).

Erigeron canadensis did extremely well on the ruins and the waste left by the air raids.

16. Daisy. *Bellis perennis* L. 112, H 40

Local names. BABY'S PET, Som; BAIRNWORT, Yks; BAIYAN-FLOWER, Lancs; BANEWORT, Yks, Cumb, N'thum; BANWOOD, Yks; BENNERGOWAN, Dumf; BESSY BANWOOD, Yks; BESSY-BAIRNWORT, Yks; BILLY BUTTON, Som, Wilts, Shrop; BONE-FLOWER, N Eng.

CAT POSY (cf. German *Katzenblume*), Som, Cumb; CURL-DODDY (i.e. 'curly head'), SW Scot; DAY'S EYE, Som; DOG DAISY, Lincs, Lancs, Yks, Cumb; EWE-GAN, Yks; EWE-GOWAN, EWE-GOLLAN, N Eng, Scot; EYE OF DAY, FLOWER OF SPRING, Som; GOLLAND, Derb, N'thum, N Eng, Caith; GOWAN, Berw, Mor, Inver; GRACY DAISY, Dev.

INNOCENT, Som; KOKKELURI, Shet; LITTLE OPEN STAR, LITTLE STAR, Som; MARY GOWLAN, N'thum, Berw; MAY GOWAN, Berw, Ang; MISS MODESTY, Som; NAILS, Wilts; OPEN EYE, Som; SHEPHERD'S DAISY, SILVER PENNIES, N'hants; STAR, TWELVE DISCIPLES, WHITE FRILLS, Som.

The Daisy was a medicinal herb – particularly a wound-herb – yet 'also so well known to almost every Child,' wrote Culpepper, 'that I suppose it is altogether needless to write any Description.' The universal favourite of the cropped meadow, the 'day's eye' (OE *daeges-eage*). Two quotations may be allowed, out of hundreds – Chaucer, faithful to his books,

Save, certeynly, whan that the month of May
Is comen, and that I here the foules singe,
And that the floures ginnen for to springe,
Farwel my book and my devocioun!
  Now have I than swich a condicioun
That, of alle the floures in the mede,
Than love I most these floures white and rede,
Swiche as men callen daysies in our toun.
To hem have I so greet affeccioun,
As I seyde erst, when comen is the May,
Than in my bed ther daweth me no day
That I nam up, and walking in the mede
To seen this flour agein the sonne sprede,
Whan hit upryseth erly by the morwe;
That blisful sighte softneth al my sorwe.

Prologue : The Legend of Good Women.

Chaucer in the fourteenth century; in the nineteenth century Gerard Manley Hopkins, writing on 20 April 1874, of being surrounded by fields of a 'deep green lighted underneath with white daisies'.

17. Hemp Agrimony. *Eupatorium cannabinum* L. 102, H 40

Local names. ANDURION, Lancs; BLACK ELDER (*scawen du*, in Cornish), Corn; CROW ROCKET, Donegal; FILAERA, Berw; JACK O' LANTERN, Dor; RASPBERRIES-AND-CREAM (colour of flowers), I o W; VIRGIN MARY, Corn; THREAD-FLOWER, Som; WATER AGRIMONY, Ches.

' 'Tis Epatick and Vulnerary. 'Tis chiefly used for an ill Habit of Body; for Catarrhs, and Coughs; for Obstructions of Urine, and the Courses. It cures the Jaundice' (John Pechey, in *The Compleat Herbal*, 1694). For healing, drying, cutting, and cleansing thick and rough 'humors of the Brest', Culpepper held the Hemp Agrimony 'inferior to but few Herbs that grow' (49). A plant of no very strong personality or appeal, though mile after mile of roadside will be filled with the raspberry-and-cream of its flowers through the late summer and early autumn. In Irish it is *sceachóg Mhuire*, Mary's Little Bush. In his *Kreuter Buch* the great Hieronymus Bock (1498–1554) called it *Herba Sanctae Kunigundis*, the herb of St Cunigond, the sainted empress who died in 1033.

18. Stinking Mayweed. *Anthemis cotula* L. 79, H 35

Local names. BALDER'S BRAE, N'thum; CAMOVYNE, Scot; DEVIL DAISY, Wilts; DOG BANNER, DOG BINDER, Yks; DOG DAISY, Dev, Kent, Shrop, Yks; DOG'S FENNEL, Som, War, Midlands; DOG FINKLE (i.e. fennel), Yks, N Eng; DOG STINKS, Lakes; DOG STINKERS, Yks; HORSE DAISY, Bucks; JAYWEED, Suss.

MADDERS, Dor, Wilts, Hants; MAGWEED, Suss; MAISE, Dor, Shrop, Lincs; MAITHEN, Dor, Wilts, Hants, I o W, Glos, Suss, Worc, Shrop; MAITHEWEED, N Eng; MARG, Hants, Suss; MARGON, Hants; MATHERS, Dor, Som; MAYWEED, Dev, Hants, Suss, Kent, Mddx, Ess, Herts, Suff, Beds, Heref; MEADEN, Dor; MORGAN, Hants, I o W, Suss; MURG, Hants.

PIG'S CRESS, Som; PIG'S DAISIES, Dor; PIG'S FLOWER, Dor; POISON DAISY, Som, Suss.

In the days of hand husbandry few weeds were more hated. 'Dogge-fenell and mathes is bothe one, and in the commynge up is lyke fenell and beareth many white floures, with a yelowe sede: and is the worst wede that is, excepte terre [tare, *Vicia hirsuta*], and it commeth moste commonly, when great wete commeth shortly after the corne is sowen' [Fitzherbert's *Boke of husbandrie*, 1523). 'The May weed doth burn' – any farmer who read that in Tusser's *Five Hundred Pointes of Good Husbandrie*, 1573, knew what was meant: the leaves and the seeds blistered the hands, the feet, the ankles, and the bare chests of the men in harvest, whether they were sickling and scything or building the corn into stooks. The pain could be so bad that men were often unable to work for days (21). 'An unprofitable weede among corne,' as Gerard put it, 'and raiseth blisters upon the hands of the weeders and reapers.' The Anglo-Saxon farmer knew *Anthemis cotula* as *maegthe* – Mayfield in Sussex was the open ground where *maegthe* grew, Maytham in Kent the *hamm*, or close, of the *maegthe* (66). Turner, in 1538, called it 'Stynkynge maydweede', the *d* was dropped, and this plant of the harvest became the Mayweed, as though the plant of May.

Name after name is contemptuous – except for Balder's Brae, which apparently does mean the 'brow of Balder', the Scandinavian god. 'So fair and dazzling is he in form and features,' says the Prose Edda, 'that rays of light seem to issue from him; and you may guess his beauty when I tell you that the whitest of all plants is called Balder's Brow (*baldrsbra*).' *Balders braa* in Danish is *Anthemis cotula*, *Balderbraa* in Norwegian and *baldrsbra* in Icelandic is *Matricaria inodora*. The name (193) is not likely to be an old one. Hateful or no, Stinking Mayweed can make a fine pure sheet of blossom; it can whiten the stubble after the harvest has been cut and carried.

It crossed to America with the colonists. 'Some of our *English* Housewives call it *Iron Wort*, and make an Unguent for old Sores' (John Josselyn, in *New-Englands Rarities*, 1672).

19. Chamomile. *Anthemis nobilis* L. 57, H 21

Local names. CAMEL, CAMIL, Corn, Som; CAMMANY, Lancs; CAMOMINE, Shrop, Scot; CAMOVYNE, Scot.

The good herb after the evil one, taken to be the *anthemis* of Dioscorides, which 'others call *chamaimelon* [ground apple], since it has an odour of apples'. (There is more apple scent to the Wild Chamomile, *Matricaria chamomilla*.) 'It is of force', wrote Gerard, 'to digest, slacken, and rarifie ... wherefore it is a speciall helpe against wearisomnesse, it easeth and mitigateth paine, it mollifieth and suppleth.'

The Chamomile flowers of commerce, for Chamomile tea (delicious), shampoo, and poultices (all in the 1949 edition of the *British Pharmaceutical Codex*), are mostly from the double variety *flore-pleno*. Flowers of the wild plant are scarcely less good. Pick them off in July and August when the bracts are still green and the white ray-florets are just beginning to bend backwards. You can make your tea of the flower heads fresh or dried.

20. Yarrow. *Achillea millefolium* L. 112, H 40

Local names. ANGEL FLOWER, Som; BUNCH O' DAISIES, Dor; CAMEL, CAMMOCK, Dev; DOG DAISY, N Ire; GOOSE TONGUE (cf. German *Gänsezunge*), Som; GREEN ARROW (i.e. yarrow), Suff; HEMMING-AND-SEWING, Hants; HUNDRED-LEAVED GRASS, Berw; MELANCHOLY, Shet; MOLEERY-TEA, Caith; MOTHER OF THOUSANDS, Dor; NOSEBLEED (cf. French *saigne-nez*), Som, Suss, Suff, N'thum.

OLD MAN'S MUSTARD (i.e. devil's mustard), Lincs; OLD MAN'S PEPPER (i.e. devil's pepper), Som; SNAKE'S GRASS, Dor; SNEEZEWORT, Glos; SNEEZINGS, Glos; STANCH-GIRS (i.e. 'grass'), Scot; SWEET NUTS, Dor; TANSY, Ches; THOUSAND LEAF (translation of *millefolium*), Dor, Som, Ches, Lancs; THOUSAND-LEAF CLOVER, Berw; THOUSAND-LEAF GRASS, Staffs; TRAVELLER'S EASE, Wilts.

WILD PEPPER, Berw; WOUNDWORT, Som; YALLOW, Lancs; YARRA-GRASS, Ess; YARREL, Suff; YARROWAY, Norf, E Ang.

In medicine principally a wound-herb, as the English knew from the *Herbarium* of Apuleius Platonicus and from Dioscorides. 'It is said that Achilles the chieftain found it; and he with this same wort healed them who were stricken and wounded with iron,' says the Anglo-Saxon version of the *Herbarium*. 'For wounds which are made with iron, take this same wort, pounded with grease; lay it to the wounds; it purgeth and healeth the wounds' (42). *The Grete Herball* of 1526 says of *Achillea millefolium* (which it calls Yarowe, Carpenter's grasse, Bloudworte, and Myllefoly) that 'it is good to rejoyne and soudre wounds'.

But this Yarrow, this *gearwe* of the Anglo-Saxons, had a potency in other ways: it was powerful for or against evil. In France, in Ireland, it is one of the herbs of St John (see *Hypericum perforatum*). On St John's Eve the Irish hang it up in the houses to avert illness (137, 188). Alexander Carmichael, in *Carmina Gadelica* (vol. ii, 1928), gives several Gaelic incantations for plucking yarrow. Here is one of them – a woman's incantation – translated by Kenneth Jackson:

'I will pick the smooth yarrow that my figure may be sweeter, that my lips may be warmer, that my voice may be gladder. May my voice be like a sunbeam, may my lips be like the juice of the strawberry.

'May I be an island in the sea, may I be a hill in the land, may I be a star in the dark time, may I be a staff to the weak one: I shall wound every man, no man shall hurt me' (106).

An elf's herb, a devil's herb, giving also the power of divination. Put the leaves up your nose, and they can make it bleed, as Gerard remarked. So it foretold love in East Anglia:

> Yarroway, yarroway, bear a white blow
> If my love love me, my nose will bleed now.

21. Sneezewort. *Achillea ptarmica* L. 112, H 38

Local names. ADDER'S TONGUE, Aber; GOOSE TONGUE, N'hants, War, Shrop, Yks; HARDHEAD, Ayr; MOLEERY-TEA, Caith; OLD MAN'S (i.e. devil's) PEPPER-BOX, Som; PEPPER-GIRSE, Shet; SHOLGIRSE, SJOLGIRSE, Shet; SNEEZE-WORT, Som; WHITE WEED, N Ire; WILD-FIRE, Donegal.

Gerard supplied himself with Sneezewort from the heavy, low-lying meadows of Kentish Town. His was the first botanical record for the British Isles, and he maintained that the smell of *Achillea ptarmica* was enough to make you sneeze. But Sneezewort was chiefly employed against toothache, as a native substitute for the Pellitory of Spain, *Anacyclus pyrethrum*. The root had much the same effect. It tastes sharp and hot, and causes a flow of saliva.

The brown, or brownish-grey, and white of the flower heads make an unusual combination of tones.

22. Corn Marigold, Gold. *Chrysanthemum segetum* L. 111, H 40

Local names. BIGOLD, Som; BOODLE, Herts, N'hants, Suff, Norf; BOTHAM, Dor, Hants; BOTHEN, Hants; BOTHERUM, BOTTLE, Dor; BOZEN, Hants; BOZZEL, Wilts, Hants, Herts; BOZZOM, I o W; BUDDLE, Herts, Norf, E Ang, N'hants; CARR GOLD, Lancs; CORN MARIGOLD, Cumb.

FAT HEN, Hants; FIELD MARIGOLD, Shrop; GEAL GOWAN, GEAL-SEED, GIL GOWAN, Ire.

GOLD, Som, S of E Midlands, N'hants, Eastern Counties; GOLDEN CORNFLOWER, GOLDEN DAISY, GOLDEN FLOWER, Som; GOLDINGS, N'hants, Ches, Notts, Lincs; GOLLAND, GOWLAN, Yks, N Eng, N'thum; GUILLS, Som; MANELET, Scot; MARIGOLD, Ches; MARIGOLD-GOLDINS, N Ire; MARY-

GOWLAN, N'thum; SUNFLOWER, N'hants; TANSY, Glos; WILD MARIGOLD, N Ire.

YELLOW BUSSELL, Berks; YELLOW BOTTLE, Kent, Yks; YELLOW BOZZUM, I o W; YELLOW GOLD, Cumb ('yelwe golds' in Chaucer's *Knight's Tale*); YELLOW GOWAN, Caith, N Scot; YELLOW GULL, Cumb; YELLOW HORSE DAISY, Corn; YELLOW MOONS, Dor, War, Worc; YELLOW OX-EYE, Som, Yks.

A loathly weed with a flower of clear and exquisite yellow. 'Golds', wrote Fitzherbert in the *Boke of husbandrie*, 1523, 'hath a short iagged lefe, and groweth halfe a yarde hygh, and hath a yelowe floure, as brode as a grote, and is an yll wede, and groweth commonlye in barleye and pees.' The tale of the gold goes back much further, to court rolls of the fourteenth century ordering tenants to uproot '*quandam herbam vocatam gold*' ('a certain plant called Gold'), and to Anglo-Saxon place-names, such as Goldhanger in Essex, Golding in Shropshire, Goldor in Oxfordshire, Goltho in Lincolnshire, or Gowdall in the West Riding. Eilert Ekwall, in his *Concise Oxford Dictionary of English Place-Names*, takes the OE word *gold* to mean Marsh Marigold, *Caltha palustris*, which seldom fits the place-names.

Going back further still, it seems that *Chrysanthemum segetum* was anciently introduced, and that its home is western Asia (and possibly the area of the Mediterranean). It has been found in Scottish neolithic deposits, but no earlier, and so probably arrived with the neolithic introduction of agriculture.

John Parkinson, in his *Theatrum Botanicum*, 1640, said that Orpine leaves and Corn Marigold flowers were made into midsummer garlands (presumably for St John's Eve and St John's Day) and hung up on houses. Notice, by the way, not only the charming green and yellow of the Corn Marigold, but the oddly attractive scent of the flowers. The delightfulness of Golds hasn't gone unnoticed. See, for instance, the Gaelic love-keen *'S dubh a choisich mi'n oidhche*:

> Your hair was like the flower that grows in the barley,
> When you put a comb in it the sheen of gold could be seen.
> (Derick Thomson, *Introduction to Gaelic Poetry*, 1974).

23. Ox-eye Daisy, Moon Daisy. *Chrysanthemum leucanthemum* L. 112, H 40

Local names. CATEN-AROES, Lancs; BILLY BUTTON, Shrop; BOZZOM, I o W; BULL DAISY, E Ang, Ches, Yks, Cumb; BULL'S EYE, Corn, Som, Ches; BUTTER DAISY, Dor, Som; COW EYES, Corn; COW'S EYES, Som; CRAZY BETT,

Wilts; DEVIL'S DAISY, Mddx; DOG DAISY (cf. equivalent names in German, Dutch, Danish, Swedish), Dev, Som, Wilts, Bucks, Herts, Worc, Ches, Leic, Lincs, Lancs, Yks, Dur, Cumb, N'thum, N Ire; DOG FLOWER, Cumb; DRUMMER DAISY, DUN DAISY, DUNDER DAISY, DUNDLE DAISY, Som; DUTCH MORGAN, I o W; ESPIBAWN (Irish *easbog bán*, 'white bishop'), Ire; BISHOP'S POSY, Donegal.

FAIR-MAIDS-OF-FRANCE, Som, Wilts; FIELD DAISY, Dev; FRIED EGGS, Som, Wilts; GADJEVRAWS (Cornish *caja vras*, 'great daisy'), Corn; GIPSY DAISY, Dor, Som, Norf; GOLLAND, GOWLAN, N'thum; GRANDMOTHERS, Notts; HARVEST DAISY, Dor; HORSE BLOB, N'hants; HORSE DAISY, Corn, Dev, Dor, Som, Wilts, Suss, Kent, Bucks, Hunts, N'hants, Heref, N'thum; HORSE GOLLAN, HORSE GOWLAN, N'thum, Scot; HORSE PENNIES, Derb; LARGE DICKY DAISY, Ches; LONDON DAISY, Som.

MAISE, Lincs; MAITHEN, Wilts, Glos; MARGARET, Yks; MARGUERITE, Som; MATHER, Heref; MAUDLIN, Wilts; MAYWEED, Suff; MIDSUMMER DAISY, Som, Suss, War; MONNIES, Som; MOONS, Wilts, Glos, Berks, Bucks, Oxf, Ess, Norf, Camb, War; MOON DAISY, Som, Wilts, Glos, Surr, Berks, Oxf, Shrop, War, Worc, Ches, Yks; MOON FLOWER, Som, Worc; MOON PENNY, Ches; MOON'S EYE, MOTHER DAISY, MOWING DAISY, Som; MUCKLE KOKKELURI, Shet; OPEN STAR, Som; POOR-LAND DAISY, N'hants; POVERTY WEED, Ches.

RISING SUN, Som; SUN DAISY, Dor; THUNDER DAISY, Dev, Som; WHITE GOLDS, Lancs, Cumb; WHITE GOWLAN, N'thum; WHITE GULL, Cumb.

Moon, Moon Daisy, or Dog Daisy are more authentically English names. Ox-eye, or Oxe-eye Daisy, began as a book name in the sixteenth century, because this plant appeared to be the *bouthalmon* of Dioscorides. Only the later herbalists found medicinal uses for it. John Pechey, for example, after remarking that the flowers 'cast forth Beams of Brightness', had this much to say: 'The whole Herb, Stalks Leaves and Flowers, boyl'd in Posset-drink, and drunk, is accounted an excellent Remedy for an *Asthma*, Consumption, and Difficulty of Breathing. 'Tis very good in Wounds and Ulcers, taken inwardly, or outwardly applied. A Decoction of the Herb cures all Diseases that are occasion'd by drinking cold Beer when the Body is hot' (141).

Yet the Moon Daisy is not a favourite entirely upon the force of its appearance. Blooming at midsummer, it is one of the notable plants of St John in other countries, with names to match (such as *Johanneskraut*) in German, French, Dutch, Polish, Russian, etc. In the Tyrol, and in various parts of Germany, these daisies were hung upon houses, round the doors, on the roof, on the hay lofts, to keep away lightning; Austrian names for the plant include *Sunnawendl*, *Sunnwendbleaml*, 'solstice', 'solstice flower'. Elsewhere in Germany it is *Gewitterblume*, 'storm flower' (129). English names suggest we had the same beliefs – Dunder Daisy, Thunder Daisy, certainly; and perhaps Midsummer Daisy, Rising Sun, Sun Daisy.

For the plants of St John, see *Hypericum perforatum*.

24. Feverfew. *Chrysanthemum parthenium* (L.) Bernh. 108

Local names. ARSESMART (see *Polygonum hydropiper*), Yks; BACHELOR'S BUTTONS (see *Melandrium rubrum* and *Knautia arvensis*), Som; BOTHEM,

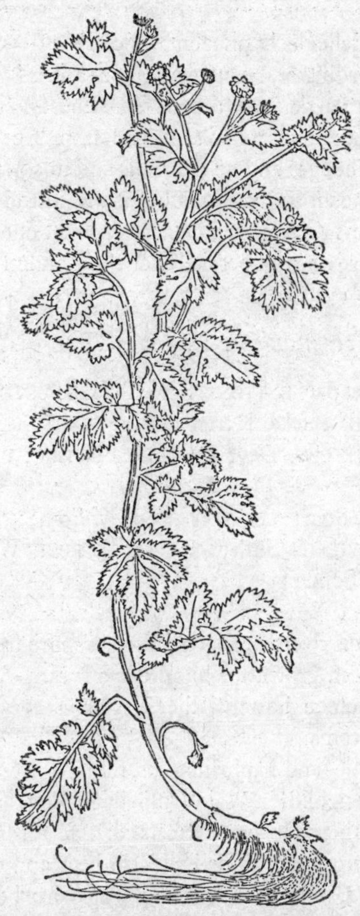

71 Feverfew *Chrysanthemum parthenium*

Corn; BUTTONS, Som; DEVIL DAISY, Som, Wilts; FEATHERFEW (in general use), FLIRTWEED, Dev; MADRON, Dor; MICHAELMAS DAISY, Dev; MIDSUMMER DAISY, Dev, Som; NOSEBLEED, Kent; STINK DAISIES, Som; WHITEWORT, I o W. The Cornish name was *les-derth*, 'ague plant'.

An old physic herb which was very widely grown throughout the British Isles and throughout Europe, partly as a feverfew – *quasi fugans febres*, for putting fevers to flight – partly as an emmenagogue and a plant against headaches, which, like aspirin, was required in every family – a herb doing 'all a bitter Herb can do' (141), in fact, the aspirin of the herbal era (see page 277).

The leaves are deliciously pungent. Once Feverfew has a hold, it persists in spite of weeding and contempt (though it is by no means an ugly plant). It survives particularly around farmhouses, in the English lowlands or in the mountains of Wales, and from Cornwall to Scotland. Gerard knew it in hedges and on old walls and in corners of the garden – precisely as one finds it today. No doubt it was introduced in the Middle Ages. Miller, in his *Gardeners Dictionary*, mentioned it as one of the medicinal plants grown commercially for the London market.

25. Tansy. *Tanacetum vulgare* L. 112, H 40

Local names. BACHELOR'S BUTTONS, Som; BITTER BUTTONS, Mor; BUTTONS, Yks; GINGER, Kent, Bucks, Herts, Lancs; GINGER-PLANT, Kent, Bucks, Herts; GOLDEN BUTTONS, Dev; HINDHEEL, N Eng; PARSLEY FERN, Dev, Som.

SCENTED DAISIES, Som; SCENTED FERN, Dev, Som; STINKING ELSHANDER, Perth; STINKING WILLIE, Suth; TRAVELLER'S REST, Wilts; WEEBO, Scot; YELLOW BUTTONS, Som.

Sometimes native on river banks, etc., though more frequently an escape, since Tansy has been grown for hundreds of years. And it is still worth growing for the golden buttons, the fern-like leaves, and their warm, refreshing, spicy aroma.

Vulnerary, uterine, and nephritic – useful on many counts. 'Let those Women that desire Children,' said Culpepper, 'love this Herb, 'tis their best Companion, their Husband excepted.' He recommended it either bruised and laid on the navel, or boiled in beer and drunk to stay miscarriages. *Per contra*, the poisonous Oil of Tansy has been taken hopefully to procure abortions (123, 49).

Thomas Cogan explained in *The Haven of Health*, 1584, that there was good cause for eating Tansy with fried eggs at Easter time, since it purged away the phlegm 'engendered of fish in the Lent season', and killed the worms to which this phlegm gave rise. However, Tansy puddings of Tansy

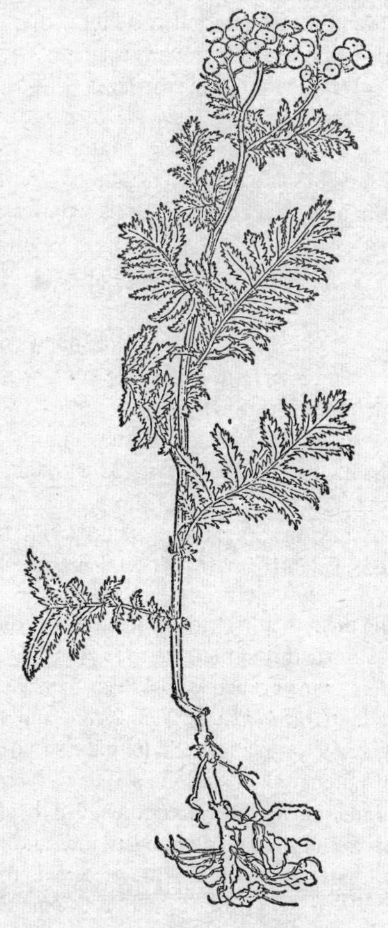

72 Tansy *Tanacetum vulgare*

leaves, milk, eggs, and flour, a survival of the 'tansies' eaten on Palm Sunday and Easter Sunday, are not to be recommended. Perhaps there is more to be said for an Irish 'drisheen', a sausage of sheep's blood and milk with a tansy flavouring. For its scent and bitterness Tansy has been used as a repellent. Highlanders laid it in the corn to keep away the mice (31b). It has been rubbed on meat to keep off the bluebottles, and on corpses to save them from earthworms or corpse worms – which, at least, is how Linnaeus explained *tanacetum*, the 'deathless' plant, from the Greek *athanasia*; it was the plant which made bodies deathless. Gerard explained the name by the flowers, which 'do not speedily wither'.

They could not do without Tansy in New England – perhaps to be sure of peopling the New World without miscarriages. The settlers introduced it, according to John Josselyn's *New-Englands Rarities*, 1672, and it is now widely naturalized up and down the eastern states. It was introduced into the Faeroes, where it is found, like medicinal herbs elsewhere, in churchyards (139).* For the name Bachelor's Buttons see *Melandrium rubrum*.

26. Mugwort. *Artemisia vulgaris* L. 112, H 40

Local names. APPLE-PIE, Ches; BOWLOCKS, Scot; BULWAND, Scot; DOCKO, Berks; DOG'S EARS, Pemb; FAT HEN, Bucks; GALL-WOOD, Scot; GREEN GINGER, Lincs; GREY BULWAND, Shet.

MIGWORT, Dor; MOGVURD, Som; MOOGARD, Caith; MUGGER, Scot; MUGGERT, Cumb, Scot, Ire; MUGGERT KAIL, Mor; MUGGONS, Scot; MUGGURTH, Ire; MUGWEED, Ches; MUGWOOD, Shrop, Yks, Dur, Cumb; MUGWORT, Corn, Bucks, Ches, Yks; OLD UNCLE HARRY, Som; SAILOR'S TOBACCO (leaves often smoked), Hants; SMOTHERWOOD, Lincs; WORMWOOD, Bucks.

A denizen, along the dusty road verges, or a native? Either way a medical and magical herb known throughout Europe, the *Mater Herbarum*, or Mother of Herbs. German, French, and Welsh names include *Johanniskraut*, 'St John's plant', *couronne de Saint-Jean*, and *llysiau Ieuan*, St John's herb – it is, in fact, a plant of St John's Eve no less famous, important, and full of potency than St John's Wort, *Hypericum perforatum* (q.v.). It was picked and purified and strengthened in the smoke of the bonfires on St John's Eve, and then made into garlands and hung over doors, etc., to keep off all the powers of evil. This practice was known in fourteenth-century England. Where Mugwort is in houses 'na elves na na evyll thynges may com therin, ne qware herbe Jon comes noyther' (Wright and Halliwell, *Reliquiae Antiquae*, 1845, 1, 53). Banckes's *Herball* of 1525 makes a similar statement – 'yf it be within a house there shall no wycked spyryte abyde' – deriving from *Agnus Castus*: 'Gif this herbe be in a mannys hous ther schal dwelle non wycked gost ne non wycked spiritus'

* But for churchyard plants see also page 372.

(20). Mugwort was used in this way in Wales in the sixteenth century (161), in the Isle of Man in the nineteenth century (Joseph Train, *Historical and Statistical Account of the Isle of Man*, 1845), and in Ireland. In the Pitt-Rivers Museum at Oxford there are scraps of 'Muggurth' from Co. Cork, which had been held over the St John's Fire, and then hung over byres and dwelling houses.

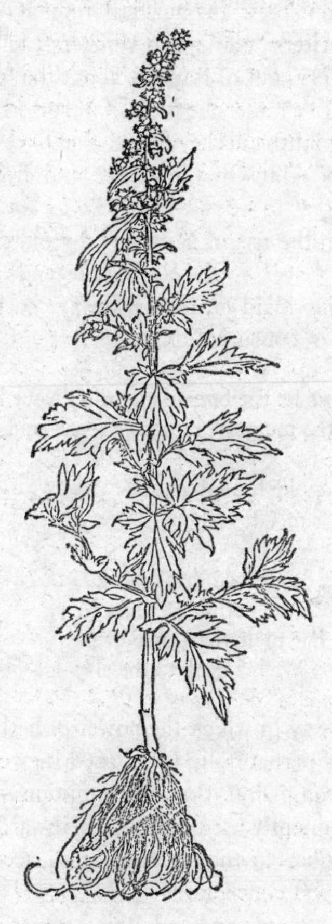

73 Mugwort *Artemisia vulgaris*

English evidence for *Artemisia vulgaris* in magic and medicine goes back to the eleventh-century manuscript of the *Lacnunga* and the pre-Christian 'Lay of the Nine Herbs' against the venoms which fly through the air (42, 169, 88). The first herb appealed to by the incantation is Mugwort:

Have in mind, Mugwort, what you made known,
What you laid down, at the great denouncing.
Una your name is, oldest of herbs,
Of might against thirty, and against three,
Of might against venom and the onflying,
Of might against the vile She who fares through the land.

Gerard and Parkinson knew the magical reputation of Mugwort, but scorned fantastical writers 'tending to witchcraft and sorcery', although Parkinson tells the story, out of Bauhin, about the 'coals' to be found at midday on St John's Eve under the roots of Mugwort – which 'are used as an Amulet to hang about the necke of those that have the falling sicknesse, to cure them thereof. But Oh the weake and fraile nature of man!' (*Theatrum Botanicum*, 1640 – see *Plantago major* for a similar tale).

So it goes on, from the use of Mugwort, by old classical authority, to prevent weariness in travel, to the Scottish legends of the mermaid and the Mugwort – the mermaid on the Galloway coast who told the lover how to cure his lady of consumption,

Wad ye let the bonnie may die i' your hand
And the mugwort flowering i' the land,

or the mermaid who spoke too late, as the young girl's funeral procession passed along the Firth of Clyde:

If they wad drink nettles in March,
And eat muggons in May,
Sae mony braw maidens
Wadna gang to the clay

– tales typical of the way in which the power of herbs is so often held to have been revealed supernaturally (35). Yet Mugwort was chiefly given for less formidable complaints than consumption – for nervousness in women, etc. ' 'Tis frequently used by Women, inwardly and outwardly, in all the Diseases peculiar to them' (141); and, according to Banckes's *Herball* (1525), it helped conception.

Like Hawthorn and Maple and Oak, this Mother of Herbs was carved in churches in the thirteenth century. There are fine sprays on the roof bosses of Exeter Cathedral, for instance, illustrated in C. J. P. Cave's *Roof Bosses in Mediaeval Churches* (1948). The author calls them 'wormwood', which is misleading since they are obviously leaves of *Artemisia vulgaris*, powerful in church as well as elsewhere.

It is tempting to call *Artemisia vulgaris* a scruffy, mean plant, as one sees it along the roads at the end of summer, and to ask why such a species ever became so distinguished. The flowers are small and brown. Crush the leaves: they are only faintly aromatic compared to the scent of *Artemisia absinthium* or *Artemisia maritima*. Yet, looked at more closely as the mediaeval sculptor looked at it, Mugwort does prove to have some attraction – especially in its leaves, deeply and delicately cut, dark green and glistening above, white below. In a vase it looks surprisingly agreeable. Still, that does not explain why this and allied species have become magical plants throughout Europe, Asia, and China (see 'Mugwort Lore', Edward A. Armstrong, *Folklore*, lv. i, 1944). Mugwort itself is a disappointing name, meaning no more in its original OE form than the 'midge plant' – because it was believed to have the power of repelling midges, like the *apsinthion* of Dioscorides?

As for the tale of Mugwort and the weary traveller ('quo-so go ony weye and he bere this herbe on hym he schall noght ben wery in hys gate' – *Agnus Castus*), every time the weary traveller stopped and looked about him, he would be apt to find Mugwort; which is no less true of continental than of English roads. Both in England and abroad, Mugwort spread happily, also, from the verges into the bomb sites.

27. Wormwood. *Artemisia absinthium* L. 81, H 20

Local names. MUGWORT, N Eng; WERMOUT, Pemb; WERMUD, N'thum; WORMIT, Dev, N'thum, N Eng; WORMOD, N'thum.

Aromatic, bitter, and, in its silvery leaves, quietly beautiful. 'It strengthens the Stomach and Liver, excites Appetite, opens Obstructions, and cures Diseases that are occasion'd by them; as, the Jaundice, Dropsie, and the like. 'Tis good in long putrid Fevers, it carries off the vitious Humours by Urine, it expels Worms from the Bowels, and preserves Clothes from Moths. The Juice, the distill'd Water, the Syrup, the fixed Salt, and the Oyl of it are used; but the Wine or Beer seems to be the best' (141).

Like Tansy, Wormwood is pleasant to have in the garden for plucking a silvery leaf and crushing out its aroma, even if it is no longer needed for fleas:

> White wormwood hath seed, get a handful or twaine,
> to save against March to make flea to refraine
> Where chamber is sweeped, and wormwood is strowne,
> no flea for his life dare abide to be knowne. (188)

28. Burdock. *Arctium lappa* L. 65, H 7

(The names also cover *Arctium vulgare* (Hill) A. H. Evans, 83, H 15, and *Arctium minus* (Hill) Bernh., 106, H 38.)

Local names. BACHELOR'S BUTTONS, Dev, Som; BARDOG, Shet; BAZZIES (the burrs), Kent; BEGGAR'S BUTTONS, BILLY BUTTON, Dev; BOBBY BUTTONS,

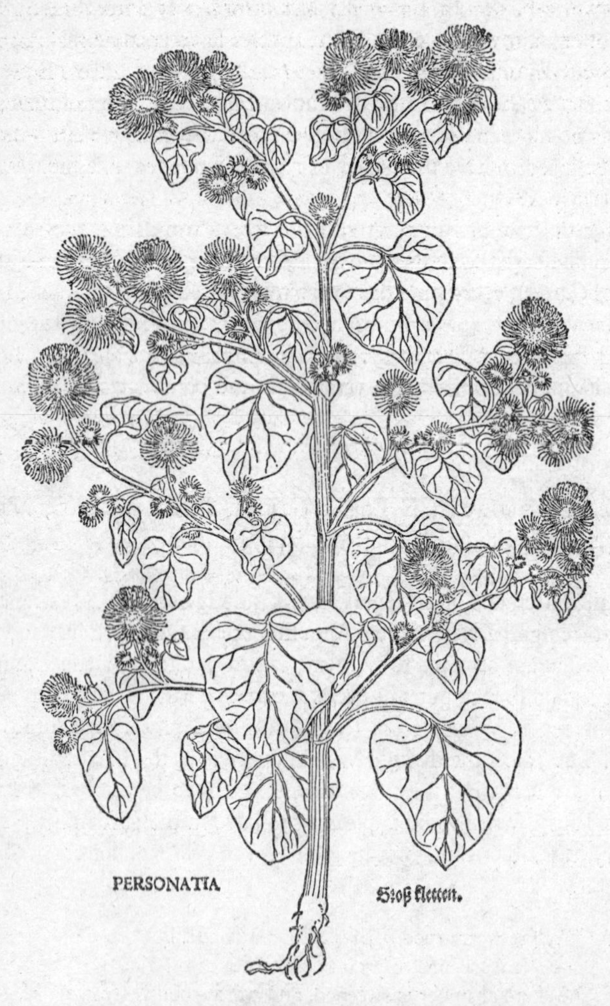

74 Burdock *Arctium lappa*

Som; BUDDY-WEED, N'thum, N Eng; BURDOCK, Dev; BURDOCKEN, Yks, Scot; BUTTER-DOCK (leaves used for wrapping butter), Corn.

CLEAVERS, Som; CLITCH-BUTTON, Dev; CLITE, Som, Glos; CLOUD-BURR, Yks, Cumb; CLOG-WEED, Wilts; CLOT-BURR, Som, Cumb, Yks; CLOTE, I o W; CLOTS, CLOUTS, CLUTS, Ches, N Eng, Yks; COCKLE-BELLS, Corn; COCKLE BUTTONS, Corn, Dev, Dor, Som; COCKLES, Dor, Wilts, Hants; COCKLE DOCK, Corn; COPPY-MAJOR, I o M; CRADAN, CREDAN, Donegal; CUCKOLDY-BURR-BUSSES, Cumb; CUCKOLD BUTTONS, Som; CUCKOLD DOCK, Corn, Dev, Som; CUCKOLD, CUCKOO BUTTONS, Som; CUCKOW, Dor.

DONKEYS, Som; EDDICK, Ches; FLAPPER-BAGS, Scot; GIPSY COMB, Berks, Notts; GIPSY'S RHUBARB, HAYRIFF, Som; HURR-BURR, Som, Shrop, Leic; KISSES, KISS-ME-QUICK, LOPPY MAJOR, Som; OLD MAN'S BUTTONS (i.e. devil's), Dev; PIG'S RHUBARB, Dor; SNAKE'S RHUBARB, Dor; SOLDIER'S BUTTONS (cf. French *bouton de soldat*, German *Soldatenknöpfe*), Wilts, Som; STICKY BUDS, Wilts; STICKY BUTTONS, Dev; STICKY JACKS, Som; SWEETHEARTS, Dor; TOUCH-ME-NOT, TURKEY RHUBARB, Som; TUZZY-MUZZY, Dev, Som; WILD RHUBARB, Som.

'It is so well known,' wrote Culpepper, 'even to the little Boys, who pull off the Burs to throw and stick upon one another, that I spare to write any description of it.' Do boys also throw the burrs in the air to catch bats, which mistake the burr for a moth and so entangle themselves, or is this a tale?

Burdock has had its medicinal virtues, summed up as 'Drying, Pulmonick, Diuretick, Diaphoretick, Cleansing, and somewhat astringent' (141); Gerard mentioned that the young stalks, peeled and eaten raw with salt and pepper, or boiled in meat broth, are pleasant to eat, and increase seed and stir up lust. *Arctium lappa* has enjoyed a greater distinction. For its sturdy habit and wide leaves, which fill up holes in the foreground, it must have been more painted than any other plant – by Claude Lorraine and by his landscape followers in England. George Stubbs, for example, is the Master of the Burdock – Burdock in the foreground, an oak sloping across the middle distance, and a racehorse in between.

For Bachelor's Buttons, Cuckold Buttons, see *Melandrium rubrum* and *Knautia arvensis*.

29. Musk Thistle. *Carduus nutans* L. 85, H 6

Local names. BUCK THISTLE, Yks; QUEEN ANNE'S THRISSEL, Berw; TEASER, Som.

'The floures growe on the tops of the branches, consisting of a flockie downe, of a white colour tending to purple; of a most pleasant sweete smel, striving with the savour of musk' (Gerard). His 'white colour tending to purple' refers to the spiny involucre rather than the purple florets.

Thistles were little regarded, or regarded only as a nuisance. Only three English kinds were recorded and distinguished before Gerard. He added three more, including the Musk Thistle, but still left four common species undescribed.

30. Spear Thistle, Boar Thistle. *Cirsium vulgare* (Savi) Ten. 112, H 40

Local names. BELL THISTLE, War, Yks; BIRD THISTLE, BLUE THISTLE, Worc; BOAR THISTLE, Som, Hants, Kent, Bucks, Herts, E Ang, Worc, Shrop, Staffs, Ches, Lincs; BULL THISTLE, Dor, Som, N Ire; BUR THISTLE, Ayr; BUCK THISTLE, Yks; CUCKOO BUTTONS, Som; HORSE THISTLE, Dev, Som; PRICKLY COATS, Dor; QUAT-VESSEL, Hants; SCOTCH THISTLE, Worc.

'The speare Thistle hath an upright stalke, garnished with a skinnie membrane, full of most sharpe prickles: whereon do grow very long leaves, divided into divers parts, with sharpe prickles: the point of the leaves are as the point of a speare, where of it tooke his name' (Gerard again, who gives the first English record). Spear Thistle translates the *carduus lanceolatus*, which Tabernaemontanus gave to this plant in 1591.

Which of the Thistles is the badge of Scotland and the Stuarts and of the Most Ancient Order of the Thistle, which James II founded in 1687? The claimants are *Onopordum acanthium* and *Cirsium vulgare*; the flower heads on badge and star suggest rather the military *Cirsium*, the militant Spear Thistle, though the motto, *nemo me impune lacessit*, 'no one attacks me and gets off scot-free', fits either of them. However, it is an academic argument, *ex post facto*. The kings of Scotland chose a thistle, not a particular thistle, for the prickliness and the gallant colour. They were not botanists.

31. Stemless Thistle, Picnic Thistle. *Cirsium acaule* (L.) Scop. 46

Local names. GROUND THISTLE, War, Worc; POD THISTLE, N'hants.

Odd and beautiful, but well armoured, spreading like a desert plant over the turf of the downs or the ramparts of an Iron Age camp. 'Picnic Thistle' it might be called. Before the flowers are out to warn you of the spiny rosettes, down you sit – and up.

A herb alexipharmic, sudorific, and diuretic, according to herbalists of the seventeenth century, able to prevent and cure the plague. Philipp Melanchthon the reformer (1497–1560) cured himself with the stemless Thistle of the pains of melancholy and depression (141).

32. Melancholy Thistle, Fish Belly. *Cirsium heterophyllum* (L.) Hill. 55

Local names. CARL DODDIE (i.e. 'curly head'), Angus; FISH BELLY (from the white underside of the leaves), Cumb; TAZZLE, Yks.

A thistle with a difference, unarmed, slender, belonging to the Highlands and the north of England. It looks pretty, if a little sad, in the mountains – for instance, around the base of Ingleborough, against the grey limestone. At first the single flower heads droop and hang; and the mountain wind turns the Fish Belly leaves, which reveal their under-surface of white.

Did this melancholy look suggest that here was a plant good against melancholy? So it was used: 'The Decoction of the Thistle in Wine being drunk, expels superfluous Melancholy out of the Body, and makes a man as merry as a Cricket ... 'tis the best Remedy against all Melancholy Diseases that grows' (49) – Dioscorides abetting, since he had mentioned one thistle-like plant which was a useful amulet against melancholy or sadness.

It was from Ingleborough that the Melancholy Thistle was first recorded – and not merely for Britain. It was found there by Thomas Penny, who in 1581 communicated the find to Charles de l'Écluse, who in his turn recorded it in 1583 in his *Stirpium Pannonicarum Historia*, calling it the *Cirsium Britannicum* of the 'Montis Englebrow'.

33. Milk Thistle. *Silybum marianum* (L.) Gaertn. 60, H 25

Local names. BLESSED THISTLE, War; LADY'S THISTLE, Som, Lincs; MILKY THRISSEL, N'thum; OUR LADY'S THISTLE, War; VIRGIN MARY'S THISTLE, Hants, Bucks, E Ang, Ches; VIRGIN'S THISTLE, Norf. In Welsh *ysgallen Fair*, 'Mary's Thistle'.

The leaves, wrote Gerard, 'are without downe, altogither slippery, of a light greene and speckled with white and milkie spots and lines drawne divers waies' – and he goes on to a silent display of Protestantism and no

Popery. He gives the names *Carduus Mariae* and Our Ladies Thistle; he must have known the legend very well (that the white veins were marked by the Virgin's milk, when she suckled the child), and he must have known that this sacred Milk Thistle was held to increase the flow of milk. But not a word. There must be no superstition. Culpepper and Pechey observe the ban in their herbals as well, though John Evelyn mentions the Thistle (in his *Acetaria*, *A Discourse of Sallets*, 1699) as a 'great breeder of milk' and a proper diet for wet nurses.

It is an old introduction from southern Europe, cropping up unexpectedly in waste places. In the sixteenth century it seems to have been more common, no doubt because it was more frequently cultivated. 'It groweth upon waste and common places by high waies, and by donghils, almost every where,' according to Gerard.

He also remarks, like most authors after him, that the tender leaves can be eaten, once you have removed the prickles. They can be boiled, they can be blanched as a winter salad, and the young stalks can be eaten like rhubarb, though they need peeling, and then soaking to remove the bitterness (124). You can also eat the bracts of the involucre, like artichoke.

Lady's Thistle is not to be confused with the old physic herb *Cnicus benedictus* L., the Blessed Thistle, another plant of southern Europe, with yellow instead of purple flowers, but also with milky veins, though on the underside of the leaves.

34. Cotton Thistle, Scotch Thistle. *Onopordum acanthium* L. 74

Local names. PIG-LEAVES, Yks, N Eng; QUEEN MARY'S THISTLE, N'hants ROUGH DASHLE, Dev; SCOTCH THISTLE, Berw.

This is what the Englishman means by the Scotch Thistle, a plant he has grown himself and still grows now and then in the garden, a plant which was probably introduced, and which is not at all common north of the border. Gerard knew it along the roads, and said that the greatest quantity of the soft cotton or down from the stem and the leaves was collected for stuffing pillows, cushions, and beds. The rich upholsterers mixed this cotton with feathers – 'which deceit would be looked unto'. The bracts of this species as well can be eaten, one by one, like the bracts of artichoke; and the young stalks can be boiled, after blanching them and peeling them (124).

William Turner called it Cotten Thistle, Gum Thistle, and Otethistle 'because the seedes are lyke unto rough otes'. He had noticed it in gardens in England, 'but never a brode' – i.e. wild. That was in 1548. In his

Herbal, in 1562, he mentions it as a wild plant near Sion House on the Thames – where perhaps it had escaped from the gardens. In Gerard's time, 1597, it grew along the roads and the ditches 'almost every where'. For some reason north-country Freemasons carried the Cotton Thistle in their processions (109).

35. Bluebottle, Cornflower. *Centaurea cyanus* L. Rare

Local names. BACHELOR'S BUTTONS (see *Melandrium rubrum* and *Knautia arvensis*), Som, Derb, Yks; BLAVER, Berw, Rox; BLAWORT (cf. French *bleuet*, *blavet*, Danish *blaurter*), Berw, Rox, Angus, N Scot; BLUE BLAW, N'thum; BLUE BLAWORT, Aber; BLUE BLOW, Dor; BLUE BOBS, Hants; BLUE BONNETS, Corn, Som, Staffs, Stir, Aber, Mor; BLUEBOTTLE, Wilts, Oxf, Shrop, Lincs, Yks, Berw; BLUE BUTTON, Som; BLUE CAP, Kent, N'hants; BLUE JACK, BLUE POPPY, BOBBY'S BUTTONS, Som; BOTTLE-OF-SORTS, Yks; BROOMS AND BRUSHES, Dor; BRUSHES, Stir.

CORN-BLINKS, Dev; CORNBOTTLE, Dev, N'hants; CUCKOOHOOD, Scot; KNOBWEED, Notts; KNOTWEED, N'hants; LADDER LOVE, Som; LOGGERHEADS, N'hants; MILLER'S DELIGHT, Dor; PIN-CUSHION, Suff; THUMBLE (i.e. Thimble), Scot; WITCH BELLS, N Eng; WITCH'S THIMBLE, N Eng.

One glory of colour has gone out of the English scene. John Clare, a hundred and fifty years ago, in his boyhood, marked 'the lighter hues of wheat and barley intermixed with the sunny glare of the yellow charlock and the sunset imitation of the scarlet headaches (poppies); the blue cornbottles crowding their splendid colours in large sheets over the land and troubling the cornfields with destroying beauty'. The seed-corn now comes well cleaned from the merchant, and the Bluebottle is dwindling to extinction (in contrast to the Poppy or Charlock). Since it gave a cornfield, and not a woodland blue, a domestic or pastoral blue and not one of the wilderness, the Bluebottle was at one time far more popular than the Bluebell, with its washes of azure and silver. So in the eighteenth-century poem, *An Invitation to Daphnis*, by Anne Finch, Countess of Winchelsea (1666–1720):

> Rich colours on the vellum cease to lay
> When ev'ry lawn much nobler can display,
> When on the dazzling poppy may be seen
> A glowing red exceeding your carmine;
> And for the blue that o'er the sea is born,
> A brighter rises in our standing corn.

Before the flood of exotics began to pour into the English garden, the Bluebottle was much cultivated. 'It is sowen in gardens, which by cunning looking to, doth often times become of other colours, and some also double' (Gerard). We grow it still. But it is one thing to raise a little patch of Bluebottle or Cornflower, another to see it by the acre, as in French cornfields. Bluebottle used to be taken against plague, poison, wounds, fevers, inflammations (49).

36. Knapweed, Hardheads. *Centaurea nigra* L. 111, H 40

Local names. BACHELOR'S BUTTONS, Ire; BLACK SOAP, Dev, Glos; BLUEBOTTLE, BLUE JACK, Som; BLUE TOPS, Worc; BOBBY'S BUTTONS, Som; BRUSHES, Dor; BULLHEADS, BULL THISTLE, BULLY HEADS, Som; BUNDS, E Ang; BUNDWEED, Norf, E Ang; BUTTON-WEED, Suss.

CHIMNEYSWEEP, CHIMNEYSWEEP'S BRUSHES, Som; CLOBWEED, Glos; CLOVER KNOB, Notts; COCKHEAD, N Eng; DROMEDARY, Wilts; DRUMMER-BOYS, Dor; DRUMMER HEADS, Som; DRUMSTICKS (cf. German *Trommelschlegel*), Som, N'hants; GNAT FLOWER, HACKYMORE, HAIRY HEAD, Som; HARDHACK, Wilts; HARDHEADS (cf. German *Hardkopp*), Corn, Dor, Wilts, Glos, Worc, Shrop, Ches, Lincs, Lancs, Yks, Cumb, N'thum; HORSE-HARDHEAD, Dev, Som; HARDINE, HARDIRON, Staffs, Ches, Notts, Lancs; HORSE BUTTON, Donegal; HORSE KNOBS, Dor; HORSE KNOBS or KNOPS, N'hants, Lancs, Yks, Cumb, N'thum; HORSE KNOT, Scot, N Eng; HORSE KNAP, Dev; HURT-SICKLE, Worc.

IRON KNOBS (cf. French *tête de fer*), Ches; IRON-WEED (cf. German *Eisenkraut*), Som, N'hants; KNOB-WEED, N'hants; KNOT-GRASS, Herts; LADY'S BALLS, Som, Wilts; LADY'S CUSHION, Kent; LOGGERHEADS, Som, Glos, Bucks, Oxf; LOGGERUMS, Wilts; MATFELLON, Yks; NIGGERHEADS, PAINTBRUSHES, Som; SHAVING BRUSH, Shrop; SWEEPS, Derb; TARBOTTLE, Oxf; TASSEL, Berw; TOP-KNOT, Som.

Most of the names are explained by the hard, knobbly heads, the bottle-shaped involucre, and the toughness of the plant, a universal species of English grassland. Knapweed was used for wounds and ruptures, bruises, sores, scabs, and sore throat, etc. (49). The heads have been made to foretell the future of love, in the same way as the heads of *Plantago major*. The expanded florets are picked off, and girls put the head inside their blouse. After an hour they look at it again, hoping that the unexpanded florets will have blossomed, as a sign that love will come their way from the right quarter (for this, and the name Bachelor's Buttons, see also *Melandrium rubrum* and *Knautia arvensis*).

37. Saw-wort. *Serratula tinctoria* L. 70

The leaves, wrote Gerard, are 'somewhat snipt about the edges like a sawe' – *serra* in Latin – 'whereof it tooke his name'. And it was owing to the name, no doubt, that Saw-wort was 'woonderfully commended to be most singular for wounds, ruptures, burstings, hernies, and such like', a companion plant to the Sanicle. Abroad, if not in England, Saw-wort was much employed as a dye plant (German *Färbkraut*, *Färbedistel*). Alum and the leaves were placed together in the dye bath, and the leaves then gave a good green-yellow to woollens (12) – one of the best yellows next to that of *Reseda luteola*.

38. Chicory. *Cichorium intybus* L. 72, H 31

Local names. BLUE ENDIVE, Som; BUNKS, Suff, Norf; STRIP-FOR-STRIP, Som.

Probably, but not certainly, a native: even then a plant too local to have played much in English life and sentiment. German folklore has several pretty tales to explain the blue-eyed Chicory by the roadside. The young girl wept for her dead lover, and would only stop weeping, she said, when she was turned into a flower by the road. So she became the *Wegwart* or Chicory (93).

Careful English wives grew Chicory among their herbs. It was good for purging and for the bladder, and by sympathetic magic it was believed that the water distilled from the round blue flowers, which open to the sun and shut when the sun disappeared, worked against inflammation and dimness of the sight (141).

'It has been discovered that the fine blue colour of the petals is convertible into a brilliant red by the acid of ants; Mr Miller the Engraver assured me, that in Germany the boys often amused themselves in producing this change of colour by placing the blossoms in an ant-hill' (Curtis, in the *Flora Londinensis*).

39. Nipplewort. *Lapsana communis* L. 112, H 40

Local names. BALLAGAN, Ayr; BOLGAN-LEAVES (a *bolgen* is a swelling), Scot; CARPENTER'S APRON, War; HASTY ROGERS, Dev; HASTY SERGEANT, Dor,

Som; JACK-IN-A-BUSH, MARY ALONE, Glos; SWINE'S CRESS, WORMWOOD, Som.

Prussian apothecaries, according to the Nuremberg physician and botanist, Joachim Camerarius, in his *Hortus medicus et philosophicus*, 1588, called this plant *Papillaris*, 'because it is good to heale the ulcers of the nipples of women's breasts'. So John Parkinson translated Papillaris into the English 'Nipplewort' (140). Sympathetic magic, once more, because the naked buds have a nipple-like appearance.

40. Jack-go-to-bed-at-noon, Goat's Beard. *Tragopogon pratensis* L. 94, H 25

Local names. GO-TO-BED-AT-NOON, E Ang, Midlands, War, USA (where it is naturalized); JACK-BY-THE-HEDGE, Suss; JACK-GO-TO-BED-AT-NOON, Dor, Som, Wilts; JOHN-GO-TO-BED-AT-NOON, Som, Wilts, Glos; JOHN-THAT-GOES-TO-BED-AT-NOON, N'hants; JOSEPH'S FLOWER, Som, Suss; NAP-AT-NOON, Shrop, Lancs, Midlands, Cumb; ONE O'CLOCK, Dev, Som; SHEPHERD'S CLOCK, Yks; SLEEPY-HEAD, TWELVE O'CLOCKS, Som.

A 'clock plant', closing at noon (cf. *Anagallis arvensis* and *Ornithogalum umbellatum*). It was called Goat's Beard by William Turner in 1548 as a translation of the shop name *barba hirci*, which goes back to the *tragopogon* in Theophrastus and Dioscorides. The beard is the long silky pappus. 'Joseph's Flower', because Joseph, the husband of the Virgin, is always bearded in pictures of the Nativity. 'Joseph's Flower' was apparently borrowed from the German. In France – in Savoie – it is called *barbe à Jean*, St John's Beard, and it is one of the herbs of St John picked before St John's Eve (see *Hypericum perforatum*).

41. Sow Thistle. *Sonchus oleraceus* L. 112, H 40

Local names. DINDLE, Norf; DOG'S THISTLE, Som; HARE'S LETTUCE, Dev; MILK THISTLE, Dor, Som, War, Lincs; MILKWEED, Som; MILKWORT, Dor; MILKY DASHEL, MILKY DASSEL, MILKY DISLE, Corn, Dev; MILKY DICKLE, Dev; RABBIT'S MEAT, RABBIT'S VICTUALS, Som; SOW BREAD, Kent; SOW DINGLE, Lincs; SOW-FLOWER, Wilts; SOW THRISTLE, Berw; SWINE THISTLE, Lancs, Yks, Cumb, Scot, Donegal; SWINIES, Berw.

Sonchus oleraceus and *Sonchus asper* are two of the world-conquering 'anthropophytes', plants that go around with man and flourish wherever he disturbs the soil. *Sonchus oleraceus* is happy in town or in country – indeed almost anywhere and everywhere, highland and lowland, inland and by the sea, on a wall or on the burnt soil where there has been a bonfire. It was 'one of the earliest plants to colonize completely new terrain raised from the sea in Hawkes Bay, New Zealand, after the earthquake of 1931'. Botanists believe it a native species, probably, on sea cliffs, from which it spread into the human context.

Older legend connected Sow Thistle with the hare rather than the sow. The pseudo-Apuleius said that it gave strength to the hares when they were overcome with heat. The *Herbarius zu Teutsch* of 1485 and *The Grete Herball* of 1526 tell how *Sonchus oleraceus* is the 'hare's house' or 'hare's palace', or 'hare's bush'. The German book says that when the hare is under the hare's house, it is not afraid and believes itself quite safe, for this herb has power against melancholy, 'and there is no animal so completely melancholic as the hare'. The mediaeval *Agnus Castus* (20) calls it *Lactuca leporica* or harys thystle – 'the vertu of this herbe is quanne an hare is wood [mad] in somer he wele etyn of this herbe and thanne schal he ben hol'. By sympathetic magic a plant with milky juice obviously increases milk. *Sonchus oleraceus* was therefore given to nursing mothers (Gerard, etc.).

William Coles, in 1657, fancied it was Sow Thistle because sows knew by 'a certain natural instinct' that it would increase the flow of their milk when they had farrowed (45). John Ray remarked in 1690 that some people used *Sonchus oleraceus* as a winter vegetable with salad. '*Nos cuniculis et leporibus eum commanducandum relinquimus*' – 'We leave it to be masticated by hares and rabbits' (154).

42. Dandelion. *Taraxacum officinale* L. 112, H 40

Local names. BITTER AKS, Shet; BUM-PIPE, Banff, Lanark; BURNING FIRE, Som; CANKER, Glos; CLOCK, Som, Bucks, Ess, Norf, Camb, N'hants, Leic, Notts, Lincs, Yks, Ire; CLOCK FLOWER, CLOCKS AND WATCHES, COMBS AND HAIRPINS, Som; CONQUER MORE, Dor; DEVIL'S MILK-PLANT, Kirk; DEVIL'S MILK-PAIL, Som; DINDLE, E Ang; DOG-POSY, Lancs, Yks; DOON-HEAD CLOCK, Scot; EKSIS-GIRSE, Shet.

FAIRY CLOCKS, FARMER'S CLOCKS, FOUR O'CLOCK, GOLDEN SUNS, Som; HEART-FEVER GRASS, N Ire; HORSE GOWAN, Scot; IRISH DAISY, Yks; LAY A-BED, LION'S TEETH, Som; MALE, Dor; MESS-A-BED, Som; MILK GOWAN, Scot; MONK'S HEAD, Som, Wilts.

OLD MAN'S CLOCK, Dev; ONE O'CLOCK, Dev, Ches, Lancs; ONE, TWO,

THREE, PEE-A-BED, Som; PISHAMOOLAG, Donegal; PISS-A-BED (cf. Dutch *pisse-bed*, *pis in 't bed*), Eng, Scot, Ire, USA; PISSIMIRE, Lakes, Yks; PITTLE BED, Suff; PRIEST'S CROWN, Lincs; SCHOOLBOY'S CLOCK, SHEPHERD'S CLOCK, Som; SHIT-A-BED, Corn, Wilts, Hants; STINK DAVIE, Clack; SWINE'S SNOUT, TELL-TIME, TIME FLOWER, TIME-TELLER, TWELVE O'CLOCK, WET-A-BED, Som; WET-WEED, Wilts; WHAT O'CLOCK, Som; WISHES, Wilts; WITCH GOWAN, YELLOW GOWAN, Scot.

The Lion is a noble beast of the sun, Dandelion flowers stare up at the sun and follow it around, they are yellow, and each of them is a sun in miniature. So much is easily felt. But it does not explain why *Taraxacum officinale* was made *dens leonis*, tooth of the lion, in mediaeval Latin, *dent de lion* in French, and Dandelion in English. There is no satisfactory explanation. The lion's tooth may be the tap root, the jagged leaf, or the parts of the flower.

Blanched Dandelion leaves, Dandelion wine in the cottages, Dandelion beer for steel workers – these are all relics of the plant in medicine. It is diuretic and cleansing, still 'Piss-a-bed' up and down the country, though the name has been politely ignored (unlike the French *pissenlit*). You tell the time, and foretell the future, by blowing on the seed-head. You leave the receptacle pitted and bare: in the Middle Ages of monasticism that suggested '*caput monachi*', the 'monk's head', bald, and pitted, the 'prestis croune', a name that goes back to the fourteenth century (99).

Dandelion is a brave plant. Far away from its European home and its proper fields, you see Dandelions on a scrap of waste land under the factories of Detroit or Philadelphia, still spreading their leaves and opening their small suns after the bitter frost has killed everything else.

There is one thing for which Dandelions are useless: pick them and arrange them in a bowl, and they soon close up. They need the full sun out of doors.

LXXXVII. Alismataceae

1. Water Plantain. *Alisma plantago-aquatica* L. 103, H 40

Local names. DEIL'S SPOONS, Scot; GREAT THUNDERBOLT, UMBRELLAS, Som; SLANLIS, Donegal.

Gerard wrote that the leaves of the Water Plantain – 'faire great large leaves like the lande Plantaine (i.e. *Plantago lanceolata*), but smoothe, and not so full of ribbes or sinewes' – were good for dropsy. Water Plantain is native also to North America, and the settlers were pleased to find it there. They too applied the acrid, blistering leaves to dropsical legs, 'to draw out

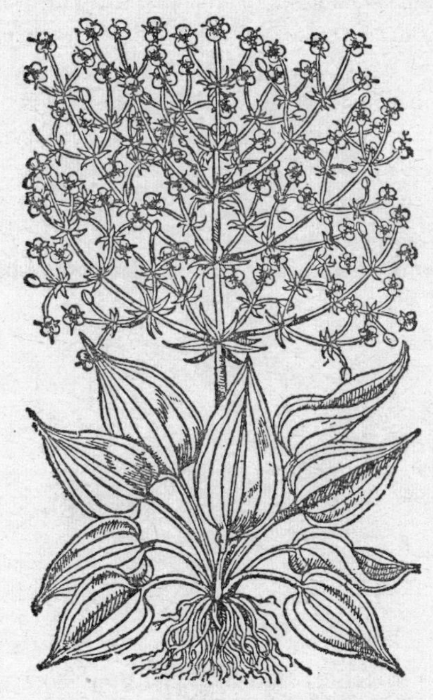

75 Water Plantain *Alisma plantago*

water between the skin and the flesh', and they called it 'Water Suck-leaves' and 'Scurvy-leaves' (John Josselyn, *An Account of Two Voyages to New-England*, 1674), two names which have disappeared from English.

In America, too, Water Plantain was applied if you were bitten by a rattlesnake.

2. Arrowhead. *Sagittaria sagittifolia* L. 62, H 23

Local names. ADDER'S TONGUE, Dev; MOSES-IN-THE-BULRUSHES, Som.

Pick the three-petalled blossoms of Arrowhead, and every petal will have dropped before you get home. A fine plant, growing in the water, with 'large and long leaves, in shape like the signe *Sagittarius*, or rather like a bearded broad arrowe heade: among which riseth up a fat and thicke stalk, two or three foot long, having at the top many pretie white flowers, declining to a light carnation' (Gerard).

There are several other things to notice about the Arrowhead – the way in which the anthers as well are arrow-shaped, the milk which exudes from the 'fat and thicke' stalk, and the tubers – 'winter-buds', in fact, enabling the plant to persist through the winter; which are unexpectedly blue and spotted with yellow. But these last objects have to be searched for, under the water and in the mud, on the strings of the rhizome. Also they are edible, though the Arrowhead is too scarce a plant for trying food experiments. Gerard knew of no virtues for Arrowhead, which is surprising for a plant with so many suggestive features. In Devonshire a strengthening tea was made, always out of *nine* leaves (192).

Arrowhead is a good name, though Gerard also used a better one which has dropped from use – Water Archer.

LXXXVIII. Butomaceae

1. Flowering Rush, Pride of the Thames. *Butomus umbellatus* L. 68, H 19

Local names. PRIDE OF THE THAMES, Som, Dor; RAXEN (OE *rixen*, rushes), Som.

One of the scarcer beauties, known by a name altogether inadequate, which we owe to Mathias de l'Obel (58). He described it as *Juncus cyperoides floridus paludosus*, the rush with leaves like a *Cyperus* and with flowers, which grows in marshy places; though in fact it grows out of the water. Gerard called it Water Gladiole, from another botanical description as *Gladiolus palustris*, though he preferred a name of his own invention, the Lily Grass. Why not Pride of the Thames, first and always? De l'Obel gave the first British record from the Thames. Gerard also mentioned it growing 'by the famous river *Thamesis*, not far from a peece of ground called the Divels neckerchiefe neere Redriffe by London'.

Butomus umbellatus has been introduced as a garden plant into the United States and has escaped into the St Lawrence and established itself along the Great Lakes (131). Notice how the loveliness of the flower depends not only on the pink segments, but on the bold red of the stamens and carpels.

LXXXIX. Hydrocharitaceae

1. Water Soldier. *Stratiotes aloides* L. 21

Local names. CRAB'S CLAW, Som; WATER PINE, Ches.

You have to search for this scarce and odd water plant in June, July, and August, when it comes up to the surface to flower, like a petrel or a shearwater coming in from the ocean to breed. It has saw-edged military leaves like an Aloe (*aloides*). Growing in a rosette, these bayonet leaves suggested that it was a Houseleek which had exchanged roof for water.

The apt name of Crab's Claw comes out of Gerard (who did not know *Stratiotes aloides* as an English plant). Among the leaves, which are 'set round about the edges with certain stiffe and short prickles', there come forth 'divers cases or huskes very like to crabs clawes; out of which when they open growe white flowers, consisting of three leaves [petals]'.

Likened to Houseleek, it was given the same properties: 'This Housleeke staieth the blood which cometh from the kidneies, it keepeth green woundes from being inflamed, and it is good against Saint Antonies fire and hot swellings' (Gerard).

2. Canadian Pondweed. *Elodea canadensis* Michx. 87, H 37

Local names. CAT'S TAILS, Ire; DRAIN DEVIL, Worc; RAAVE, N Ire; THYME-WEED, Lincs; WATER THYME, Berks, Oxf, Camb, War, Yks. Also AMERICAN WEED, GLORUS, SWAN WEED, Donegal.

In Ireland this invader from the New World was first seen in 1836. In Scotland it was detected on 3 August 1842 in the lake of Duns Castle in Berwickshire. It spread to ponds, ditches, lakes, canals, and slow rivers with a startling rapidity during the next twenty to twenty-five years. Claridge Druce, in his *Flora of Oxfordshire*, tells how one small scrap was introduced in the Lower Aquarium of the Oxford Botanic Gardens in 1849. Taking a Sunday walk four years later, an Oxford botanist was surprised to find a ditch near the Thames filled with it – an 'uninterrupted dense mass from one end of the ditch to the other'. Soon after he found it by the Canal, in Port Meadow, and in the river. By 1858 this Drain Devil was thick in the Thames as far as Reading. For a while it blocked the Thames here and there.

It reached its maximum between 1866 and 1874, then began to diminish, Druce thought because of the absence of male plants (the Pondweed maintains itself vegetatively by winter-buds). He recorded it (in his

Flora of Berkshire) in a dew pond on the Ridgeway, nearly 800 feet up, to which it was carried, no doubt, by birds; and it climbed to ponds on the Wiltshire chalk, above Pewsey.

XC. Potamogetonaceae

1. Broad-leaved Pondweed. *Potamogeton natans* L. 112, H 40

Local names. BUTTER-DOCK, Ches; DEVIL'S SPOONS, Scot; FISH-LEAVES, Suff, War; FLATTER-DOCK, Ches; PICKEREL-WEED, Suff, E Ang; PLATTER-

76 Pondweed *Potamogeton natans*

DOCK, Ches; TENCH-WEED, E Ang. Several names are recorded in Donegal: BLACKWEED, DULAMAN, SLANLIS, LADY WRACK, LIAGH-RODA.

The leaf was believed to conceal a fish. In East Anglia it was believed that young pike, or pickerels, were actually bred from the floating leaves.

2. Curled Pondweed. *Potamogeton crispus* L. 103, H 40

Local names. SMALL FROG'S LETTUCE, WATER CALTROPS, Cumb.

With leaves like the most delicate, translucent, wavy-edged hack-saw, this is by far the most beautiful of English pondweeds.

XCI. Liliaceae

1. Bog Asphodel. *Narthecium ossifragum* (L.) Huds. 98, H 40

Local names. LIMMERICK, Shet; MAIDEN'S HAIR, Lancs, Yks; MOOR-GOLDS, Yks; MOOR-GRASS, Herts; TAE-GIRSE, Shet; YELLOW GRASS, Scot. Names in Donegal include CRUPPANY GRASS (from giving sheep cruppany or bone stiffness), FAR-FIA, GLASHURLANA.

This was taken to be a miniature Asphodel, and it was called *Asphodelus luteus* and *Asphodelus Lancastriaē*, Lancashire Asphodel. Note the mixture of colours, the red of the anthers against the yellow of the perianth segments (which are green on the lower side), and the way the flowering stem changes, when the flowers are over, to a dark saffron colour.

In Shetland the Bog Asphodel has been used as a substitute for saffron, in medicine and in dyeing. In Lancashire, in the seventeenth century, women gathered it off the moors to use as a hair dye (81).

2. Lily of the Valley. *Convallaria majalis* L. 69

Local names. DANGLE BELLS, FAIRIES' BELLS, Som; INNOCENTS, Dor; LADDER TO HEAVEN, Som; LADY'S TEARS (cf. French *larmes de Ste-Marie*, and similar names, in German, Rumanian, and Italian), LILY CONFANCY, Som; LILIES AND VALLEYS, Bucks, Beds; LINEN BUTTONS, LITTLE WHITE BELLS, Som; MAY BLOSSOMS (cf. German *Maiblume*), Dev; MAY LILY (cf. *Mai-lili* in north Germany), ST LEONARD'S LILIES, in the St Leonard's Forest area of Sussex, having sprung from the blood of a dragon killed by St Leonard the Hermit, whose cult was perhaps introduced by the Norman lords of this forest; WHITE BELLS, Som.

'*Ego sum flos campi et lilium convallium*,' says the Song of Solomon, or in the Authorized Version: 'I am the Rose of Sharon, and the lily of the valleys.' The *lilium convallium* and the enclosed garden of the song – 'A garden enclosed is my sister ... Thy plants are an orchard of pomegranates, with pleasant fruits' – were taken as symbols of the Virgin, and the *lilium convallium* was thought to be *Lilium candidum*, our Madonna Lily, and in the north *Convallaria majalis*, our Lily of the Valley. The white blossoms and the fragrance of both plants stood well for the innocent, the precious, and the rare. However, *lilium convallium* was probably no lily, but the fragrant *Hyacinthus orientalis*, which is common throughout Palestine (132).

When he wrote his *Names of Herbes* (published in 1548), William Turner did not know of *Convallaria majalis* as a wild plant in Great Britain. He said that German apothecaries – and it was common in Germany – called it *Lilium convallium*, that the French and German names were *muguet* and *Maiblume*, and that in English it might be called May Lily. Laurence Andrew in 1527, in his version of Hieronymus Braunschweig's *Liber de arte distillandi*, had called Lilies of the Valley both May Flowers and Pache Flowers (i.e. Easter Flowers). By Lyte's time *Lilium convallium* was pleasantly corrupted to 'liriconfancy', and Lily of the Valley was in use. Gerard's names were Lily in the Valley, or Lily of the Valley, May Lily, Convall Lily, and Liriconfancie. He knew it as a wild plant in several places, and especially upon Hampstead Heath, in 'great abundance'. The wild Lily can still be found in plenty of dry, chalky, or limestone woods from the Cotswold escarpment up to Yorkshire. Its strangest habitat of all is upon the clints or pavements of limestone around Ingleborough, in the West Riding, where the leaves rattle and rustle in the wind.

According to German folk-tales the Lily of the Valley originated from the tears which the Virgin shed at the foot of the cross, or tears shed by Mary Magdalene when she found Christ's tomb empty after the Resurrection.

Lily of the Valley has been employed a good deal in medicine. Gerard advised the use of it against palsy and apoplexy, weak memory and inflammation of the eyes, and gout. The dried inflorescence, for action similar to that of Foxglove, has a place still in the *British Pharmaceutical Codex* (1949). French investigators found the extract of Lily of the Valley was so active a poison that four drops injected into its veins would kill a dog in ten minutes (46).

3. Solomon's Seal. *Polygonatum multiflorum* (L.) All. 51

Local names. DAVID'S HARP (curving like the neck of a harp), Som, Hunts, N'hants, Shrop, Derb, Lincs; JACOB'S LADDER (cf. old apothecaries' name of *scala coeli*, 'ladder of heaven'), Som, Wilts, Berks, Leic; JOB'S TEARS, Corn; LADY'S LOCKETS (from the hanging flowers), Som; LILY OF THE MOUNTAIN (to distinguish it from Lily of the Valley), War; SOW'S TITS, Dor.

It has also been called Our Lady's Belfry.

In a garden Solomon's Seal – either *Polygonatum multiflorum*, or the hybrid (more usual) between *P. multiflorum* and the much rarer *P. odoratum* – frequently has no room to reveal its points. In a wood Solomon's Seal stands in colony after colony, uncluttered, altogether more slender and delicate, holding its leaves higher and showing more of its greenish-white flowers.

During the late Middle Ages, the question was whether to call the plant Solomon's Seal, with its suggestion of magic, or the Seal of St John or the Virgin. 'Sigillum sancte marie or sigillum Salamonis is al one herbe that is calld Salomons seale or our ladies seale,' said *The Grete Herball.* From Josephus, the great Solomon was known as conjuror, enchanter, and philosopher. His seal was the magic pentacle, the Star of David, made of two interwoven triangles, which had great power as a *fuga daemonum* – in putting evil spirits to flight. The true pentacle of five points was supposed to touch and point out the five wounds of Christ (8). By mediaeval notions then, Solomon, who was wiser than all men, and who 'spake of trees, from the cedar tree that is in Lebanon even unto the hyssop that springeth out of the wall' (I Kings, iv. 33), had set his approbation upon the roots of *Polygonatum multiflorum*; which had also been known to Dioscorides. 'The roote is white and thicke, full of knobs or joints,' wrote Gerard, inheriting the past, 'which in some places resemble the marke of a seale, whereof I think it tooke the name *Sigillum Salamonis.*' Three pages later, as though he realized that the resemblance is very slight, he added: '*Dioscorides* writeth, that the rootes are excellent good for to seale or close up greene wounds being stamped and laide thereon: whereupon it was called *Sigillum Salomonis*, of the singular vertue that it hath in sealing, or healing up wounds, broken bones, and such like. Some have thought it

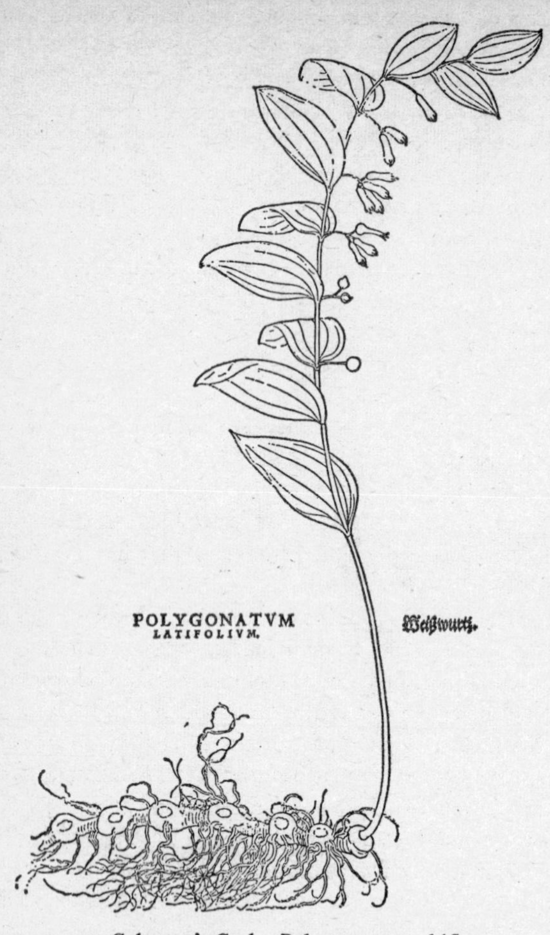

77 Solomon's Seal *Polygonatum multiflorum*

tooke the name *Sigillum*, of the markes upon the rootes: but the first reason seemeth to me more probable.'

Turner as a Protestant, humanist, and reformer who detested superstition, disliked names after Saint or Virgin or Solomon, and he did not care for the other name *scala coeli*, 'ladder of heaven' (from the arrangement of the leaves). 'It is called in English,' he wrote in his *Herbal* in 1568, 'scala celi. The Duch men call it Weiss Wurtz [the "Duch men", i.e. the Germans, still call it *Weisswurz*], the herbaries calle it *Sigillum Salomonis*. It were better to call it by an English name taken out of Duch (from whence our English sprang first), whyte wurt then scala celi; for so shall men lern better to know it, and to remembre the name of it'. Luckily no one listened, and Solomon's Seal it remains.

Gerard, and those who came after him, praised Solomon's Seal very highly, especially for soldering and glueing together broken bones, when the root was taken inwardly in ale, or applied outwardly as a poultice. The common people of Hampshire used it for themselves and their cattle. Gerard also observed that an application of the fresh root 'taketh away in one night or two at the most, any bruse, blacke or blew spots gotten by fals, or womens wilfulnes, in stumbling upon their hastie husbands fists, or such like'. Culpepper was no less enthusiastic; all of which explains the ubiquity of Solomon's Seal in old gardens around the country.

In the Lake District *Polygonatum odoratum* (Mill.) Druce, 22, was known as Vagabond's Friend, for its similar employment against black eyes, broken noses, bruises, etc.

4. Asparagus. *Asparagus officinalis* ssp. *prostratus* (Dum.) E. F. Warburg. 9, H 3

This is the native Asparagus, prostrate and not erect, which John Ray discovered on the cliffs of the Lizard in 1667. The garden Asparagus, Sperage, Sparrow-grass, *Asparagus officinalis* ssp. *officinalis*, was introduced; and it was grown already in Tudor gardens (Tusser in the *Five Hundred Pointes of Good Husbandrie*, 1573, includes Sperage among the 'herbes and rootes for sallets and sauce'), a delicate vegetable which had the reputation, according to Gerard and Culpepper, of increasing seed and stirring up lust. In several places this garden Asparagus was already escaping and establishing itself in meadows and marshes.

5. Butcher's Broom. *Ruscus aculeatus* L. 41

Local names. JEW'S MYRTLE (as the material of the Crown of Thorns), Kent; KNEE HOLLY, S Eng, Kent, Ess; KNEE HOLM, Hants, I o W, Suss, Kent; KNEE HULL, Ess; SHEPHERD'S MYRTLE, I o W.

Always a little hard to realize that this prickly shrub, so often condemned to slow, dusty starvation in the shrubbery, is a member of the aristocratic family of Lilies. Turner, in his *Libellus* of 1538, called it Butcher's broome or Petygrew.

How much did it have to do with butchers? William Coles, in his *Adam in Eden*, 1657, wrote 'now it is used by few unless it be *Butchers* who make cleane their stalls, and defend their meat from the flyes therewith', which does not suggest he had seen the Butcher's Broom in the hand of a butcher. However, W. A. Bromfield declared in the *Flora Vectensis*, 1856, that butchers in his time decorated 'their mighty Christmas sirloins with the berry-bearing twigs', which were also used in the Christmas decorations of church and home. Identified with the *mursine agria* or Wild Myrtle of Dioscorides, it was reputed to be 'of a gallant clensing and opening quality' (Culpepper, etc.). It has also been said that the young shoots can be eaten, but this again is derived from Dioscorides' account of *mursine agria*. He says that its young shoots are edible, but a trifle bitter, exciting urine.

6. Snake's Head, Fritillary. *Fritillaria meleagris* L. 22

Local names. BLOODY WARRIOR (each one having grown from a drop of Dane's blood. See *Sambucus ebulus*), Berks; CHEQUERED LILY, Som; COWSLIP (probably for the same reason as with *Primula veris*), Hants; CROWCUP, Bucks; DAFFODIL, Hants; DEAD MEN'S BELLS, Shrop; DEATH BELL, Cumb; DOLEFUL BELLS OF SORROW, Oxf; DROOPING BELL OF SODOM, Dor; DROOPING TULIP, Ches.

FALFALARIES, Yks; FIVE-LEAVED GRASS, Oxf; FRITS, Berks; FROCKUP (frog cup, or from the segments of the perianth which hang like a frock and then turn up?), Bucks; GUINEA-HEN FLOWER, Cumb; LAZARUS BELL (Lazarus rising from the dead), Dev; LEOPARD'S LILY, Dev, Som.

MOURNFUL BELL OF SODOM, Som; OAKSEY LILIES (from growing in Oaksey parish), Wilts, Glos; PHEASANT'S LILY, Cumb; SHY WIDOWS, War; SNAKE'S HEAD, Som, I o W, Surr, Mddx, Berks, Oxf, Herts; SNAKE'S HEAD LILY, Bucks, Oxf; SNOWDROPS (when white, or strictly white tinged with green), Hants; SOLEMN BELLS OF SODOM, Som.

TOAD'S HEAD, Wilts (Minety parish); TURKEY'S EGGS, Berks; TURK'S HEAD, War; WEEPING WIDOW, N'hants, Staffs; WIDOW WAIL, Shrop; WILD TULIP, Berks, N'hants, War.

Of Fritillaries seen near Oxford in 1864, Gerard Manley Hopkins wrote in his diary: 'Snake's-heads like drops of blood. Buds pointed and like snake's heads, but the reason of name from mottling and scaly look.' The names show how the Fritillary has suggested blood, death, snakes, and sorrow. Yet the Elizabethans, who were connoisseurs of death, felt no such feelings. When Lyte translated the herbal of Dodoens in 1578, he wrote: 'The nature and vertue of these flowers, are yet unknowen, nevertheless they are pleasant and beautifull to look upon.' So Gerard, nineteen years later: 'Of the faculties of these pleasant flowers there is nothing set downe in the ancient or later writers, but (they) are greatly esteemed for the beautifieng of our gardens, and the bosomes of the beautifull.'

Observe that neither Gerard nor Lyte, nor any English writer of the sixteenth or seventeenth century, knew of wild Fritillaries in Great Britain. Gerard's Fritillaries came from France: 'These rare and beautifull plants grow naturally wilde in the fields aboot Orleance and Lions in Fraunce, from whence they have been brought into the most parts of Europe. The curious and painfull Herbarist of Paris *John Robin*, hath sent me many plants thereof for my garden, where they prosper as in their owne native countrey.'

Fritillaries have a patchy range from Sweden to France and Yugoslavia. They are accepted as English natives, which is geographically possible. Yet the first record of them as wild plants is hardly more than two centuries old. It was noted in 1736 that they had been growing for more than forty years in Maud Fields near Ruislip Common, in Middlesex. 'Wild' Fritillaries were first recorded in the Oxford neighbourhood, '*in pascuis humidis prope Oxford*', 'in damp meadows near Oxford', as late as 1780, with a more detailed record of 1785 that they grew in Magdalen College Meadow. Claridge Druce wrote in his *Flora of Oxfordshire*: 'It is not a little singular that the Fritillary, so conspicuous a plant of the Oxford meadows, should have so long remained unnoticed by the various botanists who had resided in or visited Oxford.'

It seems not only singular, but impossible. The only explanation is that the Fritillaries were not there, that they are not native, that they spread from foreign plants set in the Tudor or Jacobean garden. Here, after all, is a plant of the water meadow, not of the mountain or the cliff, a plant which flourishes under man's nose in a man-made habitat, along with Cuckoo-flower and Cowslip. It could never have been overlooked by Turner, Gerard, Parkinson, Johnson, Ray. It is always worth scrutinizing the neighbourhood of fritillary fields for evidence of a large garden at some time in the past. Thus at Oaksey, in Wiltshire, the acres of 'Oaksey Lilies' almost certainly originated in the large grounds of two country houses, now vanished, but marked on a map of 1773.

The Elizabethan botanists had to invent, or rather borrow, their names for *Fritillaria meleagris*. To continental botanists and gardeners it was Meleagris, i.e. guinea-fowl, because each segment of the perianth 'seemeth to be the feather of a Ginnie hen'. So it was called 'Ginny hen flower' or 'Turkie hen flower' (Gerard). Meleagris was a name first used by Dodoens. *Fritillaria* was coined by de l'Obel from *fritillus*, a dice-box. He seems to have been thinking less of the shape of the perianth than of the look of a dice-board divided up into squares, which was how *fritillaria* was understood by Gerard – 'of the table or boord upon which men plaie at chesse, which square checkers the flower doth very much resemble'. Hence his third name, 'Checkerd Daffodill' (cf. the French *damier*, literally 'draughtboard').

Things to notice about Fritillaries are the way in which early morning or evening sun shines through the perianth, the fertilizing bumble-bees which slip in and out, and the nectaries, which glisten low down on the inner surface of the perianth segments, and are sweet to the tongue. These snaky, deathly beauties are in fact poisonous. They contain a poisonous alkaloid not unpleasantly named Imperialine (123).

The car is doing the 'wild' stock of Fritillaries no good. People drive to the fritillary fields, park their cars, and pick bundles big enough to fill a vase in every room of a mansion. In his article 'The Snake's Head' (*Lily Year Book*, 15, 1951–2), Dr W. B. Turrill, of Kew, says that between 1941 and 1945 the number of Fritillaries around Oxford had so greatly decreased (except in Magdalen Meadows where they are carefully protected) that hardly a blossom was to be found by Sandford Lock and in the meadows between Oxford and Iffley.

7. Yellow Star of Bethlehem. *Gagea lutea* (L.) Ker-Gawl. 46

This engaging little creature was added to the list of British plants by Mathias de l'Obel, who had gathered it in Somerset (*Stirpium Adversaria Nova*, 1570). He described it as a yellow Ornithogalum, or Star of Bethlehem. 'The flowers of this are on the backside, of a pale yellow stripped with greene, on the inside of a bright shining yellow colour' (Gerard); but after a journey to one of its rare meadow or woodland stations in March or April, the trouble often is to find, not a plant, but one with flowers.

8. Star of Bethlehem. *Ornithogalum umbellatum* L. 58

Local names. APOSTLES, Som; BETTY-GO-TO-BED-AT-NOON, Shrop; ELEVEN O'CLOCK LADY (cf. French *dame d'onze heures*), Som; JACK-GO-TO-BED-AT-NOON, Som, Ches; JOHN-GO-TO-BED-AT-NOON, Shrop; MORNING STAR,

Hants; NAP AT NOON, Shrop, USA; NOON PEEPERS, Wilts; ONE O'CLOCK, Dev; OPEN AND SHUT, Wilts; PEEP O' DAY, Shrop.

SHAMEFACED MAIDEN, Wilts; SIX O'CLOCK FLOWER, Midlands; SIX O'CLOCKS, Bucks, N'hants; SIX O'CLOCK SLEEPERS, Lincs; SLEEPY DICK, Lancs, USA; SNOWFLAKE, Dev, Wilts; STAR AND GARTERS, Wilts; STAR FLOWER, Bucks, Berks; STARRY EYES, Som; STARS, Berks; SUNFLOWER, Dev; TEN O'CLOCK LADY, USA; TWELVE O'CLOCKS, Corn, Som, Dor, Oxf; WAKE AT NOON, Wilts, I o W.

'Stars of Bethlehem grow in sundrie places which lie open to the aire,' wrote Gerard; and among the grass *Ornithogalum umbellatum* can easily be overlooked. The flowers, each like a rayed star, and so recalling the star of Christ's nativity and the journey of the Three Kings, shut early in the day, and shut always in dull weather, and the back of their white perianth segments are striped with green. The leaves are striped with white, but they are grassy and inconspicuous.

An old and charming 'clock flower' of the garden. Turner (*Names*, 1548) wished to call it 'Dogges Onion' or 'Dog leke', after the German, but Star of Bethlehem was the name used by Gerard and the name which has been used ever since. Tusser, in 1573, had included a 'Star of Bethlehem', perhaps the same plant, among the 'herbs, branches, and flowers, for windows and pots' (187), and it might be charming to plant *Ornithogalum umbellatum* in this way.

In the United States, known as Sleepy Dick, Ten o'clock Lady, and Nap at Noon, *Ornithogalum umbellatum* is a nuisance, a weed which has been introduced and which has escaped, presumably, from the garden (131). With us it may or may not be a native, though it looks native over the sandy stretches of the Breckland in East Anglia. According to Sowerby's *English Botany* (3rd ed., 1899), the bulbs are edible and are much eaten in the Near East, where they are, or were, dried and taken on journeys, especially on the pilgrimage to Mecca; though fresh or in hay this innocent seeming plant is poisonous to cattle (134).

9. Bath Asparagus. *Ornithogalum pyrenaicum* L. 10

Local names. BATH ASPARAGUS, FRENCH ASPARAGUS, PRUSSIAN ASPARAGUS, WILD ASPARAGUS, FRENCH GRASS, Som, Wilts, Glos.

Confined to seven counties, Somerset, Gloucestershire, Wiltshire, Berkshire, Sussex, Norfolk, and Bedfordshire. Near Bath and Bristol, in the

Avon valley, it is very common in the copses and on the banks. The slack, bluish-green leaves emerge like bluebell leaves which have no strength. In May the long flowering stems push up from the centre of the leaves, and in June the flowers open, greenish-white, close, and delicate.

Before the flowers expand, bunches are picked and taken into the shops at Bristol and Bath; and they are good eating after the manner of the garden asparagus – though a dainty of the spring which could only be forced into the stomach of a moral field-botanist.

10. Spring Squill. *Scilla verna* Huds. 31, H 7

Local names. SEA ONION, I o M; SWINE'S BEADS, Ork; SWINE'S MURRIKS (i.e. roots), Shet.

11. Autumn Squill. *Scilla autumnalis* L. 11

Neither the Spring Squill nor the Autumn Squill impressed our forefathers. They are too local, though they wash their blue across acres of sea turf. Lousewort and Spring Squill are often intermingled, giving whole slopes of bright pink and papery pale blue.

The flowers of the Spring and the Autumn species are all the more affecting in detail by the contrast of the violet anthers (*S. verna*), or the dark violet, almost black anthers (*S. autumnalis*), with the open blue of the perianth. These plants do not produce the squills of the chemist's shop, which come from the foreign *Urginea maritima*, the true Sea Onion, or French Onion, as it was called by William Turner.

12. Bluebell, Wild Hyacinth. *Endymion nonscriptus* (L.) Garcke. 110, H 40

Local names. ADDER'S FLOWER, Som; BELL BOTTLE, Bucks; BLOODY MAN'S FINGERS (see *Orchis mascula*), Glos.

BLUEBELL, Corn, Dev, Dor, Som, Glos, Kent, Bucks, Oxf, Herts, War, Worc, Shrop, Ches, Derb, Leic, Rut, Lincs, Lancs, Yks, Dur, Cumb, N'thum; BLUE BONNETS, Som; BLUE BOTTLE, Dor, Som, Wilts, I o W, Kent, Surr, Bucks; BLUE GOGGLES, Wilts; BLUE GRANFER-GREYGLES (cf. *Orchis mascula*, *Melandrium rubrum*), Dor, Som; BLUE ROCKET, Ire; BLUE TRUMPETS, Som; BUMMACK, BUMMUCK, Donegal.

CRAKE-FEET (i.e. 'crow feet'), N Eng; CRAWFEET, Lancs, West; CRAW-TAES (i.e. crow toes), Berw, Scot; CROSS-FLOWER, Dev; CROW-BELLS,

Wilts; CROW-FLOWER, Dev, Som, Wilts, Hants; CROWFOOT, Lincs, Lakes, West, Rad; CROWPICKER, Donegal; CROW'S LEGS, Wilts; CROWTOES, Ire.

CUCKOO, Corn, Dev; CUCKOO-FLOWER, Corn, Dev, Som; CUCKOO'S BOOTS, Dor, Shrop; CUCKOO'S STOCKINGS, Staffs, Derb, Notts; CULVERS (i.e. wood pigeons), Oxf, Ess; CULVERKEYS, Som, Kent, N'hants; FAIRY BELLS, Som.

GOOSEY-GANDER, Dev, Dor, Som; GOWK'S HOSE, Scot; GRANFER-GREGORS, GRAMMAR-GREYGLES, Dor; GRANFER-GRIGGLES, Dor, Som; GRANFER-GRIGGLESTICKS, Som; GREYGLES, Dor, Wilts; GRIGGLES, Som; HAREBELL, Dev, N Ire; LOCKS AND KEYS (cf. Culverkeys), PRIDE OF THE WOOD, Som; RING O' BELLS, Lancs; ROOK'S FLOWER, Dev; SINGLE GUSSIES, SNAKE'S FLOWER, Som; SNAPGRASS (from the rubbing, clicking noise of the stalks when gathered), Kent; WOOD BELLS, Bucks.

Names in Welsh include *botasen y gôg*, 'cuckoo's boots', *croeso-haf*, 'welcome summer', and *glas y llwyn*, 'blue of the wood'. In Gaelic, also, *brog na cubhaig*, 'cuckoo's shoe'; in Irish, *fuath muice*, 'pigs' hate'; in Cornish *blejen an gucu*, 'cuckoo flower'.

The Bluebell has an Atlantic or Oceanic distribution. It belongs to the western fringe of Europe, so for the early botanists it was a plant with no history and no warrant from Greece and Rome. They attempted, it is true, to make it into a hyacinth, but it was *hyacinthus non-scriptus*, a hyacinth *not* inscribed with *AI*, *AI* on the petals, not the flower which sprang from the blood of Hyacinthus, carrying those letters of grief. For centuries it had raised its wet blue in the oak forests of Great Britain. But there is no record of it at all before William Turner's brief entry in *The Names of Herbes*, in 1548:

'The commune Hyacinthus is muche in Englande aboot Syon and Shene, and it is called in Englishe crowtoes, and in the North partes Crawtees. Some use the rootes for glue.'

No more. In his *Herbal*, in the third part published in 1568, he adds that this Crowtoes, Crawtees, or Crowfote is a remedy 'against the bytinges of a felde spider' (an item lifted from the account of *huakinthos* in Dioscorides, which was taken to be *Hyacinthus orientalis*), and he amplifies the glue: the boys in Northumberland where he was born – 'scrape the roote of the herbe and glew theyr arrowes and bokes wyth that slyme that they scrape of'.* Lyte in 1578 calls the plant 'hyacinthe or Crowtoes', Gerard in 1597 calls it *Hyacinthus Anglicus*, Blew English Hare-bels, or English

* The 'slime' does, in fact, make a strong glue.

Jacint – English, 'for that it is thought to grow more plentifully in England than elsewhere', which is true, since it is found all through England, Wales, Ireland, Scotland, the Scilly Islands, etc. (though not in Orkney or Shetland). He appreciated the 'naked or bare stalks, laden with many hollow blew flowers, of a strong sweete smell', repeated Turner on the glue, and added that the bulbs gave a starch second only to Starch from Wake Robin (*Arum maculatum*).

There is a good deal to suggest that Gerard's *Herbal* was read by Shakespeare, so helping him to objectify his own experience of plants. To Gerard's entry we may owe 'the azured hare-bell' in *Cymbeline* (IV. ii. 222). Arviragus promises the supposedly dead Fidele (Imogen in disguise) the pale primrose which is like her face, the 'azur'd hare-bell' which is like her veins, and the eglantine which was no sweeter than her breath – 'wench-like words' which annoyed Guiderius, and which show the Bluebell beginning to exist independently in the English mind, and for its own sake. Actually 'Bluebell', in the sense of *Endymion nonscriptus*, does not appear in print until 1794 in Martyn's edition of Rousseau's *Botany*, according to the *Oxford Dictionary*; although it occurs much earlier as a name for *Campanula rotundifolia*. It is not really until the nineteenth century that Bluebells begin to be much noticed and often mentioned – e.g. Keats on the 'shaded Hyacinth, always Sapphire Queen of the mid-may', which is a bit of cockney affectation, or Tennyson on Bluebells being like the blue sky breaking up through the earth, or William Barnes on the wind coming from the copses and carrying the scent 'O' graegles wi' their hangèn bells'.

It is Gerard Manley Hopkins whose verbal analysis and synthesis of Bluebells is most extraordinary of all. In his journal for 1871 he describes Bluebells coming 'in falls of sky-colour washing the brows and slacks of the ground with vein-blue' (a reminiscence of Shakespeare in *Cymbeline*). 'The stalks rub and click . . . making a brittle rub and jostle like the noise of a hurdle strained by leaning against.' He noticed their 'faint honey smell'. In May 1873 he wrote: 'Bluebells in Hodder Wood, all hanging their heads one way. I caught as well as I could while my companions talked, the Greek rightness of their beauty, the lovely – what people call – "gracious" bidding one to another or all one way, the level or stage or shire of colour they make hanging in the air a foot above the grass, and a notable glare the eye may abstract and sever from the blue colour – of light beating up from so many glassy heads, which like water is good to float their deeper instress in upon the mind' – instress being a special word which Hopkins used for the whole flush, truth, force, and reality of things, that which is deeper and more total than the mere design or pattern.

In his poems there are two extraordinary bluebell passages. One of them,

> Or like a juicy and jostling shock
> Of bluebells sheaved in May,

reshapes the entry in the journal on bluebell stalks, rubbing, clicking, and jostling.

> And azuring-over greybell makes
> Wood banks and brakes wash wet like lakes,

in another poem, sets down Hopkins's own 'instress' of *Endymion non-scriptus*, in which he includes Shakespeare's 'azured hare-bell' and Barnes's 'graegles wi' their hangèn bells'.

Names in common, and other circumstances as well, suggest that there was a folk relationship between *Endymion nonscriptus* and *Orchis mascula*, which often grow together. Names they share are Snake's Flower, Adder's Flower, Bloody Man's Finger, Crow-flower, Crowtoes, Cuckoo, Cuckoo-flower, Locks and Keys, Goosey-ganders, Single Guss, Granfer Griggles – for which error or confusion is too simple an explanation. Possibly both these juicy plants of springtime, and not merely *Orchis mascula*, symbolized generation and sexual power. In support of this, look at the Unicorn tapestries, in the Cloisters, New York, at the final tapestry, probably made for the marriage of Francis I of France in 1514. The unicorn, as I have mentioned, is tied to the pomegranate of fertility, and is surrounded by plants of sex, including *Viola odorata*, *Vinca minor*, *Arum maculatum*, *Polygonum bistorta*, and *Orchis mascula* – with the Bluebell immediately alongside (1).

Though Bluebells, as if in obedience to old symbolism, are far too greedily picked every year, independent tests by Dr T. R. Peace in Bagley Wood, near Oxford, and Dr John Gilmour in Kew Gardens have shown that the bulbs are not weakened either by mere picking at ground level or by pulling the stems up long and white. Damage is done by treading the leaves without mercy, which prevents the plant from manufacturing its food (*New Phytologist*, 48, 1949).

13. Grape Hyacinth, Starch Hyacinth. *Muscari racemosum* (L.) DC., non Mill. 5

A charming flower to be able to count among the natives, though it belongs almost entirely to the dry side of England, to Norfolk, Suffolk, and Cambridgeshire, a plant of central European type, peering through the grass of the Gogmagog Hills or growing on the sandy surface of the Breckland. Sir John Cullum, a member of a Suffolk family with a passion for plants in generation after generation, lit upon this little beauty for the first time in 1776 (W. M. Hind, *Flora of Suffolk*, 1889).

It was called Starch Flower because the scent of the tight blue flowers (said to cause sickness or headache) was likened to the smell of wet starch (191).

14. Crow Garlic. *Allium vineale* L. 93, H 15

Local names. AARON'S BEARD, Wilts; CROW GARLIC (cf. German *Kräjelook*, Dutch *kraailook*), Herts, N'thum; CROW ONION, War, Worc; RUSH LEEK, USA; WILD ONION, War.

This weed of the heavy clays which grows tall, stiff, and hard in the corn, has been familiar, named, hated, and persecuted for centuries – the *crāwan lēac*, 'crows' leak', no doubt, of the Anglo-Saxon farmer. Garlic is not, or should not be, a general term for different kinds of *Allium*; it is the one kind, the condiment, *Allium sativum*, the *gār lēac*, or 'spear leek', with a long spathe pointed like a spear. A leek fit for crows distinguishes the wild and useless and harmful leek from the 'spear leek' of the gardens.

Aaron's Beard is an apt name for the common variety of Crow Garlic, *A. vineale* var. *compactum* (Thuill.) Boreau, in which there are bulbils and no flowers. The green points from the bulbils wiggle upwards into an untidy beard.

Crow Garlic can give a garlic nastiness to milk and butter simply if the cows inhale the vaporized substances it gives off (134).

15. Snow Bell, Triquetrous Garlic. *Allium triquetrum* L. 11, H 3

A beautiful plant from the western Mediterranean, with a triangular flowering stem and drooping white bells marked with green. Along the roads of Penwith, the Land's End district of Cornwall, it blossoms through April and May like a white Bluebell. Cross to the Isles of Scilly, and there, even more, it lightens the spring landscape, growing in the walls, on the grass verges, on the banks among the brambles and the sprouting bracken, and in the churchyards.

It was in Cornwall that *Allium triquetrum* was first noticed (for England) in the 1860s or the early 1870s, though in the Channel Islands (where those who allow it to grow are fined), it had been known some twenty years earlier. It can be a noxious weed, but one can be glad that it goes on spreading in the south-west – outside the market-gardener's plot.

16. Ramsons. *Allium ursinum* L. 109, H 40

Local names. BADGER'S FLOWER, Wilts; BRANDY BOTTLES, Dor; DEVIL'S POSY (cf. *Tüfelschnoblech*, 'devil's garlic', Switzerland), Shrop; GARLIC,

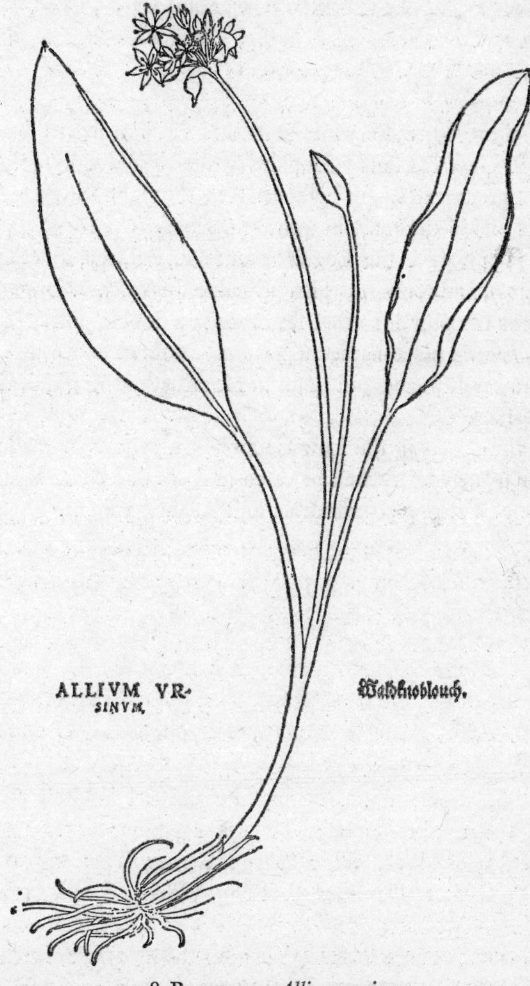

78 Ramsons *Allium ursinum*

Glos, Surr, War, Ches; GIPSY'S GIBBLES (i.e. young onions), Som; GIPSY'S ONIONS (cf. German *Zigeuner Knoblauch*, *Zigeunerlauch*), Som, I o W, S Eng; IRON FLOWER, Som; MOLY (*Allium ursinum* was at one time *Moly Hippocraticum*), Dev, Som; ONION FLOWER, ONION STINKERS, Som.

RAMPS (*rams* in Swedish, Danish, Norwegian, German), Lancs, Cumb, Lakes, N'thum, Scot, Ire; RAMSDEN, I o W; RAMSEY, Corn, Dev, Norf; RAM'S HORNS, Glos; RAMSON(S) (see below), general dialectal use, Eng, Scot, Ire; ROMMY, ROMS, Yks; ROSEMS, Staffs, Yks; SNAKE'S FOOD, SNAKE PLANT, STINKING JENNY, STINKING LILIES, Som; STINK PLANT, Lincs; WILD GARLIC, Dev, Glos, Surr, War, Worc, Ches; WATER LEEK, Som; WILD LEEK, Berw.

Not to be despised, these white stars and viridian leaves, because of a garlic smell. Gerard Manley Hopkins wrote of a wood 'curled all over with bright green garlic' (in his journal in 1871), and in blossom or leaf Ramsons is one of the most beautiful floorings.

Turner (1548) gave the English names as Ramsey, Bucrammes (i.e. buck rammes), and Rammes. The OE name was *hramsa*. *Hramsan*, giving Ramson, was the plural, so that Ramsons is a double plural. There are a good many *hramsa* place-names, e.g. Ramsbottom in Lancashire, 'Ramson valley', Ramsey in Essex and again in Huntingdonshire, meaning 'Ramson island' (66).

Gerard wrote that in the Low Country fish sauce was made from the leaves, which 'maye very well be eaten in April and Maie with butter, of such as are of a strong constitution, and labouring men'.

17. Autumn Crocus, Naked Ladies. *Colchicum autumnale* L. 55, H 3

Local names. AUTUMN CROCUS, Som, War; DAGGERS, Som; FOG CROCUS (from growing in fog, or coarse grass), Yks; GO-TO-SLEEP-AT-NOON, Som; KITE'S LEGS, Kent; MEADOW CROCUS, Yks; MICHAELMAS CROCUS, Som, Wilts.

NAKED BOYS, Som, Wilts, Hants (see Gilbert White's *Journal*, for 28 August 1789), Norf, Heref; NAKED JACKS, Som; NAKED LADIES, Dor, Som, Wilts, Glos, War, Worc, Shrop, Lancs, Yks; NAKED MAIDENS, NAKED MEN, Dor; NAKED NANNY, Som, Wilts, Glos; NAKED VIRGINS, Ches.

POP-UPS, Som; PURPLE CROCUS, Yks; SNAKE-FLOWER, Som; STAR-NAKED BOYS, Norf; STRIP-JACK-NAKED, Dev; UPSTARTS, Som.

The closed purple segments of the flower are like a snake's head in the dry autumn grass. The naked flower slides up a little more, the segments open – a deeper purple on the inner surface – and they reveal the orange anthers. The purple segments and the white tube, which carries underground into the corm a trio of white styles, are as shiny and sleek as a glass flower in the Botanical Museum at Harvard.

Never was flower so naked – not merely naked of the leaves, which do not emerge until the spring. Naked Boys, Naked Jacks, Naked Ladies, Naked Nannies, Naked Maidens, Naked Virgins. So also in French, *dames nues*, *dames sans chemise*, *cul tout nu* ('naked behind'), and in German *Nackende Jungfrau*; also, whether from a Lutheran horror of nakedness or the poison of *Colchicum autumnale*, or both, *Nackende Huren*, 'naked whores'. Other German names picture the flower as the female private parts (129). In Czech it is *naháč*, 'the naked one' (masculine). The associations are older still. An allied *colchicum* appears to have been known to the Assyrians, from the shape and structure of the flower, by a name meaning 'Come, let us copulate'. An Arabic name for *colchicum* is *sirāj al-ghūlah*, 'lamp of the ghoul' (30).

This Naked Lady offers a moral lesson, since she is dangerous in all her parts. 'It is good to knowe this herbe,' wrote Turner (who called it Meadow Saffron), in the third part of his *Herbal* in 1568, 'that a man may isschewe it, it will strangell a man and kyll him in the space of one daye, even as some kindes of Tode Stolles do . . . if any man by chaunce have eaten anye of this, the remedye is to drink a great draught of cowe milke.' This had been lifted direct from Dioscorides, out of the description of *kolchikon* which he included only as a warning. The early herbalists and physicians knew that *Colchicum autumnale* had been given against gout by some of the Greek and Arabian doctors, but they were afraid of it, with every reason, and with all the authority of Dioscorides – negative in this case – to back them. Hieronymus Bock (1498–1554), one of the fathers of the new botany, had uttered a condemnation. Dodoens called it *perniciosum Colchicum*, and Lyte in his version of Dodoens wrote that 'Medow or Wilde Saffron is corrupt and venemous, therefore not used in medicine'. Gerard quoted the Alexandrian physician Paulus Aegineta (who flourished *c.* A.D. 640) on 'meade saffron' as proper for the gout, but he gave the customary warnings of death and danger.

Sir Theodore Mayerne (1573–1655) is said to have dosed King James I with *Colchicum* and the powder of unburied skulls (72), but it was not until the end of the eighteenth and the beginning of the nineteenth centuries that *Colchicum* was properly investigated and regularly administered

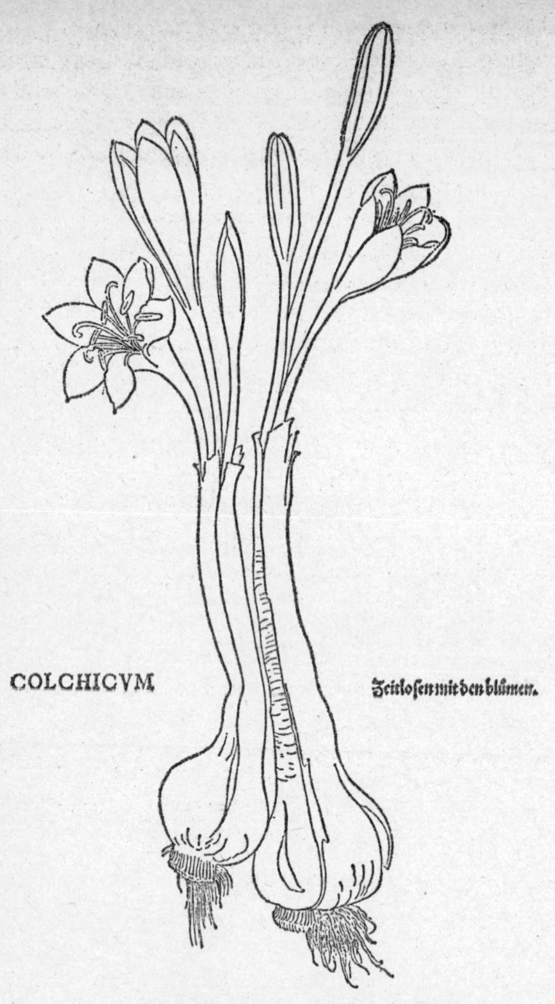

79 Naked Ladies *Colchicum autumna*

to stop the horrible throb and heat and pain of gout – in the shape of the essential and palliative *Tinctura Colchici*, made from the seeds and the corm, which is still in the British Pharmacopoeia and is likely to remain in use, as one of the last of the pure plant remedies. Oxfordshire and Gloucestershire used to supply London with *Colchicum autumnale*, though modern supplies are imported.

Archdeacon Paley, in his *Natural Theology*, 1802, used the Autumn

Crocus as evidence of design in the world, and so of an intelligent designer, i.e. God: 'I have pitied this poor plant a thousand times. Its blossom rises out of the ground in the most forlorn condition possible; without a sheaf, a fence, a calyx, or even a leaf to protect it: and that not in the spring, to be visited by summer suns, but under the disadvantages of the declining year.' However, looking closer into the structure of the Autumn Crocus, he concluded that the way its styles reach underground and its seeds appear above ground in the spring was a divine compensation for flowering so late in September – though Paley does not explain why his Designer, in the first place, designed it to flower so late.

Guillaume Apollinaire (in *Alcools – Poèmes 1889–1913*) has a poem *Les Colchiques* about the Autumn Crocus (which purples a vast acreage of autumnal France) – about the woman with crocus-coloured eyes, the cows which slowly poison themselves, and his own life which slowly poisons itself as well, on account of, too, crocus-coloured eyes. '*Le pré*', begins the poem, '*est vénéneux mais joli en automne.*'

The first British record was William Turner's in 1551, that the Meadow Saffron grew (as it still grows) 'in the West cuntre besyde Bathe'.

XCII. Trilliaceae

1. Herb Paris. *Paris quadrifolia* L. 77

Local names. DEVIL-IN-A-BUSH, Perth; FOUR-LEAVED GRASS, Som; TRUE-LOVE, Som, Berks, Herts, War, Dumf; TRUE-LOVERS' KNOT, Som, Yks, Cumb.

Herb Paris attracted notice by the numerical harmony of its parts – *herba paris*, herb of equality. A usual specimen has an ovary with four cells and four styles, four inner and four outer segments to the perianth, subtly different in their greeny-golden tone and their outline. There are twice four stamens, and there are four leaves. At first the stem upholds a greeny-golden star, then one single black shining berry.

Clearly a powerful herb. Pierandrea Mattioli (1501–77), the Italian botanist, recorded the use of the berries against witchcraft, or the mental consequences of witchcraft. You became half out of your senses, or you had an attack of epilepsy: the black berries of Herb Paris put you right, but they needed to be given, from this herb of equality, in unequal numbers – three, five, seven, or nine, said the German doctor Martin

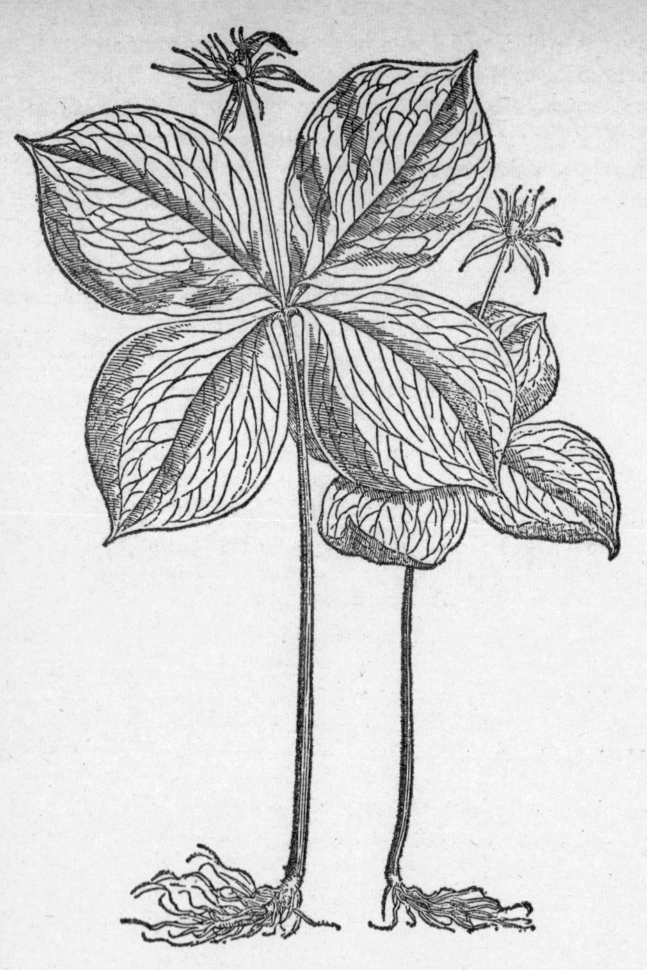

80 Herb Paris *Paris quadrifolia*

Blochwich, reporting the practice of matrons who valued Herb Paris (14). The berries were taken against poison and the plague, the roots gave a medicine against colic, the leaves were laid upon wounds, sores, and tumours (141, 49). Here was a herb, wrote Culpepper, 'fit to be nourished in every good Woman's garden'.

Par, in Latin, means not only 'equal' but a 'pair', and the arrangement of the leaves suggested the pair of lovers, or rather the endless True Lovers' Knot. Gerard was familiar with the name 'Herb Truelove' as well as with Herb Paris, and William Turner's invention of One Berry. 'It is much in Northumberland,' Turner had written in 1548, 'in a wodde besyde Morpeth called Cottingwod. It hath foure leaves lyke unto great plantaine, and in the overmost top a litle blacke bery lyke a blacke morberry, but blacker and greater.' He invented another name for Herb Paris – 'Libardsbayne', leopard's bane, from *pardalianches*, the alternative name in Dioscorides for an aconite which Turner and Fuchs took to be *Paris quadrifolia*.

Herb Paris is in fact poisonous – 'emetic, purgative, intensely acrid, and narcotic' – though it is not among the killers (123).

XCIII. Amaryllidaceae

1. Loddon Lily, Summer Snowflake. *Leucojum aestivum* L. 13, H 11

Local names. LODDON LILIES, Berks; MOUNTAIN SNOWDROP, SUMMER SNOWDROP, War.

Imagine, if you have never seen the wild Loddon Lily, a black swamp on the edge of the Thames, alders or willows overhead, a swamp which quivers and soggs and stinks. In the gloom, not the more usual light of Marsh Marigolds, but white flowers hanging in a severe purity from the end of the long stems. One thinks at first of an extra long Snowdrop, then of a new garden escape, except that there is no garden.

William Curtis discovered *Leucojum aestivum* apparently wild – 'undoubtedly wild', according to him, 'betwixt Greenwich and Woolwich . . close by the Thames side, just above high water mark, growing (where no garden, in all probability, could ever have existed) with *Arundo phragmites*, *Caltha palustris*, *Oenanthe crocata*, and *Angelica sylvestris*.' This was in, or before, 1788. He found it, too, on the Isle of Dogs on the other side of the river; and since it differed essentially from the Snowdrop and was generally called Great Summer Snowdrop in gardens, he took the opportunity of coining the new English name of Snowflake (50).

One discovery led to another: along the Thames and its tributaries and elsewhere, along the Shannon in south-west Ireland, on riverside mud and in wet meadows. In Berkshire it is called the Loddon Lily from growing by 'The Loddon slow, with verdant alders crown'd', which Pope celebrated in *Windsor Forest*. Claridge Druce, in his *Flora of Berkshire*, drew attention to its 'great luxuriance and beauty' by this tributary of the Thames, and (in his *Flora of Oxfordshire*) to the quantity of it on a small wooded island in the Thames itself opposite the mouth of the Loddon.

Is it truly wild? One might fear for the answer of a legally trained botanist, strict in the weighing of evidence. Curtis asked 'how so ornamental a plant, growing in so public a place, could have escaped the prying eyes of the many Botanists who have resided in London for such a length of time'. It seemed strange, but he was 'perfectly satisfied' of the Snowflake's native standing, and so is present authority. Yet it is as much open to suspicion as the native standing of *Fritillaria meleagris* (q.v.), for the same obvious reasons. A plant remarkable and excitingly conspicuous, a plant cultivated in gardens since the sixteenth century (see Gerard on *Leucoium Bulbosum maius*, 'the great Bulbed violet', grown in 'our London gardens' for 'many yeeres past'), the seeds of which are dispersed by water, a plant suddenly observed to be wild *by the Thames, below the gardens*, some two hundred years later, and supposed to be truly wild further up the Thames in a score of places where the seeds could have floated down from other gardens; a plant overlooked all that long while by generation after energetic generation of botanists? Impossible. Natural conditions would allow the Loddon Lily to be native, common sense and logic allow nothing of the kind. Stranger things, too, could have happened than the deliberate planting of *Leucojum aestivalis* along the Loddon in honour of Alexander Pope, and to mark his celebration of the river in one of the most famous of all landscape poems – flowers, in fact, for the nymph in Pope's lines who was saved from Pan by being transformed into the cold Loddon:

> The silver stream her virgin coldness keeps,
> For ever murmurs, and for ever weeps.

Botanists, one may suspect, have *wished*, and still wish, both *Leucojum aestivalis* and *Fritillaria meleagris* to be as native as the Primrose.

The case for *Leucojum aestivum* as an Irish native, in the marshes of the Shannon, in Tipperary, Offaly, Waterford, Leix, Tyrone, and Antrim, is a little stronger, and has been much argued (see R. Ll. Praeger's *Botanist in Ireland*, 1934). Yet how was the wild plant overlooked in Ireland until so recently – so much more recently – as the eighteen-nineties? Living botanists must sometimes think poorly of their predecessors' curiosity, energy, and eyesight.

2. Snowdrop. *Galanthus nivalis* L. 60

Local names. CANDLEMAS BELLS (cf. Flemish *lichtmisbloem*), Wilts, Glos; DEATH'S FLOWER, DEWDROPS, DINGLE BELL, DROOPING BELL, DROOPING HEADS, DROOPING LILY, EVE'S TEARS, Som; FAIR MAIDS, Hants, Norf; FAIR MAIDS OF FEBRUARY, Som; FEBRUARY FAIR MAIDS, Som, West.

NAKED MAIDENS (cf. Dutch *naakte wijfjes*, German *Nackte Jungfrau*), PIERCE-SNOW, SNOW-PIERCER (cf. French *perce-neige*), SNOWBELLS (cf. German *Schneeglöckchen*), Som; SNOWDROPPERS, Glos, Bucks; WHITE BELLS, WHITE QUEEN, Som.

Again, one of the plants which may or may not be native, and which grew in Elizabethan gardens, the *Leucoium Bulbosum praecox* or 'Timely flowring Bulbus violet' of Gerard's *Herbal.* Snowdrop does not seem to have been a common word until the end of the seventeenth century. In 1659, in his *Garden Book*, Sir Thomas Hanmer still talks of 'Bulbous Violets', of which *Galanthus nivalis* is 'the EARLY WHITE, whose pretty pure white bellflowers are tipt with a fine greene, and hang downe their heads'. It looks as if 'Snowdrop' had been borrowed from the German *Schneetropfen* (cf. Swedish *snödroppe*), though the usual German name is *Schneeglöckchen*, 'snow bell'. Snowdrops were Candlemas Bells, since they blossom in February: Candlemas, the Feast of the Purification, is 2 February; they were Death's Flower, since in several counties – Shropshire, Staffordshire, Derbyshire, Worcestershire, and Sussex among them

Leucojum aestivum was recorded as a wild plant in the 1780s: 'wild' Snowdrops were first recorded in the 1770s – in Gloucestershire and Worcestershire.

3. Wild Daffodil, Lent Lily. *Narcissus pseudonarcissus* L. 69

Local names. AFFODIL, AFFRODIL (Mediaeval Latin *affodilus*), Ches; BELL-FLOWERS, Dor, Som; BELL-ROSE, Som; BUTTER AND EGGS, Dev, Som, N'hants; CHURN, Lancs; COWSLIP (see *Primula veris*), Dev; CUCKOO-ROSE, Dev, Som.

DAFFODIL – in general use Eng, Scot, Ire. The commonest form from Somerset to the north is Daffydowndilly. Notice also DAFFYDILLY, N'hants; DILLYDAFFS, Som.

EASTER LILY, Dev, Som; EASTER ROSE, Som; FAIRY BELLS, Dor; GIGGARY, Dev; GOLD BELLS, Wilts; GOLDEN TRUMPETS, GOOSEFLOP (cf. Cowslip, above), Som; GOOSE-LEEK, I o M (where it is unlucky in the house till the goslings have hatched. The Manx name is *lus-ny-guiy*, 'goose herb'); GRACIE DAISIES, Dev, Som; GRACIE DAY, HEN AND CHICKENS, Dev; HOOP PETTICOATS, Dor; JULIANS, Herts; KING'S SPEAR, Som; LADY'S RUFFLES, Wilts.

LENT-COCKS, Dev, Som; LENT-LILY, Corn, Dev, Dor, Som, I o W, Glos, Suss, Kent, Surr, Suff, War, Ches, Derb, Lincs, Yks, West; LENT PITCHERS, Dev, Som; LENT-ROSEN, Dev, Som; LENTS, Corn, Dev, Lancs; LENTY CUPS, Som; LENTY LILY, Corn; LILY, Scot; QUEEN ANNE'S FLOWERS, Norf; ST PETER'S BELL, Wales; SUN-BONNETS, Som; WHIT SUNDAY, Dev; WILD JONQUIL, Yks; YELLOW MAIDENS, Som.

The one certain native among the Daffodils; and still as excellent a plant as any hybrid, improved and lengthened and fattened in cultivation. On 1 May 1871 Gerard Manley Hopkins looked with his extra-Pre-Raphaelite eye at Wild Daffodils in Lancashire. 'The bright yellow corolla is seeded

81 Daffodil and Snowflake *Narcissus pseudonarcissus/Ornithogalum umbellatum*

with very fine spangles,' he noted in his journal, 'which give it a glister and lie on a ribbing which makes it like cloth of gold.'

The Daffodil's form and the Daffodil's cloth of gold delighted sixteenth-century Englishmen. They enjoyed the Wild Daffodil for itself as we do (although it did enter into the books as a purge and a vomitive and a cure for erysipelas and the palsy). Gerard wrote that the 'yellow English Daffodil' grew 'almost every where through England'. He is not always reliable on matters of that kind, and did not travel a great deal himself. Yet it may well have been more common, and it may have suffered since then in a number of ways – transplantation, draining, and improvement of heavy grass land, the clearing of woods. Travel into Gloucestershire to see Wild Daffodils at their best, in wood or in meadow (and now carefully and necessarily protected), on the hills towards Herefordshire and the Forest of Dean. Newent is a base for this daffodil country.

Narcissus pseudonarcissus owes a debt also to its English name, which goes back, through the Mediaeval Latin *affodilus* and Latin *asphodilus*, to the Greek *asphodelus*, name of that plant which grew across the meadows of the underworld and which belonged to Persephone, the Queen of Hell.

XCIV. Iridaceae

1. Blue-eyed Grass. *Sisyrinchium angustifolium* Mill. sec. Fernald H 10

It was in 1845, in the damp, mountainy south-west of Ireland, that the Canadian Grass or the Blue-eyed Grass was discovered in a wild context, in a wood in Galway. Some specimens past flowering were exhibited in 1846 at a meeting of the Botanical Society in London. Many years later it was found in Co. Kerry. Here, then, Ireland has a plant which does not belong to next door or the next ditch, but to North America; and authority has pronounced it to be a native, partaking of a natural community. If there is no sun (and there is often no sun in the south-west of Ireland), the blue eyes remain shut, and this plant merges with the grass in the wet meadows and on the wet gravels. It could easily have been overlooked (148, 164).

Arguments quieten the sceptic, even though he realizes that by 1845 gardeners had already known the Blue-eyed Grass for a long while, even though it does escape easily from gardens elsewhere, if not in Ireland, even though plants recognized late as native species are often unusually attractive and have often had a previous history in the garden. But though other American species grow in the western fringe of Europe, there is one thing, in this case, which revives suspicion: and that is yet another 'Irish' species of *Sisyrinchium* – this time, *S. californicum*, the Golden-eyed Grass, with bright yellow flowers, which comes from California (whereas the Blue-eyed Grass grows in the eastern states). The Golden-eyed Grass is inexplicably present (introduced, it is agreed), and naturalized in Ireland, near Rosslare, where it was found in 1896. And like the Blue-eyed Grass, it had been known to gardeners since the eighteenth century.

An odd coincidence.

2. Gladdon, Stinking Iris. *Iris foetidissima* L. 54

Local names. ADDER'S MOUTHS, Som; BLOODY BONES (? from its use in fractures), Dur; BLUE DEVIL, Som; BLUE SEGGIN (i.e. sedges), Ayr; DAGGER-FLOWER (from the leaves, cf. Latin *gladiolus*, 'little sword', for Iris), Dev, Som; DRAGON-FLOWER, Dev; DRAGON'S TONGUE, Kent; FIELD LILY, Dor; FLAG, Dev; GLADDING-ROOT (the medicinal root), Ire; LEVER, Corn; POISON BERRIES, Dev, Som.

ROAST BEEF (smell of the leaves), Som; ROAST BEEF PLANT, Dev, Dor; SEGGIN (i.e. sedges), Cumb, Scot; SNAKE FIDDLES (? vittles, victuals), I o W; SNAKE'S FOOD, Dev, Som; SNAKE'S MEAT, SNAKE'S POISON, Dev.

(The Scottish and Irish names are presumably from Gladdon grown in the physic garden.)

A maligned species. It does not stink so much as to deserve the superlative *foetidissima*. Its curious smell of raw beef – raw rather than roast – it keeps to itself, or in itself, until you crush the strap-like, dagger-like flags. The pencilled lead-blue or lead-purple of the flowers is snakish or dragonish; the capsule opens, and the seeds are small blood oranges, set in rows, clear and formal in the brownness and the decay of November. Gerard described the tint of the flowers as 'an overworne blewish colour, declining to grayish, or an ash colour'.

Since it was equated – not unconvincingly – with the *xuris*, to which Dioscorides allowed a good many virtues, Gladdon was prominent in the herbals and in the physic garden. It provided a purge chiefly. 'This herbe is called in the yle of Purbek Spourgewort, because the juyce of it purgeth' (William Turner in the second part of his *Herbal*, 1562). 'Heereof the countrey people of Sommersetshire have good experience, who use to drinke the decoction of this roote. Others do take the infusion thereof in ale and such like, where with they purge themselves, and that unto very good purpose and effect' (Gerard, 1597). An old practice, this purge by Gladdon rhizome and ale. In the Anglo-Saxon *Lacnunga*, the root of *glaedene* is prescribed in purging draughts with other herbs steeped or boiled in ale (88). Now and again one may find Gladdon flourishing in a herbal context – for instance, in an old churchyard with Alexanders, Danewort, etc.

'Gladdon' originally was an Iris without specifying the kind, the OE *glaedene*, going back to the Latin name *gladiolus*, 'little sword', from the leaves.

3. Yellow Flag. *Iris pseudacorus* L. 112, H 40

Local names. BUTTER AND EGGS, Bucks, Oxf, N'hants; CHEEPER (from making a cheeping noise with the leaves), Rox; CRANE BILL, Som; CUCUMBERS (the capsules), Dev; DAGGERS (from the leaves), Dev, Som, Lancs; DRAGON-FLOWER, Dev, Som; DUCK'S BILL, Corn, Som; FLAGGER, Som, Ire; FLAGGON, Ire; FLAG LILY, Som; FLAG-PLANT, Lincs; FLAGS, Corn, Dev, Som, Hants, I o W, Bucks, Lincs, Notts, Yks; FLIGGER, E Ang.

JACOB'S SWORD, Aber; LAISTER (Cornish *elester*), Corn; LAVERS (OE *lǣfer*, not uncommon in place-names), Corn, Dev; LEAVERS (OE *lǣfer*), Corn, Dev, Dor, Som, I o W, Ire; LEVVER, Dev; MEKKINS, Lancs, Yks, Cumb; POND LILY, Dev; QUEEN-OF-THE-MARSHES, QUEEN-OF-THE-MEADOW, Som; SAGGON, Lanark, Ire; SEGGEN (i.e. sedges), Yks, Cumb, Scot, N Ire; SHALDER, Dev, Som; SHEEP-SHEARS, SHEEP-SHEARING FLOWER, Som; SKEGG, Yks; SOLDIERS AND SAILORS, SWAN BILL, Som; SWORD-FLOWER, Dor; SWORD-GRASS, Wilts, N'thum, Scot; SWORD LILY, SWORDS, Som.

TRINITY PLANT, Dor; WATER-LILY, Dev, Dor, Som, Donegal; WATER-SEGG, Yks; WATER-SKEGG, Scot; YELLOW DEVILS, Som; YELLOW FLAG, Herts, War, Ches.

'Camps of yellow flag flowers blowing in the wind, which curled over the grey sashes of the long leaves' (Gerard Manley Hopkins again, in his journal for 1873). The Yellow Flag, the Wild Iris, had become a poet's plant by the nineteenth century. So with Hopkins's friend Patmore, from *The Angel in the House*:

> Not in the crisis of events,
> Of compass'd hopes, or fear fulfill'd
> Or acts of gravest consequence,
> Are life's delight and depth reveal'd.

The day of days was not the day;
  That went before, or was postponed;
The night Death took our lamp away
  Was not the night on which we groan'd.
I drew my bride, beneath the moon,
  Across my threshold, happy hour!
But, ah, the walk that afternoon
  We saw the water-flags in flower!

82 Yellow Flag *Iris pseudacorus*

In France the Yellow Flag is a special herb of St John's Eve, in Ireland bunches are hung outside the doors on the Feast of Corpus Christi. So it is apotropaic: it averts evil. More practically, by our way of thinking, the rhizomes give a black dye and a black ink; and a Fenland doctor, more than a century ago, maintained that the roasted seeds make an excellent and healthy drink like coffee. Doctors once used the Yellow Flag for its astringency, in stopping the flow of blood (188, 137, 191, 141).

Turner's names were Gladon, flag, yelowe floure de lyce, though he mentioned that his people of Northumberland called it a 'seg', while it was a 'lug' to the men of Ely and the Fens (1538). Gerard called it Water Flags, Bastard Flower-de-luce ('lily flower'), or Water Flower-de-luce.

4. Wild Gladiolus. *Gladiolus illyricus* Koch 2

The garden Gladiolus which everyone likes today is of African and South African descent. Before it reached its super-cinema magnificence or became a blend of a leading boy in a pantomime and a very late nude by Renoir, gardeners were happy with the older European 'Corn Flags', or 'Corn Sedges', or 'Corn Gladins', plants of the sixteenth- and seventeenth-century border which had simplicity, slenderness, grace, and intensity enough of colour.

They were not indigenous; but at last, in 1856, a clergyman found *Gladiolus illyricus* in the New Forest, a species near to the *Gladiolus communis* of the old gardens. It was in two places, two miles apart, each 'at least a mile from any house', not a long way from Lyndhurst. The Gladiolus was growing among the bracken 'which overtops it before it comes into flower' – so, like the Blue-eyed Grass (q.v.), it was a native which had been overlooked (*Annals and Magazine of Natural History*, xx. 158, 1857). Carefully hiding its magenta blossoms in this way, more of *Gladiolus illyricus* was discovered in the Forest and in the Isle of Wight.

To be suspected? Perhaps, but like the Blue-eyed Grass or the Loddon Lily of the Thames and its tributaries, it does fit into a natural community. The Hampshire Basin, moreover, is the island's southerly verge, inhabited by other plants of continental Europe or the Mediterranean, and by a few creatures as well – Bechstein's Bat, the Sand Lizard, and the Smooth Snake, and among the butterflies the Lulworth Skipper and the Glanville Fritillary – which just retain a hold on this side of the Channel.

XCV. Dioscoreaceae

1. Black Bryony. *Tamus communis* L. 70

Local names. ADDER'S MEAT, Dev, Som; ADDER'S POISON, Dev; BEADBINE, Hants; BLACK BINDWEED, Som; BROYANT (from rubbing on animals sick with 'broyant'), Mont; CHILBLAIN-BERRY, Som, I o W; DEVIL'S BERRY, Corn; ISLE OF WIGHT VINE, I o W.

MANDRAKE (distinguishing it from the 'Woman drake' of *Bryonia dioica*, q.v.), Lincs, Yks; MURREN-BERRIES, I o W; OXBERRY, Heref, Worc, Shrop; POISON-BERRY, Dev, Som, Wilts, Suss, Kent; POISONING-BERRIES, Yks; ROLL-BERRY, ROW-BERRY, RUE-BERRY, Dev; SNAKE-BERRY, Suff; SNAKE'S FOOD, Dev, Som; SNAKE'S MEAT, Dev, Dor, Som; WILD VINE, I o W, Ches; WITHYWINNY, Dev.

In Gerard, who gives the first English record of *Tamus communis*, it is called Black Bryony, Our Lady's Seal, and Wild Vine – the last name because it was taken to be the *ampelos agria* of Dioscorides (*Bryonia dioica*, White Bryony, was the *ampelos leuke*). In France, in the late Middle Ages, it was *sigillum Beatae Mariae* (still called *sceau de la Vierge*, *sceau de Notre-Dame*), as if it had been a snake-plant, a poison-plant, an enchanter's herb, which needed reformation. The lovely glittering fruits *are* poisonous and they have caused the death of farm animals (123). Culpepper described *Tamus communis* and *Bryonia dioica* as 'furious Martial Plants'. The roots purged the belly with great violence – 'troubling the Stomach and burning the Liver', and though valuable they needed correcting by 'an abler hand . . . than most Country-people have'.

Dioscorides said that the berries would remove freckles and other spots or blemishes. The use was extended: 'The berries do not only clense and remoove such kinde of spots, but do also very quickly waste and consume away blacke and blewe marks that come of bruses and drie beatings, which thing also the roots performe being laid upon them' (Gerard). So in France Black Bryony was also *herbe aux femmes battues*, 'the herb for beaten wives'.

In the Isle of Wight the berries were laid on chilblains, as a counter-irritant; and to save them for the winter they were preserved in gin or in brandy (21).

XCVI. Orchidaceae

1. Lady's Slipper. *Cypripedium calceolus* L. 4?

Is it extinct, this most tropical-seeming of all English flowers, or still growing somewhere high in a limestone corner of the West Riding, the North Riding, or Durham?

Gerard called it Our Ladies Shoe, Our Ladies Slipper, from de l'Obel's name *calceolus Mariae*, but to his knowledge it flourished only in the mountains of Germany, Hungary, and Poland. Parkinson, by the time he published his *Theatrum Botanicum* (1640), knew it was an English plant growing in Yorkshire, near Ingleborough. He had plants from Yorkshire, and the depredation began.

What a flower! Four long, burgundy-coloured segments, and in the middle a shoe, or a sabot (*sabot-de-Vénus* in French), or more exactly a winkle-shell of clear pale yellow. Altogether too desirable. Other stations were discovered, most of them in similar woods upon limestone, but the orchid was rare, and rapidly became rarer still. In the last years of the eighteenth century Lady's Slippers were sold in the market-place at Settle (126). Botanists, though, and gardeners were all to blame. 'The singular structure, and peculiar elegance of this plant, render it a desirable acquisition for the flower garden; the more so since the indiscreet zeal of simplers to possess this beautiful rarity, and the ravages of certain gardeners, impelled by filthy lucre, have nearly exhausted several of its favourite haunts.' That was written more than a hundred years ago. Mr J. E. Lousley (*Wild Flowers of Chalk and Limestone*, 1950) calls it 'one of the rarest and most elusive of British plants'; and it is one of the very few that this king of field botanists has never seen wild. *Calceolus Mariae*, though, may not be extinct. There is evidence that it is very shy of flowering, and that it fluctuates in number.

Another botanist, Mr V. S. Summerhayes, writes more mysteriously (*Wild Orchids of Britain*, 1951). He does not say that he has never seen it wild. He does say (i) that the most recent record, *in print*, for the Lady's Slipper flowering in Wharfedale is 1937; (ii) that 'it is being carefully preserved in the few places where it is known to occur'.

In Germany rare plants are protected as we protect birds. There the *Frauenschuh* is one of the plants it is illegal either to pick or to remove (122). In Norway it is called not only *Fruesko*, 'Virgin's Shoe', but *St Olafs bolle*, 'St Olaf's bowl'.

2. Autumn Lady's Tresses. *Spiranthes spiralis* (L.) Chevall. 63, H 18

Local names. LADY'S TRESSES, Dev, Yks; LADY'S TRACES, Hants, Yks.

For Orchids in general see *Orchis mascula*. Not many kinds were firmly identified and located as English plants before 1600. William Turner, though, had recognized 'a certeyne ryghte kynde' which grew near Sion House, opposite Kew Gardens, and was already known and named: 'it bryngeth forth whyte floures in the ende of harveste, and it is called Lady traces' (1548). That is the first record. Gerard included Lady Traces under Sweet Cullions (cullion = testicle: see *Orchis mascula*. 'Sweet' for the almond scent of the blossoms). They were not often used in physic, he said, though 'the full and sappy rootes of Ladie traces eaten or boiled in milke and drunke, provoke venery, nourish and strengthen the bodie, and be good for such as are fallen into a consumption or fever Hectique'.

The white blossoms twist spirally around the stem. So the stem looks like a braid of hair – or of the Virgin's hair.

3. Twayblade. *Listera ovata* (L.) R. Br. 111, H 40

Local names. ADDER'S TONGUE, Wilts; MAN ORCHIS, SWEETHEARTS, Som; WILD TULIP, Donegal.

'Sweethearts' from the pair of broad leaves, set against one another, 'Adder's Tongue' and 'Man Orchis' from the labellum, forked like an adder's tongue, or forked like a man. Indeed *Listera ovata* is as good a 'man plant' as the proper Man Orchis, *Aceras anthropophorum*.

Turner gave the first record in 1548, calling it Martagon. Lyte invented Twayblade, Gerard called it Twayblade and Herbe Bifoile, and described each flower as 'resembling a gnat, or a little gosling newly hatched'. He used it in ointment and balsam for healing wounds.

4. Greater Butterfly Orchid. *Platanthera chlorantha* (Cust.) Rchb. 104, H 40

Local names. NIGHT VIOLET, Wilts; WHITE ANGEL, Som.

'That kinde which resembleth the white Butter-flie,' said Gerard in the first English record, 'groweth upon the declining of the hill at the North ende of Hampsteed heath . . . as ye go from London to Henden a village thereby' – which would be somewhere between Jack Straw's Castle and Golder's Green, on one side or the other of the Bull and Bush. As Curtis long ago remarked in the *Flora Londinensis*, 'the English name of *Butterfly Orchis* is scarcely warranted by the appearance of the flowers' – which is exceptional in the naming of orchids.

Gerard placed this orchid, and the Fly Orchid and Bee Orchid, among the Foxstones, of which there was no great use in physic. They were 'regarded for the pleasant and beautifull flowers, where with nature hath seemed to plaie and disport hir selfe'. It is curious that he should not have mentioned the voluminous and possessive fragrance of the Butterfly Orchid.

5. Bee Orchid. *Ophrys apifera* Huds. 61, H 34

Local names. BEE-FLOWER, Dor, Som, Wilts, I o W, Kent; BUMBLE BEE, Dev, Som; DUMBLE DOR, Surr; HONEY-FLOWER, Kent; HUMBLE-BEE FLOWER, Dor.

Gerard again gives the first record, from the Kentish hills and from Hertfordshire, near St Albans: the 'Humble Bee Orchis'. We have dropped the Humble, though an accurate name really requires it.

6. Green-winged Orchid. *Orchis morio* L. 70, H 22

Local names. BELDAIRY, Aber; BLEEDING WILLOW, Bucks; BLOODY MAN'S FINGER, Ches; CRAWFEET (i.e. 'crow-'), Yks; CROWFOOT, Lincs, Yks, Cumb; CUCKOO, Ess; CUCKOO-FLOWER, Dev, Hants, Herts, Ess, E Ang; DANDY GOSLINGS, DANDY GOSHER, Wilts.

GIDDYGANDER, Dor, I o W; GOOSE AND GOSLINGS, Som; GOOSEY GANDER, Dor, Som; KING FINGER (originally applied to *O. maculata*, the old 'Royal Satyrion' or Finger Orchis?), Bucks, N'hants, War, Lincs; KINGFISHER, War; NUNS, Notts.

PARSON'S NOSE, Dev; QUEEN'S FINGERS, War; RAM'S HORN, Suss; SINGLE CASTLES, Isle of Portland.

In Dioscorides the first orchid to be described as *kunosorchis*, the 'dog testicle' or 'Dog's Stone', important in matters of sex and fertility, though a venereal power was allowed to the other kinds. By the sixteenth-century botanists, the Green-winged Orchid was taken to be one kind of *kunos-orchis*, or rather the female of one kind – of *Cynosorchis morio*, the

Fool Stone, so called and distinguished because the perianth segments come together 'like to a fooles hood or cocks-combe'. It was the female, because the stones or tubers are smaller and more compact. *Cynosorchis morio foemina*, then, was the Green-winged Orchid, and the Male Fool Stone, *Cynosorchis morio mas*, with larger stones, was our Early Purple Orchid, *Orchis mascula* L.

Male and female, these two Fool Stones were looked upon as hot and moist in 'temperature', and venereal (see *Orchis mascula*), in the spirit of Dioscorides. The tubers were collected, but *Orchis morio* is less common. The orchid *par excellence*, powerful, venereal, easily come by from end to end of the country, easily recognized by the speckled leaves, was the other Foole Stone, the next species.

7. Early Purple Orchid. *Orchis mascula* L. 110, H 39

Local names. AARON'S BEARD, Berw; ADAM AND EVE (from 'male' and 'female' tubers, cf. Faeroese *Adam og Eva*), Corn, Dev, Som, N'hants, N'thum; ADDER'S FLOWER, Corn, Som, Hants; ADDER'S GRASS, Som, Ches, Berw; ADDER'S MOUTHS, Som; ADDER'S TONGUE, Dev, Dor, Ches; BALDEERI, BOLDEERI, Shet; BELDAIRY, Aber; BLOODY BONES, Som; BLOODY FINGERS, Glos; BLOODY MAN'S FINGERS (i.e. devil's fingers), Glos, Ches, Worc; BLUE BUTCHER, Som; BOG HYACINTH, Kirk; BULL'S BAG, BULL-SEGG, Scot; BUTCHER BOYS, BUTCHER FLOWERS, BUTCHERS, Som.

CAIN AND ABEL (the two tubers), Berw; CANDLESTICKS, Dor; CLOTHES PEGS, Som; COCK-FLOWER, Hants; COCK'S KAME, Berw; COWSLIP, Rut; CRAKE-FEET (i.e. 'crow-'), Yks, N Eng; CRAWFEET, Yks; CRAWTAES (i.e. 'Crow toes'), Cumb, Berw, Banff, Scot; CROSS-FLOWER, Dev; CROWFOOT, Lincs, Lancs, Yks, Dur, Cumb, Berw, Banff, Scot; CROW-FLOWER, Dev; CUCKOO, Corn, Dev, Som, Bucks, Herts, E Ang, Rut; CUCKOO-BUD, N'hants; CUCKOO COCK, Ess; CUCKOO-FLOWER, Corn, Dev, Som, Hants, Herts, Ess, E Ang; CUCKOO-PINT (see *Arum maculatum*), Hants, Bucks; CURLIE-DODDIE (i.e. 'curly head'), Shet.

DANDY GOSLINGS (? from the bird shape of the labellum), Wilts; DEAD MAN'S FINGER, Dev, Som, Glos, Suss, Kent, Shrop, Berw; DEAD MAN'S HAND, Som, Hants, Glos, Suss, Worc, Berw; DE'IL'S FOOT, Berw; DOG STONES, Som; DUCKS AND DRAKES (from 'male' and 'female' tubers), Dor; FOOL'S STONES, Ork; FOX STONES, Dor, Som; FRIED CANDLESTICKS, Dev; FROG'S MOUTH, Som.

GANDER-GAUSE, GANDIGOSLINGS, Wilts; GETHSEMANE, Ches; GIDDY-GANDER, Dor, I o W; GOOSE AND GOSLINGS, Som; GOSLING, Wilts; GOOSEY-

GANDERS, Dor, Wilts, Glos, Oxf; GOSSIPS, Som; GRAMMER GRIGGLES, Dor; GRAMFER-GRIDDLE-GOOSEY-GANDER, GRANFER-GOSLINGS, Wilts; GRANFER-GRIGGLES (cf. *Endymion nonscriptus* and *Melandrium rubrum*), Dor, Som; GRANFER-GRIGGLE-STICKS, Som; HENS, HEN'S KAMES, Berw.

JESSAMINE, War; JOHNNY COCKS, Dor, Som; JOLLY SOLDIERS, Dev; KEEK LEGS, Kent; KEET LEGS (i.e. 'kite legs'), Som; KETTLE-CAP, I o W; KETTLE-CASE, I o W, S Eng; KING-FINGERS, Bucks, War, Leic; KITE'S LEGS, Kent; KITE-PAN, Wilts.

LADY'S FINGERS, Som, Kent, Berks, War; LOCKS AND KEYS (from the tubers), LONG PURPLES, Dev, Som, Suss; LORDS AND LADIES (from the 'male' and 'female' tubers), Dor; MOGRAMYRA, Ire; POISON MORE (i.e root), Dev; POOR MAN'S BLOOD, Kent; PRIEST'S PINTEL (cf. *Arum maculatum*), War, Ches, Cumb; PUDDOCK'S SPINDLES (i.e. kite's legs), Perth.

RAM'S HORN, Suss; RED BUTCHER, Kent; RED GRANFER-GREGORS, Dor; RED ROBIN, Herts, Ess, Norf, Camb, Pemb; REGALS, Dor; SAMMY GUSSETS, Som; SINGLE CASTLES, Dor (Isle of Portland); SINGLE GUSS, Som, Wilts; SNAKE FLOWER, Som; SOLDIERS, Dor; SOLDIER'S CAP, Som; SOLDIER'S JACKET, Dor; SPOTTED DOG, STANDING GUSSETS, Som; UNDERGROUND SHEPHERD, Wilts; WAKE ROBIN (cf. *Arum maculatum*), Ches.

Dig up an Early Purple Orchid and you find two root-tubers in which food is stored, a new, firm one, which is filling up for next year's growth, an old, slack one, which is emptying and supplying the present needs. The symbolism of the kinds of *Orchis* with undivided tubers could hardly be overlooked. Sympathetic magic made them venereal, for which there was support even in the sober Dioscorides, spring-head of all wisdom in the *scientia scientarum* of botany. *Orchis* meant 'testicle'. Of *kunosorchis*, 'dog testicle', 'dog stone', Dioscorides mentioned that Thessalian women gave the tubers in goat's milk, the full one for exciting desire, the slack one for restraining it. Also that if men ate the large tuber, male children were born, if women ate the smaller one, female children.

Through Europe and through the centuries, the Orchids, and especially *Orchis mascula*, became a prime source of aphrodisiac food and medicine. Here was the plant used by the satyrs, the lustful, shaggy divinities of the wood (one kind of Orchis was known to Dioscorides as *saturion*, the 'satyr plant'). The author of the twelfth-century *Circa instans* summed up the matter by writing that *satyrium* '*virtutem habet attrahendi a remotis partibus, unde et coitum adiuvat*' (176). Three centuries later Hieronymus Braunschweig, in the *Liber de arte distillandi*, 1500 (which Laurence Andrew translated in 1527 as *The vertuose boke of Distyllacyon*), gave instructions for making and taking the 'water of Satyrion': 'In the mornynge and at nyght dronke of the same water at eche tyme an ounce and a halfe causeth great hete, therefore, it geveth lust unto the workes of generacyon and multiplycacyon of sperma.' Here was the plant for love potions such as were being made till recently in Shetland and in Ireland. John Josselyn, in 1672, records the way in which this old belief went to New England, where women collected native orchids. 'I once took notice of a wanton womans compounding of the solid roots of this Plant with Wine, for an Amorous Cup; which wrought the desired affect' (111). Robert Turner (*Botonologia*, 1664) remarked that enough orchids grew in Cobham Park in Kent to pleasure all the seamen's wives in Rochester; and the Pharmacopoeia of the College of Physicians in London included a 'Diasatyrion', or electuary, often used 'for a Provocative to Venery'.

This was made of Orchid tubers, dates, bitter almonds, Indian nuts, pine nuts, pistachio nuts, candied ginger, candied eryngo root (see *Eryngium maritimum*), clover, galingale, peppers, ambergris, musk, penids (barley sugar), cinnamon, saffron, Malaga wine, nutmeg, mace, grains of Paradise, ash-keys, the 'belly and loins of scinks', borax, benzoine, wood of aloes, cardamoms, nettle seed, and avens root (141). Salep (Arabic *Khusy ath 'lab*, 'testicles of the fox'), a food made from the dried tubers, was still highly regarded at the end of the eighteenth century (191) – 'a mild and wholesome nutriment superior to rice'.

Food, medicine, art, folklore – all were affected by this venereal heritage from Greece and Rome, based upon an older folklore; though in fact the tubers of *Orchis mascula* are altogether without the power attributed to them. In Forfarshire the dried tubers of 'Bull's Bags' were carried as love talismans. In the seventh scene of the Unicorn tapestries (see *Endymion nonscriptus*, *Viola odorata*, *Polygonum bistorta*), which was woven about 1514, *Orchis mascula* stands up long and purple against the white flank of the unicorn – a striking symbol of generation. In *Hamlet* Shakespeare knew exactly what he was about when he included in the garlands of the drowned Ophelia the 'long purples', to which the liberal shepherds give a grosser name, but which cold maids do Dead Men's Fingers call (IV. vii. 170–2). The grosser name may have been Priest's Pintel (cf. the French name *testicule de prêtre*), or one of the Cuckoo names.

Contrariwise, some names show an effort to supplant the usual associations. Gethsemane and Cross-flower come from the etiological legend that *Orchis mascula* grew under the cross, where the leaves were spattered with the blood of Christ. In Germany Marsh Orchid and Spotted Orchid are called *Kreuzblume*, and to explain the spots on the leaf, Early Purple

Orchid and Spotted Orchid have also been called by German names equivalent to Our Lady's Tears (10, 82).

8. Spotted Orchid. *Orchis fuchsii* Druce 85, H 33; *Orchis ericetorum* E. F. Linton 85, H 36

Local names. ADAM AND EVE, ADDER'S FLOWER, Som; ADDER'S GRASS, Berw; BALDEERI, BOLDEERI, Shet; BUTCHERS, CHOOGY-PIG, Som; CROW-FOOT, Yks; CROW'S FLOWER, Som; CURLIE-DADDIE (i.e. 'curly head'), Shet; DANDY GOSLINGS, Wilts; DEAD MAN'S FINGER, Som; DEAD MAN'S HAND, Berw; HEN'S KAMES, Berw; KETTLE-CASE, Dor; KITE-PAN, Wilts.

NIGHTCAPS, Derb; OLD WOMAN'S PINCUSHION, Wilts; PRIEST, Yks; QUEEN'S FINGERS, War; RING-FINGER, Bucks; SCAB-GOWKS, Dur; SKEAT-LEGS, Kent; SNAKE'S FLOWER, Dor, Som, Hants, Cumb.

The Spotted Orchids are William Turner's Hand Satyrion, or Royal Satyrion (1548). 'Royal Satyrion or finger Orchis', wrote Gerard, 'is called of the Latines Palma Christi,' because it has large roots which are 'knobbed, not bulbed as the others, but branched or cut into sundrie sections like an hand.' The 'Dead Men's Fingers' of *Hamlet* in one way fits the Spotted Orchids or the Marsh Orchids better than *Orchis mascula*, but their spikes are not 'Long Purples'.

9. Marsh Orchid. *Orchis strictifolia* Opiz. 93, H 40; *Orchis praetermissa* Druce 67, H 5; *Orchis purpurella* T. & T. A. Steph. 47, H 7

Local names. ADAM AND EVE, Berw; BALDEERI, BOLDEERI, Shet; BLUE ROCKET, BRUSH, SWEET WILLIE, Donegal; CAIN AND ABEL, Yks, Dur, N'thum, Scot; COCK'S KAMES, Berw; CUCKOO-FLOWER, I o W; CULLIONS, Donegal; CURLIE-DODDIE (i.e. 'curly head'), Shet; DEAD MAN'S FINGERS, Berw; DEIL'S FOOT, Berw; DODGILL-REEPAN, Kirk, Wigt; MEADOW ROCKET, Dumf.

Hand or Finger orchids, not 'stone' orchids. See Spotted Orchid and Early Purple Orchid.

XCVII. Araceae

1. Sweet Flag. *Acorus calamus* L. 40

Local names. GLADDON, Norf; SWEET FLAG, Ches; SWEET RUSH, War, S Eng; SWEET SEG, S Eng.

When Gerard grew the aromatic *Acorus calamus* in his garden towards the end of the sixteenth century, it was a new, or not a very old plant, which had been introduced from Turkey into Europe in 1567.

By 1666, Christopher Merret noted in his *Pinax*, the Sweet Flag was growing wild; and in 1668 the great Sir Thomas Browne wrote to Merret about this 'elegant plant', which was then so plentiful in the river at Norwich that one of the churches in the suburbs was sometimes 'strowed all over with it' (23). Later it was used to carpet the cathedral, and Browne also wrote that it had been transplanted to 'marish places of the countrey' – perhaps by his own medical hand (see *Urtica pilulifera*). Visiting Lady Clarendon's garden at Swallowfield, in Berkshire, John Evelyn in his *Diary* for 1685 describes how the waters in the garden were 'all flag'd about with Calamus aromaticus'. Lady Clarendon had hung a closet with it, which retained the scent very perfectly.

'*Kalamos aromatikos* grows in India,' Dioscorides had written. Long before his time it had been known to the doctors of Assyria, and the root is still one of the commonest bazaar medicines through India – used much as the Assyrians had used it, for bruises, rheumatism, and rubbing on a child's chest when he has a cold (30). ' 'Tis hot and dry. 'Tis chiefly used in Obstructions of the Courses, Liver and Spleen, and in the Cholick; it also provokes Urine. The Root of it candied, tastes very pleasantly, and is grateful to the Stomach' (John Pechey, in *The Compleat Herbal of Physical Plants*, 1694).

The scent of Sweet Flag, and especially the root, is like orange peel, warm and pungent, and unexpected.

2. Lords and Ladies, Cuckoo-pint. *Arum maculatum* L. 91, H 40

Local names. ADAM AND EVE, Som, Leic, Lincs, Yks; ADDER'S FOOD, Som; ADDER'S MEAT (cf. German *Schlangenbeeren*, 'snake berries'), Corn, Dev, Som; ADDER'S TONGUE, Corn, Som; ANGELS AND DEVILS (cf. German *Engelcher und Deiwelche*), Som; ARON, Scot; ARROWROOT, Isle of Portland.

BABE-IN-THE-CRADLE (cf. German *Kinneken in der Wieg*), Som; BLOODY

FINGERS, Hants; BLOODY MAN'S FINGER (i.e. devil's finger), Som, Worc; BOBBIN AND JOAN, N'hants; BOBBIN JOAN, Corn; BULLOCKS, Som; BULLS, Dor; BULLS AND COWS, Som, N'hants, Lincs, Lancs, Yks; BULLS AND WHEYS ('whey' or 'quey', a heifer), Yks, West.

CALVES' FOOT (translation of the old botanists' name *Pes vituli*), Som; COCKY BABY, I o W; COWS AND CALVES, Dev, Dor, Som, Wilts, Glos, Bucks, N'hants, War, Worc, Shrop, Notts, Lincs, Yks, Lakes; COWS AND KIES, Yks; COW'S PARSNIP, Som; CUCKOO COCK, Ess; CUCKOO-FLOWER, N'hants; CUCKOO-PINT (i.e. pintle, penis. See also *Orchis mascula*), Suss, E Ang, N'hants, Leic; CUCKOO-POINT, Yks.

DEAD MAN'S FINGERS, Worc; DEVILS AND ANGELS (see above), Dor, Som; DEVILS, LADIES AND GENTLEMEN, Denb; DEVIL'S MEN AND WOMEN, Shrop; DOG BOBBINS, N'hants; DOG COCKS, Wilts; DOG'S DIBBLE, Dev; DOG'S SPEAR, DOG'S TASSEL, FAIRIES, Som; FLY-CATCHER, Wilts; FROG'S MEAT, Dor.

GENTLEMEN AND LADIES, Oxf; GENTLEMEN'S AND LADIES' FINGERS, GENTLEMAN'S FINGER, Wilts; GREAT DRAGON (see below), Suss; HOBBLE-GOBBLES, Kent; JACK-IN-THE-BOX, Som, Bucks, N Ire; JACK-IN-THE-GREEN, Som; JACK-IN-THE-PULPIT, Corn, Som, Lincs; KINGS AND QUEENS, Som, Lincs, Dur; KITTY-COME-DOWN-THE-LANE-JUMP-UP-AND-KISS-ME, Kent; KNIGHTS AND LADIES, Som.

LADIES AND GENTLEMEN, Som, Wilts, Kent, N'hants, Shrop; LADIES' LORDS, Kent; LADY'S FINGER, Wilts, Glos, Kent; LADY'S KEYS, Kent; LADY'S SLIPPER, Wilts; LADY'S SMOCK, Dor, Som, Hants; LAMB-IN-A-PULPIT, Dev, Wilts; LAMB'S LAKENS (i.e. 'toys'), N'hants, N'thum, N Eng; LILY, Wilts; LILY GRASS, Suss.

LONG PURPLES, War; LORDS AND LADIES, general from Cornwall to Lakes and Yks; LORDS' AND LADIES' FINGERS, War.

MANDRAKE, Yks; MAN-IN-THE-PULPIT, MEN AND WOMEN, Som; MOLL OF THE WOODS (cf. *Anemone nemorosa*), War; NIGHTINGALE (i.e. knight *in galeâ*, in a helmet), Ess; OLD MAN'S PULPIT, Som; OXBERRY, Worc; PARSON AND CLERK, Dev, Som; PARSON-IN-HIS-SMOCK, Lincs; PARSON IN THE PULPIT, Dev, Dor, Som, Ches, Yks; PARSON'S BILLYCOCK (i.e. pintle, see *Orchis mascula* and Shakespeare, *King Lear*, III. iv. 74–5), PREACHER-IN-THE-PULPIT, Som; PRIESTIES, Lancs; PRIEST-IN-THE-PULPIT, Som; PRIEST'S PILLY (i.e. pintle), West; PRIEST'S PINTLE (i.e. penis), Derb, Lincs, Dur, Cumb; POISON-FINGERS, Dor; POISON-ROOT, Wilts; POKERS, Som.

RAM'S HORN, Suss; RAMSON, Cumb; RED-HOT POKER, Som; SCHOOL-MASTER, Suss; SILLY LOVERS, Som; SMALL DRAGON (see below), Suss; SNAKE'S FOOD, Dev, Som; SNAKE'S MEAT, Dev; SNAKE'S VICTUALS, Wilts,

Glos; SOLDIERS, Som; SOLDIERS AND ANGELS, Dev; SOLDIERS AND SAILORS, Som; STALLIONS, STALLIONS AND MARES, Lincs, Yks; STANDING GUSSES, Som; STARCHMORE (i.e. starch root), STARCHWORT, Isle of Portland; SUCKY CALVES, SWEETHEARTS, Som; TOAD'S MEAT, Corn.

WAKE ROBIN, Corn, Dor, Suss, Berks, War, Worc, Ches, Yks, Scot, N Ire; WHITE AND RED, Dor; WILD LILY, Dev.

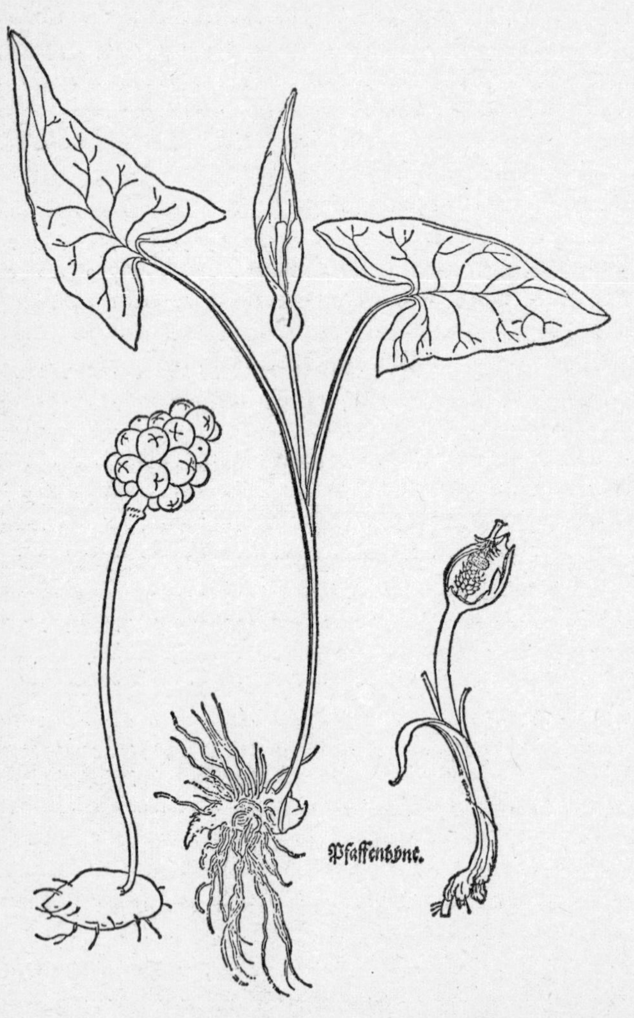

83 Lords and Ladies *Arum maculatum*

Obviously, as with *Orchis mascula*, the form dictated the names and reputation of *Arum maculatum*. The name Lords and Ladies and others like it are excellently explained once more by the seventh tapestry in the famous series at the Cloisters, New York, which tell the story of the hunt and capture of the unicorn. The tapestry is supposed to have been made for the marriage of a lord to his lady, the marriage of Francis I of France in 1514. Inside the fence which surrounds the captured unicorn, itself both a sexual and religious symbol, which is chained to the pomegranate tree of fertility, the designer has placed all the appropriate herbs, *Polygonum bistorta*, *Endymion nonscriptus*, *Orchis mascula* (q.v.), etc., and *Arum maculatum*, the form of which – the spadix in the spathe – stood for copulation. So the many male plus female names, human and non-human; and so the actual virtues of the plant. Lords and Ladies was believed to be the *aron* of Dioscorides, which was also *drakontia*, the dragon plant, and its virtues were those as well of the *drakontia megale*, the Large Dragon, including the excitement of venery. No plant had a more obvious signature. Thus in John Lyly's pastoral play, *Loves Metamorphosis* (1601): 'They have eaten so much Wake Robin, that they cannot sleep for love.'

The Grete Herball (1526) gave Cuckowe pyntyll as one of the names for *Arum maculatum*, and also, with more charity towards ecclesiastics, the name 'prestes hode' – 'for it hath as it were a cape and a tongue in it'. Lyte, in 1578, adds Wake Robyn and Priestes pyntill, to which there are equivalents in Dutch, German, Danish, Swedish, and French – *vit de prêtre*, *membre de prêtre*, *membre d'évêque* – and Manx – *bwoid-saggart*. French names include *vit de chien*, but also (and compare our own Lady's Smock, Lady's Finger) the more respectable and purifying *manteau de Ste-Marie* and *manteau de la Vierge*. The tubers were made into a food like salep (see *Orchis mascula*), and no doubt for the same reason – that it would be venereal and strengthening. Later this food became popular as Portland Sago, since large quantities were sent from the Isle of Portland to the London dealers, and it was looked upon as a substitute for arrowroot.

The mediaeval *Agnus Castus* (20) calls *Arum maculatum* cokkowyl pyntyl, which at first sight looks like cuckold's pintle (Old French *coucuol*, Middle English kokeweld, etc.). There seems some folk-etymological confusion between cuckoo, cuckold and cuculle = hood, cowl, Latin *cucullus*, in the name Cuckoopint: here's the penis of the lecherous cuckoo, who cuckolds the birds, and here's the penis inside a hood, or in a cowl – that is to say, the penis of the priest, or monk or friar who goes round cuckolding husbands, like the man who reads the gas meter. The robin of the name Wake Robin may also be explained as 'penis on the alert', a use of Robin as a pet name for the penis (cf. Dick, John Thomas).

Gerard also gives Starchwort, since 'the most pure and white starch is made out of the rootes of Cuckowpint; but most hurtfull for the hands of the laundresse that hath the handling of it, for it choppeth, blistereth, and maketh the hands rough and rugged, and withall smarting'.

XCVIII. Lemnaceae

1. Duck Weed. *Lemna minor* L. 109, H 40

Local names. BOGGART, War; CREED, Wilts; DIG MEAT (i.e. 'duck' meat), Ches; DUCK'S MEAT, N'hants, War, Ches, Derb, Scot, Donegal; DUCK-POND WEED, Cumb; GROVES, GROZENS, Som; JENNY GREEN-TEETH, Som, War, Lancs; MARDLENS (the weed on a 'mardle' or watering pond), Suff; TOAD SPIT, Lincs.

In Cheshire, Shropshire, Derbyshire, Lancashire, and Yorkshire children were scared away from dangerous ponds by talk of Jenny Green-teeth, a pond boggart or elf, whose presence under the surface was shown by the green water-carpet of the Duckweed.

Bibliography

The titles quoted in the text are not necessarily repeated in this list. Some books of outstanding interest or value for the purposes of this volume are marked with an asterisk.

1. ALEXANDER, E. J., and WOODWARD, C. H., *The Flora of the Unicorn Tapestries*, New York, 1950
2. AMHERST, Alicia, *History of Gardening in England*, 1895
3. *ARBER, Agnes, *Herbals: Their Origin and Evolution*, 1939
4. ARMITAGE ROBINSON, J., *Two Glastonbury Legends*, 1926
5. AUBREY, John, *Miscellanies*, 1890
6. AUBREY, John, *Natural History of Surrey*, 1718
7. AUBREY, John, *Natural History of Wiltshire*, 1685, ed. J. Britten, 1847
8. AUBREY, John, *Remaines of Gentilisme and Judaisme*, 1686–7 (1881 ed.)
9. *BANCKES, Richard, *Herball*, 1515. Edited by Larkey and Piles in Scholars' Facsimiles, New York, 1941
10. BAUHIN, J., *De Plantis a Divis Sanctisve Nomen Habentibus*, 1591
11. BÉALOIDEAS, *Journal of the Folklore of Ireland Society*, VII, 1937
12. BERTHOLLET, C. L. and A. B., *Elements of the Art of Dyeing*, trans. A. Ure, 1824
13. BLACK, W. G., *Folk-Medecine*, 1883
14. BLOCHWICH, Martin, *Anatomia Sambuci: or, the Anatomie of the Elder*, 1670
15. *BOULTON, E. H. B., and JAY, B. A., *British Timbers*, 1947
16. BOWLES, W. L., *Banwell Hill*, 1829
17. BRAUNSCHWEIG, Hieronymus, *Liber de arte distillandi*, 1500, trans. as *The vertuose boke of Distyllacyon* in 1527 by Laurence Andrew
18. British Calendar Customs, 1936–40 (Folklore Society)
19. BRITTEN, J., and HOLLAND, R., *Dictionary of English Plant-Names*, 1886
20. BRODIN, Gösta, *Agnus Castus: A Middle English Herbal*, vol. vi. Essays and Studies in English Language and Literature, English Institute, University of Upsala, 1950
21. BROMFIELD, W. A., *Flora Vectensis*, 1856
22. BROMFIELD, W. A., Catalogue of Plants . . . in Hampshire. *Phytologist*, 1849–50
23. BROWNE, Sir Thomas, *Letters of Sir Thomas Browne*, ed. G. Keynes, 1931
24. *Brut or Chronicles of England*, ed. F. Brie, E.E.T.S., 1906–8
25. BRYANT, C., *Flora Diaetetica*, 1783

26. BUDGE, E. A. WALLIS, *The Divine Origin of the Craft of the Herbalist*, 1928
27. BULLEIN, William, *The Booke of Simples*, 1562
28. CAMDEN, W., *Annales of Elizabeth*, 1625
29. CAMERON, John, *Gaelic Names of Plants*, 1883
30.*CAMPBELL THOMPSON, R., *Dictionary of Assyrian Botany*, 1949
31a.*CARMICHAEL, Alexander, *Carmina Gadelica*, vol. ii, 1928
31b.*CARMICHAEL, Alexander, *Carmina Gadelica*, vol. iv, 1940
32. CARTWRIGHT, William, 'The Ordinary', in *Comedies, Tragi-comedies, with Other Poems*, 1651
33.*CAVE, C. J. P., *Roof Bosses in Mediaeval Churches*, 1948
34. CAVE, C. J. P., *Mediaeval Carvings in Exeter Cathedral*, 1953
35. CHAMBERS, R., *Popular Rhymes of Scotland*, 1847
36. CHILD, F. J., *English and Scottish Popular Ballads*, 1882–98
37. Choice Notes from *Notes and Queries:* Folklore, 1859
38. CLAPHAM, A. R., TUTIN, T. G., and WARBURG, E. F., *Flora of the British Isles*, 1952
39. CLARK, A. (ed.), *Life and Times of Anthony Wood*, 1891–1900.
40.*CLARK, J. G. D., *Prehistoric Europe*, 1952
41.*CLARKE, W. A., *First Records of British Flowering Plants*, 1900
42.*COCKAYNE, O., *Leechdoms, Wortcunning, and Starcraft of Early England*, 1864–6
43. COGAN, Thomas, *The Haven of Health*, 1584
44. COLES, William, *The Art of Simpling*, 1656
45. COLES, William, *Adam in Eden*, 1657
46.*CORNEVIN, Charles, *Des Plantes Vénéneuses*, Paris, 1887
47. CORNISH, Vaughan, *Historic Thorn Trees in the British Isles*, n.d.
48. CRANE, M. B., and LAWRENCE, W. J. C., *The Genetics of Garden Plants*, 1938
49. CULPEPPER, Nicholas, *The English Physitian Enlarged*, 1669 ed.
50.*CURTIS, William, *Flora Londinensis*, 1777–91
51.*CURWEN, E. C., *Plough and Pasture*, 1946
52. DAVIES, Hugh, *Welsh Botanology*, 1813
53. DAVY, F. H., *Flora of Cornwall*, 1909
54. DAWSON, W. R., *A Leech Book of the Fifteenth Century*, 1934
55. DEARMER, P., VAUGHAN WILLIAMS, R., SHAW, M., *Oxford Book of Carols*, 1948
56. DEERING, C., *Catalogus Stirpium*, 1738
57.*DELATTE, A., *Herbarius: Recherches sur la cérémonial usité chez les anciens pour la cueillette des simples et des plantes magiques*, 1938
58. DE L'OBEL, Mathias, *Stirpium Adversaria Nova*, 1570
59. DERHAM, William, *Physico-Theology*, 1713
60. DIOSCORIDES, *De Materia Medica*, ed. C. G. Kühn, Leipzig, 1830
61. DIVERRES, P. (ed.), *Le Plus Ancien Texte des Meddygon Myddveu*, Paris, 1913
62. DODOENS, Rembert, *A Niewe Herball, or Historie of Plantes*, trans. Henry Lyte, 1578
63. DRUCE, G. C., *Flora of Berkshire*, 1897
64. DRUCE, G. C., *Flora of Oxfordshire*, 1886

65. DRUCE, G. C., *Comital Flora of the British Isles*, 1932
66.*EKWALL, Eilert, *Concise Oxford Dictionary of English Place-Names*, 1951 ed.
67. EKWALL, Eilert, *Studies in English Place-Names*, 1936
68. ELWES, H. J., and HENRY, A., *The Trees of Great Britain and Ireland*, 1906–13
69. EVANS, Estyn, *Irish Heritage*, 1949
70. EVELYN, John, *Sylva, or A Discourse of Forest-Trees*, 1664
71. EVELYN, John, *Acetaria, A Discourse of Sallets*, 1699
72. FERNIE, W. T., *Herbal Simples*, 1914 ed.
73. FLÜCKIGER, F. A., and HANBURY, D., *Pharmacographia*, 1874
74. FOGG, G. E., *Journal of Ecology*, 38, p. 415, 1950
75. FOLKARD, Richard, *Plant Lore*, 1892
76. *Forest of Dean, National Forest Park Guides*, 1947
77. FRÄNGER, Wilhelm, *The Millennium of Hieronymus Bosch*, 1952
78. FRAZER, Sir J. G., *The Golden Bough*, 1936
79. GARDENER, John, 'The Feate of Gardening', *Archaeologia*, liv
80.*GERARD, John, *The Herbal or Generall Historie of Plantes*, 1597
81. GERARD, John, *The Herbal ... Enlarged and amended*, by Thomas Johnson, 1633
82. GERTH VAN WIJK, H. L., *Dictionary of Plant-Names*, The Hague, 1911
83.*GILBERT-CARTER, H., *British Trees and Shrubs*, 1936
84.*GILBERT-CARTER, H., *Glossary of the British Flora*, 1950
85. GILLET ET MAGNE, *Nouvelle Flore Française*, 1883
86. GILPIN, William, *Remarks on Forest Scenery*, 1791
87. GOOD, Professor Ronald, *A Geographical Handbook of the Dorset Flora*, 1948
88.*GRATTAN, J. H. G., and SINGER, C., *Anglo-Saxon Magic and Medicine*, 1952
89. GREGORY, Lady, *Visions and Belief in the West of Ireland*, 1920
90.**Grete Herball, The*, 1526, 1529
91.*GUNTHER, R. T., *Early British Botanists*, 1922
92. HALLIWELL, J. O., *Popular Rhymes and Nursery Tales*, 1849
93.**Handwörterbuch des deutschen Aberglaubens*, 1927–41
94. HARINGTON, Sir John, *The Englishman's Doctor. Or the Schoole of Salerne*, 1607
95. HARRIS, R., *A Primitive Dye-Stuff*, 1927
96. HAYWARD, I. M., and DRUCE, G. C., *Adventive Flora of Tweedside*, 1919
97. HELBAEK, Hans, Botanical Study of the Stomach Contents of the Tollund Man. *Aarbøger*, 1950
98. HELBAEK, Hans, Early Crops in Southern England. *Proceedings of the Prehistoric Society*, 1952
99. HENSLOW, G. (ed.), *Medical Works of the Fourteenth Century*, 1899
100. HOGAN, F. E., *Luibhleahbrán, Irish and Scottish Gaelic Names of Plants*, 1900
101. HOWARD, H. W., and LYON, A. G., Nasturtium Officinale, Nasturtium microphyllum. *Journal of Ecology*, 40, p. 228, 1952
102. HUNT, Robert, *Popular Romances of the West of England*, 1922 ed.

103. HURRY, J. B., *The Woad Plant and its Dye*, 1930
104.*HYDE, H. A., *Welsh Timber Trees*, 1935
105.*JACKSON, Kenneth, *Early Celtic Nature Poetry*, 1935
106. JACKSON, Kenneth, *A Celtic Miscellany*, 1951
107. JESSEN, Knud, and HELBAEK, Hans, Cereals in Great Britain and Ireland in Prehistoric and Early Historic Times. *Kgl. Danske Videnskabernes Selskab*, 1944
108. JOHNSON, Charles, *British Poisonous Plants*, 1861
109. JOHNSTON, G., *The Botany of the Eastern Border*, 1853
110. JONES, Gwynn, *Welsh Folklore and Custom*, 1930
111.*JOSSELYN, John, *New-Englands Rarities*, 1672
112.*KITTREDGE, G. L., *Witchcraft in Old and New England*, 1929
113. LANG, Andrew, 'Fairies', in *Encyclopaedia Britannica*, 11th ed., 1910–11
114. LATHAM, M. W., *The Elizabethan Fairies*, New York, 1930
115. LEES, E., *The Botany of Worcestershire*, 1867
116. LE HÉRICHER, Édouard, *Philologie de la Flore de Normandie et d'Angleterre*, n.d.
117. LE HÉRICHER, Édouard, *Essai sur la Flore populaire de Normandie et d'Angleterre*, 1857
118. LEVER, J. W., Three Notes on Shakespeare's Plants. *Review of English Studies*, April 1952
119. LIGHTFOOT, John, *Flora Scotica*, 1777
120. LINDLEY, J., and MOORE, T., *Treasury of Botany*, 1874
121. LINNAEUS, *Critica Botanica*, trans. Sir Arthur Hort, 1938
122. LÖHR, Otto, *Deutschlands Geschützte Pflanzen*, 1938
123. LONG, H. C., *Plants Poisonous to Livestock*, 1924
124. LOUDON, J. C., *Encyclopaedia of Gardening*, 1848 ed.
125. LOUDON, J. C., *Arboretum et Fruticetum Britannicum*, 1844
126. LOUSLEY, J. E., *Wild Flowers of Chalk and Limestone*, 1950
127. LYTE, Henry. *See* Dodoens
128. MCNEILL, N., *Colonsay*, 1910
129.*MARZELL, Heinrich, *Wörterbuch der Deutschen Pflanzennamen* – in progress; 1937, etc.
130. MILLER, Philip, *Gardeners Dictionary*, 1741 ed.
131. MOLDENKO, H. M., *American Wild Flowers*, New York, 1949
132. MOLDENKO, H. M., and A. L., *Plants of the Bible*, New York, 1952
133. MOLONEY, M. F., *Irish Ethnobotany*, 1919
134. MUENSCHER, W. C., *Poisonous Plants of the United States*, 1951
135. MURRAY, R. P., *Flora of Somerset*, 1896
136. OHRT, F., Herba, Gratia Plena, *Folklore Fellows Communications*, No. 82. Helsinki, 1929
137.*Ó SÚILLEABHÁIN, S., *The Handbook of Irish Folklore*, 1942
138. OSTENFELD, C. H., Flora of Greenland and its Origin. *Kgl. Danske Videnskabernes Selskab. Biolog. Meddel.*, 1926
139. OSTENFELD, C. H., and GRONTVED, J., *Flora of Iceland and the Faroes*, 1934
140. PARKINSON, John, *Theatrum Botanicum*, 1640
141.*PECHEY, John, *The Compleat Herbal of Physical Plants*, 1694
142. *Pepys Ballads*, ed. H. E. Rollins, 1929

143.*PEVSNER, N., *The Leaves of Southwell*, 1945
144. *Philosophical Transactions*, 1698
145. PITCAIRN, R., *Criminal Trials in Scotland*, 1833
146. PLINY, *Natural History*, trans. W. H. S. Jones, 1951
147. POMET, Pierre, *Compleat History of Drugs*, 1712
148.*PRAEGER, R. Ll., *The Botanist in Ireland*, 1934
149. PRIOR, R. C. A., *Popular Names of British Plants*, 1879 ed.
150. RAPIN, René, *Of Gardens. A Latin Poem, Englished by James Gardiner*, 1706
151.*RAVEN, C. E., *English Naturalists from Neckham to Ray*, 1947
152.*RAVEN, C. E., *John Ray*, 1942
153. RAY, John, *Catalogus Plantarum Angliae*, 1670
154. RAY, John, *Synopsis stirpium Britannicarum*, 1696
155. RAY, John, *Historia generalis Plantarum*, 1680–1704
156. RAY, John, *Memorials of John Ray*, 1846
157. REID, Clement, *Origin of the British Flora*, 1899
158. RIDDELSDELL, H. J., HEDLEY, G. W., and PRICE, W. R., *Flora of Gloucestershire*, 1948
159. RIDLEY, H. N., *Dispersal of Plants throughout the World*, 1930
160.*ROLLAND, Eugène, *Flore Populaire*, Paris, 11 vols., 1896–1914
161. SALESBURY, William, *Llysieulyfr Meddyyginiaethol* (Welsh herbal of the sixteenth century), ed. W. S. Roberts, 1916
162. SCOT, Reginald, *The discoverie of Witchcraft*, 1584
163. SCRIBNER, F. L., *Ornamental and Useful Plants of Maine*, 1875
164. SCULLY, R. W., *Flora of Co. Kerry*, 1916
165. SÉBILLOT, Paul, *Le Folklore de France*, tom. iii, 'La Faune et la Flore', 1906
166. *Shetland Folk Book*, 1947
167. SOWERBY, J., *English Botany*, 1899 ed.
168. Stockholm Medical MS. *Anglia*, 18. 325. 1896
169. STORMS, G., *Anglo-Saxon Magic*, The Hague, 1948
170. STOWE, John, *The Chronicles of England*, 1603
171. STUKELEY, William, *Family Memoirs of the Rev. William Stukeley*, Surtees Society, 1882
172. STUKELEY, William, *The Medallic History of M. A. V. Carausius*, 1757–9
173. *Tales and Traditions of Tenby*, 1858
174. THEOPHRASTUS, *Enquiry into Plants*, trans. Sir Arthur Hort, 1916
175.*Thompson, Stith, *Motif Index of Folk-Literature*, 1932, etc.
176.*THORNDIKE, Lynn (ed.), *The Herbal of Rufinus*, 1945
177. THRELKELD, Caleb, *Synopsis stirpium Hibernicarum*, 1727
178. THURSTAN, Edgar, *British and Foreign Trees and Shrubs in Cornwall*, 1930
179. THURSTAN, Violetta, *The Use of Vegetable Dyes*, 1949
180. TOWNSEND, F., *Flora of Hampshire*, 1904
181. TURNER, Robert, *Botonologia: The British Physician*, 1664
182. TURNER, William, *Libellus de re herbaria nova*, 1538, ed. B. D. Jackson, 1877
183.*TURNER, William, *The Names of Herbes*, 1548, ed. J. Britten, 1881
184.*TURNER, William, *A New Herball*, 1551
185.*TURNER, William, *The seconde parte of William Turners herball*, 1562

186.*TURNER, William, *The first and seconde partes of the Herbal . . . with the third parte*, 1568
187. TURVILLE-PETRE, E.O.G., *Myth and Religion of the North*, 1964
188. TUSSER, Thomas, *Five Hundred Pointes of Good Husbandrie*, 1573
189.*VAN GENNEP, A., *Manuel de Folklore Français*, Tom I, vol. iv, 1949
190. VAUGHAN, John, *Wild Flowers of Selborne*, 1906
191. WEEKLEY, Ernest, *Concise Etymological Dictionary of Modern English*, 1952
192. WITHERING, William, *An Arrangement of British Plants*, 7th ed., 1830
193. WRIGHT, E. M., *Rustic Speech and Folklore*, 1913

Index of Scientific and English Names

Index of Names from Celtic Languages

CORNISH

Index of Local Names